A FEW GREAT CAPTAINS

A FEW GREAT CAPTAINS

The Men and Events That Shaped the Development of U.S. Air Power

DeWITT S. COPP

The Air Force Historical Foundation

Publications, Inc.
McLean, Virginia

To all the U. S. Army airmen, few in number great in spirit—the seekers, the pathfinders, the builders. They dared the heights and saw beyond their time.

Acknowledgments
The Air Force
Historical Foundation

A Few Great Captains was written under the auspices of The Air Force Historical Foundation. The Foundation was established by the Department of the Air Force in 1953 as a nonprofit, independent organization for the purpose of helping to perpetuate military aviation and aerospace history. Maintained wholly by its members, the Foundation is especially grateful for their interest and support, which made possible the writing of this book. Additional contributions were also received, and the Foundation is equally grateful to the following donors for their special assistance: P. P. Ardery, Major General T. C. Bedwell, Jr., USAF (Ret.), General M. E. Bradley, USAF (Ret.), Major General R. W. Burns, USAF (Ret.), Major General W. B. Campbell, USAF (Ret.), J. R. Cannon, Major General L. A. Curtis, USAF (Ret.), K. C. Davidson, Hon. J. H. Douglas, A. Dupre, L. Edmonds, H. L. Eiss, Lieutenant General H. L. and V. H. George Foundation, Senator B. M. Goldwater, Brigadier General B. S. Gunderson, USAF (Ret.), D. Haughton, Major General J. B. Henry, USAF (Ret.), Major R. H. Hodges, USAFR, General B. K. Holloway, USAF (Ret.), Major General D. E. Hooks, USAF (Ret.), S. A. Hopper, Major General J. T. Kingsley, USAF (Ret.), Captain J. M. Litman, USAF, Brigadier General R. S. Macrum, USAF (Ret.), Major General J. L. Martin, USAF (Ret.), Major General C. E. McCarty, USAF (Ret.), Brigadier General G. F. McGuire, USAF (Ret.), Hon. J. L. McLucas, Colonel H. O. McTague, USAF (Ret.), Dr. A. G. B. Metcalf, Major General J. B. Montgomery, USAF (Ret.), Lieutenant General D. G. Nunn, USAF (Ret.), Major General J. S. Patton, USAF (Ret.), J. D. Peeler, Brigadier General J. R. Pugh, USAF (Ret.), R. C. Reeve, Hon. R. C. Seamans, Jr., Captain H. G. Shell, Jr., USAF, W. P. Sherman, Brigadier General J. L. Stewart, USAF (Ret.), D. S. Swepston, N. D. Swessel, Major General R. Taylor, USAF (Ret.), Brigadier General P. W. Tibbets, USAF (Ret.), Lieutenant General W. H. Tunner, USAF (Ret.), F. Vera, Lieutenant Colonel J. P. West, USAF, and Lieutenant General J. G. Wilson, USAF (Ret.).

Particular note is made of the encouragement and assistance received from Dr. A. G. B. Metcalf, Lieutenant General H. L. George, Senator B. M. Goldwater and Lieutenant General Ira Eaker in launching this book effort.

Lt. Gen. John B. McPherson
President

Contents

Part III

"No army produces more than a few great captains."

General GEORGE C. MARSHALL

Overview

There were twenty-four twin-engine torpedo bombers in the formation, divided into two squadrons, one following closely behind the other. Flying southwesterly at three thousand feet, they traversed the lake's long reach, heading toward the city at the water's edge. The industrial smoke rising from the metropolis was a marker, discoloring the warm blue of a summer sky.

Far above the approaching formation, pursuit planes watched and waited, hidden by the sun's glare. There were two squadrons of them as well, thirty-eight in all, holding for the signal of their commander. He gave no sign until the last of the seaplanes passed over a preselected point on the ground below. Then, in a maneuver known to every fighter pilot, he peeled off, and in sharp precision his squadrons came plummeting down in the classic mode of attack.

They dived, however, not to do combat but to welcome; not to destroy but to greet; and in so doing they pulled up smoothly, offering honorary escort to the formation.

General Italo Balbo, at the controls of his S-55 Savoia Marchetti, immediately flashed a radio message to Italian dictator Benito Mussolini, in Rome, for public airing: "We have been joined by U. S. Army planes in close and splendid formation," he announced.

Balbo's air fleet had just crossed the U.S.-Canadian border near Detroit, Michigan, on the final leg of its unparalleled 6,100-mile venture, which had begun two weeks before in Orbetello, Italy.

The welcoming American airmen were commanded by Lieutenant Colonel Frank M. Andrews. He and his two wingmen, in their Boeing P-12's, moved forward to guide the aerial parade, his pilots fanning out to form a wide, protecting V around the sleek, twin-boomed ships of the visitors.

It was July 15, 1933.

Behind the pomp and ceremony that would greet the ninety-six Italian fliers with waving flags, speeches and celebration reaching all the way to the White House, was the fact that in a peaceful show of aeronautical daring, a long debated but little accepted military theory had been given added credibility. Those who held this belief maintained the airplane was a potential weapon of war that transcended all others—a strategic weapon whose long-range striking power could eliminate an enemy's industrial might, a weapon limited only by the state of its present development.

Regretfully for those advocates of air power, this concept was not an accepted military axiom by either the Army or the Navy, the battle over the proper use of military air power having gone on since the end of World War I. Shortly after Balbo led his armada on the longest aerial feat of its kind, the General Staff of the War Department, in air plan recommendations, denigrated the underlying military meaning of the spectacular flight, saying it had no bearing on the future. This denigration would continue in the years before World War II began. Integral to it was the equally explosive question: Who should control Army air power, General Staff ground officers or airmen themselves?

This book, the first of two, explores the lives and actions of a handful of air officers who believed they had the answer and who fought the twin battles of control and strategy against fierce odds. They were officers who laid their careers and reputations on the line, and frequently—in the pursuit of greater aeronautical knowledge—their lives as well.

At the beginning, foremost in the fight was William "Billy" Mitchell. Several months before the Armistice of November 1918, he had introduced a new air strategy in which he had organized his tactical units into a single striking force of nearly fifteen hundred allied planes. With this force, he had launched concentrated attacks on German positions. The effort was a first of its kind, and although the war ended before its effect could be properly evaluated, there was no doubt in Mitchell's mind as to the meaning. He came home from France in 1919 convinced that air power would soon make land and sea power obsolete. He also became convinced that air power must be separate from the control of land and sea forces.

A public hero, charismatic in style, Billy Mitchell was a visionary whose brilliance finally blinded his political judgment. When he attacked the War and Navy departments for their entrenched attitudes, he risked all for the cause that dominated his life.

Throughout, he was supported by a group of airmen who had done combat in the skies over France, the few like Major Carl A. "Tooey" Spaatz, who decided to remain with the Air Service at war's end.

Even before that, Mitchell had been championed by those who had taken wing soon after Kitty Hawk. Major Henry H. "Hap" Arnold was one, and he would become Mitchell's loyal standard-bearer in the fight that followed.

Horace Hickam was another. Testifying before an investigating board, he summed up the Army pilot's view when he said: "I am confident that no general thinks he can control the Navy or no admiral thinks he can operate an army, but some of them think they can operate an air force."

An airman who had come out of the war with no love for Billy Mitchell was Benjamin D. "Benny" Foulois. At one stage, back in 1910, Benny Foulois had been the Army's *only* pilot. As a brigadier general in France, his personality had clashed with Mitchell's, but he had not only recognized his adversary's capabilities as a combat commander, he had also subscribed to his tactical theories and his position on separation. Testifying in 1920 before a Senate committee examining the role of the Air Service, Foulois asked: "Is it any wonder that practical flying officers who have been risking their lives for the past eleven years in the development and use of this new military weapon are so keenly anxious to see aviation and aircraft development taken away from the Army and placed under separate control? . . .

"Is it any wonder that a few of us dare to risk the charge of insubordination to our superior authority and the possible charge of conduct prejudicial to good order and military discipline in order that our cause may be heard?"

Others were not so vocal as Benny Foulois or as popularly persuasive as Billy Mitchell, but they were there in support, their beliefs tied to his, their actions in the air and on the ground in league with all the demonstrations he conducted to prove that air power was a new force with which to contend. Major Frank M. Andrews had not reached France during the war, and after it he would be posted out of the country during much of Mitchell's battle. But, in time, he would become the principal leader for air independence and the leading proponent of the strategic use of air power. With regard to this strategy, Captain Ira C. Eaker, who witnessed both the internal and the external conflicts that brought Billy Mitchell's court-martial and downfall, would one day be called on to prove, with combat forces at his command, that the theories backing it were valid.

In the interim, there were other combats to fight, and with Billy Mitchell gone from the service by February 1926, those who had been dubbed as his "boys" continued the battle as best they could. It had been largely through Mitchell's agitation for a separate department of defense, that Congress, in the year of his resignation, approved the reorganization of the Air Service into an Air Corps. This had been done through the recommendations of the Morrow Board, named for its chairman, Dwight W. Morrow. The Congress authorized an Air Corps to be built up over a five-year period to number 1,800 planes. It was to be maintained by 1,650 officers and 25,000 enlisted men. The goal, called The Five Year Plan, was not met.

There were three basic causes: economic depression, no threat of war, and a War Department that had to keep its other branches functioning on severely restricted appropriations. But there was more to it than that.

The first congressional hearings on an Army Air Force separate from the Signal Corps were held in 1913. Testifying at that time, neither Billy Mitchell nor Hap Arnold had been in favor of such separation. Rapid aeronautical development and War Department failure to recognize its meaning later changed their thinking. Following the war, the crux of the matter lay in the outlook and military philosophy of the officers of the General Staff. Tactically, they saw air power as an adjunct and an aid to the ground forces. Only grudgingly would some admit that the air arm might have an independent role in combat, not always tied to ground control. The Army was not about to relinquish command willingly over a branch of its service. In this fight, its weapons, aside from political clout, were economic and doctrinal.

To establish a separate air arm, the War Department argued, would be extremely costly. Such separation not only would require duplication of many services but would also result in a separate budget and a separate promotion and retirement list. Further, the General Staff maintained that the approximately thirteen hundred officers of the Air Corps had neither the proper training nor experience to efficiently staff their own operations from the ground up.

A vivid measure of this belief was reflected in figures. As of June 30, 1933, of the sixty-seven general officers of the line—twenty-one major generals and forty-six brigadier—not a single one was from the Air Corps. Of the seven branches of service that made up the Army, only the Air Corps and the Signal Corps were below authorized officer strength—the Signal Corps by eighteen officers, the Air Corps by 368.

The doctrinal argument against air independence was subsumed in the standard phrase "unity of command," meaning there could be no proper coordination between ground and air forces with each operating under separate leadership.

Airmen such as Arnold and Spaatz felt they had refuted all these arguments many times over. But, with their numbers few, their political power in Congress with successive administrations was limited. Also, their own ranks were split. Not everyone thought as they did. There were those who wanted independence immediately, those who believed the separation should be evolutionary, and those—like Brigadier General Oscar Westover, Assistant Chief of the Air Corps at the time of the Balbo flight—who sided with the War Department's position.

Another factor was that all the contestants were in the same army. Many were West Point graduates. Most airmen had served in another branch of the service before transferring to aviation. In fact, it was mandatory for a West Pointer to do so. But the unbridgeable gap that lay between those who were flying officers and those who served on the ground was the air itself. Men who daily dared it had to be different in outlook from those who did not. Military men who flew were a special breed. They operated in an element whose challenge literally and figuratively broadened their own horizons, its uncharted dimension stimulating their thinking, bringing forth new tactical and strategic concepts.

Since they were so few in number, they came to know each other well as they made cross-country flights to their disparate bases, were shuttled from one small command to another, met on procurement boards, and joined in maneuvers to test their skills and equipment—always, of course, under the overall direction of ground officers. Because they flew and shared the same dangers from mechanical failure as much as from the elements, neither rank nor being a West Pointer had the same august meaning as it did in the other branches. This, too, rankled at the top and gave airmen the reputation of being undisciplined and unmilitary. This was not necessarily so, but freedom of flight created its own sense of independence, which was alien to prescribed military thinking and form.

The fact that aerial equipment, unlike any other military equipment, was determined by the pace of aeronautical development was a circumstance of continuing concern to airmen. The Springfield rifle was a standard weapon of the Infantry for over thirty years. Its date of first production was 1903, the year the Wright brothers flew at Kitty Hawk. The airplane had come a long way from that day and, although the comparison in hardware was patently obvious, there were ground officers who failed to see the difference and quite naturally resented the fact that more than 20 percent of the miserly overall Army budget was allocated to the Air Corps.

One airman who caught the essence of the problem was Hugh Knerr. From the outset, he was in the forefront of the fight for air independence and played a key role in the furtherance of strategic air power. It could be said of him that he started on the ground floor of powered flight, for as a small boy in Dayton, Ohio, he was frequently in the Wright brothers' bicycle shop.

In 1933, Knerr was chief of the Field Service Section of the Air Corps Materiel Division, at Wright Field. Long afterward, he succinctly stated the gut issue confronting military airmen at that time and, perhaps, at any time:

> The worrying present could be considered that brief immeasurable instant in which nebulous oncoming events come into focus as the fixed images of historical record. Our only chance to influence the shape of these images lay in our capacity to appraise our responsibilities in terms of future possibilities and pursue a course of action most likely to stand the test of war.

However, in appraising responsibilities and pursuing a course of action, air planners had to face certain rock-hard realities. On January 1, 1933, U. S. Army strength stood at approximately 119,000 men and 12,000 officers. This was less than half the strength in enlisted men called for by the National Defense Reorganization Act of 1920, and one third in officers. General Douglas MacArthur, U. S. Army Chief of Staff, in testifying before the House Military Affairs Committee on the proposed military budget for fiscal 1934, pointed out that the United States was sixteenth in military strength while being fourth in population and first in wealth.

In March 1933, Congress had approved a military budget nearly $44 million less than requested by the War Department. Whereupon President Roosevelt's budget director, Lewis A. Douglas, in the interest of economy, had lopped off an additional $80 million, which came close to cutting the total figure in half. MacArthur termed the act "a stunning blow to national defense." Accompanied by Secretary of War George H. Dern, he called on the President to protest.

Dern was not the sort of man to do battle with Franklin D. Roosevelt. By nature retiring, he was also in poor health. He quietly informed the President that to allow U.S. defenses to be further weakened, in view of German and Italian rearmament and Japanese aggression in the Far East, was a serious mistake. Roosevelt scoffed at him and, with biting sarcasm, left his Secretary of War white-faced and silent.

MacArthur stepped into the breach and a verbal donneybrook took place. It came to a climax when the General told his Commander-in-Chief that when we lost the next war and an American boy was lying in the mud with an enemy bayonet through his belly, he wanted the name in the soldier's last curse to be *Roosevelt,* not *MacArthur.* The President became livid. "You can't talk that way to the President!" he shouted. The General, emotionally wrung out, agreed. He apologized and offered his resignation, figuring his career was ended. Roosevelt cooled off swiftly, a characteristic of his self-control. "Don't be foolish, Douglas," he said. "You and the Budget must get together on this."[1]

They did to the extent that some funds were restored. Even so, the War Department faced the future with about a third less than what it needed to maintain the status quo of its skeletonized, poorly equipped forces. And if that was not enough, the General Staff was locked in a long running battle with the Navy over the use of air power.

For more than a decade, the two services had been disputing who was responsible in the air for what, with what and where. On the surface, it seemed clear enough. The Army would protect the land and the Navy the sea-lanes, and the air arm of each would function accordingly. But there were several major barriers. By international agreement, U.S. naval strength wasn't large enough to have a fleet in each ocean. Also, there was the matter of protecting far-flung possessions in the Pacific and the Panama Canal Zone. To overcome the problem, the Navy, which had been permitted to build shore bases, supposedly for training and experimentation, added bombers and pursuit planes, assuring itself of a coastal defense role. With the depression, the battle for the defense dollar became acute. The Congress saw needless expense in duplication of aircraft, and the Air Corps feared that the Navy had enough political support to put it out of business. Thus, when General MacArthur and the newly appointed Chief of Naval Operations, Admiral William V. Pratt, announced an agreement in April 1931 clarifying the issue and assuring the Air Corps's mission of coast defenses, its principal officers saw the Chief of Staff as somewhat of a hero.

The significant portion of the agreement read: "The Naval Air Force will be based on the fleet and move with it as an important element in solving the primary missions confronting the fleet. The Army air forces will be land-based and employed as an essential element to the Army in the performance of its mission to defend the coasts both at home and in our overseas possessions, thus assuring the fleet absolute freedom of action without any responsibilities for coast defense."

However, by 1933, with Pratt retired and the Navy's air arm under the command of Admiral Ernest J. King, the agreement had been pretty well junked. In the eyes of Air Corps Chief Major General "Benny" Foulois—and almost every other Air Corps officer—the Navy was out to usurp control of the air, and with a new President, whose desk was cluttered with ships but not one Army aircraft, the future looked bleak indeed. The fact that Balbo's ocean-spanning aircraft were seaplanes didn't help a bit.

It is against this broad backdrop that the protagonists of this book—Frank Andrews, Hap Arnold, Ira Eaker, Benny Foulois, Tooey Spaatz and their compatriots—waged their battles for independent thought and command. What follows is their combined and interwoven story, recounting their careers from the early days of powered flight to the onset of World War II, in September 1939.

Part I

Part 1

One

The Daring Young Men

At West Point, Cadet Henry Harley Arnold had been somewhat of a hell raiser. His nickname hadn't been Hap then, but "Pewt." He was the leader of the Black Hand Gang, a secret group of bold cadets who operated only at night, adopting their style of covert action from the popular "Pulpy Shute" series appearing in *The Saturday Evening Post*. Other gang members had equally uneuphonious names, but it was Pewt, as gang leader, who laid the plans—such as hoisting the reveille cannon up onto the barracks roof or rolling twenty-pound cannonballs down three flights of stairs in the dead of night.

Just before his graduation, in 1907, Eleanor Pool, his wife-to-be—something neither of them knew at the time—visited West Point and briefly viewed Pewt at bay. His deviltry had caught up with him. The gang had executed its greatest caper, smuggling fireworks onto the reservation and releasing same in a glorious midnight panorama of sound and fury that brought the troops running as bugles sounded, squads formed, and brave men prepared for the fray. The stunning climax was a pinwheel display on the barracks' rooftop spelling out *1907 Never Again*. And neatly silhouetted amid the fiery whirl stood Cadet Henry Harley Arnold, if not hoist on his own petard certainly well illuminated by it.

A few days later, Eleanor, who was known to all her friends as Bee, caught a glimpse of the gang leader's harried figure. Confined to his room, he waved sadly, a lion heart held captive in a castle tower.

Actually it was Cadet Arnold's older brother, Tom, who had been slated to go to West Point. The idea behind the decision was that middle-class families who couldn't afford to send their young men to Ivy League colleges could, with political help, get a son into either of the service academies, where a sound education and an assured career could be obtained.

Tom had not wanted to go, and seventeen-year-old Harley, as he was then known, did, even though his mother wanted him to become a Baptist minister.[1]

The Arnolds and the Pools were close friends and neighbors in Ardmore, Pennsylvania. Dr. Herbert Arnold, Harley's father, had been the Pools' family physician for many years. So it was that, in 1909, Second Lieutenant H. H. Arnold, returning from his first foreign duty as an Infantry officer in the Philippines, met Eleanor Pool in Europe and later saw something in the air that changed the course of his life.

The Pools were well-to-do, Mr. Pool (an Englishman by birth) a successful banker. Mrs. Pool, her two daughters, Bee and Lois, and the twelve-year-old twin boys were spending the summer of 1909 in Lucerne, Switzerland, when the Lieutenant telephoned.

"Bee and her sister are shopping in Lucerne," said Mrs. Pool. "Why don't you meet them at the ferry?"

When the two girls debarked from the ship, they saw a very thin, pale young man in a gray suit approaching them. "I'm Harley Arnold," he said with rather stiff formality. "Your mother has invited me to go down to your hotel for a few days."

They were a very pleasant few days, and some of his notorious shyness around the fair sex wore off. The interlude was ended by an unexpected cablegram saying Dr. Arnold was near death. Harley departed swiftly. It would be another four years before he and Bee were married—she spending three of them in Germany becoming proficient enough in the language to teach it.

However, in their brief encounter something had reached out and joined between them. The day would come when he would propose to her in Lafayette Park, across the street from the White House. But, for the moment, family duty called him home.[2]

He paused briefly in transit in Paris, and it was not so much the Eiffel Tower that caught his eye as an aeroplane. It was not soaring in the blue but suspended, stationary, made fast above the Champs Élysées. It was the wonderful flying machine in which Louis Blériot had crossed the English Channel only a few weeks before. In the mind of Second Lieutenant Arnold, it was not so much a thing of beauty as a matter of military recognition: a twenty-mile barrier of water had been bridged, he perceived.

But just as his course to the altar with Eleanor Pool was no Blériot-like leap, neither was his transition from the sight of his first aircraft to becoming one of the Army's first pilots. It was somewhat paradoxical, for impulsiveness was a strong force in his character. Although his interest in aviation had been sparked by what he saw in Paris, and he later followed the precarious development of aeronautics from his new post at Governor's Island, it was the War Department that extended the hand of dubious opportunity in the spring of 1911. He never knew if his name had been pulled out of a hat. He was bent on becoming an Ordnance officer for the purpose of promotion to first lieutenant when the official query arrived: Would he be willing to volunteer for flight training with the Wright brothers at Dayton, Ohio?

Somewhat puzzled, he took the dispatch to his CO, who studied it, shook his head and said, "Young man, I know of no better way for a person to commit suicide."[3] The pronouncement struck him as a challenge, something like the pinwheels going off on the barracks' roof. The Arnold blood was up! His father might be an austere, grim, overly serious practitioner of medicine who thought that fun was sin, and his mother was certainly a soft-spoken, wondrously patient and loving woman with a broad mind in a family of narrow minds, but between them they had a son willing to dare the unknown. He had a questing nature, swift to adapt and, in adapting, eager to reach out and find new ways to do the job. Anyone who flew in those days had better have that kind of nature, with a great deal of luck larded in.

The course Lieutenants H. H. Arnold and Tommy DeWitt Milling took entailed instruction in aircraft maintenance at the Wright brothers' factory, followed by flight training at Simms Station, a cow pasture nine miles out of Dayton. The aerial instruction lasted all of eleven days. The Wrights did not believe in flying on Sunday or it would have been ten.

At the conclusion of their training, the pair had about four hours of flying time apiece. Arnold's instruction had been given by civilian pilot Al Welsh and Milling's by Cliff Turpin, and at the moment, the two Lieutenants were the Army's only active rated pilots. Not that the Army didn't have other pilots: Lieutenant Frank P. Lahm, the balloonist who had helped stimulate the War Department into becoming air-minded and became its first airplane pilot, in October 1909, had also stimulated his superiors into relieving him from flight duty, it was said, for having taken the wife of Army Captain Ralph H. Van Deman flying. Lieutenant Frederic Humphries, who had been taught to fly by Wilbur Wright at the same time as Lahm, had been recalled by the Corps of Engineers because he wasn't "performing proper engineering duty." Lieutenant Benjamin D. "Benny" Foulois, the Army's third pilot, had also been rotated back to a desk job, and for one reason or another, the sprinkling of other officers, such as Captain Paul Beck, who had taken up the art, had been temporarily grounded. Two had already been killed in crashes: Lieutenant Thomas Selfridge and Lieutenant George E. M. Kelly. The list would grow and grow.[4]

But now, in the summer of 1911, Arnold and Milling were ordered to the Army's first regular flying field and flight school, at College Park, Maryland. Four years earlier, the Signal Corps had established an Aeronautical Division, whose War Department mandate was to take "charge of all matters pertaining to military ballooning, air machines, and kindred subjects." Chief of the Aeronautical Division was Captain Charles deF. Chandler, and one of Arnold's first tasks was to teach Chandler how to fly while Milling taught Chandler's Adjutant, Lieutenant Roy C. Kirtland. But aside from teaching, their mission was broad and vague: to develop their planes into a military weapon. Beyond that, the War Department did not wish to be specific. The attitude, a mixture of indifference and unknowing, gave Arnold and his companions freedom to explore the realm of flight unencumbered by confining military restrictions.[5] The first aviation appropriation by Congress, of $125,000 in March 1911, had permitted the purchase of a few Wright-Burgess and Curtiss aircraft, and with these planes Arnold, Milling, and the Army's other first pilots discovered the laws of aerodynamics by laying their lives on the line every time they sat down in a wicker chair between flimsy, cloth-covered wings and grasped the control levers.

In the beginning, the only instrument was a piece of string that indicated whether the plane was flying straight or was in a skid. This was soon followed by the tachometer, which gave the propellers' speed in revolutions per minute. Being hit in the eye by a bug while he was landing, temporarily produced blindness and a near crash for Arnold. For protection, he quickly brought goggles into being.

The helmet to go with the goggles was invented in much the same way as the seat belt. Some poor pilot in making a rougher than usual landing plowed up the turf with his head. Benny Foulois claimed to have created the seat belt after he was nearly catapulted into the wild blue by a downdraft. He at least invented the Army life saver, but the Navy was airborne too. Lieutenant John Towers, as a student, survived a crash in which his instructor was flung from the plane in mid flight. After his release from the hospital, Towers developed the Navy version.

Between Arnold and Towers there developed a lifelong friendship, and perhaps if they had been left to handle aerial relations between the two services things would have been a lot smoother and less costly for everyone. At a formal naval affair at Annapolis to which Arnold and Milling had been invited by Towers, Arnold and Towers decided they should swap uniforms. What better way to unify? Alas, the brass was not amused at this impromptu effort toward interservice melding.

It was during this rare time of genuine pioneering—of developing maneuvers and tactics by trial and error—that Lieutenant Arnold gained notoriety and got capped with the nickname "Hap."

The notoriety came as a result of his many flights. In one he set an altitude record of over four thousand feet, and in another he won the first Mackay Trophy. The award was for a rugged, forty-two mile endurance record on a triangular reconnaissance mission. He was described as white-faced and exhausted when he landed.

In September 1911, following a Long Island air meet, given by the Aero

Club of America, in which he flew, a movie company making the first
flying film ever—*The Military Air Scout*—asked him if he'd double for the
hero and be their "stunt man." He was happy to do so. What they'd pay
him would augment his salary of $124 a month. His famous smile, that
eye-twinkling grin, brought the sobriquet. Oddly, it did not stick until he
took command at March Field, California, many years later. His family
called him by his middle name, Harley, while Bee would call him Sunny
Jim, but one day all the world would come to know him as Hap, short for
Happy, thanks to the director of that early flying film.

He met a great many notables, military and civilian, in that proving
time. The most important was a thirty-two-year-old, non-flying captain,
the youngest officer ever to be assigned to the General Staff. His name was
Billy Mitchell.

Mitchell had recently returned from Alaska and the Orient, where he
had noted that the Japanese Air Force numbered a dozen planes and was
larger than our own. Several years earlier, he had written a paper maintaining that the air would become a theater of operations. Now he came
out to College Park to talk to Arnold and Milling about the use of the airplane as more than an artillery spotter—as a bomber, a fighter, a means of
long-range reconnaissance. He was going to deliver a paper on the future
of military aviation to the Army War College and, as flying officers, he
wanted their opinion. His enthusiasm was infectious, his grasp of the subject stimulating. Hap Arnold was impressed.

Additional notoriety came to Arnold from attending and competing in
civilian air meets. With his 29th Infantry buddy Lieutenant Jake E. Fickel
riding shotgun, they flew and fired in competition against the English team
of Sopwith and Campbell. Old Jake was a deadeye afoot or aflight, and six
hits in the dinner plate at two hundred feet won the brass ring.

But it wasn't all fun and games, by a long shot. Between June 14, 1911,
and November 30, 1912, Hap Arnold flew nearly 120 hours. He'd been
present for flight duty 357 days, and in that time he had made 638 flights.
The figures were simply a matter of record. They could say nothing about
the crashes on land or in the water. They could say nothing about the
wear and tear on mind and nerves and body—nothing about pusher-type
planes whose propellers could tear loose, whose engine chains could snap,
whose clumsy controls could fail to function. They could say nothing of
the buffeting of winds against man and craft, of machines that came crum-
pling down out of the sky shedding parts and pieces and killing pilots.
They could say nothing of a raw, wind-wracked day in November over
Fort Riley, Kansas.

Hap had been conducting spotting exercises for the Artillery with an observer, using a one-way wireless, sending messages down to a gun battery.
Tommy Milling's observer had used smoke signals, but the smoke, although released under pressure in blobs to form Morse code dots and

dashes, was often not readable, dissipating too quickly in the slipstream.

Previously, Arnold had taken Lieutenant Follett Bradley up as his radio operator to practice sending messages to ground stations from a primitive air-to-ground wireless set. They flew at one thousand feet and would have gone higher, but it was bitter cold in the open blast. The half-hour flight was a success, their messages all received on the six-mile course from Fort Riley to Ogden and back. They experimented next with a gun battery on a dry run, reporting the target's location by smoke and wireless. Then they were ready for the real show.

Arnold's observer, Lieutenant Alfred L. P. "Sandy" Sands, of the Field Artillery, brought along his camera. The day was dark and so were the target and its background, but they had no trouble in spotting the target and signaling the battery, correcting its line of fire. The exercise went well, and it beat dropping weighted messages through a stovepipe contraption or blurping unpredictable smoke daubs all over the sky.

Returning to land, Arnold flew the Wright C at about five hundred feet. There were pay lines of soldiers watching from below as he throttled back and began a gentle descending turn. Suddenly he was in trouble. The Wright C would not come out of the turn, and the bank began to tighten even though he now had the rudder control hard over in the opposite direction. As he sought to roll out and level off, the plane snapped into a dive, which swiftly became vertical. He threw a glance behind to see if one of the propeller chains had snapped. It hadn't. The plane plunged downward. He checked all his controls, could see nothing wrong, could only see the earth rushing up, could only haul back on the yoke helplessly, desperately, while Sandy, oblivious to the approach of eternity, was struggling, against the rising thrust of pressure, to take a picture.

Lieutenant Joseph O. Mauborgne,[6] the inventor of the one-way aircraft radio, was standing on the opposite side of the wireless tower and saw the plane heading straight down. He began to run toward the trees beyond which the craft would hit. The men in the pay lines broke ranks and began to run too.

Lieutenant Arnold never knew how he got the plane's nose up and out of its plummeting dive. Nor would he know what put the Wright C into it in the first place. He only knew, after he'd managed to land and climb out of his seat, that he was shaking inside and out.

"Come on, Sandy," he rasped, "let's walk over to the barracks."

"Walk!" Sandy by now didn't figure you walked anywhere if you could fly. Then he saw the color of his pilot's face and heard a strange sound. Turning, he saw Lieutenant Mauborgne and the men from the pay lines come charging out of the trees.

In his report to the commanding officer of the Signal Corps on the testing of an "aeroplane in connection with artillery fire," Lieutenant Arnold stated, as a result of his very near miss: "At the present time my nervous system is in such a condition that I will not get in any machine. . . . From the way I feel now I do not see how I can get in a machine with safety for the next month or two." And finally, in the last paragraph: "Lt. Milling does not care to fly No. 10 [Arnold's plane] to Leavenworth by himself. I personally do not care to get in any machine either as a passenger or pilot for some time to come."

Lieutenant H. H. Arnold had removed himself from flying duty, and he had no hesitation in saying why. In fact, he had the courage to say why, and then awaited reassignment back to Washington.

Within a very short span, eight out of fourteen qualified military aviators had been killed in crashes, flying tricky Wright B and Curtiss C pushers. "These accidents," said Orville Wright in a letter written on December 13, 1913, to Lieutenant Colonel Samuel Reber, then in charge of aeronautics, "are the more distressing because they can be avoided." He continued:

> I think I am well within bounds when I say that over ninety percent of them are due to one and the same cause—"Stalling." Stalling is a term generally used to designate that condition of a flying machine when its speed has dropped to a point where it becomes unmanageable. Recovery is possible only by regaining speed. When in this stalled condition the machine will dive downward in spite of every effort of the aviator to stop it. Many of these dives would not result seriously if the aviator had but the courage to cause the machine to make an even more fearful dive till it recovers its normal speed. . . ."

This, no doubt, was an explanation of what had happened to Arnold in his near-fatal plunge. There had been enough altitude for the plane to recover from the stall, enough lift produced so that he had recovered control.

Regardless of the causes of the crashes and Orville Wright's aerodynamically correct answer for many of them, the public and political outcry for action to end the shocking death rate was loud and clear. Within the service there was near mutiny upon the part of some of the young officers involved. They *demanded* safer aircraft and to be commanded by officers who understood flying. One major result of their criticism helped to bring about, in 1913, the first congressional hearings on the efficacy of separating the air arm from the Signal Corps and making it an independent branch of the Army.

Arnold, in testifying on the bill before the House Military Affairs Committee, repeated his desire not to continue flying, admitting to the committee's chairman, Congressman James Hay, Jr., that he had been relieved from flying duty at his own request. His forthcoming marriage to Eleanor Pool, he said, was an added circumstance affecting his decision.

Billy Mitchell also testified. Both he and Arnold were in favor of keeping aviation in the Signal Corps. The activities of both services complemented each other, said Lieutenant Arnold. Aviation, said Mitchell, was at best a reconnaissance tool. Its offensive value was yet to be proved, although dropping bombs and carrying guns were being tried out. It was Mitchell, no doubt with some input by Arnold, who recommended that "no man should be taken in [to flying] who is more than thirty years of age or married, to start with." And that's the way the new law was written, although it was rescinded three years later.

Lieutenant Henry H. Arnold and Miss Eleanor A. Pool were married on September 10, 1913. It was a big Philadelphia wedding. Captain Douglas MacArthur couldn't make it. Billy Mitchell could.

The Arnolds were strict teetotalers, and even though father Pool kept a very full cellar, he had deferred to the doctor's wishes, and accordingly the formal reception "downstairs" at the hotel was a very dry affair. Not so the upstairs room, where the groom had had the foresight to provide an oasis for his thirsty brother officers. To heighten the festivities, best man Tommy Milling and his sister jumped into the bridal car and drove away as a joke. The bride and groom, not anxious to start married life waiting on the curb, commandeered the second-best car, and off they went amid farewell shouts, confetti and the rattle of tin cans.

After a brief tour with the 9th Infantry in Kentucky, they sailed away to the Philippine Islands for duty with the 13th. There Arnold found that aviation had reached the outermost U.S. possessions through the efforts of Lieutenant Frank P. Lahm, who had been restored to flight duty. There were three hydroplanes and three pilots besides Lahm. Captain Charlie Chandler, Lieutenant Herbert "Bert" A. Dargue, who was to become one of Hap's closest friends, and Lieutenant C. Percy Rich, soon to be killed in a crash.

Hap did no flying, but there were several memorable encounters that marked the Arnolds' two-year tour. They lived first at Fort McKinley in houses, built on stilts, "with cockroaches big as lizards." There were ants, too, red and white, by the zillions, and when they weren't eating the porch supports they were eating each other.

At first Bee found the racket made by the rain pounding on the tin roof of their quarters nerve-wracking. But once she became accustomed to the sound, she grew to like it. Her husband did not like the sound of reveille at five-thirty every morning, and neither of them liked the separations troop maneuvers brought. Then he would write to her daily, calling her his "darling Beadle" or "my darling Bee Bee," signing his letters "Sunny."

Next door to them for a time lived a senior first lieutenant, George C. Marshall, and his ailing wife. Even so, the Marshalls went out of their way to help the newlyweds get settled. Bee thought they were wonderful, and Arnold had some thoughts on his neighbor too. During an important series of maneuvers centering on a supposed Japanese attack on Batangas Province, Marshall, due to the sudden illness of a ranking officer, became chief of staff for the counterattacking force to which Arnold as a company commander was detailed. Observing Marshall in action, he later commented to Bee, "You know, I've just been working with our future Chief of Staff. George Marshall is going to have that job someday." Off duty, he and Marshall got in the habit of taking long hikes together.

There were other friends: Sure-shot Jakie Fickel and his bride, Gertrude Lahm and her aloof husband, Bert Dargue and Charlie Chandler were constant callers. They'd get together and talk aviation. The pusher type of aircraft, with the propellers behind, was giving way to the tractor design, with the prop and engine in front. It made good sense. The pilot was safer and better protected from the winds. Bert Dargue had a hydroplane over at Corregidor, and Hap would journey over to the island to have a closer look. Back home, the Curtiss Jenny had come into vogue.

In December 1915 a memorable event occurred when Lois, the Arnolds' first child, was born, at Fort Stotsenburg. It was during this time that Hap learned that Billy Mitchell had left the General Staff and was now a major in the Aviation Section of the Signal Corps, deputy to its Chief, Lieutenant Colonel George O. Squirer. Some months later, Mitchell, at the age of thirty-six, went through flight training at the Curtiss Aviation School at Newport News, Virginia, paying for the instruction himself. Hap said nothing to Bee about the news, but with the war in Europe and the airplane very much a part of it, he was longing to put down the rifle and climb back into the cockpit. His opportunity came suddenly and unexpectedly.

The Philippines tour ended; the Arnolds were aboard ship, homeward bound. In the Hawaiian Islands, there was a cablegram waiting, from the Adjutant General. Would Lieutenant Arnold be willing to volunteer for duty with the Aviation Section, or if so detached, would he object?

Somewhat puzzled, he cabled for clarification. The answer was if he volunteered he'd come in as a captain. If he was drafted, so to speak, he'd still be a first lieutenant. In the offer, he easily detected the fine strategic hand of Billy Mitchell.

His decision was automatic, but Bee, desperately ill with a vicious mix of tropical ailments, had a negative reaction. *Oh, gracious, no!* was her first thought. Then, on calmer reflection, she realized that what he wanted he was to have, and she knew how badly he wanted it, and that if they had not been married he would have returned to his first love long ago.

Her best memory of that long voyage home was a young lieutenant named Millard F. "Miff" Harmon. She was too ill to leave her cabin and take Lois for her exercise. Daily, Miff would stick his head through the cabin window and in his high, squeaky voice ask, "May I have your baby?" And then off he would go with Lois—a baby-sitter par excellence.

Back in the States, on May 26, 1916, Lieutenant, soon to be Captain,

Arnold received orders to report to the Signal Corps Aviation School, at Rockwell Field, North Island, San Diego, California.

He reported in June and took over as Supply Officer. Aside from the staff, there were twenty-five aviators and fifty-one students. Bert Dargue was there with his new bride. Frank Lahm was the Executive Officer. Hap Arnold felt as though he was the only nonflying officer on the island. That had to change, but he didn't want Bee to know about it. She was going to have another child, and she still wasn't all that well.

On October 18, he took the big step. Helmeted, goggled, suited, he climbed into a Jenny with Bert as his check pilot. He'd never flown in a cockpit plane with a prop up front. The controls were different, simpler, ailerons on the trailing edge of the wings, the craft far more responsive, easier to handle. The OX5 engine offered more horsepower than he had ever known. On that twenty-minute flight, he knew he was back where he belonged. His only worry was Bee.

When he returned to quarters, she greeted him brightly, "Well, did you have a good trip today?"

He was amazed. "How did you know?"

"I've got awfully good eyes." She had seen the impression of the goggles on his cheeks.

"I didn't want to tell you because of the baby."

"It's perfectly all right. If you want to do it, go ahead and do it. That's your life." That was Bee Arnold. She had what it took, what a man who looked to the sky wanted and needed in a wife.

He went on flying, and by the end of the year he'd been up twenty-eight times, fourteen as pilot. Then, as he was steadily building up his hours and experience, he was in trouble. From the outset and throughout his life, he had difficulty in accepting the traditional method of working through prescribed channels when swift action was needed.

On the tenth of January, 1917, Lieutenant William A. "Robie" Robertson took off in a Jenny with Field Artillery Colonel Harold G. Bishop for Yuma, Arizona. They did not arrive, and it was soon apparent they were down somewhere, lost in the mountains or the desert. Dargue, who was in command of flight training, and Lieutenant Byron Q. Jones, who headed up the Aircraft Repair Section, went to Lahm, the Executive Officer. They volunteered to launch a search and rescue mission. The CO, Colonel William B. Glassford, a sixty-four-year-old former balloonist, refused the offer. He wished to proceed with care even though he had more than thirty planes at his disposal to begin the hunt. It was said he had decided the pair had deliberately flown across the Mexican border, and with relations between the United States and Mexico at a low ebb, he was not about to move aggressively.

Robertson and Bishop had been reported missing on a Wednesday. By Sunday not only were the pilots at the field up in arms over Glassford's refusal to act, but from Washington angry congressional telegrams were coming in along with press demands for action.

Following difficult negotiations with the Mexican authorities, Glassford's action was to send Lahm and a truck detachment, carrying aviation gas across the mountains to Black Butte, eighteen miles south of the Mexican border near Calexico, to set up an auxiliary base from which a search could be launched. It was much too slow a process for Hap Arnold in a situation that demanded immediate movement. He, Bert and B. Q. Jones huddled and then, against orders, they got into their planes and took off on their own search mission.

The searchers had no luck in locating the lost pair. But word of their attempt passed quickly back to headquarters. Gertrude Lahm stormed over to the Arnolds' quarters to complain to Bee.

"Wouldn't you want someone to go out and look for your husband if he was lost?" asked Bee.

"Not against orders!" responded Gertrude in true-blue Army style.

Nine days after their disappearance, Robertson and Bishop were found. They had gone down in the Sonora Desert, in Lower California, and had begun walking. After six days of wandering, Bishop couldn't continue but Robertson had gone on and been discovered by Mexicans. But Rockwell Field was in the grip of a heated turmoil over the incident. Black marks went on personnel records. Young pilots who had protested too loudly against headquarters' failure to act were transferred out. The story got into the newspapers and ill feeling permeated Rockwell Field, a line drawn between those who criticized Glassford's method of search and those, like Lahm, who supported it. The personal effect in the Arnold household was that Bee gave birth two months early to her second child, a son for a proud father, Henry Harley, Jr.

On the very next day, January 30, 1917, Arnold was ordered to Washington. He was not sure what for, but he was soon to write Bee the good-bad news. The good news was that he was off to the Canal Zone, where, once a suitable location for an airfield could be found, he would become the Panama Department Aviation Officer. In addition, he would have his own squadron—possibly three squadrons! His immediate superior would be the commanding general of the Department, and he would serve on his staff.

He wrote Bee that it was the biggest job he'd ever had, "or expect to have for years to come." There would be balloons, aero-squadrons, hydro-squadrons, and he hoped that he could handle it all, particularly when he learned that his appointment had not met with unanimous approval at the War Department. "I stand in well with some," he observed, "and not so well with others. Before I leave I hope to stand in well with all. I am trying to be very conservative." Another piece of good news was that his promotion to major dated back to September. The bad news was that they were to be separated for an indeterminate time. He missed her. He missed Lolee (Lois), his four-year-old golden girl, and he missed his newborn namesake, to whom he had barely had time to say hello. Before his departure for the Canal Zone, on February 20, aboard the transport *Kilpatrick,* he learned that the General Staff had asked for all the papers in connection with the Robertson-Bishop incident. He had already noted that there would be "no change in San Diego unless people there get their troubles to Washington through official channels via an inspection."

When he arrived in the Canal Zone and reported to the Commanding General, Major General Clarence H. Edwards, he found that the troubles at Rockwell Field would not go away. They had, in fact, preceded him. "Don't say anything," he cautioned Bee, "but Col. G---- gave me an awful efficiency report." He wondered what kind of report Glassford had given Jones and Dargue.[7] The next day, when he asked General Edwards if he could see what the Colonel had said about him, Edwards refused. "He told me," reported Arnold, "that it was so rotten it made me stink. Hence Col. G---- must like me very much. I hope he gets tried by a g.c.m. [general court martial] before he gets out."[8] As for Bee's relations with the wives of those involved, he commented, "You cannot make me mad if you never see Mrs. L--- or Mrs. G--- again." It was obvious that behind Glassford's damning report Arnold saw Lahm's knifelike thrust.

None of it made any difference to General Edwards, in whose house Arnold had been invited to live while he was trying to locate a field site. The General developed a great liking for the young Major and the feeling was fully reciprocated. In Arnold's mind, Edwards was "a real man, the first real General I have met."

In the end, it was not efficiency reports, good or bad—and Arnold had a few of the former—nor personalities at Rockwell Field or in the Canal Zone, that terminated his mission after five weeks of fruitless searching. No one in authority—Army, Navy, Panamanian—could agree on the site for the air base, and Edwards suggested his house guest return to the States and get further instructions from Chief of Staff General Leonard Wood.

It was on the voyage home that word came: the United States was at war with Germany.

Two

The Aerial Front

Unprepared as the U. S. Army was for war, no part of it was more unprepared than its Aviation Section. In the decade since the War Department had contracted for its first Wright Flyer, American military aviation had barely gotten off the ground, and in 1917 it was ranked fourteenth in strength among the nations of the world. There were two basic reasons for this: a continuing failure among the General Staff's twenty officers to appreciate the potentials of the airplane as a weapon and a concomitant lack of funding from the Congress.

When Major Arnold returned from the Canal Zone to Washington that spring to assume a post that would eventually make him the number-two man in the War Department's Air Division—as well as a temporary colonel —U.S. air strength was practically nonexistent. There were fewer than 250 aircraft. None could be rated above a trainer. There were no combat aircraft, and of the 131 officers who were airmen, only twenty-six were qualified pilots. To service and maintain the planes there were about one thousand enlisted men and two hundred civilian mechanics.

Equally as significant, no real tactics or strategy had been worked out within the Signal Corps or the War Department concerning the employment of air power, even though Arnold and his fellow pilots had experimented with the airplane in its various uses as a bomber, an interceptor and a photo-reconnaissance craft. It was generally agreed that the plane was there to aid the Infantry Division Commander as he saw fit, and since he had no real knowledge of what air power might do besides observe what the enemy was doing, that was as far as his thinking went.

For the moment, however, it was numbers and not aerial tactics that occupied the War Department, thanks in large measure to Major Billy Mitchell. Fortuitously, Mitchell had arrived in Paris just four days after the United States declared war on Germany. In a month's time, his impact was felt on both sides of the Atlantic. Swiftly, by spending ten days at the front and then talking with French and British commanders, he sized up the importance of air power in a conflict that was grimly locked in the unceasing slaughter of trench warfare. He recognized that aircraft production was a basic essential to maintaining an air offensive, and that such an offensive if carried to the full would give weight to a method by which the deadlock in the mud below could be broken.

It was Major General Hugh "Boom" Trenchard, Commander of the Royal Flying Corps, who set fire to Mitchell's thinking on the strategic employment of aircraft, not hovering over the trenches but reaching deep behind the enemy lines with bombers to strike critical targets of "supply, substance and replacement," as Trenchard put it.

Mitchell saw too that the doctrine of the offensive had been extremely costly to both the French and the RFC in men and aircraft. He believed the United States could supply both in large quantities. Working with a French military staff, he made recommendations on what U.S. production requirements should be in order to regain air supremacy from the Germans. These recommendations were incorporated in a message from French Premier Alexandre Ribot to President Wilson. They called for a force of forty-three hundred bombers and pursuit planes with two thousand replacement planes a month to be ready for an offensive planned for the spring of 1918. To man and service the planes it was estimated that five thousand pilots and fifty thousand mechanics would be needed.

At the time, the U.S. aircraft industry consisted of a dozen small, not very prosperous companies whose annual output was fewer than fifty military and civilian planes. Since Kitty Hawk, the industry had produced about one thousand aircraft.

In the French Premier's request, Arnold detected Mitchell's influence, for Billy had been bombarding the War Department with similar proposals and reports—which had been largely ignored. Now, however, the French call for help was met by a great outpouring of press and public enthusiasm, which soared skyward in the upsurging patriotic euphoria of the season. The Ribot message had arrived on May 26, and the War Department, dropping a somewhat more realistic plan advanced by the National Advisory Committee on Aeronautics (NACA), asked its air planners for an aviation program centered around the French proposal. By early June they had it, thanks to a small team of airmen headed by Major Benny Foulois and assisted by Hap Arnold.

The plan called for an unheard-of total of 22,625 planes, 44,000 engines, and spare parts enough to equal another 17,600 aircraft, all to be produced in a year's time at an estimated cost of $640 million. It was ridiculous, but at the moment no figure seemed impossible, no cost too large. Never mind that in three years of war the French had not produced that many planes. The War Department might be bowled over by the cost figure, Foulois and his group expecting it to be cut substantially, but not Secretary of War Newton D. Baker, nor the Congress. The requested sum, the largest single appropriation ever considered by Congress, sailed through passage in both houses, and on July 24, President Wilson signed it into law.

Arnold and Foulois and their minuscule band of fellow airmen had more money to build an air force than they had ever dreamed of! Less than ten years before, Benny Foulois's total yearly appropriation to maintain the Army's *only* aircraft had been $150. The difference might have boggled most minds but not Benny's.

In March 1917 he had been transferred to Washington from Fort Sam Houston, Texas, where he had been serving as Aeronautical Officer for the Southern Department. Chief of the Signal Corps Aviation Section at the time was Lieutenant Colonel John B. Bennett, a nonflying officer. Foulois, who, among other things, became Chairman of the Joint Army and Navy Technical Aircraft Committee of the War and Navy Departments, would soon succeed Bennett and be jumped from major to the temporary grade of brigadier general. He was senior as an airman to Arnold, but not by much, for aerially speaking the two were practically birds of a feather, Foulois having taken to the air just two years before Hap. Beyond that, their personalities and manner were vastly different.

In 1898, too short to be accepted by the Navy, Benny Foulois had enlisted in the Army as a private at age seventeen. By the turn of the century, he had risen through the ranks to a commission via the guerrilla campaigns in Puerto Rico and the Philippines.

A physically small man with a streak of fearlessness as wide and long as his jaw, if he was not the Army's first officer to fly, he was the first—after ninety minutes of instruction with Orville Wright—to teach himself *how* to fly! Orville hadn't had time to instruct him on the technique of landing, and Benny, in Texas, would crack up his Wright Flyer making the attempt, then write to the Wright brothers in Dayton, explaining what he thought he'd done and asking them to tell him what he should have done.

Later, Foulois would take great pride in listing some of his early aerial feats: operator of the first U. S. Army dirigible balloon; first military observer on a cross-country flight with Orville Wright; first military test pilot; first to fly more than one hundred miles nonstop; designer and user of the first radio set in an airplane; carrier out of the first air reconnaissance problems with troops; etc.

The fact was that from late 1909 through 1910, Benny Foulois was the Army's *only* pilot. Like Hap Arnold and Tommy Milling, he laid his life on the line every time he flew. But he had luck and he had guts, and during Army maneuvers at San Antonio in 1911, he convinced his old outfit, the 17th Infantry, that he had entirely too much persistence.

He was in command of a brand-new Wright C aircraft, the only plane west of the Mississippi, its predecessor having been cracked up beyond repair. Benny, flying it before the reveille bugle, had managed to disturb the sleep of many of his former comrades. Several of them decided to take him out of a Saturday night to make sure he wouldn't be in shape to bother them on Sunday morning. Present in this group were two outsiders, Lieutenants Horace Hickam, with the 11th Cavalry, three years out of West Point, and H. Conger Pratt, class of '03, serving as the Division Commander's aide.

The festivities began in the barroom of the Menger Hotel, where the object was to reduce the establishment's liquor supply. Good progress was being made in that direction when, suddenly, the assembled were attracted by a war cry. There on a balcony at the end of the barroom, high above the multitude, perched Horace Hickam. It was obvious he had been taken with the idea of powered flight, for he flapped his arms like wings, crowed like a rooster and took off.

The flight was brief, but the landing was remarkable, two points and no nose over. Horace had soloed without benefit of aircraft. He was roundly cheered and brought quickly back to the bar to quench his thirst.

By the time the evening had wended its way to the dawn's early light, via a grand tour of San Antonio's red-light district, Foulois, Hickam and Pratt had memories they could all share.

But, while the revelers repaired to their tents to sleep off the effects of the gay night, Benny rolled out his combat force and was up with the sun's rays, a wicked bee buzzing over the tent tops.

Possibly by his daring, Foulois was out to make up for his lack of a West Point class ring and the absence of a blue-blood military background. In marrying Ella Van Horn, the sister of Captain Richard Van Horn, balloonist, he entered into a family of the military elite. The marriage ended in divorce, and it was felt by some that Foulois had treated his wife shabbily—a slight that was not forgotten in the upper reaches of the War Department.

Meantime, he continued to circle skyward. In 1915, he organized and commanded the Army's 1st Aero Squadron, and the following year, in March, he led it into Mexico to support General John J. Pershing in the so-called punitive expedition against Pancho Villa's marauding guerrillas. Foulois's squadron, the first U.S. aerial combat force, consisted of eight JN-2 Curtiss Jennies. They saw no combat, but a great deal of hardship. And although all of Foulois's pilots survived the invasion, the planes, due to the elements and mechanical failure, did not.

Now, two years later, Foulois was determined to seek the cannon's mouth in France and not be nailed to duty in Washington, no matter how high-sounding or important his duties in the capital might be. Certainly an experienced air leader was needed by the AEF, and he was able to convince those making the decisions in the War Department that he was the right man to go. And so, in the fall of 1917, sporting his new general's star, he arrived in Paris and promptly ran into a war of a different sort.

Several months previously, when Major General John J. Pershing and his staff of the American Expeditionary Force had come to set up operations in France, Pershing had made Major Billy Mitchell Aviation Officer of the AEF. At the same time, he had taken the control and direction of America's air effort away from the Signal Corps, organizing in its stead the Air Service as a branch of the AEF. The move was necessary because the Signal Corps was simply not equipped to handle the explosive growth of air power as an adjunct to its other duties. Pershing's act was not appreciated in Washington, and it was a year before President Wilson would approve the separation at home as well as abroad. In the interim, the move added the confusion of divided loyalties and commands to an already confused state of affairs, which became endemic at the time to the entire U.S. air effort.

Just before Benny Foulois's arrival as the first Chief of the newly made Air Service, Mitchell was promoted to colonel and appointed Commander in the Zone of Advance, which meant he was in charge of combat air training. In January 1918 he would become Chief of the Air Service, First Army, the top aerial combat position, and by that time he and Benny were locked in a dogfight of their own.

With the exception of belief in air power, everything about them was at odds: looks, manner, thinking, approach. Although Mitchell was not a West Pointer and, like Foulois, had enlisted as a private in the Army in 1898, he had been born to the purple and came from a noted Milwaukee family. His father, John L. Mitchell, had been a U.S. senator, and his grandfather a railroad tycoon. An equestrian and polo player of note, Mitchell was also fluent in French, which was of great importance, enhancing his popularity in the councils of the mighty and with the public in a land where few Americans spoke the language. At thirty-seven, a year younger than Foulois, he looked bold and hawkish, exuding confidence and exhibiting a style that complemented the zest with which he approached the multitude of problems confronting the struggling Air Service. It was through his own efforts that so much had originally been promised. It was much too much, for grandiose projections in France had run up against the hard realities of organizational and production difficulties at home. It was a characteristic of Billy Mitchell's that his thinking often outstripped what was possible, his enthusiasm and conviction carrying others along in his wake.

Foulois, a Mutt in looks to Mitchell's Jeff, lacked the color, the background, the flare of the extremely popular Colonel. He was an Army airman through and through, not very articulate but dedicated and informed, tough and adroit, his ego less subtle than Mitchell's but just as prideful. As the Army's top airman in the combat zone, his interests encompassed exactly those of Mitchell's in getting American pilots trained and in the air to do combat with the Boche. And in April, just a year after the United States had entered the war, the first American pilots began aerial patrols against the Germans, flying not American planes, however, but French.

It was in the area of tactics that Foulois, who was more a technician than a theorist, came to see eye to eye with Mitchell on the issue of utilizing the bomber in a strategic context. He was introduced to the concept by Lieutenant Colonel Edgar S. "Nap" Gorrell. Gorrell, who had been one of Foulois's handful of intrepid pilots of the 1st Aero Squadron in Pershing's expedition, was in command of the Technical Section of the Air Service. It was his job to recommend how American air strength would be employed, and he sent to Foulois a proposal on the best way U.S. bomber squadrons could be used strategically. Wrote Gorrell: "The object of the strategic bombing is to drop bombs upon commercial centers and lines of communications in such quantities as will wreck the points aimed at and cut off the necessary supplies without which the armies in the field cannot exist. . . ."

Foulois was enthusiastic over Gorrell's prospect. The plan was the first ever by an American airman who foresaw the use of air power as a strategic weapon, to be used on an "around the clock" basis attacking the industrial heart of an enemy, independent of the actions of ground forces.

At Christmastime, Benny and Nap, like Mitchell before them, visited Royal Flying Corps Headquarters and discussed the idea with RFC Commander Trenchard. The big, deep-voiced Englishman, who had gained his nickname "Boom" from his tonal quality, had already begun strategic operations from RFC fields around Nancy, and he offered to cooperate in every way. But, like so many good ideas, the plan became mired in the mud of General Staff thinking. In January 1918, Gorrell was relieved as officer in charge of Strategical Aviation. AEF Headquarters did not like that designation anyway, and the title was changed to GHQ Air Service Reserve. It was a harbinger, for the change was made to make sure there would be no misunderstanding. The proposed bomber force of fifty-five squadrons would not be independent of Army control, as was Trenchard's operation from British Army control. As Gorrell later put it: "The Air Service failed to secure the approval of the General Staff of its plans for the employment of this aviation and consequently suffered from the fact that its plan for the use of the Strategical Air Service was not synchronized properly, especially from a mental point of view of its employment, with the ideas of GHQ."

As for Foulois, like Mitchell, he was impressed by Trenchard's thinking and that of the Italian bomber builder Count Gianni Caproni, who was in turn a voluble spokesman for his countryman Colonel Giulio Douhet. Douhet, jailed for having criticized his country's conduct of the war, was a strategist who believed that air power alone could win wars, by bombing an enemy's industrial heartland and breaking the will of the population to resist. Foulois, quick to grasp the potential, adopted the concepts, although he was soon in no position to try them out.

To Mitchell, of course, Gorrell's thoughts matched his own, and it made no difference to him that Benny Foulois also agreed. Further, he resented Foulois's promotion to general while he remained a colonel. He complained to Pershing about Foulois's inefficiency, which meant he didn't like Benny's way of doing things. But it was true that in the spring of 1918, the AEF's Air Service was in a mess bordering on chaos, with thousands of newly trained pilots milling about waiting for promised planes that never arrived.[1]

Mitchell and Foulois clashed on this problem and many others, and due to Foulois's rank, Pershing had to step in to prevent Mitchell from being returned to the United States. The fact was that the General became fed up with both men and would have gotten rid of them, but he did not want to make a bad situation worse. Instead, in an effort to straighten things out, he brought in his good friend Major General Mason M. Patrick, of the Engineers, over Foulois to command the Air Service. Beyond that, Pershing valued Mitchell's combat qualities.

Oddly enough, so did Benny Foulois. To his superiors he could be obsequious or he could stick his jaw out and show his stubbornness. With Pershing and Patrick, possibly to protect his own damaged position, he exhibited the former trait, for after witnessing Mitchell's performance at the front, he admitted that Billy's combat talents were greater than his own. He then did something neither Mitchell nor most officers would have done. He recommended that the high-flying Colonel be appointed to the position he had been assigned by General Patrick as Air Chief of the First Army. In so doing, he stepped back to become Assistant Chief of the Zone of Advance, which dealt with training and supply. Whether he made the move as a tactical retreat in the face of what he realized were superior forces or because he believed his talents were better suited to solving the difficulties of training and supply is not known. Poker was a game Benny Foulois played well.

Far across the ocean, Hap Arnold was playing another sort of game and not liking it one damned bit! Try as he might to get a combat command, he was roped to a desk in Washington. It was a bitter pill for him, for he was frustrated not only by his inability to get overseas with the likes of Mitchell and Milling but also by the requirements of his many duties, which centered on the unholy mess of aircraft production. Not only could the originally projected numbers of planes and engines and parts not be met, but it soon became apparent that, all too often, those who kept making promises lacked the knowledge to fulfill them. Arnold saw also that too many with whom he had to deal were simply out to make a buck. Added to that was the almost total absence of previous training of personnel, from the machine shop right on up. Yet, in spite of "impossibility," within a year the United States had combat pilots in the air over France even if they weren't flying U.S. planes. And the Liberty engine would be the single most important aeronautical contribution to the air production effort.

Through it all, from his office near Union Station and his rented home off Dupont Circle, Hap Arnold fought a two-front war against military bureaucracy and industrial obfuscation. The nonflying brass that commanded the Aviation Section of the Signal Corps after Brigadier General Benny Foulois left for France, had little grasp for and less understanding of aircraft, training, procurement and production.

As Executive Officer and then as Assistant to the Director of the Aeronautical Division, Arnold was at the vortex of a vast, swirling glue pot. He not only took his work home with him, he brought civilian production people along too. Frequent heated arguments took place in his den, filling the small hours with sound and fury and cigar smoke. *And goddammit, if you can't do it, we'll find someone who can!*

His anger was up. He knocked heads. He locked horns. He got things done, or else! If anything incensed him it was cupidity and phoniness. The DH-4 was a case in point. It was agreed that American manufacturers would produce the two-place British plane. The plans for the plane arrived in August 1917. The following winter, before a senatorial committee, the question was asked, "How many combat planes have been shipped abroad?"

The witnesses finessed the answer, saying they didn't have the figures but they'd bring them the next day.

As soon as the hearings adjourned, a call went out to Dayton: Get a DH-4 shipped to New York at once by express. The crated plane arrived in the morning and was rushed to the dockside. It was aboard ship by the time the hearings resumed and the report was made: "We have shipped combat planes to France."[2]

On top of it all, the Arnolds did not like the Washington scene "worth a damn!" Military protocol with its tradition of having to make formal calls did not appeal to them. Neither did political gossip and backbiting as the principal indoor sports. His grin in those days was hard to come by. Bee stepped on the cat and their third child, William Bruce, arrived July 17, 1918, two weeks earlier than expected.

In the end, at the end, Colonel Arnold did get to France. But, before that, out of the confusion, out of the trials and errors in learning how to build an air arm from scratch, he had absorbed knowledge and gained invaluable insight into the requirements and needs of training and production. Twenty years later, the experience would be of immeasurable benefit to his country. As he was to say: "After World War I, the lessons of the failure of aircraft production remained uppermost in my mind."

What finally got him to France was anything but failure. Charles "Ket" Kettering, a scientific wizard and the president of Delco, had, in working with the Sperry Company, come up with a pilotless aircraft. It was dubbed "The Bug." The Bug was a twelve-foot-long papier-mâché bomb with cardboard wings. It could fly forty miles and, with its Sperry guidance system, hit within one hundred yards of a target. It carried a payload of three hundred pounds of explosives and took off from a short track powered by two 40-hp engines built by Henry Ford. Its cost was four hundred dollars, and after several successful tests, Arnold invited a flock of Washington brass to Dayton to observe the performance.

The devilish device apparently did not take to brass, for after a smooth lift-off, it climbed skyward, where it executed some unprogrammed aerobatics and then suddenly nosed over and dived on the assembled. Decorum and rank went by the boards. All fled from The Bug's wicked intent. Except for some skinned knees and ruffled feathers, there were no casualties. Ket made some modifications to tame his brainchild's erratic behavior, and at the next test The Bug performed as secretly advertised.

General Pershing, upon being informed of the weapon's potential, wanted to know more about it, and Colonel Arnold wangled the assignment to bring the details to him. It was October 1918. Unhappily, he journeyed horizontally to England, down flat with pneumonia via flu. By the time he had recovered from the illness and reached France, the Armistice was at hand.

So was Billy Mitchell. He was full of energy, bursting with ideas, some of which he had gotten to try out in the final campaigns of the war—at St. Mihiel and the Meuse-Argonne. In the fall of 1918, with fewer than seven hundred planes under his command, he was able to draw more than equal the number from Allied forces, so that when the attack began he had nearly fifteen hundred aircraft to launch against the enemy. He gained vital control of the air, which greatly aided the AEF's first major push. Mason Patrick, who had brought order to the running of the Air Service, was as impressed as Pershing and recommended that Billy be promoted to brigadier general.

In the Meuse-Argonne Offensive, which followed in late September, General Mitchell, in conjunction with the British RAF, had a chance to try out the theory of strategic bombardment.[3] In one mission, he dispatched a force of two hundred bombers against German troop concentrations behind the lines. This was followed by British bombers attacking similar targets. In the joint mission, seventy-nine tons of explosives were dropped, which was more than half the total tonnage dropped by the Air Service during the war.

The Armistice put an end to further employment of intended strategic air strikes far behind enemy lines but as Mitchell told Hap Arnold, not an end to his plans. He foresaw an interallied independent air force, seemingly oblivious to the fact that with peace there were to be no more wars and therefore little need for an air force, interallied or not. Already, Secretary of War Baker had warned that the Air Service had better not get any ideas about having a bomber force aimed at attacking cities or industrial targets. Further, as Arnold was well aware, there were few air-power proponents in the War Department, and Mitchell himself had claimed the only ground officer in the AEF who understood what air power meant was Major General Hunter Liggett, Commander of the First Army.

That might have been so, but Assistant Secretary of War Benedict Crowell had an open and interested mind on the issue of air power, and in the spring of 1919 he led an official commission to Europe to investigate some of the claims Mitchell had been advancing. Made up of Army, Navy and aircraft-industry representatives, the commission spent three months talking to air leaders in England, France and Italy. When Secretary of War Baker got wind of the enthusiasm being generated among some of the delegates for the establishment of a U.S. department of aeronautics, he cabled Crowell informing him that his commission was a fact-finding group and its conclusions were neither sought nor wanted; it was up to the Army and the Navy to determine what they did with aviation, and the Joint Army-Navy Board was perfectly capable of doing just that. In spite of Baker's admonition, when Crowell and his delegation returned, the commission—less its naval members—filed a report supporting the setting up of a department of aeronautics. The Secretary of War rejected the recommendations out of hand. So, of course, did the War Department, in addition to all further attempts made that year in the Congress by Mitchell to put U.S. aviation under a single agency.

Hap Arnold went home from his brief, unrewarding visit to France anxious for Mitchell to follow him so that Billy could grab hold of the reins and become the first peacetime Chief of the Air Service.

Benny Foulois came home too, and quickly sized up what he felt the War Department had in store for military aviation. He, like every air officer in France, had developed a deep-rooted resentment of two kinds of ground officers: those who all too frequently had been given command of air units not knowing the first thing about what was involved, and those in the upper echelons of GHQ who viewed the airplane with less than enthusiasm. Called in October 1919 to testify before the Senate Committee on Military Affairs, which was concerned with the future of the Air Service, Benny let go with both barrels: "The General Staff of the Army," he declared, "either through lack of vision, lack of practical knowledge, or deliberate intention to subordinate the Air Service needs to the needs of other combat arms, has utterly failed to appreciate the full military value of this new military weapon, and, in my opinion, has failed to accord it its just place in our military family."

Behind Foulois's blast, Mitchell's eagerness to command and Arnold's concern for the future, a rebellious desire was groping for formulation. Independence was its name. It had found fertile soil for gestation in the turmoil and upheaval of war. Peace and penury would bring it to the fore.

Three

Something for the Birds

In the meager times that followed the war, Arnold served on scattered posts in a variety of commands, struggling unendingly to keep alive an Air Service that at times seemed headed for extinction. In this endeavor, as the CO of Crissy Field, at the Presidio, in San Francisco, he was assisted by a lean-jawed, red-haired combat veteran, Major Carl "Tooey" Spaatz.[1] They had first met in 1917 at Kelly Field, and their friendship was to become one of the most important relationships in the development of military air power.

Opposites in temperament, Spaatz was a man of few words, sardonic, direct and always in control; Arnold was voluble, impulsive, his mind ever in flight. There was between them a connection that would endure. To Arnold, Spaatz was always the man he could depend on.

As a West Point cadet, Tooey—who had been tagged with the nickname as a plebe because a former redheaded upperclassman had also been so named—had wanted to take up flying as soon as he graduated, in 1914. He had been snared watching Glenn Curtiss cruising up and down the Hudson River in his pusher. But since the Army did not permit its new officers to join the Aviation Section until they had served in another branch, he went off to Hawaii as an infantryman.

The CO of the 4th Cavalry there was Colonel Ralph Harrison. He had a pretty daughter named Ruth who wished to become an actress. When Tooey first met her, he thought she was a cute kid. Several years later, when he had already become a pilot, they met again at Kelly Field and his thoughts turned to love. They were married the summer of 1917, just before he departed for France aboard the good ship *Baltic,* bound for Southampton, England.

Colonel Harrison, whose nickname was "Death" because he had a lean face that resembled a skull, had told his future son-in-law in no uncertain terms, "There's no future in flying!" And there were times after the war that the laconic airman might well have agreed.

Back in November 1915 he had begun flight training at Rockwell Field, San Diego. The course was simple enough: usually two to five hours of dual with Albert D. Smith or Oscar Brindley in a Martin tractor-type aircraft. But Oscar let Tooey solo after fifty minutes of instruction. The idea was if the student and the Martin survived, he kept at it, getting in about twenty minutes of air time a day. More time was spent on learning how to take the engine apart and put it back together and how to balance the prop so one didn't fly around chewing up the air going nowhere. As for ground school, there was no class in the theory of flight, the idea being that, like the sky, the subject was wide open for exploration.

When he graduated and became a Junior Military Aviator in the spring of 1916, Spaatz was ordered to join Captain Benny Foulois and the 1st Aero Squadron in Mexico. His plane was a Curtiss JN-2, commonly known as a Jenny. Its ceiling was less than six thousand feet, and as Spaatz put it, he "flew by ear" from San Antonio to join the troops. It was his first meeting with Foulois, also with Lieutenant Walter "Mike" Kilner, whom he later served under in France. Another meeting in France, of more strategic importance, was in the offing. At Cazaux he lived for a time in the same chateau with Colonel Billy Mitchell.

In September 1918, Mitchell would cite Major Spaatz for bravery, recommending him for the Military Cross.* Ruth Spaatz, working in Washington, D.C., would read the story on the front page of the New York *Times* during her lunch hour. The headline attracted her: FLYING OFFICER BRINGS DOWN THREE PLANES—TWO GERMAN AND HIS OWN. No name was mentioned, but she thought to herself, *That must be Tooey.*

He had started out the war actively enough, going on his first night bombing mission with the French in late October 1917 and continuing to fly observation and reconnaissance patrols over the lines. Then, less than a month later, he was made Officer in Charge of the 3rd Instruction Center, at Issoudun, far from the front. The months went by and Spaatz soon had his fill of running a training facility away from the scene of action. Worse, Mitchell wanted to send him home, because the new pilots coming over were poorly trained and Tooey was the one to improve the situation.

Spaatz, who usually didn't talk much, talked fast this time, and got permission from Mitchell to go to the front for two weeks. The pursuit pilots of the 2nd Pursuit Group and Charlie Biddle's 13th Squadron, who were second and first lieutenants, did not take at all kindly to having an inex-

---

* He received the Distinguished Service Cross and the irreversible designation of Military Aviator.

perienced brass-hat major showing up to give them orders. He gave none. He took off his oak leaves and became one of them.

His skill was quickly recognized, and Biddle soon made him a flight commander. In combat on September 20, he shot down an enemy fighter. The two weeks passed. The St. Mihiel-Verdun Offensive was under way. Spaatz stretched his stay for another week, and on September 26, he and his patrol became engaged in a fierce scrambling action. At one point there were six German Fokkers chasing his tail. Charlie Biddle saved him, but in the plunging, turning melee he shot down two of his stalkers and chased a third far behind the lines. On his return, the plane ran out of gas, and he managed to slip down and crash-land on the side of a shell-ravaged hill. When the dust settled, he found he was unhurt and quickly got clear of the wreckage, ready to go into hiding. Then he heard voices speaking French and knew he was in friendly territory. His new-found friends decided a celebration was in order, and when they finally returned him to home base his old friends there felt the celebration should be continued, by all means. But Billy Mitchell, and with him General Mason Patrick, had come calling, and while the drinks were being poured, Spaatz was congratulated and given hell for disobeying orders, having remained at the front for longer than authorized. He was promptly sent back to the United States with the promise that when he returned he'd have his own pursuit group. The end of the war put an end to that, and then the question was: What next?

What came next was trying to keep the Army's air arm from either withering away or being cut off altogether. Billy Mitchell, who had returned home in March 1919 not to become Chief, as hoped, but Director of Air Service Training and Operations, led the crusade.

A contribution by Spaatz was to command what was known as the Far West Flying Circus. This was for the avowed purpose of a Victory Loan drive, but in putting on stunting and dogfighting exhibitions over a large swath of the country, Spaatz was generating interest in the Army Air Service.

One grand innovation put together by Billy Mitchell that captured public and congressional attention was the Transcontinental Reliability Test. It took place in October 1919, and as the first air race of its kind, it was far more than just a gala sporting event. The idea was that thirty planes on each coast would take off and fly round trip between New York and San Francisco. Since none of the planes had a range that exceeded three hundred miles and there were no air routes and no airports along the way, it was a rugged challenge for Spaatz and the other participants. As many as forty refueling stops were needed to make the continental circuit in an average elapsed time of about thirty-nine flying hours—that is, for those who completed the course. Altogether there were nine fatal crashes, which took the bloom off the event, but Mitchell was out to illustrate two major points: that the country should establish airways and that America's vaunted belief in isolation should be thrown in the discard. His first point was swiftly adopted, his second for the most part ignored.[2]

The planes that took part in the race were mostly DH-4s, but Tooey Spaatz flew his favorite SE-5, and in December he received a letter of commendation signed by Major Oscar Westover, Executive to Air Service Chief Major General Charles T. Menoher. The General asked Oscar to convey to Tooey his "hearty recommendation" on the Transcontinental Reliability Endurance Flight, in which Spaatz was the winner in elapsed time West to East.

In a two-year period, Spaatz served on many posts, briefly with Arnold at Rockwell Field and then with him at Crissy, where they joined forces to keep military aviation in the public eye. Democracy may have been saved forever, but the public had to be constantly reminded of the need and role of the Army Air Service.

CO at the Presidio was General Hunter Liggett, who had commanded the First Army in France. A gifted officer, he understood and encouraged Arnold's and Tooey's aeronautical ventures, particularly when Hap brought into being the first aerial forest-fire patrols over California, Washington and Oregon.

Regardless of their threadbare service, the Arnolds and the Spaatzes loved the life at Crissy. Since there was no night flying, Bee and Ruth would walk out near sundown and stand by the fence at the edge of the field, waiting for their men to come down from the sky before twilight faded into darkness.

The Arnold zest for public relations was at its wacky best when, in a highly publicized test of flying machine over wildlife, he challenged a flock of homing pigeons to race him and his trusty DH-4 from Portland, Oregon, to San Francisco. Bets were placed. Excitement ran high. The start was a fizzle. It took forty-five minutes to get the cold Liberty engine fired up, and by then the pigeons were long gone. But they tarried somewhere and damned if Arnold didn't arrive first!

"Hello, Bee," he telephoned after landing. "I did it!"

"Did you really beat them?"

"Beat them? I murdered them!"

But there was more to come—an aerial first that Tooey and Hap literally *cooked* up. The event deftly characterized the Spaatz sense of humor. He had become friendly with Victor Hertzler, the popular Alsatian chef at the St. Francis hotel, in San Francisco. His plan was to fly Victor up to Petaluma, the chicken center of California, on the day of the great Petaluma Egg Festival. There, before cranking newsreel cameras, Victor would take on board the DH-4 a prize hen.

In the open cockpit with him would be an electric grill. On the return flight to San Francisco, the hen would lay an egg. Chef Hertzler would fry it on the grill, and upon landing, chef and fried egg would be rushed to the

St. Francis, and the first egg to be laid and cooked in flight would be served a la carte to the clicking of more news cameras.

Of course, no one was to know there was no place to plug in an electric grill in a DH-4, and it wasn't likely that even a prize hen would lay an egg in flight. The trick was that an egg would be prepared in the St. Francis' kitchen at the very time Hertzler made a grand entrance. It took all of Hap Arnold's powers of persuasion—which were formidable—to get Victor into the cockpit. Then it was onward into the wild blue to Petaluma. There the hen was carefully added to the passenger list, and, amid the cheers of the assembled, the intrepid Tooey headed back to San Francisco.

Alas, when they landed, Victor was both crushed and shaken. The prize hen had not enjoyed the flight and had decided to go it alone. It had flapped its way out of the cockpit and, when last seen, was spiraling down into San Francisco Bay. But, with the ground firmly under him, Chef Hertzler's recovery was swift. He rose to the occasion and promptly threw a gala party in celebration of his own survival. As the newspaper stories indicated on both occasions, the Air Service had given the public something for the birds.

It was 1922 when the Arnold family returned to Rockwell Field, where Hap was assigned as Commanding Officer. Now there were four children, Jackie having been born the year before. Lois, at seven, had golden curls. She had her father's quick temper, too. On her first birthday, he had given her a case of Napoleon brandy, which was to be kept inviolate until that day far in the future when she would marry. Of course, at the time, one of the bottles had to be opened in celebration of her birthday. And later, when coming back from the Philippines and the news was received of his forthcoming promotion to captain, naturally another bottle had to be opened. Then, in the summer of 1923, it was necessary to extract still another bottle from the case—this one to celebrate an aeronautical first of considerable magnitude.

Arnold's Engineering Officer at Rockwell was Lieutenant Lowell Smith, and the two had frequent discussions on the need for planes patrolling the Mexican border to have a greater radius of action. To enlarge the gas tanks on a DH-4B, or to add extra tanks, would affect its stability and slow it down so that even if such a redesign did work there would be little benefit. Out of these confabs emerged a new idea.

"As is usually the case, in advances that are taking place daily in aeronautics," Arnold was to report, "there were no precedents to follow. The idea itself was simple—send up one plane and send up another when needed, carrying gas, oil, water or food to be transferred to the duration plane."

The rest was not so simple. Mid-air refueling had never been attempted. It would not only require expert flying upon the part of the pilots but also modifications of the aircraft so that the necessary supplies could be transferred from one plane to the other. A forty-foot metal-lined steam hose became the connecting link. After a couple of dry runs that worked smoothly, Lieutenants Smith and John P. Richter took off early one August morning and didn't come back down again until the next day. When they landed, they had been in the air for more than thirty-seven hours and had covered over thirty-two hundred miles, breaking all existing records for endurance and speed.

The Arnold household had inadvertently become a part of the record flight. During the early hours of the second day, fog had moved in to blanket the area. The ceiling had ranged from two hundred down to fifty feet, the latter over the Arnolds' quarters, which Smith appeared to be using as a pylon on his triangular run. No one in the house got any sleep. Every time the plane made its turn, it seemed to be lower and louder, and who had ever accused a Liberty engine of being anything but a ring-tailed roarer? The master of the house needed no sleep. He was up, hopping about, calling excitedly, "They're going to make it! They're going to make it!" Bee wasn't sure the rest of the family was going to make it. By dawn the children were groggy and she was exhausted.

But after Smith and Richter did make it, out came a bottle of brandy to mark the occasion.

John Richter, after tasting his, said with a sly grin, "Hey, how about trying for a new record tomorrow?"

Although it wasn't tomorrow, two months later the pair extended their performance, flying thirteen hundred miles from the Canadian border to the Mexican, refueling planes meeting them along the way.

Forest-fire patrols, Mexican-border patrols, dropping food and fodder to snowbound Indians, mid-air refueling—these were ways Major H. H. Arnold was at work exhibiting the peacetime value and uses of a shrinking military air arm.

The National Defense Act of 1920 had foreseen an Army Air Service with a strength of over 1,500 officers, 16,000 men and 2,500 cadets. In that summer of 1923, there were 880 officers, about 8,300 enlisted men and 91 cadets. As for the aircraft, of the nearly two thousand planes of all types in the inventory, more than three quarters had been built during the war and were nearing obsolescence. General Mason Patrick, the Air Service Chief, sized up the situation before a congressional committee when he said: "The Air Service today is practically demobilized and unable to play its part in any national emergency, or even to meet the peacetime demands for service with its present inadequate strength and organization."

There were at least two major factors that kept an intelligent married man with a family in the Air Service: dedication to the cause beyond the call of duty, and a very understanding wife. Hap Arnold had both. With the former went vision and inventiveness. With the latter, appreciative children who thought their Pop, with his quick laugh, fearsome temper and masterful command of forbidden words, was the greatest.

He had ways of putting all these qualities together to prove it, and the affair of the mules, Big Red and Big Blue, was a memorable example.

The pair of stubborn devils had been given to the Arnolds as a gift. The question was what to do with them. Pop took their measure and thought he might sell them to the zoo for lion meat. The family would have none of this, and so, challenged by the mean look in the eyes of the pair, Arnold's inventive spirit soared, and he was off to the drawing board.

Down behind the wooden hangars lay a sorry graveyard of plane wreckage: broken wings, crushed fuselages, pieces and parts of what had once been flyable. Before the pile stood a tombstonelike marker with the letters I.C. They were not the initials of a downed pilot who had managed to destroy such a vast amount of government property, but stood instead for . . . Inspected and Condemned. To this desolate pile of aerial rubbish quietly journeyed the inventor, where he contemplated the battered bones of ended flight.

Came a Sunday. The children would go to Sunday school with their mother. Pop would head for the golf course. It was a ritual. Not so this day. He said he was staying home, and the rest of the family went off wondering what ailed him.

Upon their return, with grin wide he proudly led them to the field stables, and there they and their gathering friends beheld a strange and wondrous sight. Big Red and Big Blue stood hitched to a wagon bed riding on four wire-spoked airplane wheels with rubber tires. At the front of the wagon, side by side, were two metal aircraft bucket seats and in front of them an instrument panel mounted on two-by-fours.

"What is it?" came the excited queries.

"It's an air-service carriage," replied the proud inventor, "the first of its kind. Hop aboard, and we'll give it a test flight."

Not only did the Arnold children quickly accept the invitation but so, too, did a goodly number of their friends. Bee Arnold signed on as copilot.

The inventor took reins in hand and spoke to the team authoritatively but in Spanish.

"That's the worst Spanish I've ever heard," said the copilot. Big Red and Big Blue must have agreed, for they failed to move out.

"Try German," suggested the copilot, and then, suiting the suggestion to the word, she spoke to the mules in excellent German, bidding them to proceed. They responded and the air-service carriage was off on its test run to the cheers and shouts of its passengers.

Rockwell Field covered 740 acres of territory, and although it was not anticipated by the inventor that the test would cover all of it, suddenly it appeared that this might be the case—and then some. Out of absolutely nowhere a wretched cat shot from the grass in front of Blue and Red. Evidently, they both saw red, for they shied wildly and then were off at full throttle.

The inventor was on his feet, hauling back on the reins, trying to control the careening air carriage. The passengers, though in their Sunday best, were wise in the ways of survival, and they began bailing out. Shortly, there were two runaway mules bouncing one air-service carriage through acres of uncut weeds. The inventor had dropped the linguistics and was roaring in airman's English while he fought to slow their furious passage. His copilot was holding on as best she could when he thrust the reins at her and shouted for her to hold them. Then, in an exhibition that would have made movie cowboy heroes Tom Mix and Hoot Gibson weep, he climbed out on the writhing tongue of the wagon, forced his way to its head and, with one arm around the neck of Big Red and the other around the neck of Big Blue, he stepped off the tongue and let his weight drag the mules to a halt!

All the passengers had landed safely, but their Sunday clothes were pretty badly torn up. Rumor had it that Big Red and Big Blue were last seen on a truck headed for the zoo.

If 1923 was a year of high adventure for the Arnolds, it was also a year of sudden tragedy and punishing worry. All in a very short period of time, the baby, two-year-old Jackie, died of appendicitis; seven-year-old Hank—who was then called Bunky, after his stuffed dog—broke his arm in such a way that the doctor said he would never be able to use it properly, ending his father's plans for his namesake one day to attend West Point; five-year-old Billy Bruce came near death with scarlet fever; and Lois was nearly killed in an accident with a runaway pony and cart, ending up with a badly fractured leg. Almost overnight, Hap and Bee Arnold's hair turned prematurely gray.

Back in the spring of 1920, Billy Mitchell had written an optimistic letter to "My dear Arnold" saying, "I believe that a Department of Aeronautics will be passed this session. It appears now that it will include everything except the tactical air service of the Army and Navy. This is to be expected, will lead to air committees and then to an Air Force. . . ."

None of it had happened. None of it was about to. In the ensuing three years, Mitchell's actions—the demonstration bombing and sinking of the battleship *Ostfriesland,* and his public and congressional attacks on the Navy and War Department attitude toward air power—had, if nothing else, spurred the Navy to strengthen its own air arm and to move toward the building of aircraft carriers. Thus the competition between the two services over the issue of air power had begun in earnest. The fact was that the Navy Department was a better-organized, sharper and more sophisticated organization than its Army counterpart. Rockwell Flying Field, which was to become a depot, was a vivid case in point.

Army installations on North Island, San Diego, had begun before World

War I. The Signal Corps had moved in and built a raft of temporary buildings, hangars and repair shops. The Navy had moved in slowly and was well entrenched with more permanent installations by war's end. Secretary of the Navy Josephus Daniels had given the Navy the go-ahead to complete all buildings that were then in progress. Secretary of War Newton D. Baker, on the other hand, ordered that all Army construction stop posthaste. "The war's over," he said; "we'll turn the money back to the Treasury."

When Major Arnold had returned to Rockwell in 1922, he found it in a badly run-down state. With a complement of forty enlisted men, about the same number of civilians and a handful of officers, he worked to whip the place into shape, but as the Inspector General said in his April report of 1923: "It is unfair to compare this depot with that of the Navy alongside because the Navy has been able to get on with its building project and has been treated more liberally both as to money and personnel for upkeep."

Even so, the Inspector General's report was critical of Arnold's command, and Arnold took vigorous exception to the findings by the Assistant Adjutant General of the IX Corps area, Lieutenant Colonel J. O. Stegers: ". . . it is believed by me that this detrimental report is not only unfair but could not possibly have been made, had General Helmick [the Inspector General] known the condition existing at the field in October and then had made his inspection in April. . . ."

When Major H. H. Arnold felt he was being unjustly put upon, he was not inclined to take criticism from anyone, regardless of rank.

Such reports notwithstanding, most of what was said of Major Arnold by his superiors was complimentary, and in the summer of 1924 the Arnolds learned they were going East. The news was unexpected. He had been assigned to take the course at the Army Industrial College, in Washington. It was not duty he had requested, and he did not know who had recommended him.

Most probably it was General Mason Patrick, seconded by his Assistant, Billy Mitchell, who, unlike J. O. Stegers, was impressed with the Major's performance.

Patrick in his three years as head of the Air Service had come to view the problems of air power in much the manner of his vocal assistant. But whereas Mitchell took matters into his own hands, the Air Service Chief worked through the War Department and was more subtle and orderly in seeking support.

In 1923 the seven-man Lassiter Board, headed by General William Lassiter, which included Lieutenant Colonel Frank P. Lahm and Major Herbert Dargue, had recommended the things that Patrick needed in the way of money and new equipment. Or, as it found, "unless steps are taken to improve conditions in the Air Service, it will in effect be practically demobilized at an early date."

To push the Board's findings and generally to push the cause of the Air Service, Patrick had quietly encouraged Major Arnold, and no doubt his other tactical commanders, to do what they could to sell their needs to any influential public and private contacts they had. And in this regard, he must have approved Arnold's methods of salesmanship, because his idea was to bring the Major to Washington as Chief of Information as soon as the Army Industrial course had been completed.

The Arnolds were not eager to leave the sunny shores of California for the humidity and bustle of Washington, but it was obvious that the shift was not so much a lateral move as it was a vertical one. The bumpy road to peacetime rank, if not to glory, also lay in attending Army schools. In that critical period, General Patrick wanted an officer to head his Information Branch who was quick and savvy and wise in the ways of public relations.

Four

Mitchell and His Boys

Arnold brought the family to his parents' home, in Pennsylvania, in September of 1924 and journeyed on down to Washington to get organized, wishing that he had a house there for them all to move into. There was something about the city's atmosphere that rankled him. He found the political climate within the War Department on a par with the capital's high humidity. In the press of affairs, he almost forgot his eleventh wedding anniversary. His excuse to Bee: "This d--- town will make me forget that I'm human if I stay here long enough."

The one big highlight of his temporary bachelorhood was the arrival home of the Round-the-World Flyers. Another Mitchell innovation, four Douglas Cruisers had set out on April 6, 1924, piloted by two-man crews, their purpose to circle the globe. Six months and 27,553 miles later, the surviving two planes, the *Chicago* and the *New Orleans,* were coming into Bolling. Fortunately, the crews of the cracked-up *Seattle* and *Boston* had survived mishaps in Alaska and on the Atlantic that had prevented them from completing the circuit.

Arnold went out to the field to watch President Coolidge, his cabinet and assorted military and political brass welcome the globe circlers. Not among the officially invited, he stood on the outskirts of the crowd until the ceremony was over. Then, slipping past the guards, he came forward to offer his own congratulations to flight leader Lieutenant Lowell Smith and his compatriots. Smith was not comfortable receiving the adulation of the high and the mighty, and he wanted to break away and go somewhere for a drink with his former CO. Hap told him no, he was a celebrity now, and he had to behave like one. Nevertheless, Smith insisted Arnold ride back into town with him in his stylish Rickenbacker, which had been loaned by the firm for the occasion. Later they were joined at the Willard Hotel by

Lieutenant Erik Nelson, pilot of the *New Orleans,* who had been delayed by a forced landing south of Baltimore. The pair related their experiences and wished they could dispense with the official festivities now that the great adventure was over. Arnold invited them to have dinner with him the next evening, and they were only too pleased to accept.

The brief encounter was a genuine pleasure, but most of Arnold's time was spent in taking the Industrial College course and in preparation for assuming command of the Air Service Information Section. It was not a new job, for he had briefly held it during the war. What it encompassed was the handling of all Air Service public statements and publicity.

He finally found a place for the family to live, and they came down to join him in the fall of 1924, moving into a house in the District on Fulton Street. But, like their father and mother, the youngsters did not really like the scene. They longed for the West and the close-knit union that went with life on a post with other kids whose fathers were aviators. Who wanted to trade the sight and sound of DH-4s and Jennies, Le Pères, SE-5s and big Martin bombers for the rattling of trolleys, the fancy pants of Washington and trading black eyes with navy brats? But . . . if you were in the Army now, you made the best of it.

The most important thing about being in Washington, as far as their father was concerned, was serving with Billy Mitchell. They had kept in close contact, Mitchell ever on the move, proving the mobility of air power not only by his own example but also in the stimulation he brought wherever he touched down. There was an electricity to the man that inspired and lit up the thinking of his airmen and sent sparks flying among those who opposed him. He had become a convinced champion of and the leading U.S. military spokesman for an air force independent of General Staff control. To this belief Arnold, Spaatz and most of their fellow airmen had become firmly dedicated.

In 1921, Major General Charles T. Menoher, a stolid Infantry officer and the Army's first peacetime Air Service Chief, had resigned in an either-or confrontation over Mitchell's publicly stated position on the need for a Department of National Defense with the air a coequal branch. The shootout had followed the sinking by Mitchell's bombers of the "unsinkable" battleship *Ostfriesland* in a widely heralded demonstration of air power over sea power. Secretary of War John W. Weeks, at the time impressed by Mitchell, had backed him, and Menoher shouldered arms and marched off the scene.

He was replaced by sixty-two-year-old Major General Mason M. Patrick, in a repeat of Pershing's actions of 1918 in France. Mitchell had served under Patrick then and now did so once more as his Assistant, keeping his temporary rank of brigadier general.

Naturally, when Menoher resigned, Arnold again hoped that Billy would be assuming command. Mitchell hoped so too, feeling that the job should have been his in the first place. After Patrick was named, Mitchell attempted to dictate what his duties would entail, and the older man promptly let him know who was boss. Threatening to resign, Mitchell reconsidered and accepted the terms.

A man of considerable intelligence and foresight, Patrick came to understand what Mitchell was shouting about, both from the obvious conditions of the present Air Service force and also for its possible use at a future time. He did a better job of handling his free-wheeling subordinate than anyone else could have done—or perhaps anyone would have had the patience to do.

However, no one could really control Mitchell's drive and his public appeal. To sell his ideas on air power—the need for it to operate free of War Department constraint, to be a major defense force—he bombarded the press, the magazines, the editorial columns with his heretical claims. His book *Our Air Force,* published in 1921, asked his readers to consider what aeronautical development would mean to the future. He used the comparison of how swift that development had been during the war. The newly formed American Legion was an important forum, offering him a chance to speak out.

In all his efforts, he was supported and advised by a tight little coterie of free-thinking aides who—as well as officers in the field such as Hap Arnold —served him faithfully. Now, Arnold as Chief of Information was only too anxious to work hand in glove with hard-riding Billy. The goal was independence, the oft-repeated theme: Air power is the weapon of the future.[1] The battleship was obsolete! They had proved that with a brigade of Martin bombers off the Virginia Capes in 1921 and then again in 1923.

By the time Hap Arnold took up his public relations duties in the capital, Mitchell's star had reached ascendancy and was about to begin its plunge. A plethora of boards and congressional committees had been unable to agree on a plan that satisfied the struggling airmen. In September 1924, the Navy had refused to accept the recommendation of the Lassiter Board that there be a single appropriation for aviation. Anything that smacked of separation was steadfastly opposed. And anything that challenged the supremacy of the fleet or the Queen of Battle, the Infantry, raised blood pressure in the upper region of the Navy and War departments.

This was particularly so in October 1924, when Mitchell addressed the National Aeronautic Convention as the representative of President Coolidge. He spoke on the unmentionable subject of strategic bombardment and its effect on an adversary's means of production. This was the first time he had raised the theory publicly, and he was the first American airman to do so. He had already proclaimed that navies were going out of business. Now he suggested that aerial bombardment properly planned might make it unnecessary for armies to meet on the battlefield.

Naturally, the President did not subscribe to such ideas and he was annoyed that his representative did. But Mitchell was only shifting into higher gear. His address was followed by a series of five articles that, over the course of the next few months, appeared in *The Saturday Evening Post*. In these he attacked the thinking and the actions of the War and Navy departments for their failure to recognize the importance of, and to assist in building, a viable air arm.

The problem was that the cause was getting out of hand. Mitchell began to attack not only bodies but individuals. Further, General Patrick had come to fully recognize the need for the Air Service to be properly financed and to have a meaningful program. He had been quietly working within War Department bounds to bring that about. Mitchell, on the other hand, in making his charges, questioning the integrity of the Navy and the War Department in blocking aviation progress, aroused the ire of War Secretary Weeks and President Coolidge.

In his zeal, Mitchell could not realize that, while the American public enjoyed seeing and reading about aerial exploits, neither the public nor its elected representatives nor the government saw any threat of war in the foreseeable future and so could not take very seriously forecasts of their cities being bombed at some future date. And when the Secretary of War was cited for his failure to take proper cognizance of Mitchell's report on the inevitability of a future war with Japan—following the airman's official tour of the Orient in 1924—Weeks had had enough. He maintained that an officer was free to offer his opinion before a congressional committee so long as he made it clear that his opinion was his own. Mitchell's opinions, however, were reported as gospel.

In February 1925, when the New York *Times* quoted him as saying the United States had only nineteen planes fit for war, Weeks, according to Mitchell, threatened that if he didn't stop making his claims he was going to lose his job.[2] As Air Marshal Trenchard so aptly put it, "Mitchell tried to convert his opponents by killing them first." In so doing, he destroyed himself. When March came, Weeks, with the President's approval, refused to renew Mitchell's appointment as Assistant Chief of the Air Service. His rank reverted to colonel, and he was ordered to Fort Sam Houston, San Antonio, Texas, to be Air Officer of the VIII Corps Area.

Hap Arnold arranged the farewell luncheon at the Racquet Club. There, Billy's boys assembled to give their hero a proper send-off. He delighted them with a fighting speech, and the barbecue dinner that followed that night was an occasion to remember.

Mitchell's stay in Texas was a little like Napoleon's exile to Elba. He plotted and planned to continue the fight. Occasionally, Arnold would receive a cryptic note indicating that things were moving along well and to "Keep going as you are." Unlike Napoleon, Mitchell was not quiet, even out in the sagebrush. The climax came on September 5, 1925, when the

U.S. dirigible *Shenandoah,* caught in a storm, crashed in Ohio with a large loss of life, and a Navy plane attempting to fly from the West Coast to Hawaii was reported lost.

Later, Mitchell's wife, Betty, was to say that if only she had been with him at the time, she could have kept him quiet.[3] It is not likely, for his decision to act was obviously premeditated, not a sudden shooting from the hip. When he summoned reporters following the aerial mishaps, he handed them a nine-page mimeographed statement. In it he accused the Navy and War departments of ". . . incompetency, criminal negligence and almost treasonable administration of the National Defense. . . ."

Several days later, he again spoke to the press, announcing that he was going on the "warpath" to improve national defense and nothing was going to stay him from his task.

It was the White House that went on its own warpath. Two weeks after the *Shenandoah* tragedy and the safe recovery of the crew of the missing Navy plane, the War Department announced the decision to court-martial Colonel William Mitchell. It was President Coolidge who preferred the charges. However, before the court-martial got under way, the President had moved to head off Mitchell's claims and the attendant publicity. He had done so by appointing a board of noted citizens to look into the state of military aviation. The group was headed by his friend Dwight Morrow, a J. P. Morgan partner.*

The Morrow Board commenced its hearings on September 21, more than a month before the court-martial began, and the investigation was completed by the time the trial got under way. It was before this board that General Patrick presented his plans for an air corps.

From the beginning of the fight—first before the Morrow Board and then at the court-martial, Hap and Bee Arnold and all those of Mitchell's boys on the scene were closely caught up in the conflict. None more so than Tooey Spaatz.

In June 1925, he had graduated from the Air Service Tactical School at Langley Field and had been assigned as Assistant G-3 for Training and Operations in the Office of the Chief of Air Service, right next door to his buddy Hap Arnold.

While attending the Tactical School, whose avowed purpose was to train officers in the War Department's view of air power, Spaatz had been summoned to Washington to give testimony before the House Military

---

* Along with Dwight Morrow, the board was made up of Judge Arthur C. Denison, Vice-Chairman; Dr. William F. Durand, Secretary of the National Advisory Committee of Aeronautics; Senator Hiram Bingham, Military Affairs Committee; Senator James S. Parker, Interstate and Foreign Commerce Committee; General James S. Harbord, president of RCA; Admiral Frank F. Fletcher (Ret.); and Howard E. Coffin, former head of the Aircraft Production Board during the war.

Affairs Committee on the Curry bill providing for "a separate Air Service."[4]

During the train ride to the capital, the Major scribbled his thoughts on his travel orders:

> National policy as to Army based on assumption that Navy in being will keep out enemy until Army is mobilized and trained. Navy cannot. Air force can.
>
> Army will not seek increase of air force at expense of men or independently of Army. Would rather have a capable mechanic with a peg leg take care of my airplane than best disciplined soldier in Army who had no mechanic's duty.
>
> Army is different from Navy. Could not operate as part of Army.
>
> Air Service is just as different from Army and Navy as Army is from Navy.
>
> Air Service cannot operate under Army or Navy any more than Navy could operate under Army. Air force must be available at outbreak to defend coasts. If separate, can be centralized to be flown to either coast—if under Navy would be close to shore and split up.
>
> If given to Navy—stations built up along the coast no line of retreat except to flanks. Therefore, cannot defend in depth.
>
> Army still has coast defense.
>
> At outbreak of major emergency air force operates independently.
>
> As soon as armies are raised the air force would be needed by Army.
>
> If under Navy as soon as Navy fight is finished has no further need for air force. Large strength will be idle, if attached to Army will operate under conditions unfamiliar.

Spaatz testified accordingly in favor of the bill, and then got off a quick letter to William Stout, who was to become one of the principal developers of the Ford trimotor. At that time, Stout had his own aircraft company in Detroit, and Spaatz, who knew him well, was seeking not only his support for the bill but also that of "more prominent citizens like Mr. Henry Ford, to publicly espouse the cause of [a] Department of Aeronautics somewhat along the lines of the Curry bill."

"A separate Department of Air," wrote Spaatz, "headed by a Minister in the Cabinet, who can be held directly responsible by the country for the development of aviation, should provide the impetus now lacking for the proper development of the civil aviation and should place our military aviation in a more satisfactory condition to meet our defensive needs."

Stout couldn't have agreed more. The bill was a step in the right direction, and he was sure it or another like it would go through. As for Ford, he didn't think he would publicly support any bill unless a national emergency was involved. "But," wrote Stout, "we may be able to get him to see it that way even now." In the meantime, he would try to enlist the aid of others, "although the biggest of these organizations move almost as slowly as the government departments. . . ."

Indeed they did, and the Curry bill, like others of a like nature before and after it, was overwhelmed by congressional and administration support of the traditional Army-Navy position.

The following month, Spaatz, Major T. DeWitt Milling and First Lieutenant Charles B. Austin were again ordered to Washington, this time to testify before the Lampert Committee of the House of Representatives, which had been conducting an investigation, since the spring of 1924, into all U.S. aviation.[5]

But all the committees, congressional and otherwise, all the claims, and testimony far-reaching and ground-bound, could not encompass the reality of a career in the Air Service. Back at Langley Field, Tooey Spaatz was daily concerned with the tactics of flight and some of its harsher effects. On Sunday, May 10, 1925, he wrote in his diary:

Benedict was killed Friday last by colliding with a balloon he was attacking. Inference: at our age muscle reactions lag behind mental impulses. Aviators after 30 years of age must allow a large majority of safety in their flying.

In the class of thirteen with him were Majors Jake Fickel, Walter Frank, Jim Chaney, Fred Martin and Clarence Tinker.

The next day he would write:

Class for the first time since Benedict's accident. Mental effect on men like Fickel and Frank, who have had little direct contact with crashes, is terrific. My mind must be case hardened, since my reactions in these cases are not severe, a few thoughts as to whether it is worthwhile and an attempt to follow the reason for being worried or afraid of lesser affairs when I am not worried or afraid of death and then back to the normal existence of eating, sleeping and occasionally some thinking.

In this last regard, he read and reflected:

A line in *Steeplejack* [a book by an unknown author]: "The sensory periphery is more masterful than the limb of his being." Another: "Time is the glittering crest of a moment, not one of a series of beads strung out through eternity."

And then such a moment:

While cruising over the beach at Grand View in an MB [Martin bomber] intent on looking over the ground to find a road to the point, a Jenny crossed just a few feet in front of me and in the same altitude, did not see it until it was past. My heart skipped a beat from the narrowness of a collision.

And on that same day:

Led a Martin bomber formation which never did get together. Some said I climbed too fast, others said I flew too slow and the remainder said I flew too fast. Judging from the flying today the effects of Benedict's crash are still lingering in a few brains. A crash always does make one slightly apprehensive and a little more cautious for a more or less extended period. Forgot to note that yesterday Ruth and I called on Miss Naylor. She is caring for Mrs. Dorothy Benedict's three children while Dorothy Benedict is at Benny's funeral at West Point. Both Miss Naylor and Mrs. Benedict [Miss Potter] were Red Cross workers at Issoudun. Benny was O.C. of training while I was C.O. there. My recollections of Issoudun are very vague in spite of the fact that I put in almost a year at that place building up the organization for which Hi Bingham later claimed false credit.[6]

The following day, he was back to reading *Steeplejack* and quoting:

"But to pass an interval between two eternities raking in gold is simply absurd to me." My creed up until now but henceforth I aim to acquire sufficient gold to cease any worry about old age. It may be ridiculous for an aviator to worry about old age, but I do coupled with apprehension that someday I may quit my Army career either thru my own volition or otherwise.

He was right on both counts. He had seriously considered the idea several years past with his pal Davey Davison. Davey wanted to be a rancher, and he convinced Tooey that he wanted to be one too. There was certainly no future in the Army Air Service the way things were going. They took a trip down into Imperial Valley, in Southern California, and came back to their wives with glowing stories of the spread they had found. All it would take was fifty thousand dollars apiece. Tooey flew east to Boyertown, Pennsylvania, where he had been born and raised and his father had owned the local newspaper. He went to see Dan Boyer the banker, who said, sure we can take care of you. Back Tooey flew with the good news, only to learn the bad. Davey had been seriously injured in a crash. He'd never be a rancher, or a pilot again either.

Aside from the tragedy that had befallen Davey, Ruth Spaatz was not all that unhappy that her husband wouldn't be swapping a cockpit for a saddle.

As for leaving his Army career for reasons that were "otherwise," he looked forward to testifying at the court-martial trial of Billy Mitchell.

When the Mitchells arrived in Washington to appear before the Morrow Board and to answer charges at his court-martial trial, the Arnolds were there at Union Station to pry them loose from an enthusiastic welcoming crowd. After the daily hearings before the board, which heard ninety-nine witnesses, and then, later, when the court-martial proceeding adjourned for the day, along with Arnold, Mitchell's close supporters such as Spaatz, Bert Dargue, Bob Olds and Horace Hickam were with him at the *Anchorage,* going over the testimony, preparing for the day to come. They knew their cause was just, but they came to realize that their leader was making a botch of his defense and all their hopes for a separate air force.

Before the Morrow Board, Mitchell had driven his supporters crazy with his long-winded reading of his book *Winged Defense.* Arnold had wanted to shout: *Come on, Billy, put that damned book down!*

But Billy went on and on, and Dwight Morrow was too wise to interrupt him. When Senator Hiram Bingham, a former air officer, attempted to do so, Mitchell's response was, "Senator, I'm trying to make a point."

He failed to make it, but Hap Arnold and Bert Dargue did in a dramatic fashion. One of the questions before the Morrow Board was the strength of the Air Service. On an afternoon as the board was exploring the problem, the air above was suddenly filled with the roar of low-flying planes. There was a rush to the windows.

A flyover of thirty-six heterogeneous aircraft was going on, all the planes that were available to protect the eastern seaboard and the nation's capital. There would have been thirty-eight, but two of the planes could not get off the ground. Majors Arnold and Dargue had arranged the graphic display, but it was rumored the sound had awakened the President from his afternoon nap, and he was not pleased. Who was responsible? unsmiling Cal wanted to know.

The court-martial trial became a much livelier forum. Mitchell's boys were there as witnesses ready to dish it out.

"Do you think the General Staff should always listen to your recommendations?" asked General Robert L. Howze, one of the court judges.

"As the General Staff is now constructed, I do," snapped Captain Olds.

"And how would you construct the General Staff?" General Ewing E. Booth asked sarcastically.

"On Colonel Mitchell's plans," came the quick response.

When Major Arnold was called to take the stand, Congressman Reid, Mitchell's defense attorney, queried him on U.S. air strength, which Arnold said was about half that of France or England. Reid read a statement by Brigadier General Hugh A. Drum that stated that the United States, from a flying-personnel standpoint, compared favorably with all foreign powers.

In cross-examination, the prosecution asked if the Major meant to charge the General with inaccuracy.

Any major who did so, particularly if the General was an assistant chief of staff, had to have his nerve. "I mean that he [Drum] gives the impression that our air service compares favorably with other large powers' air service, and I mean to say that it doesn't compare favorably," replied Arnold.

"And that is your opinion about his accuracy?"

"It will have to be accepted as more than opinion, because the figures show that whereas we have only eight pursuit squadrons, England has thirteen, France has thirty and Italy has twenty-two."

The prosecution tried to get the Major to admit that geography was a factor in size. "You consider, then, in order to be as well off as England is, the United States must have just as many airships and personnel as England?"

"I think that makes no difference in aerial war, where distance is annihilated by a few hours."

"Is three thousand five hundred miles of salt water annihilated?"

"Yes, sir. It is today."

"In what respect?"

"Airships have crossed the Atlantic and Pacific." He went on to insist that such planes could carry weapons. On the issue of casualties in wartime, he quoted General Staff figures, indicating 23 percent replacements for the air and 7 percent for the Infantry. In presenting his figures he took a crack at the General Staff for issuing "three or four sets of figures" on casualties.

The press and the public ate it up, but it did nothing to change the foregone verdict. Throughout the trial, the Arnolds and their friends were there. By General Patrick's order, Captain Ira C. Eaker had the job of seeing to it that Mitchell had all the official Air Service documentation he requested, and they all sat together in the drafty old hall that had been purposely selected for the court.

One of the first defense witnesses had been Tooey Spaatz. Previously, General Patrick had warned him, as he had Arnold and the other Mitchell witnesses, that he had better be very careful what he said or he would be jeopardizing his career.

On the witness stand, Spaatz's figure for the number of planes in the United States modern and fit for duty was a bit higher than Mitchell's, totaling fifty-nine. He pointed out, however, that there were no properly equipped pursuit planes, and to put fifteen of them in the air would require taking all the administrative officers from their desks in his office. "It is very disheartening to attempt to train or do work under such conditions," he said.

When Mitchell's defense counsel, Congressman Frank Reid, asked Spaatz if he thought aviation was being retarded by the War Department, the prosecutor vigorously objected but not loudly enough to drown out Tooey's "I do!" The crowd applauded; the War Department took note and one of the generals on the jury asked:

"Major, you are an officer in the United States Army?"

"Yes, sir."

"Well, what do you mean by criticizing the War Department and the General Staff?"

Spaatz was silent for a moment and then said, "Well, I'd like the recorder to read back over my testimony and read exactly where I criticized the General Staff."

It was the noon hour, and after the recorder had spent about thirty minutes trying to find something that would satisfy the General's implication, he withdrew the question and everyone retired for lunch.

In his testimony and under cross-examination, Spaatz made clear, as did other witnesses, the sad state of the Air Service. Major Arnold pointed out that in less than six years, five hundred and seventeen officers and men had been killed in crashes and only twelve of these fatalities had occurred in planes manufactured since the war.

Captain Robert Olds gave a grim illustration of command decision from the ground up. His CO in Hawaii had been Major Sheldon Wheeler. Wheeler had been given hell by Colonel Chamberlin, Chief of Staff to Major General Charles P. Summerall, because Wheeler's squadron was having a great many forced landings. The General, Chamberlin said, wanted the forced landings to stop forthwith. Any pilot who had one would be made to pay for the damage to the plane.

The next day, Wheeler took off in a DH-4. The engine quit at two hundred feet. Instead of following the tried and true practice of landing straight ahead or slipping down to take the impact on the wings, the Major attempted to turn back to the field. The plane spun in, and he and his mechanic were killed. Olds blamed Wheeler's attempt to do a 180 on Chamberlin's order and thus on Summerall. The latter was to have headed the Mitchell court-martial board but had been challenged by Billy on grounds of "prejudice and bias."

News of Sam Wheeler's death had been a particularly heavy blow to Spaatz. As classmates at West Point, during the Army-Navy game of 1913, he and Sam had accomplished a feat that had to rank high in the annals of cadet derring-do. Together they had managed to swipe "from the goat herders of the Navy" the goat's vaunted blanket. No doubt the mascot keepers had their eyes too firmly fixed on the game. In any case, Tooey and Sam had divided the spoils and the glory between them, brothers in crime to the end. The sad thing was the end had come for Sam, the victim of run-down equipment and boneheaded orders.

The War Department's position was unofficially enunciated by General Charles E. Kilbourne, who stated, "For many years the General Staff of the Army has suffered a feeling of disgust amounting at times to nausea over statements publicly made by General William Mitchell and those who follow his lead."

During and after the trial, Spaatz received an assortment of telegrams and notes. "Congratulations on your testimony and nerve," said one. "Thataboy!" wired Captain Frank O'D. "Monk" Hunter, fellow warbird and pursuit specialist.

Spaatz's responses were optimistic, one reason being that he felt the case was finally bringing the kind of public attention and publicity that was needed to show how badly off the Air Service was and how badly it was being treated by the War Department. "We are doing everything possible to bring the Air Service out on top," he wrote. "Most of the evidence is in our favor and with that as background it seems as though we should be able to ride over the Army-Navy trust."

There was high drama in it—no one could deny that—and they were all caught up in the excitement, spurred on by Mitchell, quietly scoffing at the War Department's most vocal disparagers such as General Hugh Drum. Chief of Training and Operations on the General Staff, Drum's outlook was epitomized by the declaration that well-directed antiaircraft fire could bring down any attacking plane.

Of the twelve Army officers weighing the evidence, Major General Douglas MacArthur had known the defendant since boyhood. The two families were close neighbors, the relationship spanning three generations, Mitchell's Scottish-born grandfather and Judge Arthur MacArthur enjoying an association of half a century; Billy's sister Janet had been Doug MacArthur's first love.

The fact was that most of those on the jury knew Mitchell and considered him a gallant officer although a brash one. The atmosphere in the dingy courtroom was never somber; frequently it crackled with laughter.

Then suddenly something ugly happened that soured the whole thing for the Arnolds. It took the smile from Hap's face and put an angry glint in his eye. It brought worry to Bee and a tense atmosphere to the home on Fulton Street. They began receiving threatening letters and crank telephone calls. The thrust was that Arnold would get his and so would his wife and kids. He stopped singing in his off-key baritone and doing his magic tricks for them. There was only one way they could still some of the fear the letters brought. They loaded the children on the train and took them north to Ardmore to stay with their grandparents. The attack was a vicious sort of thing, indicating the depth of bitter feelings that underlay the issues involved.

On December 17, the court found Billy Mitchell guilty, as charged, of conduct of a nature to bring discredit upon the military service.

The unprecedented sentence was five years suspension from duty without pay or allowances. Coolidge reduced this to half pay. The idea behind the punishment was to force Mitchell's resignation, which it did on February 1, 1926. Mitchell maintained he could do a better job fighting for the Air Service as a civilian. Although the verdict was a foregone conclusion to his supporters, it was a hard blow combined with more to come. The only positive result for the Arnolds was that the crank mail stopped, so they brought the children back home.

As to the main issue, even before the verdict on Mitchell, the Morrow Board had filed its report. Through the prestige of its chairman and fellow members, its findings set in motion the legislative and administration forces necessary to bring about an aeronautical policy that would become the platform on which U.S. aviation progress was based for the next fifteen years.

The political premise of the Morrow Board was isolationism, the military premise that the United States was safe from attack by air. Military and civilian aviation development should be kept separate. The Army and the Navy should maintain their own air arms. There was no need for a supposedly costly national department of defense and certainly not for an independent air force. There would be an Army Air Corps in name if not in fact, a five-year buildup in aircraft for both services and additional temporary promotion for flying officers. One positive step was the recommendation that assistant secretaries for air be appointed to serve the Army, Navy and Commerce departments.

Very few of these recommendations went down well with Mitchell's followers and particularly with Major H. H. Arnold. He realized that with Mitchell gone from a position of command, he was now the leader in a very one-sided battle that pitted a handful of "undisciplined flying officers of junior rank" against the massive bulk and power of the War Department and the Coolidge Administration. It could end only in a massacre.

At the time of Billy Mitchell's resignation, the House Military Affairs Committee was holding hearings on the Morin bill, embodying the Morrow Board recommendations, and on the Patrick bill, which sought if not outright independence at least an Air Corps modeled along the lines of the Marine Corps. The War Department was completely opposed to the latter. Assistant Chief of Staff Major General Fox Connor testified that an Air Corps within the Department "would create an impossible situation."

Arnold was still furious over the scare tactics that had been used to threaten his family. He was also discouraged by the treatment Mitchell had received and by the attendant failure to gain congressional support on the collective efforts that had been made.

Since in the past General Patrick had encouraged him to use public relations as a means to sell the needs of the Air Service, and believing the Patrick bill must be passed, Arnold took matters into his own hands. It was a little like setting off the fireworks at West Point; only, now it wasn't a joke. Now he was motivated by more than youthful brashness. First he wrote a letter. Then he enlisted the aid of old friend Bert Dargue in its publication. The two conspirators drafted air reservist Captain Don Montgomery, who was on duty in the Air Service section of the War Department. There a stencil was cut and a mimeograph machine was used to crank out the finished copy. It read:

> We have tried to put across the idea of reorganization in which the Air Service can be developed and operated so that it will be able to give its maximum efficiency and effectiveness.

There are two Senators from your State and a Representative from your district. Also you must know people of prominence in your State who can communicate with the Senators and Representatives, people whose communication will be given more than casual consideration. It is to your interest that you get in touch with these people as your future in the Service will depend largely on legislation in this session of Congress. Get them to back the reorganization of the Air Service along the lines outlined herewith so that their Senators and Representatives in Washington will know what the folks back home want.

This is your party as much as it is ours. We all must get busy and do it now. Next month will be too late. We are relying on you to do your share of the work. Do not throw us down.

In order to conceal the circular's origin, no identifying marks were made on it. Enclosed with the letter went an outline of the Patrick bill, urging, as the New York *Times* said, a "separate air force." The mailing went to every Air Service officer, regular and reserve, that Arnold and Dargue felt would cooperate.

Certainly such a drive to influence congressional thinking, boldly sent out under the noses of the General Staff, could not be kept secret. As Tooey Spaatz was known to say, "It became one of those days when we had more you-know-what than there were fans."

On February 8, the New York *Times* headline and subheads ran: "OR-DERS ARMY INQUIRY INTO AIR SERVICE—Secretary Davis Takes Action on Circular Issued from General Patrick's Office—It Urged a Separate Corps—Letter Also Called for Pressure on Congress by Friends of the Service—Mitchell Attacks Move—He declares that it is another attempt to silence men of experience."

The *Times* also quoted General Army Order No. 20, which stated: "Efforts to influence legislation affecting the Army, or to procure personal favor or consideration should never be made except through regular military channels. The adoption of any other method by an officer or enlisted man will be noted in the military records of those concerned."

The Inspector General of the War Department, Major General Eli A. Helmick, was called in to investigate and find the guilty parties. In concert, General Patrick launched his own, internal inquiry. In very short order, the two culprits were identified. The word spread that Major Herbert A. Dargue, who was Patrick's favorite, had confessed and double-crossed his pal. The rumor was totally untrue. Dargue was a man of few words in any situation and of absolutely none in this case. Confession probably came from those who had done the office work of stenciling and mimeographing.

Major Henry H. Arnold was called before his commanding officer and told he had a choice. He could resign from the service or he could face a court-martial. He had twenty-four hours to make up his mind. The General's white mustache was at its bristling best. He was incensed. Essentially, Patrick's ideas on air independence had developed to run almost parallel to those of Mitchell and his followers. His own bill would have brought into being an air corps directly responsible to the Secretary of War and not to the General Staff. There was to be a single overall commander to control all air operations, plus a single budget and a single promotion list. (These were the points Arnold had stressed in the outline accompanying his letter.) But the court-martialed Mitchell getting into the act and claiming that the whole incident was another attempt by the War Department to bludgeon the Chief of the Air Service into silence before Congress had only made matters worse. Patrick didn't physically throw anything at the Major standing at attention before him. Instead, he verbally threw him out of his office.

Hap Arnold went home that day sick at heart, feeling completely defeated. He was thirty-eight years old, with the end of his military career staring him in the face—until Bee took his hand and stared him in the face.

"Sunny, they told you to do it once," she said quietly.

"I know."

"Aren't there letters in the files to prove that?"

"I suppose so."

"Well, you're not going to leave the service this way. I would just love a court. I'm ready for it!"

"Think of the children, think of your parents. Think of my parents."

"Look, you're not going to leave the service. You've worked so hard, and you love it."

"I don't know who to talk to."

"Who do you know in the Judge Advocate's Office?"

When he had told her, she said, "I want you to talk to him. Don't take my advice. Talk to him somewhere. Meet him someplace, maybe out in the park underneath the trees or something, but just put all your cards on the table, what you've done. Does it deserve a court or to be relieved?"

He did as she suggested, and the next day he reported once again to Mason Patrick. "I'll take the court-martial, sir," he said.

The General's face turned the color of his mustache. He could not help but know that a trial would bring out evidence to show that in the past he had encouraged Arnold to seek the same kind of congressional persuasion intended by this latest solicitation. After all, Arnold and Dargue had been discovered doing no more than making an end run to try to solicit support for Patrick's own bill—which was doubly embarrassing to the General. He summarily dismissed the culprit and promptly backed off.

On February 17, at 11 A.M., Secretary of War Davis held a press conference in which he told the assembled: "General Patrick submitted to the Secretary of War this morning the report of the investigation undertaken to determine the parties responsible for the unauthorized attempts to influence legislation affecting the Air Service. The matter of the necessary disciplinary action has been left by the Secretary of War in the hands of General Patrick."

A half hour later, General Patrick held his own press conference. At it, he first made clear that no one in the War Department was attempting to bludgeon him, as Mitchell had claimed, that prior to testifying before the House Military Affairs Committee, he had discussed what he planned to say with the Secretary of War, and Davis had made no objection. Further, although he believed his own position on air power was correct, there was an honest difference of opinion, and so forth.

He went on to condemn the action of officers in his office who, entirely without his knowledge and through "mistaken zeal," had endeavored to influence legislation in what he regarded as an improper manner.

The punishment for the crime, he said, was to be determined by Secretary Davis after consideration of his own report and recommendations.

"The investigation disclosed the fact that only two officers in this office were concerned in an attempt to influence legislation in which I regard as an objectionable manner. Both of them will be reprimanded, and one of them, no longer wanted in my office, will be sent to another station."

The station was Fort Riley, Kansas. The officer no longer wanted was given thirty-six hours to get out of town. Later, this was shortened to thirty-two.

When nine-year-old Billy Arnold came home from school that day, he was surprised to find his Pop sitting on the floor with his back against the newel-post. He was humming an old madrigal in his off-key baritone, tossing books into two boxes—one for those he wished to keep and the other for those he didn't. He looked up at his son, his smile back in place. The fact was the entire Arnold family was happy to see the last of Washington, no matter how short a time they had to clear out.

Only one further incident disturbed their departure, and to Arnold it exemplified his feelings about the place. He had paid some utility bills for gas and water costs in advance, and the company refused to refund the badly needed cash. He was not the kind to forget the slight, and he vowed should he ever return to the area, he would not live in the District of Columbia again.

Five

The Exile

The Arnold family's arrival at Fort Riley, the Army's famous Cavalry post on the plains of Kansas, was both an anticlimax and a measure of things to come. Major Arnold had been assigned as CO of the 16th Observation Squadron, and when the train pulled into the post station that night, the only officer there to meet them was the Flight Surgeon, Captain Fabian Pratt. After the long journey, the Arnolds were all very weary and not a little apprehensive, and then suddenly on the platform before them was this smiling medico. He was in charge of the squadron because all its other officers were elsewhere. But, more to the point, he had arranged for everything: quarters, a hot meal, transportation. Bee rightly thought he was wonderful. But she also knew that after the children were tucked in bed she and her disgraced husband had an official duty call to make that she devoutly wished could be avoided.

The commanding officer at Fort Riley was Major General Ewing E. "Barnie" Booth. Booth had been one of the judges on the Mitchell court-martial board. As Hap and Bee Arnold walked the quiet, dimly lit way to the General's home, Hap wondered if his new CO might throw him out into the street as Patrick had thrown him out of the nation's capital. They saw the house was all lit up. The General was obviously entertaining. The Major gripped his wife's hand and rang the bell.

A card party was in progress, the living room filled with officers and their wives. The callers stood blinking in the sudden light, all eyes on them. General Barnie Booth rose and came across the room. Then, with a broad smile, he thrust out his hand and put his other on the Major's shoulder.

"Arnold," he said so everyone could hear him, "I'm glad to see you. I'm proud to have you in this command. I know why you're here, my boy.

And as long as you're here you can write any damn' thing you want. All I ask is that you let me see it first."

The greeting was such that it was all Bee Arnold could do to keep from crying. The strain on her had been enormous.

The unexpected welcome by Barnie Booth set the tone, and the next two years were filled with the kind of zest and activity that the Major and his family dearly enjoyed. And if there weren't many planes to fly, there were plenty of horses to ride and a wonderful sense of camaraderie and good-fellowship all about.

He had eight officers in his squadron to man five worn-out DH-4s and some Jennies used for training reserve officers. Aside from observation duties with the Cavalry at Fort Riley, the squadron was on call to supply planes to both Infantry and Cavalry units in seven adjacent states. On top of that, Major Arnold was to become the senior air instructor at the Cavalry School. It sounded important and within the area of operations it was, but Fort Riley was known to be more than a place of exile. It was the place where they dumped second-class air officers who weren't going anywhere.

This brought troubled thoughts of the future constantly to mind. Should he stay in the service? Aside from the state of his career, the state of his bank account was equally poor. He had had to pay for the cost of the trip West, the moving of all their furniture and belongings. He was in hock up to his oak leaves, in debt to his father and to Bee's father as well. At the moment, the Air Service didn't seem to be going anywhere either, and he certainly wasn't going anywhere with it.

Bee quietly cautioned patience, of which she knew he had little. Like her mother, she felt a husband should do as he wanted to do. *It's his life. He earns the living. He's the one to choose the vocation.* But she kept a gentle though firm rein on his impatience. She knew that at bedrock he loved the Service and was loyal to it and that his belief in the future of military aviation, no matter how badly the War Department failed to understand it, was the guiding force of his thinking.

Lois and her brothers had found their Cavalry contemporaries to be just great, not the "hard nuts" Air Service kids had been led to expect. Life on the plains at Fort Riley couldn't be beat. It was there that by a cut of the cards Billy became Bruce. It seemed that everyone had a son named Billy and nobody knew whose Billy anybody was looking for, so the women got together and made a firm decision over a deck of cards. Highest card meant the mother of that Billy could keep on calling him Billy. Everybody else would start using their son's middle name instead. And if he didn't have a middle name, they could make one up. Bee drew a low card and Billy became Bruce. Henry had already changed his nickname—Bunky—to Hank, so Bruce didn't mind the switch at all. It sounded kind of different.

A day that was special for the whole family was payday. They'd pile into the car and drive into Junction City to take care of bills and buy what was absolutely necessary, such as a new pair of shoes. The three children would separate to take the money owed into the stores in question, certain of picking up a lollypop or two along the way. Always enough cash was saved for a meal at Wolfermanns, a fancy German restaurant with white tablecloths and silverware. Mom would speak German to the waiters, and the kids would feel special and proud.

They realized their father was something special too. He could roar and rant at them over their schoolwork and doing their chores and obeying their mother, and then they could catch him on April Fool's morning under the kitchen table, tying the chairs together. When he discovered that Bruce couldn't read well, he took a look at the boy's books and swore a mighty oath. "Who in hell could write stuff like this!" he said to his wife. "No wonder the kid can't read. Bee, get rid of this god-damned trash!"

Thereupon he sat down and became an author. In two months, amid all the other things he was doing, he wrote, *Billy Bruce and the Pioneer Aviators*. It became one of a series of six books his children and all the other kids on the post ate up. He sold the series to a small commercial publisher, A. L. Burt, in New York, who paid him a flat fee of two hundred dollars a book and no royalties even though the books went through several printings. He was also writing serious articles on aviation developments for a number of publications such as *Popular Mechanics*.

On another economic front he did much better. He took the U. S. Government to court and won. The issue was back flying pay, for duty during and after the war, that the War Department stolidly reneged on. The amount due was a tidy sum—over three thousand dollars—and with Cornelius H. Bull, of the law firm of King and King, fighting his case, the mighty bureaucracy was forced to pay up.

Besides those who were close to him, there were others who soon recognized that Hap Arnold was an officer with special abilities. That winning grin was one thing, but what was behind it was something else. When General Barnie Booth turned over his command in April 1927, he wrote Arnold: "My very sincere appreciation for the exceptionally able manner in which you have exercised the command of the 16th Observation Squadron. I really believe you have one of the most efficient observation units in the American Army. The wonderful spirit you have maintained in your squadron is most commendable."

Arnold also brought aviation to the horsemen by teaching a course in it at the Cavalry School. He and Major Whiting wrote a brand-new set of regulations for air-cavalry operations, and a good many cavalrymen found out what it was like to ride the rear cockpit of a bucking DH-4.

In February 1927, David Lee Arnold was born, and his father jokingly referred to him as "the afterthought." Hap's military highlight of the year was flying the mail for Calvin Coolidge while the President was vacationing in the Black Hills of South Dakota.

The regular airlines would bring the mail to North Platte on their scheduled run. From there it was Arnold's job to see that the 16th Observation Squadron carried it to a schoolhouse in Rapid City, close to the President's camp. While Coolidge had precious little interest in military aviation or its problems, and absolutely no recognition or understanding of the weather conditions involved, he wanted his mail delivered on time and no ands, ifs or buts—that went for fog, hailstones as big as golf balls, gale winds and thunderstorms. In spite of the forgoing, he got what he wanted, and in September, Major Arnold received a letter of commendation—via the VII Corps Area Commander—from Secretary of War Davis "for the efficient manner in which you performed your duty in connection with the mail service to the President during his summer vacation." There was something about the assignment and the congratulations that left the Arnolds chuckling.

So, too, did the maneuvers that took place in San Antonio that year at the Training Command. Arnold flew down with his squadron to find some familiar faces among the two hundred assorted pilots and officers gathered for the mock war. General Patrick was there to conduct an inspection. He didn't have much to say to his former Chief of Information. Three times he asked, "How many men did you bring down?" and three times Arnold told him and said nothing else, and that was about the size of it.

Frank Lahm, now a brigadier general and an assistant chief of the Air Corps, in command of training, appeared to have let the acrimony of ten years past slide into the discard. Arnold found him "almost human." Conversely, he found the Assistant Secretary for Air, F. Trubee Davison, "very human." He had a long talk with the Secretary and was impressed by his aerial knowledge and his manner of listening to what the other fellow had to say.

A great many friends and old flying buddies were also present. Assistant Air Chief Jim Fechet was one who sent word to Lois Arnold, via her father, that if he could get away he'd come to Fort Riley to meet her, and that she was one of the few girls he'd fly over a hundred miles to see.

When the "war" was over, Brooks Field held a party for the noble two hundred, but "It was a frost. The bootlegger didn't show up and the chow ran out." Still, it had been a good old time, even though the maneuvers hadn't proved much of anything.

Arnold returned to the plains of Kansas, his sense of humor intact, and even though his impulsiveness was temporarily in check it remained a strong inner force that could get him in a mess of trouble. In the case of Major Sydney F. Bingham, it very nearly did.

Twelve-year-old Ted "Rossie" Milton was a great pal of the two older Arnold boys and was considered almost one of the family. On a fine summer day, Rossie, whose father and that of his sidekick Johnny Bingham were Cavalry officers, pedaled out on their bikes to Marshall Field, near Smokey Hill River.

In that year, of 1927, when Lindbergh had just flown the ocean and the air seemed to be filled with the daring exploits of aviators, the bright dreams of twelve-year-olds were easily understood. The two boys rode up close to where Major Arnold was talking to a pilot in a wondrous-looking airplane. It was a Douglas C-1, a new Army transport with a big cockpit up front and a closed cabin with portholes behind. They watched several passengers climb on board. Then Major Arnold saw them and greeted them with a smile and a wave.

"Gee, they're lucky," said Rossie as the propeller began to turn.

"Who's lucky?" asked the Major.

"Them!" Rossie pointed.

"Would you two like to go along?" The grin was as broad as the look in his eye.

"Go along! You bet!" they chorused, not believing he could be serious.

"Hold it!" Arnold shouted at the pilot over the sound of the idling Liberty. Then, before the boys could adjust to the possibility that he was serious, the good Major picked them up and literally tossed them into the cabin.

It was a day and a ride neither boy would ever forget. When they landed, dizzy with excitement, having ridden above the earth and known the unparalleled feel of flight, they cried their thanks. Then they hopped on their bikes and pedaled furiously back to the post to spread the incredible news of their adventure. Major Arnold had let them go flying.

To Lois, Hank and Bruce, the news was particularly devastating. Their father had not permitted them to fly. To Major Bingham it was quite something else. The news infuriated him. Who did Arnold think he was, risking his son's life! He was going to demand that charges be preferred. Had Ross Milton's father not calmed his friend down and talked him out of his decision, there is little doubt that Major H. H. Arnold's career would have taken another step backward, possibly right out of the service. Instead, he was preparing to go back to school.

Even before his dismissal from the Washington scene, Major Arnold had applied to attend the Command and General Staff School at Fort Leavenworth. It and the War College were Army must schools for those who expected to climb the slow and difficult ladders of rank and position. A good many fliers shunned the schools, which pointed toward upper-echelon staff jobs. And this was one of the arguments the War Department used in maintaining that the Air Corps was not capable of commanding its own units.

In view of the black mark on his record, Arnold did not believe he would be selected to attend, but he was. There were three reasons for his acceptance over the objection of the school's Commandant, Brigadier General Edward L. King, who had been one of Mitchell's most determined adversaries: General Patrick had retired and been replaced by his Assistant, Major General James E. Fechet. Fechet liked Hap Arnold. F. Trubee Davison, the Assistant Secretary of War for Air, liked him too, and finally, the officer who had been selected to fill the vacancy at the school died. General King could not prevent Major Arnold from becoming a student, but he wrote General Fechet a confidential letter telling him that if Arnold did come, he'd be "crucified."

After two enjoyable years at Fort Riley, in 1928 the sacrificial lamb marched determinedly off to the supposed slaughter, where he would learn that the road to military victory lay not by way of the stars but via the Civil War.

"Well, we fought the Battle of Gettysburg today and guess who won again? Right, Meade did it again, but think of what Lee could have done with just one damned Wright Flyer!"

They lived in a three-story brick ghetto called the Bee Hive. The building was a block long, the quarters running from front to back, which at the rear afforded three floors of porches—all with low railings that could be stepped over, or if you were of an age, hurdled. There was much of the latter, and the arrangement made for a great deal of coming and going and the mixing of offspring. It was said that no wives committed suicide living in the Bee Hive, but murder was often just a porch railing away.

As far as the Major's threatened "crucifixion" is concerned, it did not materialize. There was no doubt that General King was out to give Major Arnold a hard time, but the latter took it all in his stride and treated the efforts as a joke, often giving back better than he received. One significant incident became known to all.

The planes available to Air Corps officers at the school to keep up their flying proficiency were few and antiquated. Once a month was about the best an officer could hope to fly, and when Major Junius Jingle "Willy" Jones managed to wipe out a DH-4, landing dead stick in the river, it looked as though even that rate would be lowered.

General King was incensed over what he considered an unnecessary accident. Why hadn't the plane landed on the field, where it belonged! He assembled in his office all the Air Corps officers at the school and read them the riot act. He raked them up one side and down the other, proclaiming that none of them really knew how to fly. When he dismissed them, Hap was the last to leave the room, and on closing the door he said loudly, "You know, that guy doesn't know a damn' thing about flying."

Instantly the door swung open and the General stepped out, grabbing the Major by his Sam Browne belt. "I want you to take that back!" he ordered.

Arnold's grin was fixed. "No, General," he said; "you don't know any-thing about flying, and you can't tell these men when and how they can fly." Which was perfectly true, but everybody there knew it had taken a lot of nerve for a major to face a general down. The thing was that King real-ized it too. From that moment on, he began to change his mind about Major H. H. Arnold.[1]

However, unknown to the General and to most of his fellow Air Corps officers, Hap Arnold had come to a very distinct and important crossroads in his life. Soon after he had taken over as G-2 under General Patrick, in 1925, he had noted that a German airline, Scadta, had established routes in Latin America. Arnold learned that the line's chief, a Captain von Bauer, had plans to extend his activities into the States, carrying mail and passengers. This last move was blocked by Postmaster General Henry A. New after Arnold informed the War Department what was afoot. In ex-ploring Von Bauer's operations, Hap came up with a plan of his own for a Florida-based airline that would service the Caribbean area and Central America. He talked the idea over with Tooey Spaatz, Major Jack Jouett, Chief of Air Service Personnel, and John K. Montgomery, a naval air re-servist. Montgomery had connections on Wall Street, where financing could be found. The line's name was to be Pan American Airways. While Mont-gomery was seeking backing, Billy Mitchell's actions overwhelmed all else.

Now, several years later, Arnold learned that Pan American Airways, Inc., had finally been established. John Montgomery was vice-president in charge of operations, but at the moment the company had no chief officer. The offer to Hap was that he would join the firm as managing director and then become its president. His starting salary would be eight thousand dol-lars a year, with three hundred shares of voting stock to be increased to fifteen hundred shares if he decided to remain. The original subscribers and stockholders made impressive reading. It was now or never.

He and Bee had discussed the opportunity often enough. His thoughts, of course, were bound up in two familiar areas: where the newly organized Air Corps was going and where he was going with it. With Billy Mitchell gone and his closest adherents all but silenced, the Morrow Board recom-mendations now a matter of law, independence was but a faint flicker on a far-distant horizon. The Navy was building up its own air arm with seem-ing smoothness and efficiency, while the War Department went plodding along, living a hand-to-mouth existence on starvation budgets.

As for himself, Arnold knew it had been clearly understood that his eviction from Washington was to be permanent. There appeared to be no chance that at some future date he might serve in a high position of com-mand. The arguments were eminently sound for getting out. Opportunity in the shape of heading a new airline had great appeal. The only real com-

petition in the area they were seeking to open up was the German airline
Scadta, whose northward push to Panama and the United States had already been blocked. So, what stood against such an offer in an undertaking
he had been the principal originator of anyway? Twenty-one years of service, little chance of promotion; less chance of ever becoming a general
officer. He had four children to raise and educate. He hoped at least Hank
would go to West Point, but it wasn't something that could be planned on.
For that matter, there wasn't anything in the service besides retirement and
a pension that could be planned on.

"I'm going to get a long-distance call," he told Bee. "I've got to make
up my mind by tonight."

"All I ask," she said, "is that you never leave the service with a chip on
your shoulder. I don't want you to be among the army of disgruntled
ex-service men. When you're on top, that's all right. But you're down now,
and you're unhappy about the way things are going. Think it over. Do as
you please, but think it over."

Their bedroom faced the long flight of stairs with the telephone at the
bottom. They were sound asleep when the phone began to ring, about 1
A.M. Arnold clawed his way toward wakefulness, rolled out of bed and,
sans slippers or bathrobe, charged down the stairs. He reached the landing, and at the exact moment he grabbed the receiver he stubbed his toe
on the newelpost. Bee, who had grown used to his blue language, had
never heard such a vast array of oaths. Amid them, she heart him shout,
"I'm staying here, goddammit! I don't want the goddam job!"

And then back up the stairs he came, ranting about being wakened in
the middle of the night by a pack of damn' fools who didn't know enough
to call at a decent hour. Bee Arnold lay in bed, silently shaking with
laughter. She never did learn whether he was going to say yes to the offer
at the top of the stairs and stubbing his toe at the bottom had changed his
mind, but she was pleased over his decision. She knew, too, that when Canadian Airways later offered him their general managership and he said
no, more calmly, that it was not likely that he would be giving up the service for any outside opportunity.

As the year of winning the Battle of Gettysburg passed, Arnold's question was, Where to next? He was in touch with General Fechet on the matter, and so was twelve-year-old Bruce, who had started off the new year
with his own request to the General. He received a prompt, straight-faced
reply:

My dear Sir,

Your letter of January 2 reached me today. I am taking this early opportunity to notify you that your request not to be sent to Washington or left in

Leavenworth has received my consideration, and I have decided to send
you to Texas.

Yours sincerely,
Maj. Gen. JAMES E. FECHET
Chief, U. S. Army Air Corps

But it wasn't to be so, even though Fechet had made up his mind to assign
Arnold to the Training Command as CO of the advanced school at Brooks
Field, San Antonio. In April 1929 the General wrote to say that a wrench
had been thrown in the works by the Commanding General of the Training
Command, Brigadier General Frank Lahm.

"He is so strongly opposed to your assignment that I have given up the
idea for the present as I believe it would be a mistake to place you under
his command," wrote Fechet.

It was one of those not infrequent conflicts of personality. Lahm had
been Arnold's French instructor at West Point, and when as a senior in
1907 Cadet Arnold had seen Lieutenant Lahm make a balloon ascension
from the parade ground at the Point, he had been mightily impressed.
Later relations with the tall, thin-faced, somewhat imperious Army air
commander in the Philippines, had certainly been equitable enough. But
the incident over the lost plane at Rockwell Field in January 1917 had
soured the relationship.

Lahm's promotion to brigadier general had come about as the result of
the Air Corps Act of 1926, which permitted the Corps to increase the
number of its general officers from one major general and one brigadier to
one major general and three brigadiers.

Lahm became one of the three, with the title of Assistant Chief of the
Air Corps. His assignment was to organize and command the Air Corps
unified training center at Randolph Field.

With duty at San Antonio shot down, Fechet offered Arnold three
choices: command of Fairfield Air Depot, Air Officer of the II Corps
Area, or the Hawaiian Islands. None of them was what could be termed a
choice assignment, and none of them appealed to the man who had recently rejected the presidency of Pan American Airways. If he could not
be involved in advanced training, he wanted a tactical command.

The 1926 five-year expansion plan of the Air Corps Act foresaw a combat wing on both coasts and in the South, as well as the Canal Zone and
the Hawaiian Islands. There was to be one combat group in the northern
part of the country and one in the Philippines. Minding an air depot, or
doing a staff job for a Corps Area CO who, more than likely, had his eyes
on the ground was not the Major's idea of aerial progress. To go to Hawaii
was to join the Navy, or the "canoe club," as he called it. Out there, the
Corps had little in the way of aircraft and less of control. The planes for

the proposed wings and the proposed buildup in strength were largely on paper.

In his letter, General Fechet had added a P.S. "It may be possible that you can go to Rockwell—but not likely."

Rockwell would have meant command of the 17th Pursuit Group, which Arnold would have liked very much, but "not likely" he recognized meant not at all. Although he had vowed to Bee that he'd never accept command at Dayton, Ohio, that's what it finally came down to. He was to command the depot at Fairfield and head the Field Services section at Wright Field; the two locations were practically contiguous.

The company of Frank Andrews, George Brett, John Curry and others of a like mind had made Command and General Staff School worth the course. They had all found the school's ideas on the use of air power on a par with Civil War strategy, and before his graduation Arnold wrote a paper suggesting how air thinking might be updated a bit. Bee had found Leavenworth a joy for the Army wife, but as an Army wife she was ready for what came next. So were the children.

The departure was typical Arnold. He wanted to be the first officer out of the place following the usual formal matriculation exercise. Bee's job was to oversee the packing and then to have the car and family waiting outside the auditorium, ready to dash when the graduate appeared. At the crucial moment, the Quartermaster Sergeant inspecting the Arnolds' quarters found there was a coal scuttle missing. A mass hunt for the lost item was frantically instituted. Nothing turned up. Time was wasting. The Major would be ready to skin the fur off a bear. Regulations came before majors or bears. A long form had to be filled out on the missing fifty-cent scuttle, for which the Arnolds would be held accountable. Bee grumped that the Major had once lost a mule in the Philippines and had to pay for that, too. Finally, they were released by the inspecting NCO, who had made them feel that the penalty for the lost implement could be summary execution.

Fearful of what she would find, Bee drove the overloaded car to the assigned meeting place. But, wonder of wonders, he was not there. No one was. The formal farewell was still in progress. Everyone sighed with great relief and relaxed. Time passed, the sun grew hot, the children restless. Could they get out and play? Could they get a drink? Where was Pop, anyway?

At length, the doors to the hall flew open and out poured the released "Civil War veterans," Major H. H. Arnold leading the charge. He reached the car, waving wildly to Bee to fire up and drive on! Drive on! Once aboard, he collapsed in the seat beside her and sighed with great feeling, "Oh, that long-winded sonofabitch!"

Six

The Pilot

During the years that Hap Arnold was at Fort Riley and then attending the Command and General Staff School, he was essentially cut off from the duties and decisions that concerned the political and aeronautical development of the air arm. Those of his colleagues who were not taking school courses or serving in distant commands were occupied in one of these arenas or the other: either by duty in Washington; with the Materiel Division, in Dayton, Ohio; or assigned to a tactical unit. Only a very few were in the unique position of being actively employed in both the political and the aerial evolution of the service at the same time. Through circumstance and capability, one officer who was able to grasp and enjoy such a simultaneous opportunity was Captain Ira C. Eaker.

Arnold had known him first in 1919, when Eaker had served briefly as Adjutant at Rockwell Field. Hap had taken note of the unobtrusive but very alert twenty-three-year-old Lieutenant then, as had Tooey Spaatz, and when, five years later, they had come to serve in Washington, the association became more meaningful.

Texas-born, with the thought of becoming a lawyer, Ira Eaker had entered the Air Service during the war. He had not gotten to France, but, following duty in the Philippines after the war, he was planning to leave the Army to go to law school, when a fellow pilot's stomach ache changed the course of his life. He had returned from the Philippines in early 1922, reporting to Major Walter Weaver, CO at Mitchel Field, Long Island. Weaver, liking what he saw in the newly promoted Captain, assigned him to the 5th Aero Squadron and then brought him into his office as post Adjutant. Lean and hard, with high cheekbones and brown eyes, Eaker had the look of an Indian and somewhat the manner as well. On the Air Service playing field he was the mainstay of the football, baseball and basketball teams. His thoughts, however, followed more intellectual pursuits and he kept thinking about that law degree.

In the fall of 1923, Congress passed a law offering a year's pay to officers of Eaker's rank to leave the service. The move was to induce a further cutback in military strength, and he saw in the offer both an indication of the future and an opportunity. With the money he would get for resigning, he could afford to enroll at Columbia Law School.

He told Weaver of his decision, and although the Major, a West Pointer and career officer, was sorry, he understood the reasoning. If the Captain would submit his application, it would be processed.

And then came the upset stomach. It was a Saturday, and Eaker was working with his boss, the two of them alone in the office. They noted a distinctive DH-4 with blue markings coming in to land and went out to see who was paying a visit.

The plane's pilot was Captain St. Clair "Billy" Streett. His passenger was Air Service Chief General Mason Patrick. Streett, a small man with a neatly trimmed mustache, was ashen-faced. It was not the air that was bothering him but a bad case of the flu. It was better he not try to fly any farther. "Farther" was the Army-Harvard football game, which the General was most anxious to attend. While Billy was eased away to the infirmary, the General made his wishes known: "Major, pick me a pilot who can fly me to Boston right now."

Weaver looked at Eaker and said, "Captain, you're elected."

And so, unexpectedly, Captain Eaker found himself flying the big boss to a football game, which he was invited to attend. The next day, they flew back to Mitchel Field and Sunday dinner with the Weavers.

During the course of the meal, the Major told the General he was going to need a new adjutant, since Captain Eaker was leaving the service. Patrick made no comment, but after Eaker had flown him to Washington, landing at Bolling Field, the General asked suddenly, "Captain, why are you getting out of the service?"

Eaker explained. The opportunity was too good to forgo. He had previously begun his law studies at the university in Manila.

"Well, if you remain in the Army," the General said, "I can send you to law school. I have authority to send 2 percent of Air Service officers to school. Of course, if you go, you'll have to study contract law. How does that strike you?"

On immediate reflection it struck very well, for it meant that he could combine the best of both worlds. He'd be living at Mitchel, going to Columbia during the day, and able to fly on the weekends and whenever else the opportunity afforded. More, he'd keep his regular commission, which was an important factor.

After Eaker had flown back to Mitchel Field, he told Weaver of the General's offer and his acceptance of it.

"Well," said Weaver, "the Chief's getting a little old. He'll probably forget it by the time he gets to the office, but I'll hold up your discharge application for a week, and we'll see."

It was in less than a week that notice came through granting Captain Ira C. Eaker permission to apply to Columbia for a year's course in contract law.

When the year of study was ended, he received orders to report to Washington for duty as an assistant to Major Mike Kilner, General Patrick's Executive Officer. He did so in June 1924 and was closely attuned to all the high drama that followed. Placed in an office adjacent to Patrick and Mitchell's he had a close-in view of the rising conflict swirling around the Chief of the Air Service and his irrepressible Assistant. He saw that the major difference between the two on air matters was in method of approach. Patrick often considered Mitchell's behavior that of "a spoiled brat," but there was no doubt of Billy's persuasiveness and his effect on the staff. It was a staff dedicatedly convinced, for the most part, that air power could never reach its potential under Army or Navy control. Eaker, who joined it first as an acolyte, was soon a firm believer and follower in the faith. Before Mitchell's exile to Texas, in March 1925, and his replacement as Assistant Chief by Brigadier General James E. Fechet, Eaker came to form a great admiration for him. It was not just the man's inspiring personality, but the way in which he was able to exert influence on the press and the public, his ability to draw attention to his ideas.

Conversely, in working for Patrick through Kilner, Eaker could clearly see that without the General's acceptance by Secretary of War Weeks and the top brass, the plight of the Air Service could have been even more desperate than it was. Most members of the General Staff liked Patrick; he was, after all, one of their own. They were willing to listen to him, whereas by 1925 their hostility to Mitchell was pronounced.

By the winter of 1926, Billy Mitchell was gone and Hap Arnold was gone, but in the summer, out of the wreckage of high hopes—from the point of view of "Mitchell's boys"—had come not a whole loaf but a halfbaked loaf: the Air Corps Act of July 2, 1926. For Ira Eaker, the Act would be something more than that. It would mean a new association at the top.

When the legislation had been drafted, it provided for assistant secretaries of air for the War, Navy and Commerce departments. There is little doubt that Dwight W. Morrow, chairman of the Morrow Board and a close friend of President Coolidge, recommended F. Trubee Davison, the politically active son of Henry P. Davison, for the post with the War Department. Behind the move was a decision by the Republican party to bring some young blood into its executive ranks. The new Assistant Secretary of Air for the Navy was Edward P. Warner, aeronautical engineer, and for Commerce, William P. MacCracken, Jr., who had flown with the Army during the war.

On a hot July morning, General Patrick summoned his Assistant Executive Officer and said without preamble, "You live near here, don't you?"

"Only a few blocks away, sir," admitted Eaker.

"Well, go over and put on your uniform. I'm going up at eleven o'clock for the swearing-in ceremonies for Mr. Davison, and I need you as my aide."[1]

Eaker already knew something of Davison's background, knew that in the summer of 1916 the family summer home had been converted into a flying field for Trubee and a dozen of his air-minded friends at Yale. Young Davison had created, through the backing of his wealthy father, Henry P. Davison, the Yale Aviation Unit, whose twenty-nine graduates in September 1918 were commissioned into the Naval Air Service. It was shortly before the graduation that Trubee had crashed into New York Harbor and been permanently injured. The lower parts of his legs were so damaged that from then on he limped badly, had to wear specially made shoes, and needed a cane to walk. The infirmity had not affected his political career, nor his interest in aviation.

Following his swearing in, Trubee Davison's first official act was to turn to General Patrick and make a request: "General, I'd like to start out with an airplane flight. Is that possible?"

Eaker was the only Army pilot present, and Patrick turned to him and said, "Accompany the Secretary to Bolling Field and give him a ride in my Blue Bird DH."

And so the introduction was made. It was the sixteenth of July, 1926, and Eaker took Davison up for a half-hour flight over the capital. After they had landed, Davison, obviously exhilarated by the experience, had another request. "I've got to go up to my home at Locust Valley this afternoon," he said. "Would you fly me to Mitchel Field?"

After they arrived at Mitchel, Davison invited his pilot to come home with him. Eaker spent the evening getting to know the family—mother, two sisters and brother—living in quiet splendor on their Long Island estate. The next day, he flew the Assistant Secretary back to Washington, and from then on he added to his duties that of being Davison's pilot.

The result was that Captain Eaker was now in close personal association with the new Air Corps's three most important individuals: General Patrick, Assistant Chief Brigadier General Jim Fechet and Trubee Davison. It could only be an enlightening experience, but in the midst of these associations he was to participate in a great aerial adventure indicating what air power was all about.

In 1919, Billy Mitchell had written to Hap Arnold saying he was "very anxious to push through a flight to Alaska with land planes," and he thought the Alaskan venture "might develop into a round-the-world flight."

He had been right on both counts. The following year, the Alaskan flight had taken place, and then, four years later, in 1924, the famous globe-circling effort had excited world attention.

Although the original idea for the global circuit had been Mitchell's, General Patrick was seen as its principal planner, and it was just two years later that he came up with an idea of almost equal magnitude. It was to become known as the Pan American Goodwill Flight. What Patrick envisioned was a small fleet of U. S. Army planes touching base with twenty-three Central and South American countries, showing a friendly flag by air, so to speak. The diplomatic purpose of the expedition was to help bolster relations with America's neighbors to the south, a somewhat continuing need. The technical purposes were to provide beneficial flight training to the crews while demonstrating the capability of the amphibian-type aircraft that would be used.

There were other, underlying considerations, however, beyond garnering publicity for the newly established Air Corps and showing that the Army was more adept than the Navy at flying long distances over either land or water. The word was around that the Germans were making inroads into South America, planning to set up civil airlines. The thought was to send down superior planes to interest South American governments in buying U.S. equipment.

Both the new Secretary of War, Dwight Davis, and Secretary of State Frank B. Kellogg liked General Patrick's proposal and recommended its acceptance to President Coolidge. Coolidge could see the diplomatic and political value of such a flight and gave it his laconic nod.

Patrick thereupon handed the organizational and operational planning over to his Assistant Chief, Jim Fechet. Captain Ross G. Hoyt concentrated on the overall details, extending from the administrative to the operational. In this last was the choice of the aircraft. The plane finally selected was a Loening, bearing the designation COA-1. The Loening's fuselage was mounted on a wood and duralumin hull. The plane's 400-hp Liberty engine was mounted inverted so that the propeller would swing clear of the hull's nose. The cruising speed of the open-cockpit amphibian was about 85 mph.

All of these points were of major interest to Ira Eaker, for he devoutly wished to be one of the ten pilots who would man the five planes on their twenty-three thousand-mile journey. The crew members were to be chosen by Patrick on Fechet's recommendation. Commanders at principal Air Corps stations were asked to submit the names of two officers, basing their choice in large part on "length of service, cross-country flying experience and practical engineering experience." Observation pilots were more in demand than either pursuit or bomber types, due to the nature of the flight.

However, to be one of the lucky ten, there was nothing like being close to the center of things. Eaker knew that Major Bert Dargue, Patrick's aide, was going to be selected to lead the flight, and the thought of two pilots being chosen out of the Chief's office was worrisome. Certainly Dargue was eminently qualified and a very active pilot. So, too, was Ira, and if Dargue was considered Patrick's favorite, Eaker had become Fechet's.

It had not always been thus. At the time Fechet replaced Mitchell, Eaker did not hold him in very high regard. His opinion dated back to the days of flight training, when the former hard-riding cavalryman had been in charge of the training command. A mess sergeant at the field had absconded to Mexico with a large amount of money taken from the officers'-club till. Fechet's method of recouping the loss was to divide the amount lost among the officers going through the school at the time of the robbery and dun them for it. Eaker's bill was nearly ninety dollars, received *after* he was no longer at the field. He thought it was "a helluva way to run a business!"

It was following Fechet's initial conference with General Patrick, after Mitchell's exile, that Eaker met the new Assistant Chief. Fechet came out of the confab with Patrick into the office Eaker shared with Major Miff Harmon and Lieutenant David Lingle.

"Can any of you guys fly?" rasped Fechet.

When Harmon failed to speak up, Ira responded, "A little, sir."

"Okay, let's go," Fechet said on his way out the door.

In going, they flew through an afternoon of low clouds and heavy rain to Wright Field, where Fechet was to command some scheduled maneuvers. Only one other plane had gotten through that day from Bolling Field, and its pilot was Tooey Spaatz. That night, the General was hungry for a game of bridge, and the three sat down to the cards to play a little cutthroat. It was the beginning of a beautiful friendship. They never flew anywhere together without a deck of cards, and the big, outgoing ex-cavalryman came to look upon "Iree" almost as a son.

Once Eaker knew he had been given the green light on the Pan American Goodwill Flight, he worked closely with Dargue in the final administrative and logistical preparations as well as the choosing of the other pilots from a list of twenty-six names that had been submitted.

It was fully understood that at least one of the two pilots on board each plane must have an engineering background. Thus Eaker's choice of a crewmate was First Lieutenant Muir S. "Santy" Fairchild. Fairchild had been a bomber pilot in the war, following flight training in France and Italy. He had flown night missions with the French, attacking targets east of the Rhine. Following the war, he had become a test pilot at McCook Field and then Wright, joining a handful of daredevils that included Jimmy

Doolittle. Flight testing was the wild-blue-yonder aspect of aeronautical engineering, and Eaker had gotten to know Santy Fairchild at Wright Field and later at Langley in both pursuits. He felt that if any pilot knew the Liberty engine, in whatever attitude of flight, it was Santy Fairchild. Aside from that, he was a savvy partner to have around in the air or on the ground.

The planners foresaw fifty-six flying days and seventy-seven delay days for diplomatic functions and aircraft maintenance. Day one was December 21, 1926, when, following a gala send-off, the five amphibians departed from Kelly Field, Texas. Each aircraft bore the name of a principal U.S. city; *New York, San Antonio, Detroit, St. Louis* and, lastly, *San Francisco,* which was crewed by Eaker and Fairchild. The course they followed also pioneered what was to become essentially the route established by Pan American Airways.[2]

The team of Eaker and Fairchild flew on the basis of a two-man partnership "in which each had invested his total assets, his reputation, his ambition, even his life." They soon learned that what was required as much as aeronautical skill was physical stamina. The flying was arduous enough with a 4 A.M. rising and a six o'clock takeoff, but it was the nightly diplomatic functions after a four-to-six-hour flight that sapped a man's strength. Every evening, there was a banquet followed by dancing, and as Captain Art McDaniel, commander of the *San Antonio,* was to remark, it felt as if they ended up dancing more miles than they flew.

Fortunately, they were all young and rugged, but that didn't make the maintenance chores any less. It took long hours to refuel in primitive landing sites where most of the population had never seen an airplane. Another problem was that Liberty engines mounted in an inverted position leaked oil through their rings, which meant constant cleaning and changing of spark plugs.

There were soon more serious mechanical difficulties. Bert Dargue cracked up the *New York* while taking off in Guatemala. The field was bordered by tall trees and, beyond them, a fifty-foot-high stone aqueduct. The plane had no brakes, and even though airborne, Dargue saw he wasn't going to clear the aqueduct. He chopped the power and managed to plop the Loening down in a cornfield, damaging its landing gear and hull. Repairs were made on the spot, and Dargue and his partner, First Lieutenant Ennis C. Whitehead, flew from a nearby lake to France Field, Canal Zone, for a complete overhaul.

The *St. Louis* was next. It developed engine trouble over very rough seas off the Colombian coast, and the others helplessly watched Lieutenants Bernard Thompson and Leonard Weddington going down into the tumbling waves. Hours later, after Dargue and his men had landed at the mouth of the Magdalena River and were housing their planes in the sheds of the German Scadta Company, the *St. Louis* came racketing in.

"No boats were in sight," explained Thompson to his mightily relieved comrades; "we could see the waves were going to break us up, and we couldn't get her ashore. So we decided we might just as well break up, trying to fly so as to hang on until she went to pieces under us. So we just gave her the gun and bounced from wave to wave till a big one threw us about fifty feet in the air. We nosed her over the next roller, and here we are."

And aside from a leaky radiator and a propeller that needed its three steel blades worked back into shape, they were ready to proceed.

It was over the Andes that Eaker and Fairchild felt the icy breath of disaster. Following three days of banquets and hard work on their planes at Chile's delightful inland capital of Santiago, they moved on down the coast, flying from a tropical zone into a temperate one and, with it, into heavy rain. It rained so hard they flew bunched together just above the surf, forward visibility practically nonexistent. It was dark and foggy when they spotted Valdivia and happily made their landing.

They found Valdivia to be somewhat of a pioneer town, populated by Germans busy clearing the forests to build their homes. The settlers told them they hadn't seen a decent day in nine years, and the travelers couldn't help but wonder why anyone would want to live in such a place. At the moment, however, they had more important considerations. Although they couldn't see them, the high Andes lay ahead. They had a 650-mile flight over them to reach the east-coast port of Bahía Blanca, Argentina.

It was just as wet taking off in the dawn as it had been in landing, but then, fifty miles inland over hilly forests, they broke out of the rain into clear air. And there before them, hiding the mountains, lay a shroud of clouds higher and thicker than they had ever seen. The only way was up and over. Dargue gave the signal to circle and climb. Each plane would make the crossing on its own.

Eaker was flying the *San Francisco,* and he went on up, paralleling the cloud bank. Twelve thousand feet was the best the Loening could give with three tons of fuel and equipment aboard. At that altitude they were just on top, but across the white cloud sea, peaks jutted through like a shark's teeth. The cold was bitter. A bottle of water froze as they proceeded, balanced delicately above the cumulus surface, the plane's shadow, haloed in the mist, winging along with them. And then, an hour into the crossing, the Liberty suddenly began to cough and spit. They knew it was caused by ice in the carburetor, with no way to clear it out and no way to hold precious altitude.

"Do you want to bail out?" shouted Ira.

"No!" came Santy's immediate response. "No point in breaking a leg in the middle of nowhere."

Or as Eaker was later to relate, "We had either to take our parachutes and jump out or nose the plane down and trust blind luck not to crash on some invisible peak. . . . Also we figured that sitting on some isolated Andean peak holding a lone parachute in our laps wouldn't be much better than lying on the same peak in a mess of airplane wreckage. So we stuck to our ship."

While Eaker kept it flying just above stalling speed, they sank down out of the horizon-reaching blue into the deadly opaqueness—no visibility beyond the nose and the wing tips. Eyes straining, their heads on a swivel, the descent seemed endless, the altimeter's moving finger irrevocably marking their loss of altitude.

"For five thousand feet we slid down through obscurity. Each second, we knew the next might be our last," related Eaker. "But, that day, the air gods were kind. Like a glimpse of paradise, through a cloud gap a lake appeared below us. . . ."

Santy Fairchild was immediately excited. He stood up in the rear cockpit and held out a piece of paper with a crudely drawn terrain sketch, a lake on it the shape of the one below. It had been given to him the night before by a British engineer who was involved in surveying a route across the Andes at a lower level, the lake marking the pass. By incredible luck they had broken out of the clouds above it. Eaker circled the lake, ready to land if the choking Liberty didn't cough up its carburetor ice. It did, and they happily headed eastward again, cruising low over the picturesque Patagonian plains, knowing that someone up there had been watching over them.

But if they had missed death over the mountains, it had hurried on ahead and was waiting over Polmar Field, at Buenos Aires. The four planes flew above the city in a tight formation, preparatory to landing, excited crowds peering up from below.[3] Flight Commander Bert Dargue and Lieutenant Ennis Whitehead led in the *New York*. *San Francisco* was off their right wing and to the rear. *Detroit,* with Captain Clinton F. Woolsey and Lieutenant John W. Benton, was in a similar position on the left, *St. Louis* closing the diamond formation. An escort of three Argentine Army planes flew alongside.

As they passed over Polmar, Dargue gave the signal to break formation. *Detroit* and *San Francisco* pulled up, turning outward. Dargue began a gliding turn to his left, his attention momentarily distracted to the right and downward by one of the escort planes crossing beneath him.

Earlier, when Woolsey and Benton had landed the *Detroit* at Mar del Plata, a landing-gear cable had snapped. Woolsey decided to fly the plane with the gear retracted. Once they reached Polmar, Benton would climb out on the wing, holding on to the brace wires, and his weight would manually dump the gear. They knew the wing walking worked, because they'd all tried it in training flights. John Benton climbed out onto the wing. While Dargue was looking to the right and down, Clinton Woolsey was looking left at Benton.

As Dargue was to recount, "It was only a matter of seconds when I glanced up to the left. I caught a flash of black and yellow slightly higher and just off the rear of my left wing.

"Then we crashed."

The two planes, locked together, fell into a violent spin. Pieces of wreckage swirled about Dargue. Automatically he flipped the catch on his safety belt and was catapulted out of the cockpit. He yanked the rip cord and was slammed into the tail of the plane. His billowing chute promptly became tangled in the spinning mass, and he plummeted down with it, trapped, knowing this was the end.

Rugged Ennis Whitehead was trapped too, with the wing of the *Detroit* folded over his cockpit. Then the two planes tore loose and, amid a rain of tools, parts and papers, he went over the side, whacking his ankle on the tail. He had seen Dargue bail out, seen him become ensnarled in the tail and waited, therefore, until the wreckage was beneath him before he opened his chute. Then Whitehead watched the two planes hit the ground, the *Detroit* bursting into flames. He looked for other chutes and saw only one, a plowed field was coming up at him, and he pulled off his shoes so as not to break his ankles.

When the planes separated, Bert Dargue was flung free of the inevitable. Two large hunks of silk had been torn from his chute and some of the shroud lines snapped, but nevertheless he reached the ground safely. Not so Captain Clinton Woolsey and Lieutenant John Benton.

When he had gone out on the wing, Benton had left his chute in the cockpit. Once the collision took place, he had no chance to retrieve it. Woolsey could have bailed out and saved his life, but he would not leave his helpless partner.

Eaker, who had observed the entire tragedy, was to say, "I have never witnessed a more courageous self-sacrifice." He wired a description of the accident in an official report to Washington, but no one could describe the awful sense of loss and desolation they all felt—the grim reality of an avoidable crash that became unavoidable.

Dargue was to write: "The people of Buenos Aires could not have been more considerate had Woolsey and Benton been two of their favorite fliers." The President of Argentina and his cabinet marched in the funeral procession.

There were no more major mishaps, but on the long return up the coast of South America, Eaker and Fairchild had another near miss. Again it was the carburetor—not ice this time, but a blockage that threatened fire. They were off Montevideo, Uruguay, over extremely rough seas and Eaker took one look at the waves and decided, "I'd as soon burn as drown." So he headed for shore, some miles away. They reached it, and then it was not fire but water and rocks that threatened their safety. They had landed and gotten the anchor out, but the rope snapped just as a wave smacked the plane's nose and drenched the engine. The flooding seas swept the *San Francisco* toward the rock-infested shallows.

Eaker worked furiously to get the Liberty fired up and succeeded long enough to steer the plane away from certain destruction. "Waves threw us up on the beach," he related, "and about one hundred natives rushed out of the brush. They got a rope and, like a long team of horses, helped us pull the plane ashore. That night, I slept on the ground under the wings of the plane, so tired that giant cockroaches crawling over my hands and face didn't awaken me. . . . Next day, after repairs, we got safely off and overtook the flight at Santos."

On May 2, 1927—175 days after departure—the grand adventure was concluded at Bolling Field. There, before a large crowd, President Coolidge and a host of civilian and military brass greeted the eight airmen. Major Dargue presented to the President messages of goodwill sent by the heads of the twenty-three Latin American countries, and Coolidge in return awarded the weary but elated fliers a newly created medal, the Distinguished Flying Cross, a measure of their superior performance. As one State Department official put it, the flight did "more good than ten years of diplomatic correspondence."

After the captains and the kings had departed, Ira, looking like a lean and tough Indian, and Santy, with his jungle bush mustache, bid each other farewell, knowing a friendship had been made, tempered out of a thousand weathers and a half year of unforgettable association.

During the latter part of the 1920s, the public was captivated by succeeding aerial feats, none more daring or acclaimed than Charles A. Lindbergh's flight from Mineola, Long Island, to Paris. Rapid aeronautical progress spurred the adventurers, and the Army Air Corps participated when and where it could for reasons that had an underlying military purpose. Ira Eaker and his fellow airmen were out to test and prove their equipment, and in so doing hold public attention.

In this endeavor, they had the aid of Trubee Davison. The Air Corps Act and Davison's appointment in 1926 had helped to deflect the strong thrust for independence on the part of Billy Mitchell's followers. His coming to office did much to calm the disgruntled and bring positive results. At least, that was how Ira Eaker viewed Davison's stewardship, for shortly after his return from the epic Pan American Goodwill Flight, he became the Secretary's military aide. Since the Captain also served as General Fechet's pilot and faithful card partner, he was in the catbird seat—right at the center of the Air Corps's heartbeat.

Immediately, he learned that Davison had some definite ideas on Air Corps public relations, a subject dear to every Army airman who knew that only through publicity could the Army keep the Navy from usurping its yet undefined role.

Davison told him he had long felt that even with Billy Mitchell, the air arm, for all its publicized record flights, lacked a proper image, particularly with Congress. All too often, its public image had been stymied by War Department restrictions stemming from either an inability to understand or simply a desire to block the airman's point of view. To aid him in trying to correct the problem, Davison hired Hans Christian Adamson, an Albany, New York, newspaperman and Air Reserve officer whom he had long known and regarded highly. He paid Adamson's salary out of his own pocket, making him his legislative aide as well.

Eaker's job was to assist Adamson, and the two worked closely together, soon sharing the expense of an apartment. While Eaker would supply the military background and data for whatever the assignment might be—a speech, testimony before a congressional committee, a General Staff conference or a proposal on Air Corps demonstrations, such as a mock surprise attack on Los Angeles by a mass of bombers—Adamson would do the lion's share of the writing, and Davison would develop the final draft and add his own personal touch.

From the beginning, the Air Secretary and his Air Corps Chief, Jim Fechet, had a close understanding. Davison would handle the major contacts with the Administration and the Congress, the War Department, the Navy and the press. Fechet would oversee the administration of the Air Corps and join Davison when necessary. The arrangement worked better than anything theretofore, and the arrival of Trubee Davison on the scene to speak for the Corps in the councils of the mighty brought a decided uplift to the morale of its officers and men.

Liaison with the principal House and Senate committees on military and financial affairs became close and informative. Frequently, Captain Eaker was airborne, carrying senators and congressmen to Air Corps bases and installations where the solons under Davison's guidance could see firsthand what the Army air arm was all about.

On the Administration side, Davison locked horns with the Bureau of the Budget, determined to get more money for aircraft procurement, for R & D, for facilities. Chief of Staff Major General Charles P. Summerall complained that Trubee Davison was continually demanding another fifteen or twenty million dollars a year. Summerall did not believe one branch of the service should receive more of the meager appropriations for military security than the other.

It was realized in very short order that Trubee Davison, as much as providing a shot in the arm to the Air Corps, was an unpluckable thorn in the side of the General Staff. It wasn't that he was obnoxious, rude or ever out of bounds, but he was persistent, and Ira Eaker was in a position both to observe and be a part of the process. He saw to it that his boss was

thoroughly briefed on the military air aspects of every question and point
of contention. There were occasions when Eaker was required to deal with
War Department boards in Davison's name. A captain putting forth positions before senior officers had damned well better know his facts and be
on his toes. But even if the Air Secretary was not in the room, his presence
was felt, and it made a very great deal of difference in attitude and attention span.

It was soon apparent, at least in Washington, that Trubee Davison was
a good man to have on your side. The politicians came to appreciate his
capabilities and the military his power. Something of his quiet determination was reflected on the tennis court. Cripple that he was, he and Ira
Eaker were doubles partners and no pushover as a team in anybody's
game. Since Davison could not move about the court, he positioned himself deep. He had developed powerful arms as a result of his injury, and
using either forehand or backhand he could make a ball smoke. With
Eaker, fleet of foot in the forecourt, they were a formidable pair.

It was during Davison's six-year term that the Air Corps began to obtain its first really new planes: Curtiss and Boeing pursuits, Keystone and
Martin bombers, Douglas and Loening observation craft, as well as
Fokker and Ford transports. He recognized early on the potential of the
long-range bomber and began to push for its development.

There were new training planes as well, and it had long been General
Fechet's desire—having spent a decade in the training command—to see a
central Air Corps training headquarters established. In his thinking, it
would bring under one administration primary and advanced flight training
as well as a school for aviation medicine. Through Davison's adroitness
the funds were appropriated. The eventual result was Randolph Field, at
San Antonio, Texas. Frank Lahm, as a brigadier general and one of three
assistant chiefs of the Air Corps, was given the job of organizing and
becoming the center's first commander.

In Ira Eaker's mind, the Air Secretary was not only the Corps's able
amicus curiae but also an unsung hero, never given the proper recognition
or credit for his many contributions to the cause of air power. That the Air
Corps's five-year legislated plan of growth and development was not successfully accomplished was a condition that could in no way be attributed
to a lack of effort on Davison's part.

If, to Ira Eaker, Trubee Davison was a boss with a gift for political
salesmanship, an official at home in the White House as well as the
Congress, his other boss, General Fechet, was something else again. Big
Jim Fechet, with his rough-hewn Cavalry bark, was a soldier's soldier. He
hated confined places, such as offices, and he found the subject of logistics
a dreary bore. "I've got a staff for that sort of thing," he'd say, "and if
they can't handle it, I'll find someone who can." He'd outline what he
wanted done, and it was up to those given the task to complete it and not

to bother him with details. He was all for operations, particularly if it meant getting up in the air.

With his large, rugged features, he looked and sounded like a fighting man. He had joined the Cavalry in 1897 as a private, was badly wounded at San Juan Hill in Cuba, and in the Philippines earned a commission in the fighting. His horsemanship put him on the Cavalry Team in 1909. While Benny Foulois was trying to get his Jennies on the track of Pancho Villa, Fechet was pounding the dusty desert trail after the wily Mexican bandit. He spent twenty years on horseback before he switched to aviation. The Air Corps took to him not only because he was the first Chief who had come up through the ranks but also because he was a pilot.

No one ever got tired of telling the story of Jim Fechet's parachute jump. He had become interested in the parachute and decided he'd like to try one out. The idea was that he'd go up lying on the wing of a Jenny, and when everything was set he'd roll off the wing. Somehow, while the Jenny was still climbing, the rip-cord ring got caught on a wing brace, and when Fechet tried to get it loose he inadvertently pulled it. Suddenly airborne, he drifted down, surprised but not at all displeased.

By whatever mode of transportation, Fechet liked to be on the go. On one memorable adventure he and "Iree" traveled from the East Coast to the West by train, playing with religious fervor the highly regarded card game of cooncan across the mountains, the valleys, the plains and the deserts in daylight and in darkness. They were really warming up to take on Tooey Spaatz and his coterie of cardsharps at Rockwell Field. Tooey had thrown down the gauntlet, suggesting they bring their checkbooks. Previously, Eaker had indicated that no mercy would be shown, whether the game was cutthroat (three-handed bridge) or something for a larger gathering, such as five-card stud. For real rest and relaxation, there was a weekend pilgrimage to Tijuana, just across the border in old Mexico. There at Paul and Alex's Place, where Major Spaatz was always an honored guest, one might hear the composer strumming on a borrowed guitar, singing his very own madrigal "Tooey, the Oysterman." Or one might go to the races, for in Agua Caliente the horses you bet on were bound to win . . . sometime! Serious Air Corps business was reserved for the daylight hours of weekdays, but once recall had been sounded, the flag lowered, the sundown gun fired, the cards were cut and the battle joined.

The aftereffect of one such venture was not a story the Old Man was anxious to have batted around the squash court. He and Eaker were ferrying back from Rockwell Field two high-wing Douglas O-28 observation planes. Fechet knew his limitations as a pilot and so did Eaker. The General would follow the Captain. They would fly only in the best of weather. On a beautiful morning with a gin-clear sky they took off from Rockwell. On reaching their refueling stop, at Winslow, Arizona, several hours later, Eaker, preparing to land, looked around for his boss and found the sky still wonderfully clear but empty of other aircraft. From time to time en route he had checked to see that the General was accompanying him. Now he was gone. Worried, Eaker landed quickly to see if those on the ground might have some report on him. The answer was no one had heard anything. Now Eaker was truly concerned. Obviously General Fechet had experienced difficulty of some sort and was down in desert country.

Eaker was just on the point of taking off to retrace his course and institute a search when someone shouted, "There he is!"

And there he was, O-28 and all, looking and sounding in A-1 shape. Eaker's sigh of relief could be heard across the field. On alighting from his plane, it was observed that the General was looking somewhat sheepish, an unusual expression for so rugged and forceful a character.

Later, out of earshot of the welcoming committee, Fechet confessed to "Iree" that the previous night's joy around the card table, piled on the previous nights of taking Spaatz's money, had been a bit too much. Somewhere in the course of the flight, he had dozed off. When he had awakened, he found he was practically on the ground and for the moment not at all sure of his location. When he had recovered some altitude as well as his senses, he had spotted "the iron beam"—a railroad track—which he thought he recognized from previous flights. Shortly after that, he was sure he was on course because dead ahead was a crater caused aeons ago by a meteor, affording a well-known and dependable checkpoint to all pilots.

After he had confessed his inadvertent snooze, the General and his aide decided it might be well to forgo the card table for the remainder of the flight, to swear off, as it were, until they had arrived safely home.

Fechet's performance before congressional committees was strictly in character. At one hearing before the Subcommittee on Military Appropriations, chaired by Representative Ross A. Collins, of Mississippi, Eaker witnessed an exchange between the General and the Congressman that was to have a long-lasting effect.

Collins was known as a quiet-mannered individual who spoke softly with a slow drawl and had a gift for sarcasm. He was also a Mitchell supporter and a strong advocate of air power, saving his best barbs for General MacArthur over the War Department's hidebound position on the needs for manpower versus the building of planes and tanks. However, when Collins spotted anything in an Air Corps appropriation that looked out of line, he was just as quick to pounce. In going over the annual Air Corps appropriation, he found such an item: funds to support thirty-six horses for riding instruction at the Air Corps Tactical School. Collins knew perfectly well that before an officer could attend the Command and General Staff School at Fort Leavenworth he had to know how to ride, and therefore the instruction was included beforehand. But he looked upon the need as totally unnecessary for airmen, an example of another General Staff tradition he was pleased to take a shot at.

Three members of the wicked Black Hand Gang, with fearless leader Cadet "Pewt" Arnold at the far left. He was also known as "Faith," and his two companions, Bill and Gus, as "Charity" and "Hope." (Credit Bruce Arnold)

Second Lieutenant Henry H. Arnold at the controls, grin in place, with his instructor, Al Welsh, looking a bit pensive, at Simms Station in May 1911 at the Wright brothers' flying field, Dayton, Ohio. (Credit USAF)

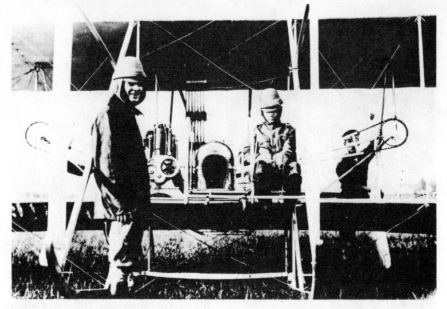

A beehivelike helmet has replaced the cap among thinking airmen, as Arnold prepares to join Second Lieutenant Tommy DeWitt Milling in a flight at College Park, Maryland, summer 1911. (Credit USAF)

At Fort Riley, Kansas, in November 1912, Mackay Trophy winner Arnold and Lieutenant Follett Bradley prepare to test a new wireless set devised by Lieutenant Joseph O. Mauborgne, of the Signal Corps. Bradley had transferred from the Navy the year before. (Credit USAF Photo)

Lieutenant Leighton W. Hazelhurst waits for Al Welsh to climb aboard the Wright C, whose engines have just been started. Arnold is to the left, his hand on the wing strut. Welsh and Hazelhurst, a bulky passenger, are about to test-fly the Wright carrying a total of 450 pounds. (Credit USAF)

Grim proof of how dangerous flying could be in 1912. At two hundred feet, the late-afternoon air of a June day was calm over College Park as Al Welsh dived to gain additional speed for his climb. The plane failed structurally, killing both men in the crash. (Credit USAF)

Arnold airborne over the parade ground at Fort Riley, Kansas, just prior to his near-fatal crash, November 1912. (Credit USAF Photo)

The Wright brothers had taught Lieutenant Benjamin D. "Benny" Foulois how to take off and fly, but there had been no time to teach him how to land before he was transferred to Fort Sam Houston, Texas. There he learned by trial and error through correspondence with the Wrights, until they sent instructor Phil Parmalee, on the right, to give him a few extra pointers. (Credit USAF Photo)

Rescue team at Black Butte, Mexico, January 1917. Efforts to find Lieutenant Walter A. Roberston and his passenger, Col. Henry Bishop, had been unsuccessful, and the mud hadn't helped. Left to right, Doc Wildman, Lieutenant Seth Cook, Captain Bert Dargue, Lieutenants A. D. Smith and B. Q. Jones, and Colonel Kenly. (Credit USAF Photo)

Major Carl "Tooey" Spaatz, Issoudun, France, 1917. He was CO of the 3rd Instruction Center in August 1918, when ordered back to the United States. He protested, saying he had not been to the front. Mitchell gave him permission to join the 13th Pursuit Squadron. In short order, Spaatz shot down three German aircraft. (Credit Mrs. Ruth Spaatz)

Major Horace M. Hickam at the controls of his DH-4. Chief of Air Service Information in the early twenties with Mitchell in the forefront, he organized international air races. (Credit USAF Photo)

Captain Frank O'Driscoll "Monk" Hunter, World War ace and pursuit specialist in the cockpit of his favorite Thomas Morse Scout at the Detroit Air Races, 1922. (Credit USAF Photo)

Brigadier General Billy Mitchell, Assistant to Air Service Chief Major General Mason M. Patrick, right, reports his pilots are ready to go at Pulitzer Trophy Aviation Meet. (Credit USAF Photo)

Mitchell's pilots of the 1st Provisional Air Brigade were ready to go after the Navy, too. Here on July 21, 1921, they make aerial history sinking the supposedly "unsinkable" German battleship *Ostfriesland* in a test of air power against sea power. In September 1923, Mitchell's bomber pilots repeated the lesson, sinking condemned naval ships *New Jersey* and *Virginia* in tests off Cape Hatteras, N.C. (Credit USAF Photo)

The route of the Pan-American Goodwill Flight covered over twenty-two thousand miles and encompassed twenty-three Central and South American countries. Five Loening amphibians took off from San Antonio, Texas, on the epic journey, December 21, 1926. (Credit USAF Photo)

Before departure. The five crews line up for the camera. Left to right, Lieutenant Charles Robinson, Captain Arthur McDaniel, Captain Clinton F. Woolsey, Lieutenant John W. Benton, Major Herbert A. Dargue, Lieutenant Ennis C. Whitehead, Lieutenant Bernard Thompson, Lieutenant Leonard Weddington, Captain Ira C. Eaker and Lieutenant Muir S. Fairchild. (Credit USAF Photo)

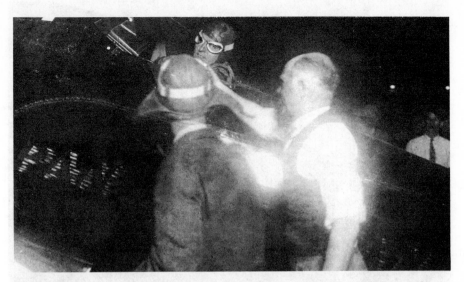

Dawn takeoff at Brownsville, Texas. Captain Ira C. Eaker attempts a dawn-to-dusk flight to Panama in his new Boeing P-12 pursuit, named the *Pan American*. (Credit Lieutenant General Ira C. Eaker)

Bucking strong winds and rough seas off the coast of South America. (Credit USAF Photo)

Ten had set out; only eight returned. May 2, 1927: President Calvin Coolidge awards the DFC to the returned airmen at Bolling Field. Major Dargue is to the President's right, Captain Eaker to his left. In the background behind Eaker can be seen Assistant Secretary for Air F. Trubee Davison. (Credit USAF Photo)

Cayenne, French Guiana: McDaniel and Robinson have found one way to keep cool in the tropics, and Ennis Whitehead is about to join them. (Credit Lieutenant General Ira C. Eaker)

New Year's Day, 1929. Five airmen get ready to test their own endurance and that of their Fokker trimotor, the *Question Mark*. From left to right, Staff Sergeant Roy Hooe, Lieutenants Elwood R. "Pete" Quesada and Harry A. Halverson, Captain Ira C. Eaker and plane commander Major Carl "Tooey" Spaatz (Credit P & A Photo)

The refueling plane is a Douglas C-1 piloted by Lieutenant Odas Moon. His crewmen were Lieutenant Joseph G. Hopkins and Andrew F. Salter. Oil and food were lowered in a basket. (Credit P & A Photo)

Lieutenant Irvin A. Woodring, copilot of the refueling plane, paints an appropriate insigne on the C-1. Woodring was later killed in a crash. (Credit USAF Photo)

January 7, 1929, nearly 151 hours after takeoff, it was the plane that gave up, not the hungry crew. Their endurance record spurred aeronautical development but left the question, Who was going to pay for the food? They did. (Credit USAF Photo)

Eaker the test pilot. He took up a fifty-four pound battery in his P-12, simulating the weight of a Signal Corps radio. The question: How would the added weight affect the plane in a spin? (Credit **UPI Photo**)

The answer was . . . not very well. Eaker had to bail out. But you can't keep a good pilot down, even with a broken leg, as Eaker proves at Bolling Field, zipping along on crutches. (Credit UPI Photo)

Talking it over at the May 1931 maneuvers that won Brigadier General Benny Foulois the Mackay Trophy: from the left, Foulois, Captain Eddie Rickenbacker and Major General Fechet. (Credit USAF Photo)

A threesome beside the Chief's official car, ready for the band to sound off. Foulois, Arnold and the man in black, Tooey Spaatz (Credit Wm. Bruce Arnold.)

Saturday reviews were an Arnold specialty, to which he invited noted aviation and film luminaries. Amelia Earhart, in white, stands with Hap and other guests on the administration building balcony watching a flyby. (Credit USAF Photo)

The Air Corps put almost every type of aircraft it had into carrying the mail. This Douglas O-25, over Great Salt Lake, Utah, is en route to Seattle, Washington, on Route 5. (Credit USAF Photo)

Martin and Keystone bombers could carry over a thousand pounds of mail, but their slow speed made it difficult to meet the schedule. (USAF Photo)

Captain Ira C. Eaker set up his route headquarters at United Airport, in Burbank, California. His chief pilot at Milford, Lieutenant Frank A. Armstrong, was to become a famed 8th Air Force bomber commander in World War II. (Credit USAF Photo)

Having a pretty girl as a dispatcher helped if you were flying an open-cockpit pursuit plane that could carry only fifty pounds of mail and wasn't designed to carry any. (Credit USAF Photo)

"General," he drawled, "I see here an item for thirty-six horses. Would you mind telling us what the hell the Air Corps is doing with thirty-six horses? Would you mind telling me what you'd do if I offered you thirty-six goats?"

There was no hesitation on the General's part. His big, gravelly voice filled the room. "Congressman," he replied, "if one of those goats was yours, I'd take it."

For an instant Eaker thought Fechet was in for trouble; then the room exploded with laughter, Collins leading the chorus. When it was quiet again he said, "General, would you mind stopping by my office when we're through here?"

When they stopped by, Collins said he understood the General liked to hunt, and he wondered if Fechet would be inclined to join him in Meridian, Mississippi, over the weekend for some duck hunting. Indeed the General would, and Eaker flew him down and was invited to join in. So, too, in time, did Trubee Davison. Because of his infirmity, they rigged a gun turret from a P-16 in the blind, which gave him the support he needed.

The relationship between the Congressman, the General and the Air Secretary became a firm and lasting one, all over a matter of horses and goats.

When the new Boeing P-12 pursuit ship was due to make its debut, in 1928, Ira Eaker had an idea on how best to show what the plane could do and at the same time exhibit its military potential. He would fly it on a dawn-to-dusk hop from Texas to France Field, in the Panama Canal Zone. Trubee Davison, General Fechet and the Boeing people liked the idea.

The point chosen for departure was Brownsville, at the eastern end of Texas' Gulf coast. The selection had more than its location going for it, for a new airport was to be opened there in March 1928. The opening would also mark the inauguration of passenger service between Brownsville and San Antonio. Combining the three events would certainly add to the news coverage of a Texas boy making good.

The event was built up as a major aeronautical happening, and on March 15 about one hundred military and civilian aircraft were gathered at the field, which was crowded with enthusiastic spectators and welcoming officials. Present also was Catherine Fechet, wife of the General, and their twelve-year-old daughter, Mary. At the proper moment, Captain Eaker appeared overhead in his spanking-new pursuit ship. The P-12 was a proud sight, bright yellow wings, red, white and blue tail fin, with *Pan American* stenciled on the fuselage. He taxied the plane up to the speakers platform, and there Mary Fechet christened it the *Pan American*. Her breaking the customary bottle of champagne on the Boeing's nose was the highlight of the ceremonies for Eaker, the Air Corps and Boeing officials.

The next morning at four, he took off in the dawn light in his bid to set a record. Very soon thereafter he ran into foul weather. Strong head winds slowed him. Heavy rain and poor visibility plagued him, and by the time he was over Nicaragua, he knew he wasn't going to make it to Panama by sundown. Disappointed, he landed at Managua and, the following day, flew on to France Field to prepare for the attempt in the opposite direction.

It was not his lucky month. Again he was off at four, and again he ran into rotten weather. In fact, it was worse than that encountered on the flight down. Near Minatitlán he was forced to descend to palm-top level, hugging the shoreline, fighting to stay in the air. He missed his gas stop at Minatitlán, and at Tampico he ran out of fuel and time, having to land in the mud in a drenching downpour. It was a miserable, wet ending after such a bright send-off. But it was not in his nature to curse the fates or moan about bad luck. When you flew, you had to accept the odds, whatever they might be. It was what came next that counted. When the rain let up, mud prevented his departure for another day. When he finally reached Kelly Field, he announced that the failure of his attempt had absolutely nothing to do with the plane or the location of the refueling sites. It was all the fault of the weather. Instruments had not yet been developed to surmount it.

It was almost exactly a month later that Eaker went on a flight of another kind, whose genesis had also filled the newspapers with headlines. With him went General Fechet, Lieutenant Elwood R. "Pete" Quesada and a German pilot, Frederick Melcheor.

Melcheor was with the Junkers Company, whose plane had been used for a record flight attempt. This one was from Baldonnel, Ireland, to Mitchel Field, New York, in the first nonstop east-to-west crossing of the Atlantic. The plane, dubbed the *Bremen,* carried a three-man crew: its gutsy pilot, Hermann Koehl; his Irish Air Corps copilot, Captain James Fitzmaurice; and the navigator and financial sponsor of the undertaking, Baron Guenther von Huenefeld.

On the morning of April 12, Koehl had barely gotten the overloaded low-wing Junkers off the ground. He had dragged its wheels through the treetops, and scraped a wing tip on the ground, turning to avoid a mountain. As Friday the thirteenth wore on, the *Bremen* was reported overdue and feared lost. Then by telegraph came the great good news that the three were safely down on tiny Greenly Island, near the coast of Labrador. They were a thousand miles off their intended course, but even so, eleven months after Lindbergh's epic flight, they had spanned the ocean in the opposite, more difficult direction.

Eaker's involvement came with a call from Trubee Davison on a Saturday morning. A representative of Junkers was in his office asking for assistance. The company felt that if its chief pilot, Fred Melcheor—who was in Washington—could get to Greenly Island, he could fly the *Bremen* off it and thus save the aircraft. Could the Army Air Corps help?

When Davison showed Eaker the plane's location, his immediate response was, "There'll be ice in there. I wouldn't be able to land. Only thing we can do is drop him in by parachute, and the only plane for the job is the Loening OC-1. Would your pilot be willing to make the jump?" Eaker asked.

The representative said that he was sure Melcheor would. The result was that Davison and Eaker took the proposal to General Fechet. The three talked the idea over and decided that since the mission would entail a certain amount of risk due to the time of year and the unpredictability of the weather, a backup amphibian should come along with a safety pilot. Eaker suggested Santy Fairchild, and Fechet got on the phone to Langley Field.

By the time Fairchild had flown up from Langley, the Air Corps Chief had decided to deal himself in. It looked like too good a game to miss, and he signed on as Santy's crewman. With Melcheor in Eaker's plane, they took off and headed north, and by that night had reached St. John's, New Brunswick. The next morning, it wasn't the weather that grounded them but Fairchild. He was in considerable pain and it appeared he was having an attack of appendicitis. A replacement pilot was needed right now. Eaker said to get Pete Quesada, and Fechet was on the phone to Major Howard C. Davidson, the CO at Bolling Field, where Quesada was stationed. But it was Sunday morning, and where the hell was Quesada on a Sunday morning? Not in bed, but out riding his horse in Rock Creek Park. Through Davidson's orders, the twenty-four-year-old Lieutenant was literally plucked from the saddle and planked on board a DH-4 bound for Boston. There he picked up an amphibian and, with a backup pilot, flew to join the waiting group at St. John's.

Eaker had recommended Quesada because he knew he was a pilot who would and could fly anything with wings. Further, the small, agile Lieutenant was a tireless worker, with a record of dependability.

The next morning, a saddened Santy Fairchild was flown back to Boston in the amphibian that had brought Quesada, and the expedition prepared to push onward. Eaker was the navigator, and Quesada, with General Fechet, was to hold formation off his wing. It was bound to happen that they would run into foul weather. Coursing up the Bay of Fundy, "the weather got bad," in Quesada's words, "and then it got worse, and then it got extremely bad." Translated, this meant that by the time they had reached the most northerly point of the bay, the overcast was on the hills, and the wind, at gale force, was in their teeth. Their fueling stop was to be Pictou, Nova Scotia, but there was no getting there at the moment, and they came down out of the muck and landed in the bay, dropping their anchors off the village of South Maitland.

They had no sooner done so than a dory came out from the headland, pulling toward them through the whitecaps. The fisherman at the oars began shouting that they should leave their planes and come ashore with him. He seemed overly excited about the tide, and in his effort to convince them to abandon their ships he slammed his own into the wing of Quesada's amphibian, ripping the fabric. A tide was a tide to the four airmen, and they shouted at him to stay the hell clear! The fisherman gave up and headed back for the rocky shore, shaking his head as he rowed.

A few minutes later, they learned what he was shaking it about. Suddenly the tide turned and began to ebb swiftly. The anchor ropes pulled taut. With the wind, the backing water quickly became a millrace. Quesada heard the sharp snap of the cleat tearing loose from the nose of Eaker's plane and, open-mouthed, saw the amphibian swept away out of control. He had been trying to stay as close to Eaker as he could, and he later reflected that the sight scared the hell out of him. He thought, *If it happened to Ira, it'll happen to me. If it happens to me, I'm up on the rocks!*

He scrambled out of the cockpit, shouting for Fechet to take his place, figuring he'd better get the engine started and free the anchor rope. He checked the rope and saw that it was going to be impossible to loosen; then he went after the engine with the inertia starter crank. As he cranked away, he heard Eaker's engine bark into life and knew that he, at least, had control. In order to free the anchor rope, Quesada had to step past the propeller, knowing, primed as it was, it could start turning without warning. He slashed the rope with his knife, stepped past the prop again, and signaled the General to pull the clutch. The engine burped a couple of times and caught hold. Back in the cockpit, Quesada taxied out to take off as a seaplane . . . and ran aground.

Ten minutes later, he and Fechet were walking around the plane on the Bay of Fundy's bottom, shaking their heads in wonder. As for Eaker, he and Melcheor had disappeared in the gloom.

Quesada's question was, What the hell am I going to do now? He left the General with the plane while he walked ashore, trying to figure out the answer. The fisherman was waiting with an I-told-you-so look. The Lieutenant climbed up the bluffs that ringed the shoreline for a look-see, thinking of finding a place where he might land if he could get off the Bay and the weather prevented him from reaching Pictou. He located a usable spot a couple of hundred yards long and left his leather jacket on a stake to mark it. The wind was blowing so hard he estimated he could land in the small area with room to spare. Then he went back down to the shore and asked the fisherman if he had a shovel he might borrow. Taking turns, he and Fechet put their backs into it and literally dug the amphibian out of the sand, first around the wheels so they could be extended and lowered, and then under the hull so the weight would be on the wheels.

Quesada could have left the fisherman's shovel and been on his way, for the light was fading fast and he was anxious to get airborne. But instead, he walked ashore again and returned the tool to its owner, remembering the Robert W. Service poem about the woodsman: if he lost his ax handle in the wilderness, he'd perish.

Fechet had the engine started by the time he returned, and Quesada taxied the groaning Loening out of the holes and took off from the rocky bottom of the Bay of Fundy. He knew that with the fuel remaining he had about twenty minutes before he reached a point of no return. He'd decided that if they hadn't gotten out of the weather by then they'd have to come back and land by his jacket. He was mightily pleased to find that while they had been shoveling, the visibility had improved and so, although he lost his jacket, they made it to Pictou. But, once there, he saw he'd lost something far more important. Eaker's plane wasn't to be seen in the Bay below. *Oh, my God,* Quesada thought, *he's hit a hill!* It was not an infrequent occurrence under such conditions.

He landed, and the look on Jim Fechet's face was a rawboned reflection of Quesada's fear. They spent an uncomfortable night laced with worry, bunked amid the tackle and gear of a ship's store on Pictou's waterfront.

In the morning the weather was on the rooftops, and Quesada spent most of the day on his back, sewing up the elongated rip made by the fisherman's dory. To do the job, he had brought the Loening up on the shore, and while he worked, the General sat beside him, huddled against the cold, straining his ears for the sound of a Liberty engine. By the time Quesada had completed his patch job, using marine glue and some linen bought from the store owner, it was getting dark, and they went up to the store together to get the chill out of their bones. The proprietor was waiting with a little brown jug. They thought it was alcohol, and although the gesture was appreciated, they said thanks but not right now.

"It's not Canadian whisky," the proprietor said. "I don't mean for you to drink it. I thought it might help you with that patch."

"Oh, the patch is fine," said Quesada. "Your glue and linen will do the trick. I've done this sort of thing before."

"Well, take a smell."

Quesada did and reacted. "Where in the devil did you get this? It's acetate dope!"

"It was left by those round-the-world flyin' fellahs a few years back when they came through here," explained the store owner. "I just remembered I'd been saving it for someone like you."

While the discovery lightened their spirits a bit, it could not in any way alleviate their unspoken worry for Eaker and his passenger.

The General and the Lieutenant settled down for a second night, hoping the weather would let them get off in the morning to look for the missing plane. At about 3 A.M. they were awakened by a pounding on the door. Quesada went to it, and in out of the dark stepped Fred Melcheor.

"Where have you been?" Quesada gasped.

"Where's Iree?" barked Fechet.

"We landed back in the hills," said the German. "When the weather clears a bit he's going to come in. I've got all the equipment from the plane out here in a farmer's wagon. You come help me with it, yes?"

While they helped him unload, Melcheor explained how Eaker had managed to bring the plane down through the overcast. The angle of the slope was such that they knew it would be impossible to take off without an assist. Eaker then told the German pilot they would have to build a ski-jumplike mound at the foot of the hill. Once they had it built, they'd remove everything they could from the amphibian to lighten its load, and then Eaker would give it a try and see if he could catapult himself into the air. With the help of the farmer, who had supplied the horse and wagon, they had spent all the previous day building the mound. Now all that remained to be seen was whether Eaker could launch himself via a downhill run. At Pictou the other three could do nothing but wait and see, and that afternoon Eaker and amphibian came chugging in out of the mist unfazed and ready to push on. To Pete Quesada's way of thinking, the performance was typical Ira Eaker. He'd get through somehow, even if he damn' well had to carry the plane.

They reached Greenly Island the following day, and as Ira had predicted, the surrounding water was crammed with ice floes. Below, on the island, they saw the *Bremen* and its waving crew. At five hundred feet Quesada and the General watched Fred Melchoor climb out of the rear cockpit and stand on the side step of the fuselage. At Eaker's signal, he fell over backward. His chute billowed, and he drifted down to a safe landing right beside the Junkers. Melchoor waved his thanks from the ground, and then Eaker pointed the nose of his plane toward home.[4]

On the return, an incident occurred that led to the famed *Question Mark* flight, the following January. Again it was a matter of weather. They got caught above it, flying from Pictou to Portland, Maine, and it all came down to a matter of fuel. Could they find an opening in the solid overcast before they ran out of it?

Quesada's experience as an instrument pilot was less than Eaker's; nevertheless he was proficient on the gauges, such as they were. But the question now was the closeness of the ceiling to the ground and what they might hit before they broke out of the overcast. The answer could be fatal. And so they sweated time and weather, hoping they'd have enough of one before they ran out of the other. Their luck held. The cloud cover broke, and down they came with a few drops of gas left in the tanks.

That night, as they settled in for some gentle cutthroat, Quesada raised the question of a limited fuel supply and what it meant. "It would have been nice to have had a gas station up there," he observed.

Eaker agreed but said he didn't know of any.

"What about the idea of in-flight refueling? Someone to come up and give you ten."

"You mean like Smith and Reichter?"

"Something like that, only longer, with a bigger plane. Why don't we conduct a refueling flight with a crew on board and stay up as long as we can in something like an O-1?"

"Hmmm," mused Ira thoughtfully, taking a firmer bite on his cigar.

"Not a bad thought," rumbled Fechet, checking his cards.

Later, Pete Quesada believed that it was through Ira Eaker that he was chosen as relief pilot on the *Question Mark* endurance flight, having planted the seed of the idea on the return from the *Bremen* mission.*

However it was done, Tooey Spaatz was named to command the undertaking. And he and his crew, which included Lieutenant Harry Halverson and Sergeant Roy Hooe, became world famous. Halverson was selected not only because of his flying ability but also because he was physically powerful and noted for his endurance, Sergeant Hooe because there wasn't any better Crew Chief in the Air Corps. The plane chosen for the endurance attempt was a Fokker trimotor, and the idea was to keep it in the air as long as they could. Because they didn't know how long that would be, they dubbed the Fokker the *Question Mark*.

The two refueling planes were Douglas C-1's. The first was piloted by Captain Ross G. Hoyt, with aides Lieutenant Irwin Woodring and Lieutenant A. C. Strickland.

Following the flight, Spaatz and his crew were awarded the Distinguished Flying Cross, as were the refueling crews. The citation, by direction of President Coolidge, read:

> For extraordinary achievement while participating in an aerial flight on Jan. 1–7, 1929, he [Spaatz] commanded the airplane "Question Mark" in a refueling flight, at and near Los Angeles, California, which remained in the air a total of 150 hrs. 40 minutes 15 seconds, a period of continuous flight longer than any previous flight ever accomplished. By his endurance, resourcefulness and leadership he demonstrated future possibilities in aviation which were heretofore not apparent and thus reflected great credit upon himself and the Army of the United States.

Aside from the obvious, there were several things about the flight that were indicative: It offered insight into thinking going on in the minds of airmen like Spaatz and his crew. It illustrated Spaatz's performance under sudden duress. And when the flight had faded from the front pages, it produced an ironic and exquisite postscript that would be difficult for any government agency to top.

Strategically, the *Question Mark* supplied answers. One purpose was to see how long aircraft engine parts could sustain uninterrupted operation. In spite of a lack of governmental funding, aircraft firms through private

---

* Quesada would later succeed Eaker as General Fechet's pilot.

investment were pushing ahead. Engines that could keep a plane aloft for a week in 1929 could double, triple, quadruple the equivalence of that performance in a decade. Airmen had no trouble understanding this, even if their superiors on the ground did. From the point of view of hindsight it was such an obvious recognition, but not so at the time. Spaatz and his crew had proved that planes would be built that could fly great distances, either by refueling in mid-air or nonstop. It meant the world was shrinking and, as Balbo and his flight of twenty-four planes would prove four years later, shrinking fast.

The *Question Mark* flew eleven thousand miles and took on over five thousand gallons of hand-pumped fuel, not to mention oil, food and water. If pilots and crew could operate within such confined quarters for nearly a week running, then, one day, bomber crews, should the occasion arise, would be capable of seeking out far-distant targets.[5] Conversely, the defenses of the United States and its possessions would not be invulnerable to future air attack.

In its six days aloft, the *Question Mark* was refueled fifty-six times, ten times at night. It was during the first daylight refueling, over the Rose Bowl, in honor of the game below, that Commander Spaatz got drenched, not with rainwater but 72-octane gasoline. He was handling the wind-whipped operation through an open hatch in the trimotor's roof. The crew had lowered the hose from the C-1, which was hovering about twenty feet above. Spaatz had the borrowed fire-department hose in hand and had inserted its nozzle into the tank. The fuel began to flow. The planes hit turbulence. The nozzle yanked free and raw gasoline spewed down on Spaatz as he grabbed for the hose, struggling to get the nozzle back where it belonged.

The crew on the C-1, seeing what had happened, cut off the flow and cranked up the hose. Eaker, also realizing what had happened, turned the controls over to Quesada and said, "Head for the ocean," knowing the air was less turbulent above the water. Then he came back to help Sergeant Hooe get Tooey's clothes off. The high-octane fuel could cause serious burns. They rubbed the casualty down with rags and coated him with oil.

"If I'm burned and have to bail out," Spaatz told them, "you keep this plane up in the air, and that's an order."

He was not burned, but he was quite a sight, stripped to the buff, wearing only a coat of oil and a parachute. When the hose came down from the C-1 again, there was Spaatz, standing naked in the breeze halfway out of the hatch, making sure this time that nothing came unstuck.

As the citation read, the flight had "demonstrated future possibilities heretofore not appreciated."

Economically, this appreciation was manifested upon the crew of the *Question Mark* in a stunning manner. They were required to pay for the cost of the meals they ate during the flight.

In Captain Eaker's case the amount was a bit more than the others, totaling $78.55. This was because after the mission had ended, at Metropolitan Field in Los Angeles, Eaker had remained until the eleventh of January, seeing to the *Question Mark*'s overhauling and repairs before flying it back to its Bolling Field base. Since Metropolitan was not an Army field, he had also put in an expense voucher for meals taken there.

But lo and behold, when Flight Commander Spaatz and his crew submitted their vouchers for payment to the War Department's finance section, the section was seized of its own question mark! It became so weighty that the consideration was passed upward until it came before J. R. McCarl, the Comptroller General of the United States.

Mr. McCarl passed judgment in a letter to Major E. T. Comegys, of the U. S. Army Finance Department, through the War Department's Chief of Finance, concluding as follows:

> Officers traveling by air are not entitled to reimbursement of subsistence expenses during a stop or delay for performance of temporary duty exceeding 72 hours. The duty performed in this case at the temporary duty station was duty devolving upon an officer of the Air Corps of the Army and no provision of law authorizes or provides reimbursement or subsistence expenses for officers of the Army performing their appropriate and required duties at their duty stations. The fact that in the performance of the duty here in question the officers could not procure subsistence in the usual manner or at their accustomed places and that it was necessary to procure subsistence by means otherwise than ordinarily procured, creates no status giving a right to have the subsistence furnished them at the expense of the United States. You are not authorized to pay the voucher and it is returned here.

In short, the crew of the *Question Mark* was required to pay for the food consumed on their unparalleled flight, which had captured headlines, earned them medals, advanced aeronautical capabilities and reflected great credit on the Army of the United States.

To airmen it was par for the course. The thought was advanced that if they'd stayed up much longer they could have gone broke. Spaatz had nothing to say. The action didn't need comment; it spoke for itself.

But on New Year's Eve 1929, Spaatz's three compatriots on the flight spoke for themselves when they sent him a telegram:

> Hope the refueling ships of Health, Wealth and Happiness will always be in contact with you and yours and contentment in life will establish a permanent endurance record at this hour. As always we think of happy hours spent with you. From three green flares at midnight . . . Ham, Ira, Slippery.*

* Harry Halverson, Ira Eaker, and Pete Quesada.

Seven

Holding Pattern

While Spaatz, Eaker, Quesada and a handful of other Air Corps pilots were probing the horizons of aeronautical development, Hap Arnold was now caught up in the business of keeping them flying. In June 1929, he had taken command at Fairfield. Fairfield was called an intermediate depot. Its job was repair and overhaul of aircraft, their engines and their instruments, whereas Wright Field, which was the headquarters for the Air Corps Materiel Division, was concerned with all the components of aircraft engineering and supply. The two stations were closely linked by more than proximity, for Fairfield was where most of the officers serving both bases lived.

The Arnolds had a farmhouse that sat in the middle of a golf course, which couldn't have been nicer for an addict trying to break 100. The job, however, was not his cup of tea. He was not oriented toward fixing up, but coming up with something better. Still, there were highlights and advantages, and as a maintenance man he gained some valuable experience and knowledge in the servicing of aircraft.

During his tour, the DH-4 was finally phased out. New aircraft with brakes and controllable-pitch propellers were phased in. Plans were going forward for a 200-mph bomber. Major Jimmy Doolittle, breaker of speed records and test pilot without parallel, took off under the hood, flew for fifteen minutes and then, with the hood still up, still on instruments, made his landing.

As for Arnold, he felt his greatest contribution to the advancement of aeronautics while at Fairfield was the destruction of the Barling bomber, a Billy Mitchell brainchild. The aerial monstrosity built in 1923 was a triplane with six engines, incapable of flying 100 mph or attaining enough altitude to get over the Appalachians. The result was that it had long sat in its own special $700,000 hangar, an ugly behemoth, going to ruin. The sight and thought of the thing was gall and wormwood in the gut of any self-respecting airman. Yet the War Department insisted upon keeping it intact because of its enormous cost of over a half million dollars, and the possibility of recurring congressional interest. Dismantled, it was stored in a nearby warehouse. In Arnold's eyes the wretched machine was a monument to those who saw no future in aviation. He put in a requisition for its elimination, not mentioning its name, only referring to its experimental designation, XNBL-1.

A few days later, at home for lunch, he looked at his watch and observed with his Merlin the Magician look, "In five minutes there's going to be a fire." The children wondered what he was talking about, but five minutes later, when the fire siren sounded, they knew. Everyone charged out of the house to chase the passing fire engine. In the distance they could see smoke rising. At the moment, they didn't know that within the burning structure was housed forty-two thousand pounds of Barling bomber.

During Arnold's first year at the depot, Benny Foulois served next door as Chief of the Materiel Division. He was a brigadier general once again, for in the decade since his return from France he had tenaciously worked his way up the rickety ladder of Air Corps command. Seniority and his pioneering flight record no doubt aided him, for even before Mitchell he had dared to speak out against what he considered the War Department's abysmal attitude toward the military air arm. Unlike Mitchell or Arnold, however, the shore he had been cast up on for his temerity, though far distant, was not one of exile. In 1920, his rank set back by the National Reorganization Act to permanent major, he was assigned as the Assistant Military Attaché at The Hague and in Berlin. The post was actually much to his liking. Far from the War Department power struggle that swirled around the head of Billy Mitchell, he did his job and he did it well, sending to the Military Intelligence Division all manner of aeronautical and political information. In this last, while swapping drinks and tales of flight with former German aces Hermann Goering and Ernst Udet, he was to claim later that he predicted the rise of German militarism.

While in Germany, Foulois maintained his personal lines of communication, corresponding with Majors Oscar Westover, Executive to the Chief, and Horace Hickam, the old birdman of San Antonio, who at the time headed the Information Branch of the Air Service. Perhaps the most notable event of Foulois's Berlin tour was that two weeks before he returned home, in 1924, with the rank of lieutenant colonel, he married again. His bride was forty-year-old Elizabeth Grant Bethel, from Philadelphia. Blue-eyed, brown-haired, with a firm chin and a determined manner, she had come over to Europe to serve as a secretary in the U.S. embassy.

However, the bridegroom did not return altogether a happy man. He had applied to go to the War College and had been turned down by the War Department. The turndown was particularly upsetting because in corresponding with General Pershing he had pointedly made mention of his desire and hoped it would be acted on favorably.[1] Later, in writing to General Patrick on his future course, Benny revealed his sense of frustration:

> I fully agree with your view that I should prepare myself for a general staff position. That has been my desire and ambition for the past 18 years! I have consistently studied and worked with that objective in view. . . . I had hoped that my pre-war and recent war record would have been sufficient to have placed me on the Initial General Staff Eligible List, but here again, I evidently failed to qualify, for some reason unknown to myself.

He closed his letter showing his pique with an implied threat: "I have had several very attractive offers to leave the Army since I have been in Germany. Whether I shall consider them or not will depend on my future prospects in the Army." He didn't mean it, of course. The Army was his whole life, and although after his return he did not receive assignment to the War College, he was posted to the Command and General Staff School. He stayed clear of the Mitchell imbroglio, his intense dislike for Billy no doubt influencing his thinking and keeping him out of harm's way. With Mitchell gone and the passage of the Air Corps Reorganization Act of 1926, Foulois's prospects brightened. Following a tour as CO at Mitchel Field, he was jumped two grades, to brigadier general, and appointed Assistant to the Chief of the Air Corps. And then, at the same time Hap Arnold took up his duties at Fairfield, Foulois assumed his as Chief of the Materiel Division.

Hap got along well enough with Benny, in spite of his having been so close to Mitchell. When Benny and Elizabeth Foulois returned to Washington, in July 1930 (he again appointed Assistant Chief), they were replaced by Conger and Sadie Pratt, one of the Air Corps's most popular couples.

Brigadier General H. Conger Pratt was a very tall man with a somewhat somber look, whereas his wife—the daughter of Major General Arthur Murray—was roundly ample and full of the joy of living. It was said you could always tell when Sadie was near, by the sound of her laughter. A graduate of the class of '03 at West Point, Pratt had been a cavalryman until 1920, when he made the switch to the Air Service. Unassuming, a gentleman of the old school, he offered ideas and sought cooperation. He believed that research and engineering funds should be aimed at greater advances in new types of planes and not used up on modifications of existing models. He believed air independence should be given a service test like an airplane, to see if it worked.

One of the things the Pratts were famous for on the post were their parties, and not just for the grown-ups. Childless themselves, they gave unforgettable Christmas and Easter gatherings for the children on the bases where they served. There were all kinds of scavenger and treasure hunts and wonderful prizes, Sadie's gay laughter a prize in itself.

In Arnold's mind, and that of a good many other officers, there was only one drawback in Conger Pratt's tenure as Chief of the Materiel Division. It was in the personality of his chief engineering officer, Major Clinton W. "Jan" Howard. No one would deny Howard's aeronautical engineering ability, particularly himself, but he was a sarcastic, acerbic individual. He could particularly infuriate the selected boards of officers who would come to Dayton to consider and test new military aircraft with an eye toward procurement. Often after a board had met and made its recommendations, comments would follow, from the Materiel Division under Conger Pratt's signature, severely criticizing the board's operations. It was recognized that Howard was the author of the attack and not Pratt, but board presidents like Major George Brett, whose fuse was short, would become exercised and want to take the matter up with the Adjutant General. Others, like Tooey Spaatz, handled Jan Howard in their stride, using their wits instead of anger to keep him in check.

With Arnold, however, it was a bit more personal, for Jan Howard had been his brother-in-law, having married Bee's younger sister, Lois. Howard had graduated from West Point in 1915 and served in the Field Artillery before transferring to the Signal Corps's Aviation Section in late 1916. It was at Rockwell Field that he'd met Lois, who was visiting Hap and Bee.

He was a complex man, irascible and often unpredictable. Lois was an emotional woman, and their marriage ended in divorce. Arnold had never liked him, and when Hap was head of the Field Services Section, they had several memorable run-ins on engineering matters. Both were majors, both outspoken, with their own ideas on how things should be done.

Only Howard's recognized brilliance as an engineering chief saved him from the wrath of a good many other officers. Arnold was sorry to be on the same post with him. Truth be known, he was sorry to be stuck in Dayton at all. The job had put him on the other side of the fence, so to speak. He found himself in the middle of a running battle between the tactical commanders, who wanted their limited aircraft back on the flight line right now and in A-1 condition, and those in his command, whose responsibility it was to maintain and service them.

The farmhouse was often the center point of heated argument, and to Bee the cacophony sounded like a stuck needle on the phonograph. *We can't ever get the equipment when it's promised! You don't deliver, Arnold! You don't give us what we want!* etc. etc!

He would go to the office at seven, an hour before the others.

"My goodness, why?" Bee would ask. "They don't get there till eight."

"I have to think. I can't think when other people are around." He was always thinking, working on ways to do it better even as he went through his daily workout, for he was a physical-fitness buff as well.

One day, he took his boys out to Simm's Station, where the Wrights' hangar still stood, and showed them where he had learned to fly. In the hangar they found a stick with ribbons on it. The ribbons had been used to signal him and Tommy Milling. It was like reaching back into the past nearly twenty years. Now, in the present, Orville Wright would come to Sunday dinner. His bushy white mustache reminded the children of their grandfather's—Daddy Doc. A kindly but taciturn visitor, the inventor thought it was a tragedy that what he and his brother had created had become an instrument of war, a device now fashioned and designed to kill thousands. His host couldn't argue. The grim fact of man's inhumanity to man was history's oldest cliché. However, if a country was strong enough and well equipped with enough airplanes, it was not likely that anyone would attack it. The answer to preparedness, however, was a matter of policy made in Washington, not at Fairfield Depot.

It was a policy that had not changed very much since Arnold had been summarily dismissed from the seat of power. What had changed was a number of the most important players. Coolidge had chosen not to run, and Herbert Hoover was President. Secretary of War Davis had been replaced by debonair oilman Patrick J. Hurley, who was a friend of General Fechet. But perhaps most important of all to the career and fortunes of Major H. H. Arnold, Major General Douglas MacArthur had replaced Major General Charles P. Summerall as Chief of Staff. Thus, all those who had had a hand in the eviction of the obstreperous Major were gone from the scene. But more to the point, the new Chief of Staff had known him well at an earlier time.

Back in the glory days of College Park, Pewt and Doug had lived at the Army-Navy Club and had gotten to know and like each other. Captain MacArthur was genuinely interested in Lieutenant Arnold's work, impressed with his eagerness and his capabilities. Arnold was then Assistant to the Chief of the Signal Corps's Aviation Section, whereas MacArthur was a Special Assistant to the Chief of Staff, Major General Leonard Wood. The two aides found they had mutual interests, and Pewt introduced the Captain to his bride-to-be.

There was only one thing about Pewt Arnold that Captain MacArthur frowned on, and that was his Belgian motorcycle. It was judged to be the noisiest two-wheeled machine in the capital. And when its eager-beaver driver fired it up at 5 A.M. every weekday, it not only awoke the sleeping brass at the club but a good many of those along the Lieutenant's thundering route. However, it was not MacArthur or the angry brass who put a stop to the awful racket, but Bee. The one thing she asked her future husband to give up was the motorcycle so that they could live in wedded bliss, and gentleman and lover that he was, he complied.

MacArthur had become Chief of Staff in the late fall of 1930. Having served as one of Mitchell's court-martial judges five years before, he knew all about Arnold's role in the event and what had followed.

In February 1931, evidence of the change in Washington's attitude toward Arnold became apparent. He received word of his promotion to lieutenant colonel. The word was certainly an occasion for celebration, and Lois's case of brandy lost another bottle. It was his first real promotion in nearly fifteen years.

"I must be out of the doghouse," he grinned at Bee as she helped him pin on the silver leaves.

"Since you're a *lieutenant* colonel, maybe you're only half out," she quipped back at him.

One thing was sure, however. Promotion had to mean promotion in command as well. Aside from the Air Corps's four general officers, a lieutenant colonel's rank was almost as senior as a flying officer could get. But then, when the word came through that he was going to get command of March Field, California, the rank became secondary in his thoughts. March had been a primary flying school. It would be his job to convert it into an operational base containing a pursuit group and a bomber group. Nothing could have pleased him more.

At the same time, he knew there were other equally significant changes in the wind. It was known in the Air Corps Chief's office that although "the Old Man" was planning to retire at the end of 1931, Secretary of War Patrick Hurley and a number of congressmen were anxious to have him serve a second four-year term. Fechet had said thanks, but no thanks. He wasn't going to stand in the way of others coming up. Speculation and bets as to who the new Chief would be began to circulate among the air staff. It was no sure thing that just because Benny Foulois was Assistant Chief he would succeed to the top job. It was thought by some that Fechet wasn't all that sold on him.

Foulois's wife, Elizabeth, was a factor as well. Intensely devoted to her husband, independent and outspoken, she made no bones as to her likes and dislikes. If it was the former, she could be gracious and kind; if not, look out. Air Corps wives such as Ruth Spaatz coined a single word to describe Elizabeth Foulois—*formidable*—whereas the wives of General Staff officers were said to have had a glacial attitude toward her, possibly never having forgiven her husband for separating from the first Mrs. Foulois.

Fortunately, there was more to the selection of an Air Corps Chief than feminine evaluation. Lieutenant Colonel Frank M. Andrews, Air Corps Chief of Training and Operations, believed that "General Foulois was the logical and probable successor to General Fechet." He offered this opinion to his former boss, retired General William E. Gillmore, and then made note of a forthcoming maneuver:

> There is a plan afoot, confidential as yet, to hold our training exercises this year in the shape of a concentration at Wright Field and demonstrations at all the principal cities on the East Coast over a period of about ten days. It looks as if it would go through. From the point of view of tactical training this is not as good as last year's exercise but it will present a most interesting problem in supply and staff work.

The organization and operation of the exercise was delegated to Benny Foulois by General Fechet with the approval of the Secretary of War and the General Staff.

The maneuver took place in May 1931 and was proclaimed "the greatest aerial demonstration the world has ever seen." The figures were impressive: 659 planes, 720 pilots, 644 enlisted men. They made up what was called the 1st Provisional Air Division. Aside from a headquarters squadron, the division consisted of a bomber wing and a pursuit wing, two observation wings and an attack group and a transportation group. For the first time on record, air cadets participated in field exercises. They were exercises that encompassed more than a half million air miles flown in a two-week period, featuring the Air Corps's largest command and staff exercises to date. The force assembled first at Dayton, Ohio, then flew on a rapid round-trip mission to Chicago. Then it was on to New York and New England, winding up over Washington on Memorial Day in an awesome, twenty-mile aerial parade.

Watching from below, Hap Arnold remarked to his wife, "Someday there'll be thousands of planes in the air. Do you realize that?"

"Well, maybe, but not while I live," Bee said.

"Yes, pretty soon," he insisted. "We're going to have lots of planes. The average person doesn't realize what's ahead of us."

No doubt he didn't, but the average person recognized that for the time it was an unusual display of air power.

Behind the scene, there were plenty of officers willing to take credit for the idea. Even before the event, Major General Frank Parker the VI Corps Area Commander, had written to Foulois stating, "As you know, this concentration is the realization of a vision which I have had since the World War and toward which I have, with such powers as given me, worked unceasingly." Undoubtedly so, but like so many ideas that succeed, this one had many pushing for it for a long time. Its overall purpose was not just to show how the fliers could fly but to direct public attention to the Air Corps. The War Department had to be in favor, because the wide-ranging display could only make the Corps look impressive and strong. A twenty-mile-long aerial parade with the newsreel cameras grinding in planes loaned to the Army by Fokker, United and the Ford Aircraft

Company, was sure to bring out the crowds and reach people via movie houses across the country. General MacArthur sent General Foulois a congratulatory telegram: "Well done 1st Air Division!" And Benny, not missing a trick, turned it over to the newspapers. It was a job well done, for in all the miles and the aggregate of more than thirty thousand hours flown during the exercise, there was not a single crack-up, not a single pilot killed or injured. Remarkable, was the word.

Even Billy Mitchell, living in retirement at Boxwood, his Middleburg, Virginia, estate, had to tip his hat to his former adversary. Mitchell had been kept apprised of the development of the rally by Frank Andrews, who had sent him photographic copies of the routes to be flown by the planes to and from the maneuvers. "Photos of the different types of ships being used are being prepared, and I'll send them," Andrews had written.

Actually, no one was in a better position to send such information. Andrews had been appointed Foulois's all-important chief of staff. This meant he was responsible for the details of the preparation of the exercise and its execution.

Every military officer who has commanded troops knows the success or failure of his mission depends in large measure on the caliber of his staff. And in this case, Benny Foulois had the best. Frank Andrews' deputy was Tooey Spaatz. It was Spaatz who acted as the principal air traffic controller. He was seen by the press as "a hooded figure in a plane on the ground, directing the movement of aircraft in the air like so many checkers, his job to send the various wings of the division to the assembly point and then toward a central rendezvous."

It was one thing to direct all the planes in flight, but it was still another to keep them flying. In this case, the officer who had that job was Hap Arnold, serving as Foulois's G-4, in charge of all the necessary supplies.

After the maneuver had ended, on Memorial Day, Foulois wrote letters of commendation to all his aides, knowing how well they had served him. It was then, too, that General Jim Fechet was said to have given the green light to the proposal that General Benny Foulois succeed him.

Although Hap Arnold was oriented more toward the tactical aspects of flight than to its repair shop, serving as G-4 for the combined air-ground maneuvers over California in 1930 and, a year later, at Foulois's grand rally, had made a lasting impression on him. This was not because of the demands of the job but because he saw that the real strength of an air force had less to do with numbers and types of planes than it did with transportation, quality of the supply, and the care of it.

A prime illustration of the problem was aircraft radios. In the early 1930s, Signal Corps radios for aircraft could not in any way compare to the voice and navigational radio equipment being used by the civilian airlines. Few Air Corps planes were equipped with two-way air-to-ground ra-

dios, and those that were, left much to be desired, particularly with regard
to clarity and reliability.

Another problem confronting Air Corps engineers was the weight of
Signal Corps radio equipment, especially when it had to be installed in
pursuit planes. At various stations—Wright, Langley, Rockwell, Bolling—
airmen were conducting their own testing of a new Signal Corps radio with
a battery weighing over fifty pounds.

On such a test on a bright summer's day in August, Ira Eaker took off
in a P-12 with a fifty-four pound battery stowed in a compartment directly
behind the cockpit. The question was, would this additional weight have
any effect on the plane's stability? Eaker learned the answer soon enough.
At five thousand feet, he put the plane in a spin to the left, and in coming
out of the maneuver, found the control responses very sluggish. He
climbed up to seven thousand feet before he tried a spin to the right. Very
shortly, the spinning pursuit's nose rose up and the plane assumed a level
attitude but continued rotating. The controls were so loose that he could
put his finger on the stick and move it all around the cockpit but the movement had no effect on the control surfaces, which were also moving without bringing any change in the plane's descending gyration. A flat spin is
now called autorotation—a term unknown at the time.

What Ira Eaker knew was that he had trouble, and nothing he tried to
do in the cockpit helped to correct it. The battery had thrown the plane's
center of gravity out of balance, producing uncontrollable instability. At
five hundred feet he knew the only way he could get himself out of trouble
was to get out of the cockpit, and right now! He went over the side and
banged into the P-12's stabilizer, skinning his nose and leg. But, at the moment, he had much larger worries. He had stayed with the plane too long.
He was too close to the ground. Would his chute open before he hit it?

The answer was yes, and no, and a very near miss. The canopy of the
parachute was only partially filled when it went over one side of the steep
roof of a house and Eaker went over the other, the risers of his chute
catching on the roof and breaking his fall somewhat. Not enough, however, to prevent him from slamming down on a cement stoop, his right leg
buckling under him. The pain was fierce, and he was rolling around on the
ground, trying to get unhitched from his chute harness when the back door
of the house opened and a woman peered out at him. She took a good
look, as though to reassure herself that he was real, then shut the door,
leaving him to agonize over his leg and the milk of human kindness. He
was also aware that his plane had torn down a hen house and was burning
up in an adjacent apple orchard. Suddenly the lady of the house appeared
again. This time she came to his aid, full of solicitude and eager to help.

"I beg your pardon, young man! I beg your pardon! Let me help you!"
she kept repeating while together they got the straps unbuckled. Then she
assisted him into the house, acting as a crutch, easing him onto a couch.

He asked if she'd mind calling Bolling Field for him, and she was only too happy to do so, but first she had a confession to make. "The reason I didn't come to help you sooner was that I had to call the newspaper first. They give five dollars to the first person who calls on an ambulance case."

A few days later, Eaker learned from his good friend newsman Ernie Pyle that, indeed, such was the custom.

Shortly thereafter, in a letter to Tooey Spaatz, Hans Adamson commented on the end result of Ira Eaker's test flight: "Came home to find Ira laid up with a bum foot as a result of a slight injury due to his parachute jump. The net effect is that he now plays cooncan with the General *all* the time instead of *some* of the time."

On radios, Adamson had a new idea for Spaatz, somewhat in advance of the technology, ". . . and that is to hear Trubee by radio from Washington directing the convolutions of your bombers in a mock attack on Los Angeles."[2]

There were other radio communication problems with which to contend, some of them teaching bitter lessons. "We were very much upset when Casey was killed several weeks ago," wrote Spaatz to Captain Hugh M. "Elmie" Elmendorf, in Washington.

> Apparently, the radio helmet had something to do with his failure to pull the parachute ring, as the helmet was plugged to the set when he jumped and the cord tore close to the plug with the plug badly bent. Parts of the helmet also showed signs of severe strain having been put upon it. I believe this force was sufficient to knock him out temporarily and prevent him from pulling the ring. We are now attempting to get all the clips on headsets in order to eliminate any possibility of such accidents. I have also instructed both squadrons that with the old-type headset they will not be plugged in below three thousand feet.

The sky above could be a dangerous realm, because all too often something went wrong with the equipment, and then it was a matter of getting down alive. Luck was a part of it, but skill was usually the greater part, and pilots like Ira Eaker had the deftness to land in backyards afoot or in the cockpit.

In a letter to Mrs. Mary Dunn, of Vanderbilt, Pennsylvania, he wrote: "I have often thought of the very hospitable way in which Mr. Dunn and yourself took me in when I so unceremoniously descended into your backyard so to speak. I have been over the mountains many times since and have always wanted to turn east a few miles and land in your field I hastily selected but have always had some pressing business that prevented."

On another flight not long afterward he was giving two West Point cadets a much needed lift when engine failure forced him into another "backyard," near Philadelphia. The cadets had to make it the rest of the way by train and arrived at the Point three and a half hours late, which meant they were AWOL for the period in question—a very serious business. They wrote to the Captain asking his help. Eaker took their plea to Trubee Davison, and the Secretary contacted the Corps Commandant, explaining the circumstances, thus preventing possible severe disciplinary action.

Of all the events of Ira Eaker's thirty-four years—the hours flown, the dangers encountered, the accomplishments attained both in the air and on the ground—none was more memorable or important to him than his friend Lieutenant Newton Longfellow's Thanksgiving party of 1930.

Longfellow and his wife, Laura, made it a really grand affair, inviting many of their friends, one of whom was Lieutenant Ronald Hicks. Hicks called Ruth Apperson to see if she could come as his date. He had excellent taste, for she was a lovely young lady, a graduate of George Washington University, engaged at the time as an instructor at the Washington School for Secretaries. From North Carolina, she had planned to spend the holidays in Chapel Hill, but Hicks convinced her she should come to the party instead, even though she didn't know the Longfellows. Ira Eaker decided he would go too.

And so they met, and poor Lieutenant Hicks had unknowingly served his purpose. She found the Captain, with his lively brown eyes and cleft chin, handsome. She liked the sound of his voice, laced with its faint Texas drawl, and the way that he looked at her. It was obvious that she in return had his undivided attention.

They were married a year later, on November 23, 1931, and spent a leisurely honeymoon, going by ship to Panama. It was much more relaxing than trying to fly there between dawn and dusk. On their return, there was no disagreement on the part of anyone that while Ira Eaker had gained a wife, the Corps had gained a beauty. Not only that, she had brains, not to mention a sense of humor.

Naturally, the Captain had introduced her to the Fechets, who in turn had invited the couple to come to dinner. Catherine Fechet, wife of the General, was a tall, angular woman, a devout Catholic, rather reserved in public but warmly agreeable once the ice was broken. An "Army brat" whose father had been a colonel, she loved to tell tales of the old Cavalry days. Life had been much simpler then, and remote Cavalry posts were really not all that lonely. Her daughter Mary, who had christened Eaker's Panama-bound plane, had an older sister, Catherine, and the four formed a close-knit family. No doubt big Jim Fechet, outdoorsman that he was, would have liked to have had a son to go with his daughters, on whom he doted. It was Iree, of course, who came close to filling the bill, for it was no secret that the Old Man looked upon Eaker with almost filial affection, and now admired his excellent taste in taking a wife.

On the occasion of that first dinner, the General rose before the festive board to carve the duck. He went at the task with typical boldness, if not with care and delicacy, and suddenly, as his knife came down, the bird was airborne! Its flight was brief and broke no records. Its landing was something else again: right in Ruth Eaker's lap.

The General turned to beet-red stone. His wife was mortified, Catherine and Mary caught with their mouths open. And for once, Captain Eaker could not come to the rescue. Ruth Eaker did, her laughter a delightful sound to fill the awful vacuum left by the act of uncontrolled flight. Then they were all laughing at the hilariousness of it as she handed the recalcitrant duck back to her host. Even so, the General's color remained high, and he felt, if nothing else, he owed the bonny young lady a dress. She charmed him back to a calmer frame of mind, and his next attack on the bird assured that it would fly no more.

Eight

The Seeker

While on the East Coast the Eakers had celebrated Thanksgiving by making the wedding bells ring, on the West the Arnold family arrived at March Field. They came fresh from a most enjoyable sea journey via New York and the Panama Canal. They also arrived flat broke, half starved and in a thunderous downpour.

Ruth and Tooey Spaatz were out on their porch with Maggie and the Arnolds' dog, Pooch, waiting to welcome them in out of the deluge. "We've had our Thanksgiving dinner," said Ruth (soon to have her third daughter), "but we want you to have what's left. Come pick the bones."

And pick them they did, Hap and Bee, Lois, Hank and Bruce and four-year-old David. The whole family at the table, the Spaatzes and their two daughters looking on.

Although the two couples had been in correspondence, and the men had seen each other at various times, they had not served together since Washington, and there was much catching up to do. While the rain beat down and they passed around the cranberry sauce and the dressing, the Arnolds related the tale of their journey.

It had begun with a farewell Halloween party given by the Pratts. Things were in an Arnold whirlwind. Parents dressing for the masquerade. Children underfoot, gathering up their possessions. Trunks strewn about, Maggie directing an aide with one hand while feeding David with the other. The uproar was not stilled by the phone ringing until Hap, unable to hear, roared for quiet, goddammit!

It was Bee's father calling. His bank, the First National Bank of Ardmore, due to the economic collapse, had been forced to close its doors. Worse, his financial officer had absconded to Mexico with a great deal of money.

It was the beginning of bad news. By the time the Arnolds had driven from Ohio to New York, visited with their families in Ardmore for a few days, then boarded ship, the Union Trust of Dayton, where they had their account, had also failed.

The news reached them aboard the *Republic*. There was a certain amount of irony in it, for their shipboard accommodations were most princely. Hap was so impressed that in a letter to his father he had drawn a picture of their cabin, the most distinguishing feature a large private bath!

"They are the most luxurious accommodations that we have ever had or expect to have," he had written. "So far everything has been lovely, fine food, good crowd, excellent weather." His only comment on the financial situation was, "Some people aboard had their money in banks which have failed. Some drew checks against their accounts in New York."

He did not add that he was one of those people and that all the bills paid between Dayton and New York were bouncing. With the failure of the Union Trust, the Arnolds were financially busted. Savings gone, education fund for the children gone—everything gone but the Army paycheck, but at least they were not unemployed, as were millions of Americans that Thanksgiving Day.

When Bee Arnold awoke the next morning and stepped out onto her patio, she saw the sun shining down on the snow-capped mountains over toward San Bernardino. The base was surrounded by fields of alfalfa, and in the near distance San Jacinto and Old Baldy were twin peaks standing guard duty over the flat green plain. The panoramic scene, the rain-scrubbed air, the smell of flowers made her feel that after the long journey and in spite of economic difficulties, everything was going to be all right. It was a sound prediction.

The decision to convert March Field into the Air Corps's major West Coast installation was in keeping with the MacArthur-Pratt agreement, brought forth earlier in the year. The joint policy hammered out by Chief of Staff General Douglas MacArthur and Chief of Naval Operations Admiral William V. Pratt had finally given to the Air Corps a defense mission on which to hang its helmet and goggles. Agreement with the Navy had been a very long ten years in coming. Fierce contention between the two services had grown out of three factors: ambiguous wording in the 1920 Army Appropriations Act as to whose air arm was responsible for what; Billy Mitchell's attack on the obsolescence of the battleship; and the economic depression. This last had not only shrunk the defense dollar to a bare minimum but had also brought congressional and press accusations of duplication of planes and equipment. Essentially, the new agreement gave to the Air Corps the responsibility for coastal defense so that the Navy would have freedom of action at sea.

That the Navy was swift to sidestep the agreement, building up its own land-based bomber and pursuit forces was a problem that did not occupy Hap Arnold at the moment.

In his assignment to convert March Field from a primary flight training school to a major tactical operation, Arnold not only had at the outset Major Tooey Spaatz as his Executive and Wing Commander but also the assistance of several other very dependable friends and airmen.

Major Joseph T. McNarney had been the CO of the primary flight school since August 1930. Now he would command the bomb group. A somewhat dour Scotsman, Joe was known more for the sharpness of his mind and his ability to make a hard, calculated decision than for any degree of levity. He had gone from the West Point class of 1915 briefly into the Infantry before making the transfer to the Signal Corps Aviation Section and flight training at Rockwell Field. He was at Rockwell at the time Hap Arnold returned to flight duty, and there he earned his wings with the rating of Junior Military Aviator in April 1917, a few days after the declaration of war on Germany.

Captain McNarney's war record was distinguished. He commanded observation groups during all the major U.S.-Allied campaigns from Château-Thierry to Meuse-Argonne. At war's end, suiting his combat experience to the word, he wrote a manual on air observation tactics.

In the decade that followed, McNarney passed through the three most important schools an air officer could attend. He spent five years at the Air Corps Tactical School and was one of the three officers who constituted the original staff.[1] After that, he put the Command and General Staff School behind and spent the next three years in the Aviation Section of the War Department's Military Intelligence Division. The Army War College was next. All of it was fine for the intellect, and Joe McNarney was a braw, canny Scot, but his flying time suffered. From being a very active pilot, he had become a very inactive one, averaging only about fifty hours a year.

In 1930, when he was finishing up at the War College, the question was where he and his wife, Helen, would be posted next. He discussed it with his friend Frank Andrews, who was then Assistant Chief of Training and Operations. Andrews brought up the lack of flying time, and McNarney allowed how if he got a flying command, he'd certainly fly. Becoming Commandant of the primary flying school at March Field had helped some, but when he came under Arnold as a bomber commander, the opportunity was of importance. Classroom theory needed application to make it meaningful, and making meaning was Joe McNarney's forte.

If McNarney lacked color and sought to conceal any inner warmth, Captain Frank O'D. "Monk" Hunter, who commanded the pursuit, had enough color and warmth to provide for them both. Monk Hunter was a fighter pilot for all seasons. His bushy mustache, his wicked roving eye, his rugged good looks and commanding voice—these were the exterior features that attracted his fellow officers and made him the paragon for the younger pilots. But behind the eight German planes shot down during the war, the five DSC's and the Croix de Guerre, Hunter was a serious pilot who knew all phases of pursuit aviation, mechanical as well as theoretical.

At the beginning of the war, he had enlisted in the Aviation Section of the Signal Corps, where he had risen to the rank of sergeant. Following flight training at Chanute Field, Illinois, he was off to France and further training at Issoudun, where he became fast friends with Major Tooey Spaatz. Eager to get into action, he was assigned as Deputy CO to the 103rd Aero Squadron. On his first combat mission, although wounded in the face, he shot down two German planes. Then, for the next seven months, Monk Hunter flew combat patrols and was soon a widely known ace.

After the war, pursuit remained his specialty, although he spent two years glumping around in observation ships for the Field Artillery. Then he got command of the 95th Pursuit Squadron, at Selfridge Field. During a part of his Selfridge tour, Tooey was his CO.

When he was posted to Washington in 1928 to serve three years in the Office of the Chief of Air Corps, the best he could say for the duty was that he got to serve on all the important pursuit boards and therefore was in on the testing and recommendations of what fighter aircraft should be procured. As for the social life of the capital, that was an entirely different matter. A southern gentleman from Savannah, a confirmed bachelor, the Captain was in great demand by the fair sex, and he did his best to honor all invitations.

In 1931 he had been assigned to Rockwell Field to serve again under Tooey Spaatz, as CO of the 95th Pursuit Squadron and then the 17th Pursuit Group. When Hap Arnold came to March Field, Monk and the 17th joined him. He'd been a captain for ten years.

A third outstanding officer who became a positive force in Arnold's command was Captain Clarence L. Tinker. A lean man with sharp features and high cheekbones, his one-quarter Osage Indian ancestry showed in his looks and direct manner. Following graduation from Wentworth Military Academy, in 1912, he had become an infantryman and did not make the switch to the Air Service until 1921. "Tink," as he was nicknamed, at one point commanded the same observation squadron at Fort Riley that Arnold had been handed in 1926. Later, as an Assistant Military Attaché in England, Tinker had been awarded the Soldier's Medal for heroism. In a 1926 crack-up, the plane he had been flying caught fire. He had gotten free of the aircraft only to realize that his Navy officer passenger was still trapped in the burning wreckage. He went back in after him and carried the injured officer to safety.

After three years' duty in the Training Command at Kelly Field, Texas, he had been made post Commander at March. This was in 1930. Now under Arnold, he would succeed Hunter as CO of the 17th Pursuit Group and then later take over the 7th Bomb Group, which in 1934 would move to a new base at Hamilton Field, near San Francisco.

There were, of course, other seasoned professionals under Arnold's command, but in a Corps that was always short of personnel with never enough to go around, their numbers were few and the demands on their services great.

Most of the pilots were either fresh out of flying school or on reserve status or both. Flying strength at March Field in mid-1932 consisted of sixty-two pursuit and forty-eight bomber pilots. The only base that had a larger complement was at Langley, in Virginia, where there were three more bombardment officers.[2] Pursuit aircraft were mostly the snub-nosed, radial-engine Boeing P-12s. They were single-seater biplanes with fabric-covered wings and a semimonocoque fuselage. Top speed at sea level was about 190, with a service ceiling of twenty-eight thousand feet. The bombers were an assortment of Keystone models. Twin-engine biplanes, they were at the time the standard Air Corps product. Fabric-covered, carrying a crew of five, they lumbered along at about 90 mph, and on a clear day with lots of lift might get up to sixteen thousand feet.

Arnold also had some Curtiss Condor B-2 transports. He would make headline use of them in dropping food to blizzard-isolated Navajo and Hopi Indians in parts of Arizona, New Mexico, Colorado and Utah during the winter of 1932–33.[3] But the main job—the main thrust of his command—was to develop, update and perfect the tactics of pursuit and bombardment. There was nothing new in the mission. What was new was the approach, the open-mindedness "with a view of getting at the facts and principles of modern aerial operations, regardless of the rules of tactical employment of the past or present."

Arnold and his confederates worked on the technique of putting squadrons at remote auxiliary stations under wartime conditions, away from the comforts and support of home base, seeking to make them self-sufficient for weeks at a time.

Night flying was stressed, and with it high-altitude operations, but in these exercises and all others dealing with combat, Arnold quickly realized there was a serious lack. He found that nowhere on the West Coast did the Air Corps have a bombing or gunnery range worthy of the name. The Navy refused to share the Pacific Ocean, evidently feeling it wasn't big enough for both services to make use of.

North of the San Bernardino and San Gabriel mountains lay the Mojave Desert. Arnold and his pilots knew its waterless topography well. Adjacent to a tumbleweed stop called Muroc, there was a vast dry lake bed. It was flat and hard and ready-made for what Arnold had in mind. He checked and found that most of the land surrounding the lake bed was government-owned. Nevertheless, he knew that if word got around that the Air Corps was interested in the territory, the effect on property values would be as though Sutter had announced another gold strike. Wearing civilian clothes and utilizing friends from the Automobile Club of Southern California for cover, he made a reconnaissance in depth into the area, supposedly seeking new routes for the club. The end result, although it took a considerable time to reach, was the acquisition of a piece of dry lake bed nineteen miles long by nine miles wide. The site was to become one of the best bombing and gunnery ranges in the world. Although final settlement on the land did not come until 1939, it was there, at Muroc, that Arnold soon began to test out the tactics of high-altitude precision bombing by daylight, using the first of the Sperry and Norden bombsights.

All these developments—organizational, tactical and aeronautical—took time, but soon the P-12 was being replaced by the Boeing P-26, called the "Pea Shooter." A low-wing aircraft, which was to be the last of the open-cockpit fighters with a nonretractable gear, it had a cruising speed of nearly 200 mph and was considered "the hottest plane around."

The clumsy Keystone bombers were giving way to new design. The Boeing B-9 was a twin-engine monoplane with an all-metal wing, its fuselage and tail surfaces largely aluminum alloy. The cockpits were still open, and it was a transition aircraft with a number of unsolvable bugs. Only nine of the planes were purchased by the Air Corps, and six of them came to a bad end. As one pilot put it, "If you leaned on the rudder too hard, you could look back and see the tail twist."

The B-9 was in competition to the Martin B-10, which was also a twin-engine, all-metal mid-wing craft with a retractable landing gear. The B-10 carried a regular crew of four, pilot and bombardier sitting in an enclosed cockpit, radio operator in a compartment aft, and gunner behind him.[4] The key feature of the B-10, aside from its 1,400-mile range, was its speed. With its twin Wright Cyclones, it could cruise at close to 200 mph.

It was this sudden surge of advance in bombers that brought into being at this time a question of unparalleled importance regarding the relationship between bombers and fighters. It was a question whose determination would have a profound effect on the course of strategy and tactics.

Arnold expressed his formative thinking on the problem to Conger Pratt in early 1933. Pratt had continued as chief of the Materiel Division, at Wright Field. The thought, Arnold said, had been bothering him for some time.

> We have been increasing the speed of our airplanes as rapidly as we could until now we find that the bombardment planes have a speed of about 200 miles an hour. Even supposing that we have a pursuit plane which will make a speed as high as 270 miles an hour, the speed differential is not sufficient to permit us continuing our antiquated pursuit tactics. Once the pursuits dive at 200 miles an hour and these planes attain their terminal velocity, their impetus will take them far below their target, and

before they can re-form the bombers will have traveled a minimum of 10
or 12 miles. Accordingly, the pursuit will have a stern chase and as you
know a stern chase is a long chase. Thus in all probability the bombers
will reach their objective with the pursuit having made but one attack. If
my premises are correct, it is obvious that pursuit tactics must be re-
vamped or the pursuit passes out of the picture.

. . . I do not know how much Tooey talked over with you when he was
back there, but it looks to me as if we will have to do something besides
thinking along that line very soon or we will have a lot of fast ships built
up for a purpose which they cannot accomplish.

He had written to Conger on his thinking because it was the Materiel
Division that had the major voice in the determination of any new plane,
and it was Arnold's opinion that, all too frequently, the Materiel Division
"stole the show" and forced tactical units to take planes that they didn't
necessarily want.[5] In spite of Jan Howard's influence, Arnold knew Conger
Pratt was a thoughtful man, and he wanted to get some of his own
thoughts on record, how changes in design that were affecting changes in
tactics should affect changes in ideas on procurement, and particularly because funding for this last was so limited. He wanted to see the best minds
in the Air Corps put to work on the matter.

In Tooey Spaatz's mind, the fighter assigned to escort a bomber would
not be as fast as an interceptor, because of the gas load required for the
former. In raising the question as to whether "it is better to attempt to
develop a distinct type of airplane for this purpose [escort] or whether it is
better to rely on the bombers to protect themselves on their long-range
mission," he laid his hand on an issue that was to become central to all
else. He could not answer the question at the moment, and the answer,
when it did come, would evolve through a number of varied conditions
and pressures.

In seeking the answer, Arnold would ask Monk Hunter to get his "old
think tank working with a view to determining just what operations we can
carry out in order to fix the role of bombardment and pursuit. . . . The
bombardment role seems pretty well fixed, its mission being to go someplace and destroy something and return with the least possible loss after
inflicting the greatest amount of damage. Now as to the pursuit, should the
pursuit furnish protection only during the take-off and landing or should it
do a certain amount of accompanying? What should be the nature of this
protection?"

The questions, and they were being raised and debated throughout the
Corps, particularly at the Air Corps Tactical School, were aiming toward a
central strategic point: Could bombers be built that could successfully defend themselves against interceptors, therefore making protective fighter
escort obsolete and unnecessary?

Arnold was looking for the answer and stimulating his commanders and confidants to seek for it too. At the time, the evidence was conflicting, influenced in part by economics, by proponents and opponents, and by the state of aeronautical art. In an exercise held at Wright Field in 1931, Major Ralph Royce had commanded a pursuit group and, over a two-week period, was unable to make contact with attacking bombers on a single occasion.

The failure had prompted Major Walter "Tony" Frank, acting as an umpire, to conclude in his official report that "due to increased speeds and limitless space, it is impossible for fighters to intercept bombers and therefore it is inconsistent with the employment of air forces to develop fighters."

This kind of reasoning drove Captain Claire L. Chennault, the senior instructor in fighter tactics at the Air Corps Tactical School, right up the wall. Chennault, who had the look and manner of a brave about to go on the warpath, was a zealous, dogmatic, determinedly inflexible proponent of fighter aircraft. He believed that fighters properly employed could dominate the skies, and that without fighter protection bombers would be sitting ducks.

He got a chance to prove his point at maneuvers held at Fort Knox, Kentucky, in 1933. Chennault had been preaching and working on air defense by utilizing a warning line of ground observers employing radio and telephone communication. Although he was to say that the board that drew up the ground rules for the Fort Knox exercise did not include a single fighter pilot, nevertheless he was free to set up his defenses accordingly.

The problem was basic: the bombers were stationed at Wright Field, in Dayton; the fighters at Louisville, Kentucky. The target to be attacked and defended was Fort Knox. Under Chennault's direction, the defenders set up a net of observation posts connected to a control center. Three mobile Field Artillery radios were manned by trained personnel. The result was that the bombers were hit by the fighters night and day and at all altitudes, sometimes twice a mission, long before they reached the target. According to Chennault, the bomber proponents cried foul and changed the ground rules. He maintained he had learned much about air-defense systems by studying information made available by both the British and the Germans. In any case, after the Fort Knox maneuver, he wrote a text on the subject titled *The Role of Defensive Pursuit*. He stressed two major points: Defending pursuit could make interception of attacking bombers before the bombers reached the target if furnished timely information and if the interception area had sufficient depth to allow for the necessary time factors. And bombardment planes flying deep into enemy territory required friendly fighter protection to prevent heavy losses if not utter failure of the mission.

Chennault was to claim bitterly that neither his paper nor his ideas were given a proper hearing at the Tactical School or anywhere else. This was hardly correct at the time insofar as Arnold and a good many other tactical commanders were concerned. Arnold had supplied a bomb group for the Fort Knox exercise, and through his own training at March Field he was fully aware of the developing need for air defense. On the matter of pursuit, he was at the moment hot for the development of an all-purpose fighter that could act as an interceptor against bombers and yet possess enough firepower and range to act as an attack bomber. As he put it to Tooey, "This plane is designed mainly to strike at enemy bombardment and not with the idea of furnishing special protection for our own bombardment, which I believe we all agree must be done by bombers themselves or by specially designed accompanying fighters."

And so with Arnold, like Spaatz, the question of an escort fighter remained open, somewhat ambivalent, awaiting aeronautical change.

There were other voices being heard. In May 1933 *U. S. Air Service* magazine had published an abstract of the air-warfare doctrines of General Giulio Douhet. The magazine had been sent to Hap by his old friend Charles deForest Chandler, who was the aeronautical editor of the Ronald Press and the publisher of the magazine. "The Douhet doctrine is attracting considerable attention in Europe," declared Chandler, "and impressed me so favorably that I made the digest and believe that you will be interested in the aggressive doctrine that requires an independent air force."

In response, Arnold thanked "Charley" very much for bringing Douhet's article to his attention. He had already seen and read it with great interest. "It contains thoughts which should be given careful consideration by all Air Corps—better all Army Navy officers."

The Italian General's twin theories on the need for air independence and the power of mass aerial bombardment to break an enemy's industrial capacity along with his will to fight was also being pushed by those at the center of things in Benny Foulois's office. Tony Frank, in the Air Corps Chief's office, wrote to Arnold, "You have probably heard about Douhet's book on air warfare. I got Dorothy Benedict to translate it from the French for me [originally in Italian] into English. George Kenney corrected the translation, and I proofread it about eight times. We have just had it mimeographed and under separate cover I am sending you eight copies. Read it over and have your henchmen read it over. Some of it you will not agree with, I know, but it will give you plenty of food for thought and it will help enlarge the vision of the youngsters."

Toward the aim of enlarging everyone's strategical vision according to Douhet, General Benny Foulois had had thirty-three copies of the mimeographed text circulated to influential contacts.

On the subject of food for thought, Hap Arnold had plenty of that whether viewing the scene through a tactical or a strategic telescope. When he had asked Monk Hunter to put on his thinking cap in regard to tactics,

Hunter was flat on his back at Walter Reed Hospital, recovering from having very nearly lost his thinking cap altogether.

It had happened in January 1933 when Monk had gone East to Patterson Field to serve on a pursuit evaluation board considering a new Consolidated twin-seater designated PY B-2. From Washington and the Air Corps Office of Training and Operations had come Captain Hugh M. "Elmie" Elmendorf also to serve on the board. Elmie Elmendorf was known for his aerial accuracy with machine guns and his high-flying pursuit tactics.[6] He and Hunter and Tooey Spaatz were close pals.

The Consolidated twin-seater that the fourteen-man board was to test-fly with an eye toward possible procurement did not have dual controls, and the board members, splitting into two-man teams, had to take turns flying as pilot and as gunner. On Friday, January 13, Elmendorf and Hunter took off with the former at the controls and Monk manning the machine gun. As they gained altitude, they were attacked by an "enemy" fighter, and for the benefit of those watching below, they rolled and dived and battled over a large hunk of sky, putting the PY B-2 through its paces in an old fashioned dogfight.

When this phase of the test had ended, Elmendorf began maneuvering the plane through a series of rugged acrobatics, finishing the workout at about seven thousand feet. He leveled off momentarily, then stalled the aircraft and, with power still on, whipped it into a violent tailspin. The ship wound up plummeting swiftly in its corkscrewing descent.

After five turns, Hunter saw Elmendorf turn his head and say something. But the only means of communication was by hand signal and voice, which certainly couldn't be heard above the engine's building scream.

When Hunter saw his friend's head tilt forward, saw no movement of his hand to ease back on the throttle and reduce power, saw no movement to halt the spin and bring the plane back to level flight, he knew something was seriously wrong. His own cockpit was open, whereas the pilot's was canopied, with a small opening at the rear. With his hands cupped he began to shout at Elmendorf through this opening, trying to get some reaction out of him. There was none. The ground was leaping up at them. Hunter knew if he didn't leave now he was dead. It was somewhere around the fifteenth turn of the spin that he managed to get clear of the plane, simultaneously pulling the rip cord of his chute. There wasn't enough altitude for the chute to open fully before he hit.

Three days later, Captain Hugh M. Elmendorf was buried with full military honors at Arlington National Cemetery, and Monk Hunter would be six months recovering from a broken back as well as other serious injuries. The last time he'd had to jump, the only damage had been to his mustache, which had taken a singeing before flames had forced him over the side. Then the hurt had been more one of vanity. This was something quite different, and the post-mortem investigation into the cause would never resolve what had happened. What had prevented Elmie from pulling out of the fatal plunge?

The best guess, with which Monk and Tooey concurred, had to do with a previous accident, in 1927, in which Elmendorf had been seriously injured. Then it had been a ground collision at Selfridge Field, and Elmendorf, who had just landed a Curtiss P-1, had suffered several broken vertebrae. The belief was that the dogfighting plus the violent acrobatics at high altitude and then the spin had affected the damaged vertebrae in Elmendorf's neck, and he had either lost consciousness or had become paralyzed, powerless to move.

It was a sickening thought and a sickening loss, and there was grim irony to be added. Six days later, Elmendorf's very close friend First Lieutenant Irwin A. Woodring, test-flying another model of the same plane, crashed near Wright Field and was also killed.[7] Woodring and Elmendorf had not only served together as CO and Operations Officer, respectively, of the 95th Squadron, but it was also Irwin who had been at the controls of the plane that had run into Elmie at Selfridge, and he, too, had been seriously hurt at the time.

The Corps had lost two fine, very popular officers in less than a week, and it hurt. Conger Pratt sent the details to Hap Arnold, who struck out at what he considered a basic error in Air Corps planning. To his way of thinking, the manner in which personnel for aircraft evaluation boards were selected was "archaic and inefficient."

He wrote accordingly to Mike Kilner, Executive to Trubee Davison.

> We have pursuit boards, observation, attack, bombardment and primary training boards. Each one overlaps into the others' sphere of action. The only permanent members on the board are representatives from the Materiel Division, and in a majority of cases, the same man from the Division doesn't sit on any two boards. The result is that we develop airplanes for use in our service without any thought to a continual development policy. Every board has its own ideas on the subject, and in some cases the board itself switches its ideas and opinions before the plane which it recommended in a prior meeting has actually gone beyond the experimental stages. This is usually caused by these boards being dominated by junior officers who have never in the past nor can they very well in the future see the whole picture, and allow themselves to be influenced by their local problems.

Arnold then went into what he termed "another side to the matter," which was that some planes could be used for more than one purpose. He felt that the new Martin bomber was such an aircraft, that it could be adapted as a long-distance reconnaissance plane, a multiseated fighter and possibly an attack plane. "Should we not then be developing further sources of planes of these characteristics . . . ?" he asked, and then returned to his original complaint.

I do not know whether or not you saw the order for the last pursuit board. As I remember it, it was composed of 14 members. Just what they expect 14 men to do in deciding on a pursuit plane, I do not know. It was ridiculous to have such a large, unwieldy board in the first place, and in the second place, there is no excuse for having some of the men on the board appointed, other than they were somebody's friends.

The point of this letter is that I am convinced that in order for the Air Corps to get a continuous progressive development of aircraft along proper lines, we must have a single board made up of the best people in the Air Corps who will decide on all types of aircraft regardless of their use. In that way and in that way only will we get away from the backing and filling and washing out of inefficient airplanes that we have had in the past. Is it not possible to conceive of a board composed of say Spaatz, Knerr and Hickam satisfactory to everyone concerning the program of development of aircraft of all types during the coming year?

The importance of Arnold's communication to Kilner was related to his thinking on the need for a better way to get better aircraft which might be adaptable to several uses. Also behind his complaint and question lay his concern for the future of the Air Corps. He knew when he wrote that the situation following the election of Franklin D. Roosevelt in November 1932 had become confused at the top.

"Things are very quiet," observed Mike at the beginning of 1933. "The ins are waiting to go out and the outs are waiting to come in. Therefore, the usual hiatus. No dope yet as to who our new Secretary is going to be and no assurance that even the job will be continued."

The matter was a question of major importance. Just who would be replacing Trubee Davison? Some months later Arnold would comment to Tooey Spaatz, who was in Washington, on the effect of no replacement. "I have heard some very disquieting bits of news which indicate that the General Staff has taken upon itself the task of running the Air Corps without acquainting the Chief with the action until after it has been taken. It looks to me as if the General Staff must then be taking full advantage of the absence of an Assistant Secretary of War for Air."

As a tactical commander far from the seat of power, the actions of the General Staff did not affect him as directly and as closely as those in the office of the Chief of Air Corps. But the control was there. Even though he exercised combat command over his forces, this did not mean that he, or any Air Corps tactical CO was really in command of them. It was the ground commanders of the nine corps areas who maintained rigid domination over all Army aviation.[8]

The procurement of aircraft and the training of air crews—that, indeed, was the prerogative of the Air Corps, but once tactical units were in being, regardless of their destination, they came under the purview of the corps area generals.

At the time, the stationing of air units in a corps area had absolutely nothing to do with defensive strategy. It was simply a matter of whether there happened to be airfields in the area. There was no system as to allocation. Consequently, as Arnold put it, "One Corps Area might have an attack group, another a bombardment group, another a pursuit group. Some Corps Areas had two or more groups, some none at all."

Beyond that, the real bone of contention was the method of control. Like so much else in the relationship between the War Department and the Air Corps, it was not a system that was clearly understood by most congressmen, and General MacArthur, as Chief of Staff, did little to clarify the issue.

In testimony before the Senate Appropriations Committee in which he was discussing the failure of the Air Corps's five-year program as a result of cuts by the Congress and the Budget Director's office, MacArthur waxed eloquent:

> Up until last July, the complete control of the Air Corps was in itself headed up with the Assistant Secretary of War for Aviation. The Assistant Secretary of War for Aviation was responsible for all its functions except some of the routine processing of feeding and so on, for everybody in the department, for the program of instruction, for the program of training and for everything else. The General Staff had absolutely nothing to do with these things until last July. To all intents and purposes within the Army, there was an independent fighting branch. The Air Corps, which had its own Assistant Secretary for Aviation, who had delegated to him by the Secretary of War, the complete authority, where the rest of the Army at large, is exercised by the ground staff. . . ."

Benny Foulois was sitting next to MacArthur as the General offered his colorful description of an all but independent Air Corps. Foulois must have had to bite down hard on his pipe stem. Later, when he received the transcript of his Chief's testimony, he made some cryptic marginal notes refuting MacArthur's claims: "P. 51. See C of S testimony re lack of *responsibility* of General Staff. Prepare reply. Show list of turn downs.

"P. 56. See C of S statement he had nothing to do with training up to July 1933. *Show continuous control of training by General Staff through Corps Area Commanders.*"

Of course, neither before, during nor after the stewardship of F. Trubee Davison had the Air Corps been "an independent fighting branch," as stated by MacArthur, who certainly knew better and was anxious to have the Assistant Secretary's position abolished.

The Corps Area Commander under whom Arnold came to serve was Major General Malin Craig. Craig came from a military family—his father retired as a colonel—and following graduation from West Point, in 1898, the younger Craig's career embodied a broad range of duties and commands. Principally a cavalryman, he had seen action in the Santiago campaign, in Cuba, and the Boxer Rebellion, in China. He had several tours in the Philippines and was there as a captain in the 1st Cavalry when newly-made Second Lieutenant H. H. Arnold was serving down the road a piece with the 29th Infantry. There is no record of their having met at the time.

Prior to the war, when Craig was not on horseback, he was passing through various Army schools, usually as an honor student, and then at the War College as an instructor. By the time war came, he was a highly regarded staff officer. During the conflict, in which he earned the Distinguished Service Medal, he served as Chief of Staff for the 41st Division and then with the First Army, participating in the major American campaigns. He remained as Chief of Staff with the American forces in Germany, returning home in 1919.

In the decade that followed, until his appointment to the IX Corps Area, Craig held eight major posts: in Washington, in the Philippines and in the Canal Zone. Among these were Chief of Cavalry and then G-3, Assistant Chief of Staff for Operations and Training in the War Department.

At fifty-five, Malin Craig was an Army officer of many military parts. Within the War Department hierarchy his capabilities were recognized, as was his membership in the Pershing clique. Quiet and somewhat retiring, there was something of the schoolteacher in his looks; perhaps it was the rimless glasses that he wore. His astuteness was essentially that of an administrative executive, conservative, circumspect, operating always within the boundaries of military thinking. Perception beyond the norm was not his forte, but neither did he have a shuttered mind. When it came to the use of air power, he was not inflexible; he could be shown. Arnold hoped to show him.

General Craig's Air Officer at the Presidio headquarters was Lieutenant Colonel Lawrence W. McIntosh. He was "Mack" to Hap, and his duty was that of liaison between the Corps Area Commander and the air units under Arnold, a sort of official mouthpiece in the umbilical cord connecting the two.

There was considerable communication between the two airmen, and a considerable amount of it appeared to reflect criticism on Craig's part. A kind of running nit-pick seemed to be the order of the day, and at the beginning of 1933 it moved toward a climax.

In January Mack wrote to accuse Hap, through Craig, of "short circuiting" and "ignoring" the Corps Area Commander. The criticism was over a presentation the previous fall of the newly established Frank Luke Trophy in Phoenix, Arizona. The arrangements had been thoroughly fouled up in Washington by Trubee Davison and Secretary of War Hurley in neglecting to inform Malin Craig of the planned proceedings. Arnold had been assured his boss knew all about the presentation and flew off to Phoenix with Tooey Spaatz and Monk Hunter to make the necessary Air Corps arrangements. Both Congressman James W. Douglas and the American Legion post had been the guiding forces behind the idea.* The result was that planes from March Field made a snappy appearance above the ceremonies, but General Malin Craig was not on hand, if he had wanted to be, nor was he aware of the affair until long after it was over.

Arnold explained the circumstances and insisted he wasn't trying to short-circuit anyone.

On top of this came Craig's complaint via McIntosh over a request for theater construction at March Field on which the Corps Area Commander had not been clued in. Arnold's response to the criticism revealed the problem of the War Department's disjointed system of command, and how simple it was to put your boss's nose out of joint:

> Now, insofar as the Constructing Quartermaster is concerned, one of the biggest kicks I have had at this station is that the Constructing Quartermaster is not under the Commanding Officer. He communicates directly with the Quartermaster General's office in Washington, and the Commanding Officer knows nothing about what goes on in the correspondence between those two offices. Accordingly, unless I have a Constructing Quartermaster who will come around and tell me everything that goes on, I will never know anything about alterations or changes in any buildings, and even at that, there is no way in which I can put an indorsement on a letter from the Constructing Quartermaster to the Quartermaster General, for that is not the channel of communication.
>
> In this particular case, the Constructing Quartermaster here took up the question of enlarging the theater.

In February Hap wrote to Monk Hunter, who was recovering from his injuries, informing him that Joe McNarney had left for a "10 day seance with the Canoe Club" (U. S. Navy) as an observer for their aerial maneuvers, and then indicated just how much authority an Air Corps tactical commander had over his own units:

> In the meantime, there is something new every day to break the monotony. First, the Corps Area had the Bombardment Group standing by for two weeks waiting to proceed on some funny mission or other, thereby throwing a monkey wrench into the training program; for they rolled the bombers out of the hangar, stood by all day waiting for orders that never came, and rolled them back in again every day during the entire period.
>
> Rumor quite positively says that the Air Corps Exercises will be held here this spring. Don't know if they will materialize or not, or if they will just give us another lesson in coordination as they did last year and at the last minute call them off.

There were other complaints and responses funneled back and forth through McIntosh, and as Arnold was later to declare to Spaatz, ". . . all training should come under the head of a commander of the Air Forces

---

* Douglas became FDR's first Budget Director.

instead of Corps Area Commanders. Even with a Corps Area Air Officer on the Corps Area Staff, they do not seem to be able to recognize that there is a big difference in the training of observation units and air force units."

Whatever a corps-area commander could or could not recognize, it took an earthquake to establish a going relationship between the Major General and his unfazed tactical air commander.

At ten minutes of six on the evening of March 10, 1933, Long Beach, California, was struck by a major earth tremor. Hap Arnold heard the news flash on his radio. He didn't sit around waiting for details. March Field was a good forty miles from the scene, but he knew someone who was close to it. Captain Ira C. Eaker was on detached duty, taking a journalism degree at the University of Southern California. He was living in West Los Angeles, near the scene of the quake, and Arnold got on the phone to him.

The Captain had been out giving his wife, Ruth, a driving lesson when the road began to tremble. At first they thought something was wrong with the car, but when they saw people running from their homes, they knew what was happening.

When they reached their apartment, they found dishes all over the floor and the piano on the wrong side of the room. They were surveying the breakage when Arnold called. He asked Ira to go have a look and report the situation to him as soon as possible. Eaker did so, reaching Long Beach while there were still pieces of masonry falling from damaged buildings. He hunted up the Mayor and the Chief of Police as well. He then got an estimate of what was needed and was back on the phone to his CO, reporting the immediate requirements of the survivors and the injured.†

Arnold knew that the commanding officer of Fort MacArthur, which was close to Long Beach, would be the responsible party should requests be made by the city for aid from the Army. But in attempting to pass the word to him following Eaker's report, he learned that the CO of Fort MacArthur could not be located. He was out of touch, out of reach.

Arnold did not wait. He put Bee on the phone to continue trying to make contact, and then he went to his office and got hold of the Mayor of Long Beach. He suggested to the Mayor that the list of his needs be forwarded to Corps Area headquarters at the Presidio and, in the meantime, March Field would set about fulfilling whatever was required.

That night, the Air Corps moved into Long Beach, setting up relief stations and field kitchens, bringing in medical supplies, bedding, tents and manpower in an effort to offer aid and comfort to a shocked citizenry.

The next day, and on following days, the press was full of praise for the promptness and efficiency with which Lieutenant Colonel Arnold's men had responded to the call for help.

The same praise was not forthcoming from Corps Area headquarters,

† There were 112 fatalities in the earthquake.

and it looked as though Hap Arnold was in serious trouble again. The missing commanding officer from Fort MacArthur had returned, and the afternoon following the earthquake word came down from the Presidio that Fort MacArthur was in charge of relief for Long Beach. Further, Arnold's prompt action was in violation of orders, and there was a good possibility that he would be charged for the cost of the entire operation. There would be an investigation into Arnold's dereliction by the Inspector General.

The first knowledge that Arnold received that he was in trouble came from General Malin Craig. On the telephone, the General gave him little chance to explain. The dressing down took him completely by surprise.

In a response to a follow-up by McIntosh, Arnold pointed out that there were twenty-four hours between the time the earthquake struck and the time that he received the order that Fort MacArthur was in charge. "What was I to do," he asked angrily, "fold my arms and ask the people who needed relief to go jump in the river?"

"I explained all this to General Craig, but who is the troublemaker? Who starts all this? What's the big idea anyhow?"

The troublemaker in this case was the CO at Fort MacArthur, trying to cover up his own absence and ineptness. Arnold was thoroughly disgusted, furious. He gave an indication of his frustration in a letter to Monk Hunter:

> The earthquake came, and it was a nightmare—not in so far as the actual damage done was concerned—although that was bad enough—but on account of the way Army procedure and custom glorifies the individual who sits back and does only that which he is told to do, and censures the fellow who tries to do the right thing and displays a little bit of initiative and intelligence. The longer I stay in the Army, the more I am convinced the man who makes the greatest success in the Army is the fellow who has no initiative, exercises no intelligence, but merely does what he is told.

He discussed the foul-up with Tooey Spaatz, and Tooey had a moderating suggestion. His father-in-law, Colonel "Death" Harrison, was an old buddy of Malin Craig's. They had served in the Cavalry together. Tooey would fly up to the Presidio, and on the strength of the family relationship, have a chat with the man. It was an idea that certainly couldn't hurt. Even so, the more Hap talked it over with Bee, the madder he got, and it seemed like an even better idea if he flew up there himself and had it out. The hell with dealing through Mack and the telephone!

Whatever calm Major Spaatz managed to spread on the troubled waters, by the time Lieutenant Colonel Arnold stormed into the Presidio his dander had been finely honed in the slipstream. His greeting to his commanding officer got right to the point: "Sir, I want to know why I'm in the doghouse!"

The query startled the General. No one had ever quite put it to him like that before, and it was obvious his subordinate had something heavy to get off his chest.

To Bee Arnold, Malin Craig was a "strict old soldier," but he was neither so strict nor so aged and hidebound that he couldn't appreciate the sincerity and dynamism of his irate airman. Somewhere behind Craig's proper military front there lurked a sense of humor, too. Lieutenant Colonel Arnold was a Daniel come to judgment to plead his own case, and the General was willing to listen. The result was that the meeting between the two completely changed their relationship, and later the change was to have a pronounced effect on the career of Hap Arnold.

He flew back to March Field, the storm cloud of trouble dissolved, his smile in place. He had invited General and Mrs. Craig to come down to March Field to attend a ground and aerial parade. They did so, and when the Air Corps files had passed in review, the General remarked that their marching was as smartly executed as any he had seen, including the Infantry. From that time on, General Malin Craig was a firm supporter of his irrepressible airman, and as Arnold was to say, life became far easier for him than it might otherwise have been.

If March was the month of upheaval through the unpredictability of nature and man, May became the month of even greater challenge, through the exigency of political and economic need.

The creation of the Civilian Conservation Corps was a Rooseveltian dream, a plan to take the unemployed youth of the country and put them to work on reforestation, flood control, national park development and a number of other worthwhile out-of-doors projects. The Congress approved. A bill was passed to set the program in motion, but the motion got jammed up amid the Departments of Labor, Agriculture, Interior and War. Roosevelt stepped clear of the embroilment and turned the mobilization and organization of CCC recruits over to his Army Chief of Staff, General MacArthur, whose capabilities he trusted.

Although MacArthur was not in favor of the Army becoming involved in any relief-work program, his alertness was such that by the time Congress had authorized the formation of the CCC, he had had the General Staff draw up a plan of implementation and was ahead of the game when Roosevelt called on him to take over. Thus the Army was given the task of organizing 275,000 CCC recruits into camps in forty-seven states by the first of July, 1933.

This new task—the greatest peacetime mobilization of manpower in the nation's history—hit March Field headquarters at the beginning of its long-anticipated spring maneuvers. Brigadier General Oscar Westover, the Assistant Chief of the Air Corps, had come from Washington to take command of the exercise. At his disposal he had about three hundred combat aircraft, which had flown in from all parts of the country. The entire West

Coast was the area of operations, and for the first time the Muroc facility would be used for bomb dropping. The focus of the maneuvers was to be on combined fighter, bomber and attack techniques in combat. It was styled as not so much a tactical exercise as a technical one, but its most important feature as far as Hap Arnold was personally concerned was that he had been selected to serve as Tubby Westover's chief of staff. Since he'd never been a chief of staff, the assignment appealed to him and he was thoroughly engrossed in it when he learned he would be wearing two hats in a big way. An unexpected query from Malin Craig arrived: *could he handle fifteen hundred CCC recruits arriving next week?*

But the maneuvers! All the combat-type aircraft in the United States were to be engaged. The General followed with a letter which he signed "your friend." In it he expressed confidence "that you will give this matter your attention until our efforts of taking care of these men become systematic, routine and successful." Craig recognized there would be a drain on Arnold's personnel, and he would try to alleviate the drain by sending in reserve officers.

"I assume," he added, "you have already selected your enlisted personnel who are to handle and go with these fellows. The whole business requires firmness, kindness and devoted attention to duty on the part of our personnel. You have a large job with your maneuvers and your other duties, but I feel certain you will put them over, as the War Department expects you to do."

Maneuvers and other duties aside, it was a very large job, once the influx began. Hordes of young men began to arrive by various means of transportation in general disarray and mostly very hungry. Under Arnold's direction the officers and men at March Field set to work with a will to accommodate, organize and instruct them in the basics of tent pitching, bunk making and cooking. Craig sent a follow-up note indicating not only his understanding of Arnold's immediate challenge but also a reminder of the mandate under which the armed services were determined to operate. "With the Air Corps tactical exercise on your hands in addition to the quota of 1,500 CCC [recruits] I readily appreciate that you will have a strenuous time, especially during the current month," he said. "However, the Army must, as it always has in the past, make good on any mission assigned it."

And make good it did, and profited in doing so. The camps were set up in California in such places as Bishop, Monmouth, and Silver Lake, among the sequoias in the northern part of the state, and across the High Sierras in Nevada, at Lake Tahoe. The young men who filled them had plenty of spare time and some asked if they might receive military training. The wish had to be denied, for pacifist groups were quick to protest any use of the military in civilian affairs and to see the dark cloud of fascism rising in any such connection. At the time, Communist fronts such as the

American League Against War and Fascism and the *American Youth Congress* were on the warpath against the CCC being run by the War Department. A bill later advanced by Congressman McSwain of South Carolina to give military training to the CCC was quickly squelched in committee. Even so, Hap Arnold did quietly permit voluntary formations to gather in the camps at reveille and retreat to raise and lower the flag—and no one was the wiser.

He liked to visit the camps, not only to see how things were progressing or how conditions might be improved, but also because he and Bee were avid fishermen and campers themselves. They loved the out-of-doors, the mountains and the trout-laden streams that coursed through them.

The Civilian Conservation Corps may have knocked the hell out of tactical operations from May to September, but there were many side benefits, as Arnold was later to record:

> The Air Corps learned a lot from handling those boys. Officers who had never before thought about anything very seriously, except flying an airplane, suddenly found themselves faced with administrative and human relations problems. As contingent after contingent streamed in, we had to find new camp sites for them, be sure the land was clear and healthy, and that good pure water was available. At the arrival of every new group there had to be enough food, blankets, tents, tent floorings and so on trucked up to the proper camp. If a truckload of lumber was dumped off in the wrong place, or the truck rolled off down the mountain, that was an Air Corps Officer's problem. The CCC outfits brought small numbers of Reserve Officers with them for administrative purposes . . . but most of them were about as bewildered as the boys when they arrived.[9]

That summer, Arnold, pro-tem chief of staff cum CCC camp maker, sent an update to Westover. "Now that the smoke has pretty well cleared away and I have my 25,000 camps operating in the woods [he was joking; the number of camps was twenty-five] I want to write and tell you how much I appreciated working for you as chief of staff and also let you know that I am sorry my other activities prevented me from giving you my full time while on the job. It was a pleasure to work with you, and I hope that someday I'll be closely associated with you again."

He had no idea as to the degree to which he would get his wish. At the time he wrote, however, he was far more concerned about losing Tooey Spaatz, who was to be transferred to Washington. All military people expected to be shuttled from one command to another; it was a way of life and a way of acquiring stature if not rank. Still, the departure of the Spaatz family was a sad farewell for the Arnolds.

Now Spaatz was to head up the Training and Operations Section in the Office of the Chief of Air Corps. His position would have the benefit of giving Arnold a small window into that office. The accent was on *small* because Tooey was as sparing with his letters as he was with his words. Still, the connection was there, and Arnold would be bombarding his friend with a wide range of comments and queries.

An area of special importance to Arnold, outside the day-to-day business of tactical command, was scientific development as it applied to flight. He had always sought out the scientific view of what could be done in the air. This went back to the days at College Park, when, on his own, he would stop in at the Bureau of Standards and question the wizards there on the forces with which he was contending above the earth. Later, it was he who sold the Kettering Bug to the brass, his mind ever alert and open to scientific input beyond the engineering capabilities found at the Materiel Division.

He had known and worked with Dr. Robert A. Millikan during the war, and when Millikan approached him in that somewhat frantic spring of 1933, he was quick to greet him. Millikan was head of the Norman Bridge Laboratory of Physics at the California Institute of Technology at Pasadena and was involved in experimenting on the mysterious ways of cosmic rays. He wished to measure the intensity of the rays at various altitudes, and that meant needing the use of an aircraft. Could Hap help him? Hap could and would. He arranged to have a plane specially modified to carry a six-hundred-pound, instrument-laden, lead ball, and Millikan and his team began their experiments.

On the surface, or above it, cosmic rays intense or otherwise appeared to have no direct bearing on the problems of tactical air command. But who could say what their effect might be on man or aircraft as planes flew higher?

There was also the relationship with Millikan, warm and stimulating. The scientist became a frequent visitor, and he and his assistant, Johnnie Mattson, spent considerable time at the Arnold kitchen table, talking half the night away. Arnold knew that activities such as Millikan's could have great reciprocal value to the Air Corps.

Millikan was of a similar view and was quick to invite Arnold to the Institute to see what was being done in the aeronautical and allied fields. "I will get together a group of men and have a little luncheon party and smoker," he wrote, "and then we will look around at the various doings on the campus that will interest you."

And then, further: "You will be interested in knowing that we got some nice records on the high flights which we made with the pursuit plane last week, reaching an altitude of 29,200 feet, which is at least 4,000 feet higher than the Wright Field fellows got last summer on somewhat similar tests. I want next week, perhaps Wednesday, if it's convenient, to get down with some lead around our scopes to repeat a couple of these high altitude flights if it will be all right to do so."

Through his interest and association with men of science, Hap Arnold was building a connection, a mutual sense of appreciation for the practical application of flight, allied to the exploration of theoretical knowledge, which was to pay large dividends in times to come.

As for the Arnold boys, Hank and Bruce were sure they were going to be pilots. How could they be anything else? And certainly David, at six, was going to follow in their footsteps. At the moment, their horizons were being broadened through the benefit of meeting and observing a parade of famous people who came to March Field as guests of their parents.

A great favorite was the German war ace Ernst Udet. On one occasion, he arrived at the field flying his specially built U-12 Flamingo, which he had designed himself. Some of the maneuvers he performed with it were considered to be aerodynamically impossible. But, like the bumblebee, Udet wasn't inclined to heed the "impossible." His specialty was a series of consecutive loops. The first was executed starting at five hundred feet, the engine throttled back, the maneuver completed with a landing. On the next go-round he'd complete two loops from five hundred feet, power off, finishing with a landing, and on the third, it was three loops while everyone held his breath. It took perfect timing and a rare skill; the thinnest margin of error would bring disaster.

Udet, a small man with a gregarious, outgoing manner, had a way of making boys feel they were adults. His English was poor, and Bee became his helpful translator as he regaled them with colorful stories of air combat over France. Only sixteen when first shot down, his feet had gotten tangled in the parachute's shrouds and he'd landed head first in a trench, knocking out his front teeth. He made it all sound like great sport. In explaining his acrobatics he constructed paper gliders, and the living room would be filled with his handiwork and laughter. In the eyes of the Arnold boys, Ernst Udet was a champion.[10]

Twice during the year, Arnold went to Washington. The first time, in March, he stayed with his former classmate who was CO at Bolling Field, Major Bart Yount, and his wife, Mildred. He made the trip there and back in ten days, also spending a night at Fairfield Depot with another old classmate, Lieutenant Colonel Augustine W. "Robbie" Robins and his wife, Dorothy. A cross-country flight was always an opportunity to touch base with old friends and pick up the latest rumors.

His second journey East was unexpected and full of sadness. During the summer, his father, Doc Arnold, had come for an extended visit. Curmudgeon that "Daddy Doc" was, with heavily tufted white eyebrows and bushy mustache to match, the children and Bee were able to penetrate his gruff wall of reserve. He stayed with them until the twenty-first of August, when they put him on a bus to return to Ardmore. There he would be living with his son Dr. Clifford Arnold. He was not too well when he departed, and a month later, Cliff informed his brother that Daddy Doc was very ill. On October 27, the word came that he had died.

Hap wired Tooey the news, adding "requesting cross country to Philadelphia. Please handle this personally for me."

Following the funeral, Arnold flew to Washington and, while there, stayed with the Spaatz family. He and Tooey talked long into several nights, catching up. He, no doubt, talked with other old friends in the Chief's office and checked in with Benny Foulois. But mostly it was the hospitality of Ruth, and Tooey's explanation of what was going on, that occupied him.

He was back home at March Field by November 6, feeling depressed not only over his father's passing but also as a result of the information Tooey had confided to him.

To Tooey he wrote: "It was good to see you and talk with you again, and I came back from the trip with a much better understanding of what is going on than I would have gained from letters. I must admit that I left Washington with a feeling of depression." The cause would be "the apparent inability of the Air Corps to put things over which are so essential for its efficient operation."

In none of these communications was any direct mention made of trying to revive a push for independence, but the thought was ever inherent, the form modifying perhaps with the years. Whatever his thinking, Hap Arnold's world was made of many parts. Through the forcefulness of his personality and the sweep of his mind, he was determined to do all he could to assure the Air Corps its place in the sun.

Nine

The Visitors

On a summer's day in 1933, Lieutenant Colonel Frank M. Andrews took command of the 1st Pursuit Group and Selfridge Field, at Mount Clemens, Michigan. He came to his post following three years' duty in Washington, having served as Benny Foulois's Executive before attending the Army War College. A close contemporary of Arnold's and long-time friend of Billy Mitchell's, he was an officer highly regarded not only by his Air Corps colleagues but also by officers on the General Staff. This last was a rarity, but Andrews' capabilities had long been recognized by his seniors, including General MacArthur. In 1929, he had been one of three pilots in the entire Air Corps to be placed on the General Staff eligibility list. Withal, he had graduated from the Command and General Staff School and, in June of 1933, from the War College. These were the Army's "must" schools for any officer who expected to ascend the leadership ladder.

His friends knew that Andrews had not only a quality of steadiness but also that rare charismatic knack of being able to make the other person feel he was doing an important job, no matter what that job was. Subordinates wanted to serve under him. His manner of speaking, with a touch of the old South in his voice, put men at ease but held their attention, for his capacity of combining knowledge and imagination was electric, stimulating. He laced it all with a dry, cajoling sense of humor, a wide grin and eyes that reflected warmth. Major General Frank Parker, VI Corps Area Commander, under whom Andrews would serve, had said he could count on one hand Air Corps officers he deemed worthy of high rank, and one finger was reserved for the new CO at Selfridge.

To Hap Arnold, Andrews was Andy, the nickname he had picked up at West Point, and to Andy, Arnold was Pewt—not yet Hap. At West Point,

Andrews had been a year ahead of the Black Hand Gang leader, his method of outwitting the school's tactical officers a bit more subtle and cagey than Pewt's; it showed the difference in their styles.

On a field exercise, Cadet Andrews' platoon had bivouacked not far from a town where it was known that a Saturday-night dance was to be held. Naturally, the affair was off limits to cadets. Andrews purposely pitched his pup tent at the end of the platoon line, and as soon as everyone had settled down for the night, he quietly arose, stowed his gear and tent in the woods and took off for town. He had reasoned that if there was a late bedcheck, tents would be checked to make sure they were occupied. Back before dawn's early light after a great time on the dance floor, he repitched his tent and at reveille rose up to meet jocund day, proud of having outwitted the enemy by using superior strategy. The incident was indicative of his nature, the ability to see beyond the obvious and then go ahead.

A Tennesseean from Nashville, Andy Andrews (whose relatives called him by his middle name, Maxwell) had developed thoughts of flying at a very early age. According to his brothers Billum (William) and David and his sister Josephine, they had assisted him in building paper balloons. They had not assisted him when on numerous occasions he had climbed out on the porch roof of the family home on Benton Avenue and attempted to take to ballooning himself, umbrella in hand. His mother, Lula's, carefully tended flowers suffered, and his father, J.D., promptly grounded him.

At West Point, he had fallen in love with horses and girls in that order, and upon his graduation, in 1906, he had literally ridden away into the sunset—his first post as a Cavalry officer in Hawaii.

Early on, he had wanted to exchange the saddle for the cockpit, but marriage in 1914 to Josephine Allen, the vivacious, brown-eyed, polo-playing daughter of his CO, Major Henry T. Allen, blocked the move. Not only did his father-in-law say he was damned if he wanted a son-in-law to make his daughter a widow, but legislation passed the previous year forbade a married man to become a pilot. It took a war to finally get Andy Andrews transferred to the Aviation Section, and by then he was a father with a two-year-old daughter, Josie (Josephine), and a new-born son, Allen.

It was not until April 1918 that he graduated from flight training at Rockwell Field and received his JMA.

From that moment on, his course was set. It was a slow beginning, however, for, like Hap, he did not get to France. Once he was awarded his wings, he was nailed to stateside command with the temporary rank of lieutenant colonel until the hostilities ended.

In that time, he got to know Arnold well. Because of her polo-playing activities, his wife had adopted the name of Johnny; and Johnny and Bee had much in common in that their husbands were caught up in the job of trying to build a combat air force out of the dreams of Billy Mitchell while faced with the limitations of production and training know-how.

Before the war, on the Army's polo-playing fields, Andy and Johnny had come to know the Mitchells, for in those days Billy was an avid equestrian. He had a great fondness for Johnny and had won a trophy at the National Capital Horse Show, in Washington, riding her conformation thoroughbred hunter *The Dargyl.*

When Mitchell returned from France in 1919 to become Director of Air Service Operations and Training, Andrews was Chief of the Inspection Division and a member of the Air Advisory Board. They had much to discuss, and while Mitchell's thrust was always dynamic, larger than life, Andrews' was low-key, direct and very steady. It was, of course, his understanding and grasp of the potentials of air power that attracted Mitchell. They rode horseback together, discussing aerial strategy and the need to build a viable Air Service.

The relationship was brief, for in 1920 the Andrews family left for duty in Germany. It was rather special duty in more ways than one, for when Andrews departed for Europe, in August of that year, he had been given the permanent rank of major and assigned as Chief of the American Air Service in Coblenz. Since the Commanding General of the American Forces, Germany, was his father-in-law, the Major's appointment had to be in the nature of a family matter. Serving as General Allen's aide was his son Henry, Jr., also present with his family. The result was that AFG became jokingly referred to as "Allen's Family in Germany" and the General's headquarters as "Allen's kingdom on the Rhine." He was a tremendously popular officer, and through his manse constantly passed persons of note. They all came to discuss matters of interest and to enjoy the hospitality of their widely informed host and hostess. The Andrews, of course, were very much a part of the courtlike atmosphere.[1]

Flying was done out of the air station at nearby Weissenthurm. The AFG's air force boasted thirteen assorted planes, and Andrews and the few pilots under his command kept them busy. He had great mobility. He could cross borders and meet the air-minded of other countries, either where they lived or when they came to visit General Allen—among them Caproni, the Italian bomber builder; Marshal Fayolle, the chief of French aviation; and British airmen Trenchard and Sykes. All seemed to agree that should war come again it would, as Trenchard had written, "inevitably open with great aerial activity far in advance of contact either upon land or sea, and that victory cannot but incline to the belligerent able first to achieve and later maintain supremacy in the air."

All the contacts, all the discussions on the place and future of air power in the scheme of national defense, broadened Andrews' thinking. And so when Billy Mitchell came to Coblenz in 1922, conducting his own investigation into the European aerial outlook, the two had much to talk over.

It had been two years since they had ridden together. Two years in which Mitchell had raised such a storm of controversy that General Patrick's underlying purpose in approving his investigation foray was to get him out of Washington. When he arrived in Coblenz, he had already been in France and Italy and was on his way to England to see Trenchard, who had just established Cranwell, the West Point of the RAF, as well as a school for aircraft maintenance. Mitchell talked with German aviators as well, and had an interview with the Commander-in-Chief of the Reichswehr, General Hans von Seeckt. In German eyes he detected the glint of a new militarism, and no doubt he indicated his feelings to Andrews. Andrews could offer insight from another segment of the population, because he also doubled as a civil-affairs officer and had come to know something about the people of the area. There was no doubt that General Allen rated as highly with the occupied citizenry as he did with the diplomats and dukes. But only the blind could fail to see that Germany, though racked by inner turmoil and economic collapse, was not a defeated nation and looked upon the Versailles Treaty as a monstrous imposition that must be overcome.[2]

Mitchell's stay in Coblenz was brief, for with his aides Lieutenant Clayton L. Bissell and Alfred V. Verville, he was anxious to talk to the British. Before his departure, however, he wished to impress Andrews' officers, and anyone else with eyes to see, on the importance of formation flying. So he and Andrews and Bissell took off in an element of three DH-4's, and with wings practically locked, they flew down the Rhine to Remagen, their wheels skimming the water. Then it was hail and farewell. In passing, his enthusiasm, his confidence in the future, the timbre of his rather high-pitched voice, left a wake of their own.

Another three years would pass before Andrews and Mitchell were in close contact again—not that they didn't see each other, for no one in the Air Service could go for long without seeing Billy coming down in his stylish DH-4 with long red-and-blue streamer extending from its tail.

When Mitchell was exiled to San Antonio, in 1925, the Andrews family was close by. They had returned from Europe in 1923, and Andy was now the CO of the Advanced Flying School, at Kelly Field. On horseback, or at the bridge table with their wives, they talked endlessly on the needs of the Air Service and how best to obtain them. Andrews was a man of great patience and calm thought, and of course, Mitchell was not. And in the end, there was no staying him.

Andrews had not been in Washington during any of the upheaval, but after it he went from one duty to the next, his keenness recognized by his fellow pilots, his circumspection appreciated by all with whom he served. When he assumed command at Selfridge, there were two major considerations intertwined in his mind, one tactical and the other strategic.

Tactically, he had come to recognize a serious lack in pilot training.

Unlike himself, most Army airmen were not capable of flying under instrument conditions. After all, the rationale went, pursuit and attack pilots did not fly in formation and engage in combat when the ceiling was down and the visibility close to zero. As for bombers and observations planes, you couldn't expect to locate and bomb a target while flying in the soup, although certainly it had been tried, with notably unsuccessful results. There was also the problem of the Air Corps not being able to afford the proper instrumentation and navigational equipment, except for a very few of its transports.

The thing was all wrong, Andrews believed. He reasoned that if the real potential of air power was to be fully utilized, then all Army pilots must be able to fly under inclement weather conditions.

Although he had made numerous transcontinental crossings, a long-distance journey he undertook in early 1932 brought his thinking on the matter into sharp focus. Serving as Benny Foulois's Executive, he sold his Chief on the idea of delivering a mass flight of some much-needed aircraft to France Field, in the Canal Zone. Foulois, who was in favor of anything that would show off Air Corps capability, and particularly over the Navy, liked the idea and obtained the necessary okay from the War Department.

The "mass" consisted of five planes—three antiquated Keystone bombers and a pair of Douglas amphibians. The departure point was Kelly Field, Texas, and Andrews, who started the show by ferrying a Keystone from Langley Field, Virginia, to Hap Arnold at March Field, led his raggle-taggle squadron on a first-of-its kind operation.

One of the pilots, Second Lieutenant Laurence S. Kuter, had not had occasion to mix with anyone as exalted as a lieutenant colonel, but he was impressed by the flight commander's relaxed and easy manner. In comparison with others under whom he'd served, Kuter found the approach refreshing.

Andrews didn't give many orders. He assumed they all knew their business or they wouldn't be present. By his attitude he engendered in them a sense of confidence in their own abilities and a liking for his style. He made them want to fly with him, to follow him. And follow him they did.

There was no attempt to travel in formation. Andrews and crew would take off first in the Douglas, a faster plane than the Keystones, and be at the designated landing point by the time the other four planes arrived. In the interim, Andrews would have seen to the necessary protocol, checking on facilities and making technical arrangements. The crews would inspect their planes, staking them down, attending to maintenance. A night in a hotel in Tejeria or Tapachula was always an experience. Orders printed in Spanish were a great help. The Marines were fighting rebels in Nicaragua, but the action on the ground did not interfere with their passage. At all stops north of Guatemala City, Andrews would wire a nightly progress report to Foulois in Washington. South of Guatemala, his report went to

Major General Preston Brown, the commanding officer of U.S. forces in the Canal Zone. The 2,200-mile flight went smoothly, leisurely; there were no mishaps. They were two weeks en route, with a total air time of about twenty-five hours. Protocol and official entertainment were causes for delay; so was maintenance. But weather was a factor on Andrews' mind. In the tropics, afternoon thunderstorms and sudden squalls were the order of the day. Therefore, the safest and best time to fly was in the morning. This was standard procedure for the thin squadrons of Army aircraft operating in the Canal Zone. To Andrews it meant that the air arm was not being used to its greatest capability as a weapon for defense.

Safely down at France Field, Andrews talked over with General Brown the latter's aviation problems, which to the infantryman centered on not having enough of anything. The General was very pleased to be receiving five additional aircraft, and he hoped the operation could be repeated on a regular basis. As for the weather and its effect on flying, there wasn't much that could be done about that, was there?

Andrews believed there was, and upon returning with his crews to the United States by ship, he obtained permission to go to Kelly Field to take the three-week blind-flying course given by the Training Command. From that point on he had sought out the weather, working to perfect his own proficiency.

The Air Corps had done considerable pioneering in the technique, and such officers as Major William C. Ocker and Lieutenants Albert F. Hegenberger and Jimmy Doolittle had brought both attention and progress to the art of blind flying. But the only organized instrument training an Army pilot received was as a flying cadet at Kelly Field. The course was rudimentary, as were the instruments: needle, ball and air speed. After graduation, most pilots learned to fly on instruments by running into bad weather. It was not a sought-after condition. But, at the time, the Air Corps did not have mandatory preflight weather briefings. As a result, a pilot on a cross-country flight would often come up unexpectedly against a storm system. Then it was a matter of experience, luck, and frequently the ability to execute the 180-degree turn, called the safest maneuver in flying. The fact was that, every year, Air Corps pilots were killed or injured by flying into bad weather. In most cases, however, when bad weather moved in, the planes sat on the ground.

Upon one highly publicized occasion in 1931, the planes not sitting on the ground had brought heartburn and anguish. Andrews had been at the center of what became known as the *Mount Shasta* affair, and it, too, had to have a direct bearing on his determination, as a tactical commander, to see that his pilots could fly by the "gauges."

The *Mount Shasta* was an old tub of a ship that the Navy had reluctantly agreed to let the Air Corps bomb, a la Billy Mitchell. The 2nd Bomb Group, at Langley Field, had been given the job. Foulois's G-2,

Lieutenant Colonel Ira Longanecker, whipped up press interest. It was going to be another sinking of the battleships, even if the *Mount Shasta* hardly qualified as such. Instead, the attempt became somewhat of a fiasco. Group Commander Bert Dargue led his bombers out to sea in a mess of bad weather. On the first day, they couldn't find the ship and became generally fouled up. On the second, they finally found the target amid the rain squalls, but out of forty-two bombs dropped there was only one hit, and it didn't do a great deal of damage. The Navy happily sank the *Mount Shasta* with its own guns, and the Air Corps had egg on its face, while the Navy rubbed it in.

Andrews had been closely involved in the planning and publicity of the maneuver, and its failure bothered him, as it did his colleagues. What the hell happened? was the sum of the questions that circulated among Air Corps compatriots.

Andrews was sure one answer was to see that his pilots at Selfridge Field became proficient instrument flyers. In the beginning, he had little chance to pursue the plan, for he had no sooner taken over command of his three squadrons than he learned that he was to be directly involved in the welcoming of General Italo Balbo and his twenty-four Savoia Marchetti seaplanes.[3] Instead, by indirection, it was the innovative Italian airman who illustrated so graphically what was behind Andrews' thinking. Balbo's crews had spent months training for their flight from Orbetello, Italy, to the World's Fair at Chicago. Their instruments and navigational equipment were the best. On the 1,500-mile leg from Reykjavik, Iceland, to Cartwright, Labrador, training and equipment had paid off. The crews had fought wing-snapping turbulence, headwinds, fog and severe icing, and they had all come through safely.

Behind the welcoming crowds, the official receptions and speechmaking, the epic venture also illustrated an aeronautical axiom. The bemedaled Balbo, with massive head, glittering eye and rakish red goatee, stated it clearly when he declared in a public report to Mussolini: "The air forces can, like the navies, confront the problem of moving squadrons. With the Atlantic flight, Italy has furnished proof of these possibilities. I believe that with this aviation policy . . . aviation can make gigantic strides in all senses whether with reference to the improvement of machines or to the preparation of the flyers or the organization of the meteorological, logistic and technical services which are still too insufficient."

Andrews and the Italian airman met at a reception, and perhaps they discussed the future, for Andrews had said essentially the same thing in a paper written at the War College titled "The Airplane in National Defense": "We often hear about the limitations of military aviation. But, year by year, these limitations are becoming less with the improvement in airplane design and manufacturing and, more important, with the improvement in aids to navigation and piloting."

He had also described the barrier the Air Corps faced in moving ahead. "Many people are prone to judge our possibilities in time of war by limitations which peacetime operations impose upon us; limitations due to lack of funds for carrying out some project in its entirety. Others, whose experience in military aviation ended with the World War, can see no improvement in military aviation, but live in the past so far as the activity is concerned."

On this last, at the time of the Balbo flight, two actions by the War Department indicated exactly what Andrews had in mind. The Italian venture had been over a year in the planning. During it, the U. S. Assistant Military Attaché in Rome, Captain Francis M. Brady, had sent back detailed reports to Army G-2, the Military Intelligence Division, on the equipment, training methods, and logistical plans the Italians were perfecting. Apparently, none of this information was passed on to the Air Corps, although Brady had specifically made note of obvious interest. It was not that G-2 was attempting to hide the information, simply that, unless there was a demand for it, the Air Corps was not on its routing list.

Following the Chicago visit and a ticker-tape parade up New York's Broadway, Balbo's visit was to be capped by a meeting with President Roosevelt at the White House and a farewell banquet at the Army Navy Country Club. The guest list included the Secretaries of War and Navy, Chief of Staff General Douglas MacArthur, Air Corps Chief Major General Benny Foulois and close to one hundred Army and Navy officers and their ladies, among whom were Major and Mrs. Dwight D. Eisenhower. It was a grand night for toasting, and the departing visitors received the kind of military send-off that could only add to their prestige and pleasure.

But, behind the smoke of good Havana cigars, there lay in this final gesture of hospitality a War Department failure to capitalize on what had to be an obvious opportunity. The Military Intelligence Division, under Brigadier General Alfred T. Smith, appeared incapable of realizing a promising intelligence opportunity when offered it.

The estimated cost for entertaining the visitors while in Washington amounted to $1,240, and the Navy asked the Army to pay half. Smith was against paying it.[4] His rationale was that the severely restricted funds of the Military Intelligence Division were "for obtaining military information."[5] Apparently, it didn't enter the Chief of Intelligence's head that entertaining the Italian Air Minister and his men offered a rare opportunity to gain added insight into Italy's military plans.

Although Mussolini's aggression in Ethiopia was eighteen months away and could not be foreseen, the War Department and its Military Intelligence Division had to be aware that the Italian dictator was building up his armed forces. From an intelligence point of view, with nearly a hundred individual prospects from whom to glean possibly valuable information, it might seem that $620 was well worth the investment and eminently justified.

Not so, said General Smith. "The funds appropriated for the contingencies of the Military Intelligence Division," he reasoned, "are for the purposes specified in the Act and there is serious legal question as to whether they would be available for such purposes as this." (Smith's underlining.)

When the festivities were ended, Andrews returned to Selfridge Field to get down to the business at hand. The 1st Pursuit Group was made up of three squadrons. Their pilots flew Curtiss P-6's, Boeing P-12's and the Berliner-Joyce P-16, a twin-cockpit fighter. The 57th Service Squadron boasted a handful of assorted aircraft, one of which was a Douglas BT-2. This last was a slow-flying two-seater that Andrews knew would be ideal for practicing blind flying under the hood.

But if the 57th had the right plane to get instrument training under way, it also had a well-deserved reputation for fouling up. Its BT-2 had been divested of its lower wings by an erring pilot who had made an unsuccessful cross-wind landing. In attaching a new set of wings, the mechanics had managed to hook the aileron cables on backward. The result was a very short test flight, a BT-2 again without wings, some red faces and the air turned a smoky blue by the 57th's CO. Since no one had been hurt, the story was the joke of Selfridge.

Of more serious note, there were only sixty-two pilots to man the planes of the 1st Pursuit, seventeen of them reserve officers on temporary duty. A year before, the pilots had numbered ninety, but a squadron had been detached to aid in forming the 8th Pursuit Group, at Langley Field, Virginia. The Army's assignment to staff and operate the Civilian Conservation Corps had further reduced the 1st Pursuit. As a tactical combat unit, it was at about half strength. Only the P-12's could be considered first-line fighters, the line now five years old. But, among them all—the pursuits, the transports, the trainers—Andrews was quick to note, there was not a *single* gyrocompass or gyro horizon, two pieces of equipment standard on commercial aircraft and tremendously helpful when flying on instruments.

Aside from no gyro equipment, Andrews knew the "compasses on the P-12's weren't worth much." Exposed to the elements, they drank water, leaked and generally gave incorrect readings. Aircraft radios produced by the Signal Corps weren't much better. Many were of the one-way type, but whatever the variety, they were not dependable.

During the summer and fall, Andrews sent letters, through channels to Washington and to his friends at Wright Field, stressing the need for improved instrumentation. He contacted at Wright Field Conger Pratt, and Major Hugh Knerr, Chief of the Field Service Section, as well as Jan Howard, head of the engineering branch. They knew exactly what Andrews was talking about and agreed with the reasoning behind his push. But they could do nothing to help. The money wasn't there, and the Division's over-all approach seemed to be that since improvements in instrumentation were coming anyway, it would be a mistake to spend what little money they had on something that would be obsolete tomorrow.

Andrews found such reasoning self-defeating. Better to have the best that was available than nothing at all. He knew from his recent duty in the Chief's office that thinking there, which he had helped to influence, paralleled his own. As Brigadier General Oscar Westover, the Assistant Chief, had previously put it, "This office considers that dead reckoning navigation will be used by all classes of aviation, and celestial navigation in addition by bombardment, long-range observation and coastal defense units."

In a later response to Andrews' position, his former boss General Benny Foulois said he fully agreed. Foulois's own effort to get pilots properly qualified on instruments had been the establishment of two small "navigation courses at Langley and Rockwell Fields," which would begin in October.[6] His thought was that from these classes there would come a nucleus of instructors who would teach their fellow pilots at all the tactical bases.

However, according to one of Andrews' young pilots, Second Lieutenant Curtis E. LeMay, whose interests were focused in the same direction as his CO's, no one at that time had a basic course in instrument flying as it should be constituted. Further, nobody knew anything about celestial navigation. You got from point A to point B by what you could see or by dead reckoning. When using the latter over water, the only guide other than the compass was a crude drift sight (which wouldn't do much good).

It was the Air Corps's need to fly over water that had stimulated Foulois in the establishment of the school he mentioned to Andrews. This had nothing to do with Balbo's feat but was the result of the Air Corps Chief's effort to support in every way the Corps's principal mission, of coastal defense. It was a mission clung to tenaciously, for without it the Corps would lack a real purpose for existence. Everything that Air Corps officers did toward aviation development—whether it was Andrews' attempting to establish uniform instrument training for his pilots, or Foulois's ongoing thrust to get more planes and men, or Arnold's working to improve pursuit and bomber tactics—everything was centered around the mission.

Part II

Part II

Ten

The Conspirators

There was no doubt in any Army airman's mind that General Douglas MacArthur, as Chief of Staff, had a far better grasp of what air power was all about than any of his predecessors. In working out the coastal defense agreement with Admiral Pratt in 1931, he had given the Air Corps its *raison d'être*.

Although the Navy was quick to step back from the agreed-upon Joint Army-Navy Board conditions ironed out by the two, particularly after Admiral Pratt retired, in 1933, the Air Corps accepted as gospel its "mission to defend the coasts both at home and in our overseas possessions. . . ." But even though all training and operations were tailored around this mandate, there was no unified combat structure in being to carry it out. What there was, was a mishmash of understrength air units situated—as Hap Arnold had said—in the nine corps areas without any real regard for defensive strategy or cohesive direction. MacArthur understood this, and when he brought forth his Four Army Plan, in 1932, discarding a much clumsier defense organization, the fifth part of the plan concerned utilization of the Air Corps should a national emergency arise.

Behind the general plan, however, of creating a defensive combat air force, lurked the central issue of control. Would the General Staff direct such a force? Would the Air Corps? Or was it better to let rebellious ideas indicating any kind of air autonomy fester where they were and do nothing? MacArthur was not inclined to accept this last school of thought, and although at the time he privately appeared ambivalent toward the use of air power, he was anxious to still congressional accusations that the Army lacked any real plan for air defense. He wrote a War Department policy letter describing the form he thought such defense should take, which involved organizing a combat arm of about eighteen hundred planes into three defensive wings, guarding the coasts and the Gulf area.[1]

Benny Foulois and his staff agreed with much of what MacArthur said, but they did not feel his plan was specific enough to their wants. They, in turn, drafted a counterproposal, which they passed on not only to the General Staff but also to Assistant Secretary of War Henry H. Woodring. The General Staff took a cold look at what it considered a very costly and unrealistic Air Corps approach to the problems of defense. War Department planners were concerned with the here and now, not tomorrow. No enemy aircraft could attack the United States, they reasoned. No carrier-borne aircraft could sustain operations off our coasts.

Air Corps planners saw it differently and so stated:

> The danger of concentrated air attacks upon nerve centers of communications, industry and government, with the object of paralyzing the nation's power to resist and thus facilitating decisive action of ground forces, is a factor which makes it imperative that the nation's peacetime strength be adequate to such an attack upon the outbreak of war.

Nonsense! was the general retort from the ground.

When efforts of the General Staff and Foulois's office failed to resolve the existing disagreements, MacArthur decided to give the Air Corps Chief his head. The defense of the United States and its possessions was based on War Department "color" plans, each color designating a variation of attack by an enemy or a combination of enemies. The Air Corps had *never* been asked to submit a proposal based on a color plan. MacArthur, through the General Staff, told Foulois to go to it and present recommendations, but to be reasonable.

Foulois, a man who had dared the cannon's mouth frequently, went overboard. The Air Corps's offering started with a direct attack on the War Department, maintaining that its color plans as written had no bearing on the use of air power. Instead, the Air Staff presented a scheme for the defense of seven critical coastal areas with air reconnaissance extending seaward three hundred miles. To fulfill the needs of the plan, a force of nearly forty-five hundred planes was required, triple the number on hand. Foulois's detailed coastal defense plan, submitted in mid-July 1933, was rejected out of hand by Brigadier General Charles E. Kilbourne, who headed the General Staff's War Plans Division. He wrote MacArthur: "The report submitted is of no value for either war planning or for a logical determination of the strength at which the Air Corps should be maintained."

Kilbourne saw in Foulois's proposal nothing but a repetition of previous proposals to increase air strength. He had no liking for the Air Corps Chief for personal reasons, and this, added to what he considered was a demand for separate war-planning privileges, angered him. Further, he felt

Air Corps officers were being indoctrinated to believe they could handle their defense mission independently and that Foulois showed "a complete lack of desire to conform to instructions issued by proper authority."

The result was that MacArthur and the General Staff rejected the Foulois submission not only because it was far too costly but also because it went far beyond any conceivable war threat.

Regarding the tactics of coast defense, the General Staff position was that air operations were to be fitted into the Army's overall strategy. Foulois's hope had been to sell the concept that there should be a special Air Corps defense program "and that organization and training for this purpose should take priority over all other considerations." Stripped of all else, it was, of course, a bid for greater freedom of action, and MacArthur, in raising considerations for the establishment of an air force, had surfaced the very strong feelings that surrounded the issue.

The idea for such a force was not new. Ten years earlier, in 1923, Major General William Lassiter and his Board had supported the air-force concept, developed by Colonel Edgar Gorrell, in which Billy Mitchell had organized his tactical units into a single striking force during the St. Mihiel and Meuse-Argonne campaigns. Not only was this 1918 strategy new at the time, but so was the name given to its command structure: General Headquarters Air Service Reserve.

In examining the sad state of the U. S. Army Air Service five years later, the Lassiter Board approved the idea of a GHQ Air Force, which would be a separate combat arm within the Air Service. It did so, of course, while stressing that the primary mission of the Air Service was to support Army ground forces as ground commanders saw fit. However, any leftover units might be organized into a GHQ Air Force, which could be employed either traditionally or independently.[2]

In the decade that followed, perennial boards set up by the War Department and Congress—supposedly to reexamine the role of military air power but more often to block the proponents of air independence—continued to accept the idea of a GHQ Air Force in theory if not in fact.

It was MacArthur, in his policy letter on the "Employment of Army Aviation in Coast Defense," who had indicated he was in favor of moving the idea from a talking point to a rough form of actuality.

At about the time Lieutenant Colonel Frank Andrews and General Italo Balbo were matching formations in the air and exchanging aeronautical opinions on the ground, MacArthur called in his Deputy Chief of Staff, Major General Hugh Drum, to consider what could be done to bring about an acceptable solution to War Department and Air Corps differences on coastal defense and the actual establishment of a GHQ Air Force. It was basic to them that should an emergency arise—M Day, it was called—time would be needed to mobilize and train ground forces, but an air force, like the fleet, had to be ready to defend the nation's shores and possessions at once. To organize combat units of pursuit, attack and bombardment into a single operational force made sound military sense. Politically, a GHQ Air Force, under the direction and control of the General Staff, would go a long way, they believed, in quieting what congressional support there was for a separate air force and would either pacify or cut the ground out from under air officers continuing to press for autonomy.

MacArthur had hopes that his Deputy would come up with recommendations that would finally resolve the controversy and still the Air Corps's outcry.

The Deputy Chief's effort went under the weighty heading of Special Committee of the War Department General Council on the Employment of the Army Air Corps. More easily, but no less pontifically, the Committee became known as the Drum Board. Of the four officers selected to assist Drum, only Benny Foulois was an airman. The other members were Major General George S. Simonds, Commandant of the Army War College; Brigadier General Charles E. Kilbourne, Assistant Chief of the War Plans Division; and Brigadier General John W. Gulick, Chief of the Coast Artillery.

Hugh Drum, at fifty-four, was a career infantryman who had fought in the Philippine guerrilla campaigns at the turn of the century and had served on General Pershing's staff in the World War. During the final days of the war, he had been partially responsible for a written order that had very nearly caused a military debacle.

In early November 1918, General Pershing became intent upon his First Army capturing the important German-held city of Sedan. In a rush to do so, the proper arrangements were not ironed out between his own forces and those of the French Fourth Army, whose boundaries abutted those of the American First. Worse, the orders sent to the units to launch the attack were not clear, and Hugh Drum added to their confusion by appending them to read: "Boundaries will not be considered binding." The result was that the night advances of elements of the First Army, principally its 1st Division, became badly entangled in the boundaries of its own forces and those of the French. In the darkness, confusion became total. Had the Germans realized what was happening in front of them, they could have inflicted a bloody defeat on the stalled attackers. Had the French opened fire on the Americans, a debacle would have ensued, with obvious political ramifications.

Fortunately, the mess got straightened out before daylight.[3] The Armistice, coming a few days later, was the smoke screen behind which an official inquiry into the massive blunder was dropped. Pershing had been at fault in his hurry to take the city. Brigadier General Frank Parker, commander of the 1st Division, became the scapegoat for the near catastrophe, and Hugh Drum, instead of facing a possible court of inquiry, was awarded the Distinguished Service Medal for his "distinguished and meritorious service as Chief of Staff of the First Army."

Now, some fifteen years later, he held the number-two job in the War Department, and in 1930 had been seriously considered for MacArthur's position. In the intervening period, a series of assignments had led him upward through division and corps-area commands: Director of the School of the Line at Fort Leavenworth, Commander of the 2nd Coast Artillery District, Assistant Chief of Staff G-3 in charge of Operations and Training, Commander of the 1st Division, etc.

In 1930 he had been promoted to major general and, after serving as the Army's Inspector General, he had become the V Corps Area Commander. Then, finally, in February 1933 he had replaced Major General George Van Horn Moseley as Deputy Chief of Staff and was moving smoothly, with solid support, toward the seat of command he so dedicatedly sought. Through background, family connections and marriage, Drum, a member of the "Pershing clique," had become one of the Army's power elite. Slightly on the Blimpish side, with smooth, hawkish features, he was described by one contemporary as a "pouter pigeon." Another said, "He gave the impression that if he wasn't the Pope, he should have been."

Withal, Drum had an agile mind and a grasp for detail. He had proved himself an efficient administrator. When testifying on War Department programs, he made a good impression before congressional committees. In 1923 he had served on the Lassiter Board, but, possessing a traditional appreciation of air power (the air must be closely tied to ground support, either via the General Staff or corps-area commanders), he was against all that Billy Mitchell stood for. As G-3 in 1925, he was considered one of Mitchell's principal enemies. During the spring and summer of 1933, he had had his hands full assisting General MacArthur in the Army's biggest peacetime challenge, organizing more than a quarter of a million young recruits into CCC camps across the country. Now, in late summer, he began his examination of the Air Corps.

On a day in the month of September 1933, when Benny Foulois sat down in the Deputy Chief of Staff's commodious office in the massive, gray-pillared keep known as the State War and Navy Building, he knew he was deep in enemy territory, outnumbered, outflanked, surrounded. The four ground officers with whom he was to work out an acceptable compromise on Air Corps coast-defense plans and War Department requirements were top professionals, strong-minded and assured in their respective commands. They carried an overwhelming amount of clout.

Foulois could light his pipe, look at the quartet around him and know that, as an airman, he had done things and been places and dared heights none of them could challenge. But he had also seen enough jungle combat and been in enough scrapes to know when to back off and lie low. Given ample warning and not faced by a sudden decision, he could check to the dealer. Deliberation was often easier for a man of action when he had decided that whatever came out of the deliberations might be used to good advantage with friends in Congress. So, instead of going down swinging, as possibly Kilbourne and Simonds expected him to do, he surprised them, and during the first meeting and those that followed he cooperated, playing along. Even if the lack of appreciation on a personal basis was mutual between himself and the phalanx of four, they were officers and gentlemen, and their Chief, Douglas MacArthur, wanted results.

A GHQ Air Force was an agreed-upon point. So were the seven vital coastal defense areas the Air Corps had cited. Never mind that the four ground officers would never agree that the Air Corps alone could protect any one of them or could prevent an invasion by hostile forces—this last was an act they considered impossible anyway. What mattered most to Foulois was that beyond the organization of a GHQ Air Force lay his determination to command it, a desire the General Staff was dead set against.

Through the weeks of discussion that followed, Foulois, mild-mannered, puffing his pipe, let it be known that he believed the essential problem had to do with lack of money, not with a lack of cooperation by the General Staff. The four were pleased with his equanimity.

To strategy-minded air officers like Andrews, Arnold and Spaatz, the Drum Board was a stacked deck. Its findings, released in the fall and signed by Foulois, supported the belief. Although recommending a GHQ Air Force of 980 planes and an Air Corps with a total complement of 2,320 aircraft, the Board stressed keeping the Air Corps under firm Army control. It did reiterate a coastal defense plan. This was seen as a compromise offer and was accepted by many airmen as such. Yes, there should be long-range reconnaissance and bombardment aircraft in support of ground forces and of the fleet when and if necessary. But in taking a crack at strategic air power advocates, lo and behold, the Drum Board used the Balbo flight to prove its point!

It would not be possible, stated the Board, for "land-based bombing planes of foreign nations [to] cross the Atlantic or Pacific, rendezvous at some selected point, deliver a concentrated attack on some vital objective and then return to home base . . . nor with land or floating bases established in route and in territory contiguous to our nation could an air force superior to our own launch a decisive attack against some vital area of the United States."[4]

In using Balbo's exploit to support its conclusions, the Board noted that the Italian armada had required eight air bases in foreign countries, eleven surface vessels and two months of preparation, and that in spanning over six thousand miles, the elapsed time to accomplish the epochal journey was fifteen days, which included forty-six hours of flying.

A key to this intended knockdown was the repeated phrase "under the present stage of development." It was as though the present stage would remain constant. Also, completely missed was the amazing logistical and cooperative success of the Balbo venture. The entire undertaking had pointed toward a rapidly approaching future which Drum and his cohorts refused to understand or to consider. Instead, they sought to ignore the future by belittling the present.*

There was something prophetic in the way the War Department General Staff had looked upon, misread, failed to take advantage of and, finally through the Drum Board, misjudged the meaning of the Balbo crossing.

There was irony, too, for a few years down the road, Hugh Drum would find himself given the responsibility of defending the U.S. East Coast against possible attack, his principal weapon aircraft and, at the outset, far too few of them.

In October 1933, General MacArthur approved the Drum Board's recommendations, and they became official War Department policy. But the Navy did not approve and, having already revoked the MacArthur-Pratt agreement, began a push in Congress on an expansion program. The program's intent was to scuttle the Air Corps mission and, in so doing, confine Army air action solely to support of troops, and thus eliminate any strategic role it might seek to develop.

On New Year's Day 1934 the Andrews held the post Commander's eggnog reception. It was a traditional affair in which the officers of the post came to pay their respects and exchange New Year's greetings with their Commander and his wife. The silver bowl was on the dining-room table, surrounded by cups and canapés. The festive air and decorations of the holiday season provided warmth against the icy chill of winter in Michigan. So, too, did the CO and his lively wife. For the moment, the blizzard of problems facing the Air Corps in the New Year were not permitted to intrude.

Never mind that the Corps seemed to be shrinking into itself: only thirty-three cadets graduated at Kelly in the fall, and publication of the Air Corps newsletter was suspended for lack of funds. These were small things best not discussed around the wassail bowl any more than the large ones such as the 15 percent cut in pay and the President's authority to cut flight pay altogether. No one was going to mention that FDR, in his infinite wisdom, had reduced the nearly forty million dollars in Public Works Administration funds allocated by the War Department for new aircraft to seven and a half million.

---

* That the RAF, unlike the U. S. War Department, fully understood the significance of the Balbo flight was evidenced during the Battle of Britain, in the summer of 1940. When the British began to put up large formations of aircraft, they did not refer to them as Wings but as Balbos, the image of the mass flight seven years before having left its impression.

The talk was focused on serious but more pleasant subjects, such as bridge and poker and how long it would be before the golf clubs could be gotten out. There was talk of riding, too, for Johnny Andrews had brought her three polo ponies from Washington to join others in the stable, and she and her ten-year-old daughter, Jeannie, rode daily along the dike that surrounded the airdrome and helped keep Lake St. Clair from flooding the field. It was a known fact that no one could talk to Johnny Andrews for long and not end up talking horses.

Her genial husband had other things on his mind. In ten days he would be leading the 24th Squadron to Miami to be present at the All-American Air Races. It was an annual affair, and Air Corps representation was a public-relations must. Otherwise the Navy would run away with all the press and newsreel coverage.

En route he would land at Nashville's McConnel Field for a visit with his air-minded father, James D. Andrews. There was considerable similarity in the looks of the two. His father was smaller, but his son had inherited his high forehead, long, straight nose and wide, slightly canted mouth.

At seventy-five James Andrews, with his thatch of white hair, had the look of an elderly frontiersman. A former newspaper editor, he had switched to real estate after a fire had destroyed his newspaper plant in Pulaski. A well-regarded citizen, he and his three sons (all army officers[5]) were frequently mentioned in the Nashville *Banner,* a paper for which he had once worked.

Semiretired, living with his daughter Josephine Sykes and her husband, Gillespie, his aim in life was to get a decent airport for Nashville, for the present site, at Sky Harbor, was inferior at best. Behind the scenes Andrews had been quietly aiding his father and was looking forward to seeing him. And so on that New Year's Day, as he greeted the sixty or so officers and wives of his command, most of them half his age, he could foresee that he would be starting the month off by touching base with members of the family. Beyond that, he could not see, could not know, that this would become the watershed year of his life.

In a month he would be fifty. It was not a milestone to which he would give much thought, nor would anyone else besides Johnny, because he looked a good ten years younger. But in this, his fiftieth year would come developments over which he would have no control and little influence. They would nevertheless set in motion forces that when spent would give rise to a call for leadership. That he would be chosen to accept the mantle was the measure of all his years of service. Now removed from the center of conflict, he heard only indirectly from friends in the Chief's office as to what was going on in Washington. He knew you had to be there to know, and then often all you knew were the latest rumors. Or so it was until after he had returned from the Miami air races, near the end of the month, and had as his guest at Selfridge, Congressman John J. McSwain.

Over the years, there had been members of both the U. S. Senate and the House who had championed the cause for air independence. None was more of a champion than John Jackson McSwain, a South Carolina farm boy who had gravitated from teaching to the law. Following service as an Army captain in France and Germany, he had returned home to run for Congress. He had been elected to the House of Representatives in 1921. Six terms in office later, he'd gained enough weight and seniority to become, in 1933, Chairman of the House Military Affairs Committee. With a hail-fellow-well-met veneer to mask his political shrewdness, he was seen by the War Department as a threat to its air policy. MacArthur privately referred to the big, ruddy-faced legislator as "McSwine," but in Benny Foulois's office he was known as a very good friend—when very good friends were not all that plentiful.

Before his visit to Selfridge, McSwain and Andrews had not met. In view of mutual interests this might seem odd, particularly since Andrews had served tours in Washington while McSwain was serving in the House of Representatives. The undoubted reason was that Andrews had a fine sense of circumspection. He was known and accepted by the General Staff, first through social entrée by his marriage to Johnny Allen and then by his performance and his ability to get along in places and situations where so many other air officers failed. Even so he and Billy Mitchell had remained close confidants after the latter's resignation, and before coming to Selfridge, the Andrews had frequently ridden with and visited the Mitchells at their Middleburg, Virginia, home. He had also kept Mitchell posted on major aerial developments, sending him maps and routes and photographs and critiques on combat aircraft. At the same time, Andrews had kept a low profile on the political front and, instead of speaking out on the principal issues, he listened and thought.

At their meeting, McSwain was obviously impressed, for on his return to Washington the first thing he did was to sit down and write Andrews a letter of appreciation and thanks. The next was to take action that sent danger signals through the War Department.

At the beginning of the year, McSwain had announced that his committee would once again consider the question of the role of aviation in national defense. At the time, the Washington *Herald* was running a series of articles giving the General Staff's point of view on its air arm, and it was thought that possibly the thrust of the articles might have moved McSwain to act.

This was window dressing, for behind the Chairman's move was concealed the fine poker-playing hand of Benny Foulois and some members of his staff.

There had been no action by the War Department on the Drum Board recommendations of October to set up a GHQ Air Force. Foulois and his confidants had grown impatient, worried with good cause at moves by the Navy to grow in strength and size while the Air Corps faced atrophy. They believed the War Department would not move on the establishment of a GHQ Air Force unless forced to do so.

McSwain's announcement brought the desired reaction. Even while he was talking with Andrews at Selfridge, his committee was hearing General MacArthur in executive session. With his predilection for colorful metaphors, MacArthur described the utilization of GHQ Air Force and in so doing was careful to declare that one of its uses would be "independent missions of destruction aimed at the vital arteries of a nation." This was the Mitchell dictum of strategic air power. The force, he said, would consist of five wings of two hundred aircraft each.

As a result of MacArthur's testimony, McSwain raised the ante by asking the War Department to submit a bill so the new force could be brought into being. Before the end of January, Secretary of War George Dern had complied. In his letter of transmittal to McSwain, the Secretary pointed out that although the Bureau of the Budget had not yet approved the funds to bring the GHQ Air Force into being, the War Department was anxious that it do so.

On the first of February, McSwain introduced the War Department's bill. On the second, he bowled over its General Staff proponents by introducing his own bill—the first of two that called for cutting the Air Corps's umbilical cord to the War Department. Complete independence was the bill's goal, with dictation and control by the General Staff eliminated from the picture. The Chief of the Air Corps would report directly to and be the immediate adviser of the Secretary of War on all matters relating to military aviation.

The bill had everything in it Army air-power advocates were seeking: separate promotion list, separate budget, increase in personnel and a new five-year program aiming at over four thousand aircraft plus a rise in rank for the Chief of the Air Corps to lieutenant general.

The War Department's response was swift. It released to the press a synopsis of the Drum Board report and, covering itself, sent a complete text to McSwain. The same day, the *Army and Navy Journal,* of John C. O'Laughlin, came out with strong support for the report.[6] The *Journal* maintained that a GHQ Air Force would offer "the benefits of an independently acting air force." Unity of command would be protected and costly duplication would be avoided. Further, the *Journal* declared that President Roosevelt would probably oppose the McSwain bill and that "the Air Corps headquarters in Washington is entirely satisfied with the recognition its arm has received from the General Staff since the practical abolition of the Assistant Secretary of War for Air." The first assumption, regarding FDR, was correct. The second was totally incorrect.

O'Laughlin's pronouncement on how the Air Corps felt must have brought some chuckles from the second floor of the Munitions Building, for the fact was that McSwain's latest bill, calling for air independence, had been secretly drawn up in the office of the Chief of Air Corps. Foulois and some of his close subordinates—no one knows who for sure—had been playing a game of cutthroat with the War Department. The game went back to MacArthur's turndown of Foulois's coastal defense plan. At that point, Foulois had gone along with Drum and Kilbourne and the others to get a GHQ Air Force. When he saw nothing happening to move its development beyond the report stage, he let McSwain know of his frustration. The wily Southerner, who at the time considered Benny Foulois the greatest military airman on earth, no doubt suggested they go one better by going all the way. Foulois probably realized that the bill he and his subordinates drafted would have little chance of getting through Congress, but its pressure might be great enough to force the War Department to bring a GHQ Air Force into being.

There were times in his career when Benny Foulois showed poor judgment, when an inner aggressiveness in his nature took hold. This was such a time: He saw Roosevelt as no friend of the Air Corps, and although he publicly genuflected toward General MacArthur, he did not really trust him on air matters. As for Drum, the dislike was thoroughly mutual. At the moment McSwain offered his bill, Foulois was puffing angrily on his pipe over three interlocking issues and a host of smaller ones.

When McSwain had asked the War Department to draft its own bill, Foulois had been hurriedly summoned to assist Drum in doing so. In the proposal, numbers of planes were not mentioned but the existing ratio of 9:5 between Army and Navy aircraft was. Implicit in the draft was the buildup of Army aviation. MacArthur approved the bill and told Foulois to present it to McSwain. But, the very next evening, Foulois was again summoned by MacArthur and presented with a new version of the bill that made no mention of the ratio. This, to Foulois, was basic. Without it, the Navy, which was already launching its new attack to command the air, would be given almost *carte blanche*. Foulois was furious over what he considered a War Department sellout to the Navy, but he said nothing to MacArthur about his feelings.

The second cause for the Air Corps Chief's anguish dealt with aircraft procurement and was of long-smoldering duration. It illustrated in another manner the type of War Department control that to airmen not only impeded progress but discouraged it as well.

Of the fifteen million dollars finally agreed upon by Franklin Roosevelt in 1933 for the purchase of new aircraft with PWA funds, half the amount was allotted to the Army. However, the selection of types and numbers of planes was determined almost exclusively by General Kilbourne and his War Plans Division. Foulois had been anxious to include in the purchase a new long-range amphibian. To keep peace with the Navy, Kilbourne had shot the plan down. Foulois and his people saw the act as the most blatant kind of interference in their own plans for coastal defense. Even in being finally permitted to recommend the number of pursuit, attack and bombardment planes to be purchased with the seven and a half million, the numbers were subject to General Staff approval. On the need for the aircraft in the first place, it was not Foulois or members of his staff who testified before the Public Works Administration but Brigadier General Robert E. Callan, Chief of the G-4 section of the War Department.

Foulois's third cause for frustration was tied directly to the other two and indicated the power block he faced. With Trubee Davison gone, he felt isolated. As long as the Air Corps had Trubee fronting for it, a certain sense of independence did prevail. Trubee was there to act as the honest broker, and the General Staff could not automatically override him. With the election of FDR, the office became vacant. The question was, Who was going to fill it? Or was anyone going to fill it?

It was not until June of 1933 that FDR announced that he was abolishing the office for economy reasons and that the new Assistant Secretary of War, Henry H. Woodring, would assume the job with his other responsibilities. The recommendation to abolish the post had come from George Dern, the Secretary of War. Dern, who had wanted to be the Secretary of the Interior, was a successful metallurgist and former governor of Utah. He was, however, a bad choice for Secretary of War. He knew nothing about the Army, was pacifistic by nature and was in poor health. Beyond that, to Foulois and others, it soon became apparent that Dern was under the strong influence of Army Chief of Staff Douglas MacArthur. Foulois saw the Secretary of War's recommendation to abolish the post coming directly out of the Chief of Staff's office, and what followed simply bore out his fears.

So it was that when it came Benny Foulois's turn to testify in executive session before McSwain and his committee, assured his remarks would be kept confidential, he cut loose and, in a rambling, somewhat disconnected and contradictory discourse, tore into the War Department's bill and supported the one for which he was secretly responsible.

"For twenty years, Mr. Chairman, at least twenty years, I have fought consistently against the Army General Staff, trying to build up aviation within the Army," he declared.

"I announced to the Chief of Staff in his office four days ago, in the presence of the Deputy Chief of Staff, that the steps that they were taking at the time—toward the building up of Army aviation—were the first steps to my knowledge, in the past twenty years, that had been initiated by the General Staff itself.

"In saying that, I'm not impugning the integrity of any particular individual member of the General Staff nor criticizing any particular member of the General Staff. The General Staff is made up of intelligent men—made up of men who have been trained in other branches of the service. Ninety-nine percent of them have no knowledge of aviation except what they read out of books.

"This is a practical problem that has to be handled by practical men trained in aviation and at the same time who have just as much knowledge, training and experience on General Staff work and organization of the Army and other branches of the service as any member of the General Staff.

"That is the basis on which I want to start in connection with building up an organization within the Army, under the War Department, that can go ahead, keep up to date, and not be hampered and hindered by the routine methods that hamper and hinder the Army now.

"We cannot operate under those methods. We cannot operate at a speed of four miles an hour, the speed of Infantry. We are hitting it up around a speed of 200 miles an hour. Today the psychology of the development of aviation requires that all other military must increase their speed of operations, and the rest of the Army must be reorganized with this in view, if it ever expects to effectively use the modern military weapons of war.

"I will give you the very latest action of the General Staff. You have had it yourself in this bill.

"The day before yesterday, in the evening, I presented to you a draft of a proposed piece of legislation which I delivered to you informally under the directions of the Chief of Staff. If you have read it, you will remember that there was a provision in that first draft that I submitted the day before yesterday that fixed a ratio between the Army and the Navy to what we felt was necessary to take care of the Army Air Corps in this general move to build up the air defenses of the United States.

"Within twenty-four hours they changed that. The new draft you have on your desk now I do not have. It was supposed to have been furnished to me, but I did not receive it. It was presented to you formally and officially over the signature of the Secretary of War. The new draft contains an ambiguous clause in it. I went before the Deputy Chief of Staff and the Chief of Staff last night, and they asked me to read it. I read it, and the very point that stood out was that the new formal draft fixed a maximum limitation of 2,000 planes. In other words inside of twenty-four hours they had changed the policy. I asked the reason for it and was advised to the effect that the amendment provided for more leeway in securing the number of airplanes desired. I had a different explanation from a General Staff officer this morning to the effect that it was put there deliberately for the purpose of ambiguity. Who are they trying to fool? You? Me? Or someone else?

"I am suspicious. As I say, I fought them for twenty years, and I think they are trying to fool you and trying to fool me. . . .

"The vital point that I am trying to develop is the necessity of having an independent Army Air Corps organization that can function efficiently outside of the present cumbersome General Staff system."[7]

In declaring that "the main blocking element in the War Department" to the growth and development of Army aviation over the past twenty years "was the War Department General Staff," he sounded like Billy Mitchell a decade earlier. And like Billy Mitchell, he went overboard and said things he would later come to regret.

Yes, he was for autonomy. Yes, the Air Corps should have its own budget, its own promotion list and an end should be put to a system by which aircraft equipment such as radios, guns, ammunition, even pay for reserve officers, was procured and managed at the discretion of outside military agencies. The thing was that the Air Corps was supposed to do a job and be responsible for the manner in which it was done, but how could it without control over its own resources?

The question was hardly new, nor were efforts by some members of Congress to break the Air Corps free from its economically strapped parent. It was, in fact, well-trod ground. During the previous session of Congress, House Majority Leader Joe Byrnes had pushed for similar action along with McSwain. Byrnes, a Tennesseean, had written to Andrews' father, in Nashville, saying that his bill for a department of defense had been defeated "by only forty votes." He was confident that an independent air force was coming "much sooner than had been expected."

As for McSwain, his moves to accomplish the desired end spanned a decade. The General Staff, as noted, was on guard against his moves and this was all the more reason for the secrecy with which McSwain's bill was put together.

Two weeks before the Congressman began his hearings, Captain Bob Olds, a former aide to Billy Mitchell and one of the brightest minds in the Air Corps, reported on a meeting he had had with Deputy Chief of Staff Drum on Saturday, January 13, 1934. The General had called Olds to the War Department to question him on the exchange of information between the Office of the Chief of Air Corps and Congressman McSwain's office.

When Olds informed him that, yes, a report containing information concerning subsidies and aid paid by foreign governments to commercial aviation operations in their respective countries had been passed directly to McSwain, Drum said such practices must stop right now!

Olds pointed out that a copy of the correspondence in question had already been furnished to the War Department, in keeping with existing policy. Impressed by neither rank nor authority, the Captain suggested that if that was to change, then the policy would have to be changed.

Drum's response was that from now on he wanted General Foulois to consult his office by telephone or otherwise in order to determine what data would be made available on congressional request.

Several days later, a request from Congressman Crosser came into the air chief's office asking for information on appropriations for aviation since 1929. Olds called Drum and said they were planning to comply. Drum said no. Send the information over. His reason: the same data had previously been supplied by the War Department to members of Congress, and Drum wanted to make sure there were no discrepancies in the figures.

It was a fair example of lack of trust, indicating General Staff care and watchfulness. This tight rein had also been reflected in a change of policy in drafting travel orders for airmen going on temporary duty. Routinely, the task had been handled in the office of the Chief of Air Corps. Now the War Department said the Adjutant General's office would manage the function. Foulois protested vigorously, infuriated by what he considered costly interference and nit-picking.

All of these factors played a part in the determination by the Air Corps Chief to secretly aid the Chairman of the House Military Affairs Committee in the drafting of a bill that, if successful, would free the airmen from War Department dictation. It was the most concerted move toward independence since the departure of Billy Mitchell.

Of the thirty-six airmen who staffed the office of the Chief of Air Corps at the time Benny Foulois testified on the bills before Congressman McSwain's House Military Affairs Committee, there were four ranking officers of particular note. They were Brigadier General Oscar Westover, Assistant Chief; Lieutenant Colonel James E. Chaney, Chief of Plans; Lieutenant Colonel Walter H. Weaver, G-2 Chief of Information; and Major Carl Spaatz, Chief of Training and Operations, G-3. All were younger than their bald chief, but all were old air hands.

Tooey Spaatz had served under Foulois during the Mexican incursion in 1916 and later in France. Walter Weaver and Jim Chaney had transferred from the Infantry to the Aviation Section of the Signal Corps in 1917 and Westover had made the switch from the Cavalry. All were West Pointers. Of the four, Westover was the most traditional and circumspect in his military outlook. It didn't make him any less enthusiastic toward air power, but he could not openly be counted in the air autonomy group. When he had been selected as Assistant Chief in 1932, the news had not been met with great joy by the activists in the Corps.

Short of stature, like Foulois, round of face with a black brush mustache, the Assistant Chief was of serious mien, with a reserved, slightly mournful aspect. A classmate of Andrews, he had garnered the nickname Tubby not because he was but because as a wrestler and athlete he had developed a powerful torso. He was also a crack marksman. He earned his wings as a balloonist in 1921, and two years later, when he attained his JMA, he was forty, rather late in the game to become a pilot.

In the former pursuit, however, he gained acclaim in two areas, first as winner of the National Elimination Free Balloon Race in 1922 and the following year as a briefly held prisoner of the Hungarians. This last occurred at Geneva during the Gordon Bennett Trophy Race. Admiral Horthy, head of the Hungarian Government, had announced that overflights of his country's territory were forbidden. Tubby Westover, caught in unfriendly weather, descended to have a look around. Hungarian peasants, thinking he was in trouble, grabbed his drag rope and hauled him out of the sky. Then he was in trouble until the guns of diplomacy could be brought to bear on the Admiral for his release—a rather unusual case of air power against sea power. This was in 1923, during Benny Foulois's tour as the Assistant Military Attaché in Berlin. Foulois had been on hand for the race and, before and after it, Westover had kept him posted on Washington developments. Westover was then serving as Executive to the Chief of the Air Service, Major General Mason M. Patrick. Foulois and Westover were on friendly terms, and in their correspondence Westover made it a habit to address Foulois as "my dear general" even though the latter had reverted to the rank of major.

More to the point, Westover, who was no extrovert and rather humorless, had a straight-line approach on the matter of air power. He had expressed it before; he would express it again. He did not deviate: "As an individual, I, of course, have my own ideas about what is best for the future, and I am willing to work for the accomplishment of my ideas provided I know definitely that they are not in contravention of the plans of my military superiors, to whom any action on my part would necessitate or appear to be one only in the chain of command and, therefore, fundamentally based upon loyalty to their policies."

These were not the words of a man who would surreptitiously aid in the writing of legislation in defiance of his superior's position. Loyalty to the War Department was Westover's most steadfast trait. There were those of his associates who would see it as a form of blindness and consider him a yes-man, solid, hardworking and unimaginative. But it all depended on which side of the line one stood on.

Lieutenant Colonel Walter "Trotsky" Weaver knew where he stood. He had been born to the military purple, so to speak, for his father had been Chief of the Coast Artillery, retiring as a major general. A turn-back at West Point, Weaver had graduated in 1908 instead of 1907. He had made the switch from the Infantry to aviation during the war, and in that period he swiftly established, in the face of towering bureaucracy, the Signal Corps's first aviation mechanics school. In 1919, as head of the Supply Group in the Office of the Director of Air Service, he had come to know Andy Andrews well. In Washington, Andrews had been Chief of Inspection in the same office. Later, they had served together under Foulois before Weaver went to the Army Industrial College and Andrews to the War College. Weaver had earned his wings in 1921 and was classified as a bomber pilot. Never all that avid a flyer, his principal specialties were training and logistics, but he had other talents of command as well. In 1922, as CO of Mitchel Field, he had converted it from a run-down mudland with deteriorating installations, into one of the best-operated air bases in the Army.

He had picked up his nickname, "Trotsky," from his golfing pals, Andy Andrews and Tony Frank. As tireless on the fairway as he was at the chessboard, Trotsky Weaver was a fierce competitor and strongly opinionated, especially when it came to the use of air power. In this regard, Frank saw him as a fiery revolutionary and so had dubbed him Trotsky. Conversely, Weaver viewed his friend as more of a compromiser and thus had returned the favor by naming him Kerensky, even though most of Frank's friends knew him as Tony. To add to the Russian confusion, they both referred to Andrews as Lenin!—possibly because they believed that one day he might lead a revolution.

Although through his family Weaver and his wife, Elizabeth, were able to circulate socially in the upper regions of the military hierarchy, his major contact at the seat of political power was his old West Point roommate Colonel Edwin "Pa" Watson. Watson had become military aide and a secretary to FDR. Trotsky and Pa had remained close friends, and in 1933, as Chief of Air Corps Information, Weaver was well informed on the White House attitude toward the Air Corps. Watson also saw to it that his friend had access to congressional sources that were friendly to the Corps. Weaver was both serious and conscientious, a hard taskmaster, who took things very much to heart. He was tall and angular, with a square face. One could read intensity in his features but little of the comic spirit. A firm friend of Billy Mitchell's and as strong a believer in strategic air power as Andrews, he had written a scathing attack on the Drum Board report, titled "Air Power Has Its Own Theater of Operations." The piece was for publication in the *United Air Services* magazine, but the editor, Earl Findley, was worried that Weaver would get himself in trouble if he signed it, so he asked Mitchell if he would. Mitchell was glad to accept and rewrote Weaver's conclusion "to give it more punch."

That Weaver would write such a broadside, quoting Douhet and others of the same philosophy, illustrated where he stood. If he was not one of the drafters of McSwain's bill, he was certainly in complete agreement with it.

Prior to the hearings, he had written to Andrews at Selfridge stressing the need to arouse public opinion to the Air Corps plight. "If we all get together," he wrote, "we might accomplish something which might be effective during this period just before Congress meets."

He had written similar letters to Hap Arnold, at March Field, California; Major Hugh Knerr, at Wright Field; and Lieutenant Colonel Horace Hickam, at Fort Crockett, Galveston, Texas. His contention to all of them was that the Air Corps was "in a rather crucial position. I don't know of anyone who is going to help it unless we do something for ourselves."

His suggested method was to seek publicity in all parts of the country. His fear was "that if we don't do this the Navy is going to run away with the show." One thing that he felt must be stressed and reiterated was the MacArthur-Pratt agreement of 1931. "The Navy seems to be trying to adopt an attitude that no such agreement ever existed."

It was only a few weeks after Weaver had written to his command associates that McSwain introduced his bills. It is not improbable that just before the move it was Weaver who suggested to the energetic Congressman that a visit to Selfridge Field and Frank Andrews might be of mutual benefit.

A classmate of Weaver's, Jim Chaney, from Chaney, Maryland, was the son of a doctor. He had traveled the Infantry route to the Air Service in 1917 via the Philippines, the Hawaiian Islands and a stint of teaching modern languages at West Point. This last was an indication that Chaney was an officer seeking broader horizons on the ground as well as in the air. During the war, serving in France on the staff of the Air Service Headquarters and then with the Army of Occupation, his capabilities of administration were quickly recognized and appreciated. In the five years following the Armistice, his duties took him from Coblenz, Germany, to Romorantin, France, to Great Britain, and to Rome, where he served as Assistant Military Attaché. Upon his return to the States in 1924, he went back to the academic life: the Air Corps Tactical School, followed by the Command and General Staff School, where he was an honor graduate in 1926.

As a result of his duty assignments abroad, attending schools and then being given command of flight training in Texas—first primary and then advanced—Jim Chaney remained clear of the bitter battles waged over air independence. Strong-featured, quiet-spoken, reserved, thought by some to be stuffy and by others to be the most polite officer on the scene, the General Staff saw him as the sort of deferential airman with whom it liked to deal. He was no Mitchell-type firebrand, and after graduating from the War College, in 1931, he served briefly in Foulois's office before being selected as Aviation Technical Adviser to Brigadier General George S. Simonds at the 1932 World Disarmament Conference in Geneva. When he returned, it was to act as Foulois's Executive and then Chief of Plans. He had known Benny since the war, had served under him in France, keeping contact during the period when both were military attachés. There is no doubt that Foulois depended on and trusted Jim Chaney. His seeming stolidness was a measure of his care. He was a thinking man. He did his homework, knew his facts, was there to answer questions of detail. At the age of forty-nine he did not appear to be the kind of officer—by either background or temperament—who would knowingly be in the forefront of a secret move to hoodwink the General Staff. The key word is *knowingly*, for on January 18, 1934, he submitted to Foulois a paper discussing an independent air organization as a department of air or a subdepartment of national defense.

The paper had been prepared as a result of conferences held by officers on Chaney's staff and for possible use by Foulois when he was called to testify before McSwain's committee. Three of these officers were Major Walter Frank and Captains George C. Kenney and Robert Olds. Frank was Chaney's assistant and Kenney and Olds were a pair of airmen who had long been engaged in the fight for independence. Both combat veterans of the war—Kenney in Observation and Olds in Bombardment —they were dedicated and loyal supporters of Billy Mitchell and believers in the theories of Douhet et al. Freedom from War Department control was their dream.

It was Kenney who had written the final draft of the paper "in a clear and excellent presentation of the various factors involved," Chaney reported to Foulois, and then added, "The attachment is in the memo in the safe."

It is possible that it was the attachment that formed the basis out of which the McSwain bill was ultimately drafted.

One who would have known the answer to that was Major Tooey Spaatz. Spaatz and Trotsky Weaver saw eye to eye on matters aerial but not on the game of golf. Tooey had given it up early. It was too slow for his taste, and the only time he really enjoyed it was when he played with Andy Andrews and they bet on everything from the color of the tee to the make of the lost ball. Squash and tennis were far more to his liking, and in his younger years he had particularly excelled at the former. Poker and bridge were his favorite indoor sports. He would play them endlessly and lose endlessly, sitting there with his scotch and cigarettes, sure he could draw two cards to an inside straight.

He epitomized what was bad about the promotion system under which the Air Corps was forced to operate. He had been a major for seventeen years, a pilot for nineteen and an officer for twenty. He was forty-three years of age. His rank was no measure of his capabilities but a prime reason why the Corps was constantly agitating to have its own promotion system.

The Army had a single promotion list. Rank depended exclusively on length of service. As a result, in 1932 there was no airman who had had enough service to rank as a full colonel. At that time, Foulois and Westover had been jumped from their permanent rank of lieutenant colonel to that of temporary general officers because their command positions required it. Added to this problem was the World War "hump." The hump was the product of sudden wartime expansion, after which a number of air officers who were not pilots decided to remain in the service. Because a

pilot went through nine months of training before he received his commission, and ground officers only three, pilots were generally the lowest on the
promotion list, often lower than those officers who had taken up flying
after the war.

It did not make for very good feelings within the Corps or toward the
War Department, which was not about to budge on the problem.

Although Spaatz did not quite fit the form, he was a victim of it. He was
not one to waste time in protesting, but one of the features of the McSwain
bill was a separate promotion list. In his usual manner, he did not have
much to say for attribution.

His lean, rather grim countenance, with close-cropped mustache and
knobby nose, softened in the eventide when he took guitar in hand and
began strumming away. The Spaatz home was always a warm oasis for
young bachelor lieutenants like Pete Quesada, made particularly so by the
hospitality of Ruth Spaatz. She had wanted to be an actress—was an actress in local productions—and her responsive nature was a nice foil for
her husband's taciturnity. They laughed at the same things. Their three
daughters, Tatty, Becky and Carla, filled the house with their own special
kind of music, while Tooey strummed and hummed and sipped his drink.

At the time, he was thinking that he would retire in six years. Possibly
due to his poker losses, he was frequently considering ways of "attaining
financial independence." He had made note of one way: "By saving $150
a month and investing at 6%, I should have about $50,000 in 1940, the
year of my retirement for 30 years service. At this time I have $300 in the
bank and owe about $2,000."

At the beginning of 1934 he had written to his good friend and former
West Point classmate, Major George E. Stratemeyer. "Strat" was an instructor at the Command and General Staff School at Fort Leavenworth
and thought Tooey should put in to join him.

"With reference to my taking the course at Leavenworth," responded
Spaatz, "I cannot see my way clear to do this. I fully expect to put in for
retirement March 1, 1940, and I see no reason to spend two of the next six
years at that place."

His antipathy toward attending "that place" was reflected in a letter he
had once received from Major Clarence Tinker, then taking the course.
Wrote Tink:

> The favorite expression of all instructors here is "the school believes,"
> and we are having an awfully hard time trying to find out what the school
> believes. . . . Sometimes I wonder just how much difference it would
> make to me where Lt. X puts his machine gun squad when I am cruising
> over his sector at 25,000 feet, and sometimes I'm really dumb enough to
> believe that Lt. X's tactical disposition of his platoon will not have a very
> great influence on my actions.
> When I finish here I think I shall apply for the Cooks and Bakers

School and the Horseshoe School, then I should be well qualified to command an air force unit. There hasn't been a single Air Service suicide yet. . . .

The camp followers seem to be able to keep themselves busy with bridge, mahjong, golf, tea, riding, tennis, gossip. . . . Madeline and myself spend our weekends in Kansas City or St. Joe, where there is less discussion of General A and more general entertainment.

In giving a brief but clear picture to Strat on the equipment situation, Spaatz in turn wrote: "There is no possibility of getting any modern equipment at Leavenworth. The equipment you have is about as modern as any except the new Martin Bomber of which we have 1 and the [Boeing] P-26 of which we have 10."

Again, Spaatz's scribbled notes reveal his thinking in the time frame of McSwain's actions. His thoughts, more than anything, illustrate that his outlook over the years had remained consistent, the implications clear:

Morale: Built up by belief that you are doing something useful or important in the scheme of things, plus confidence in the controlling agency. Knowledge that you are being treated squarely.

Lack of air doctrine—policy based on doctrine—plans for effective executive.

a. Are we a first line of defense? Are we charged with frontier defense?

b. If so why permit pursuit and tactical organizations to become ineffective for lack of equipment and personnel?

1. Movement of personnel to schools etc.

2. Use of units for CCC 1st Bomb Wing.

c. Pan American hunts for European trade—later for Asiatic trade. Air commerce requires air power, same as sea commerce sea power.

Growing realization of an Air Power theater of operations with necessity for development of air power. Vigorous leadership for tactical units cannot be obtained under present system of diversified control.

Competition with Navy for air force of USA in which competition Navy apparently gets everything and we get what is left—evidenced by their completed 5 year program and started with further development whereas we end up 5 year program 8 years later with fewer airplanes than when we started. Regardless of appropriations we will sink below strength at beginning of [new] 5 year program before we get caught up and 348 officers short. Lack of equipment is evidence of lack of interest or confidence in importance of Air Corps as an element of national defense. Failure to eliminate incompetent commanders. Inspection system believe more attention paid to administrative efficiency efforts than to combat efficiency of units.

Lack of sufficient personnel in Chief's office resulting in overwork and hasty consideration of important papers.

Belief that War Department has done nothing to develop air force or air power but on other hand has acted as though air power were being developed to fight only with Army instead of to fight air power of future enemy.

Feeling of unfairness, hopelessness of wartime 1st lts. and capts. smothered by hump due to ranking officers going to schools, etc. These same officers not attained Air Corps experience but due to hump have little hope for commands commensurate with experience.

Impression War Department has never assisted to retain flying pay and at times have even submitted data to Congress with a view of doing away with it.

Feeling that the Army believes every A. A. shell is a hit and every bomb not only a miss but a dud.

Giving up air bases in Philippines—French and Japs grabbing air bases in Far East off coast of Indo China.

In a character analysis of himself, Spaatz had once noted that one of his principal characteristics was the "tendency to false illusion that nothing is worthwhile except that done by myself."

Considering his appraisal of the War Department's treatment of air power in 1934 and matching it with his confessed tendency, it would not be difficult to surmise that Major Carl Spaatz had played a contributory role in the drafting of the McSwain bill.

His position on Army air power and its future was, of course, in keeping with most of those around him—with Andrews and allied officers throughout the Corps, and with his far-distant buddy Hap Arnold.

None of them were aware of the challenge that lay waiting to test them in the winter winds. The caliber of political thinking that had kept them beggared in equipment and personnel was about to make demands upon their limited resources that would eclipse congressional efforts to bring hoped-for autonomy. Ironically, in so doing, these actions would briefly force upon them a degree of the freedom they had been seeking.

Eleven

The Challenge

Like the bone-chilling weather that plagued most of the country during the winter of 1934, the 73rd Congress of the United States sought to make its mark. Taking office the previous year with the administration of Franklin D. Roosevelt, the legislative body became possessed of an investigation mania. Whereas the executive branch was busy establishing a host of governmental agencies aimed toward correcting the economic and social ills that beset the nation, the Congress seemed bent on proving that most of these ills were the result of a cabal created by the previous administration, of Herbert Hoover, and allied business interests.

In these investigations, the letting of contracts between the War Department and its industrial clients became fair game for the reform-minded, politically motivated congressmen. The year 1934 was, after all, an election year.

Two such investigations were of basic importance to the Air Corps. Timing was a key factor, for one inquiry was reaching a climax in a special Senate committee, while the other, in the House of Representatives, was just beginning. In the Senate, Alabaman Hugo L. Black presided, and in the House Congressman John McSwain was launching his Air Corps autonomy inquiry. Black's hearings into the letting of sea- and airmail contracts, which had been going on for a year, did not appear as a threat to the Air Corps, and certainly John McSwain's intent as Chairman of the House Military Affairs Committee was clear and friendly enough to Foulois and his staff. But very swiftly, in the first nine days of February 1934, all that would change.

Senator Black at forty-seven was a mild-appearing legislator with a deep southern drawl and a reputation as a populist. Elected to the Senate in 1927, a former member of the Ku Klux Klan, he was a champion of the disadvantaged and a strong supporter of labor, the author of a bill for a thirty-hour work week. Beneath the mildness, the pale blue eyes and his rather ascetic, unlined features lurked a sharp mind and an astute awareness of political opportunity. His committee had been established before Roosevelt's inauguration, its original intent being to focus on mail contracts that subsidized the merchant marine. Senator Key Pittman, of Nevada, however, had insisted that airmail subsidies be included.

Utilizing two years of unpublished evidence gathered by Hearst newsman Fulton Lewis, Jr., Black, after holding extensive closed hearings, came to the conclusion that President Hoover's Postmaster General, Walter Folger Brown, had used government subsidies to support large holding companies at the expense of small operators in the air industry. Twenty-four of the twenty-seven airmail contracts were controlled by three corporations.[1] To Black and the members of his committee, it looked like a matter of collusion, of rampant favoritism at the expense of the little guy. The exposure of Brown's methods in dealing with the air transport industry became of far more importance to the Alabaman than the reasons behind the former Postmaster General's actions and the end results. Brown, who had served as Assistant Secretary of Commerce when Herbert Hoover had been its Secretary, took over as Postmaster General in March 1929. He had a great and genuine interest in air service, and he found, due to inadequate legislation, that it was in a mess. As President Hoover described the situation, "We were threatened with a permanent muddle such as had resulted from our chaotic railway development with all its separation into short and long lines, duplication and waste."

Brown sought to change all that, and he did so by getting Congress to pass the McNary-Watres Act late in 1929. That Act, among other things, gave him the power he sought, and he went about using airmail subsidies as a club to force large companies into expanding their routes and passenger service while driving the less affluent short-line operators out of business. To develop an integrated system of airways, he eliminated competitive bidding, which he considered the *bête noir* of the industry, and put in its place the negotiated contract. In simplest terms, it meant he would decide among a handful of companies who would get what. At one such meeting, an uninvited independent operator was literally thrown out of the office. Brown's ruthless, overbearing manner brought him many enemies, all anxious to testify before Senator Black's special committee.

The result of Brown's endeavors was something else. By the time he left office, in March 1933, the nation had a national airline system of which it could be proud. Thoroughly instrumented transport aircraft were servicing thirty-four mail routes, covering over twenty-seven thousand miles of airways, carrying passengers and freight. The cost of airmail service had doubled under Brown's administration, but the miles flown had trebled and the cost per mile flown had been cut in half—from over a dollar to forty-two cents. Few government agencies could point to such an enviable record of progress, and Walter Brown went out of office feeling he had accomplished something positive.

Black and his colleagues hardly saw Brown's stewardship in that light. To their way of thinking, the negotiated contract was highly undemocratic regardless of its merits; the McNary-Watres Act must be repealed; and even though the new Postmaster General, James E. Farley, was having to maintain the airways on a severely cut budget, independent operators who were clamoring at his door should be permitted to bid for a piece of the airmail subsidy.

On January 9, 1934, Black's committee held its first public session. Witnesses described Brown's way of doing business in detail. According to testimony, in May 1930 Brown had held a "spoils conference" in his office to divvy up the nation's air routes among a chosen few. Five thousand miles of additional airways were added to the existing routes by a method of subcontracting to lines only carrying passengers. The subcontractors, however, were connected to the parent corporations. One after the other, independent operators told how they had been forced to sell out to the big boys, often through Post Office pressure. The testimony became daily headline news.

As a result of the revelations, Black, after meeting with officials of the Justice and Post Office departments, announced that the two departments would launch a joint investigation to determine whether there had been criminal violations during Brown's administration. Roosevelt's Attorney General, Homer Cummings, announced the investigation on January 20, and Farley assigned postal inspectors to pursue the inquiry under the supervision of Colonel Carl L. Ristine, a Special Assistant to the Attorney General.

On Friday, January 26, Hugo Black journeyed down to the White House to have lunch with FDR. They talked over the work of his committee, and he told the President "the whole system of airmail contracts emanating from the 1930 split was fraudulent and completely illegal."

In June of the previous year the Congress had given the President the authority to cancel, after hearings on sixty days' notice, any government contracts dealing with the transportation of persons or things that he felt were not in the public interest. Senator Black was one of the guiding forces behind granting this authority, feeling the provision's purpose might best be served when existing airmail contracts were reviewed. Now, over lunch, he reminded the President that the White House had the power to act.

That afternoon at his weekly press conference, Roosevelt indicated he was considering the move, but he was careful not to commit himself.

Neither was Postmaster General Farley. A very amiable and canny politician, Gentleman Jim knew all the tricks of the game, and he wasn't about to be suckered by political haymaking. Appearing before the Black Committee on January 30, he stated that he wasn't in a position to approve or disapprove the existing system. He hoped to be able to make a recommendation soon, but his investigators were still at work, and in the meantime he didn't wish to be unfair to any contractor. "We have no right to assume the contracts are not right until an investigation proves otherwise," he said.

Within the next week, the pace of political fervor outran the course of careful investigation. Not only did the headlines spur a desire for bigger headlines, but FDR's penchant for decisive and bold action also rose to the fore. Post Office Department Solicitor Karl Crowley, a vigorous New Dealer from Texas, had been appointed by Farley to head the Department's inquest. By February 5, he had compiled two thick volumes of raw evidence that he felt presented a clear case of conspiracy between Walter Brown and the airmail companies. However, he did not feel the investigation was complete, and he said so.

Farley had not read Crowley's draft report but was aware of its general thrust. Even so, he was not prepared to move until the investigation was completed and the evidence in final form. It was Roosevelt who moved instead. He sensed he could capitalize on public outrage over the apparent monopolistic control of the airways, could force legislative reform that would break the hold of the wealthy few and open up the industry to the excluded, making the airmail subsidy equitable and honest. The public had been with him since the beginning, and he saw a way to foster his popularity while sustaining his political image as a crusading innovator. He called in Farley and Crowley and asked for their recommendations. The latter was for cancellation of the airmail contracts, but not until June 1. Neither he nor his boss was anxious to flip service on the nation's airways into a tailspin.

When they called on the President again, on Friday morning, February 9, they presented their detailed plan, which was "to allow the domestic lines to continue to carry the mail until new contracts could be negotiated." The Postmaster General would then issue the cancellation order, not the President.

The only part of the recommendation FDR bought was this last. He wanted the contracts canceled right now. Farley had checked with the Attorney General and had been advised that the grounds for executive action were sufficient. The only question remaining was, Who was going to carry the mail in the emergency until new contracts could be negotiated?

It was around eleven o'clock that same morning that Benny Foulois received a telephone call from Harllee Branch, the Second Assistant Postmaster General. Branch wanted to know if Foulois could stop by after lunch and go over some aviation matters they had previously been discussing with Eugene Vidal, Director of Aeronautics for the Department of Commerce. The three of them had been designated as an Interdepartmental Board by the Secretary of Commerce, Daniel C. Roper, to iron out airway policies. Assuming this was what Branch had on his mind, Foulois said he'd be over.

At the moment, Foulois had a lot more on his mind than a discussion concerning joint airway policies. McSwain's hearings, which had begun so promisingly a week before, were taking on a broader range. Their expanded scope was tied in with the general atmosphere of reform, which had focused on a grand jury inquiry into charges of collusion in the sale of surplus government property.

Assistant Secretary of War Henry H. Woodring, whose major function was the procurement and sale of military equipment, had been before the investigating body in January, and the Washington rumor mill had it that he was not only in trouble but also on the outs with Secretary of War Dern and General MacArthur. Simultaneously with McSwain's hearings, the House Naval Affairs Committee had revealed that aircraft manufacturers had reaped profits as high as 50 percent on the sale of planes to the Navy, and before Senator Black's committee, William E. Boeing, Chairman of United Aircraft and Transportation—one of the big three airway operators —admitted his company had made a killing in selling aircraft engines to both services.

With this background, McSwain, whose committee was beginning its own probe into War Department procurement, decided to include the purchase of Army aircraft. To handle the probe, an eight-man subcommittee was formed headed by Congressman William N. Rogers, Democrat of New Hampshire.

On this same Friday morning, Rogers' subcommittee began its hearings in executive session. Foulois knew that not only would a number of his officers be testifying, such as Conger Pratt, head of the Materiel Division, and his Chief Engineer, Jan Howard (as well as Spaatz, Weaver, Chaney, Frank and Kenney), but also Assistant Secretary of War Woodring.

Like Secretary of War George Dern, Henry Woodring had not sought the appointment the new President had offered him. A former Governor of Kansas and one of the first public figures openly to come out for FDR, he had hoped to be Roosevelt's Secretary of Agriculture. When he learned otherwise, if it had not been for Jim Farley and affable George Dern, the forty-six-year-old Kansan who had also been a successful banker and political power in his home state might well have elected to remain there. But they talked to him on the basis of the appointment being a stepping-stone, and certainly the excitement of the Washington scene appealed to him. Short and balding, with puckish features, he was an affluent Midwesterner who had attained prominence through a combination of wits and a smooth, outgoing personality. His greatest drawback in taking over as Assistant Secretary of War was that, also like Dern, he knew nothing about the job and was not all that interested in it.

Late in 1933, a major difference had arisen between Woodring and Foulois that dealt with the procurement of aircraft. As with Hugo Black's Senate committee, the difference was over the negotiated contract as opposed to open bidding.

Since 1926, the Air Corps had pursued the former course in its quest for quality. It entailed an involved procedure of selection, beginning with recommendations to the Chief of Air Corps for various types of aircraft from the applicable Air Corps users. On these recommendations the Chief would issue a directive to the Materiel Division. The Division, in turn, would make its own studies and eventually send to manufacturers it deemed competent a description of the aircraft desired. The list of manufacturers could be counted on one hand, although other companies could receive the specifications by asking for them. The selected aircraft companies would then submit their plans, and from them, with the Chief's approval, a choice would be made and a contract granted for an experimental model. Sometimes two companies were given the go-ahead. If the model lived up to expectations, a production contract was negotiated. In its purchasing of aircraft, the Navy used the same procedure. At the time, the reasoning for it appeared sound. There were only a few aircraft manufacturers capable of meeting the military standards required. If purchase became a matter of open bidding, with the award being granted on the basis of low bid, quality and performance would suffer. This also applied to aircraft engines, of which there were only two manufacturers worthy of the name: Pratt & Whitney, and the Wright Aeronautical Corporation.

Henry Woodring did not understand this reasoning, because he did not understand the mechanical difference between an aircraft and a truck. Further, as a fiscal conservative, he was influenced by the need for economy and the complaints of Burdette Wright, the Washington representative of the Curtiss-Wright Aircraft Company. Wright, who had been one of Billy Mitchell's confidants, had left the Air Corps in 1931 to go with the aircraft firm.[2] He called on Woodring and complained that favoritism on the part of Foulois had excluded Curtiss from selling any aircraft to the Air Corps for nearly three years, and also that some of the performance requirements of the planes to be bought with PWA funds were unreasonable.

What concerned Woodring most about these accusations was that they revealed how much power the Chiefs of the Air Corps and the Materiel Division had in the decision of selection.

Near the year's end he informed Foulois that from now on the Air Corps was to purchase its planes by the method of open bidding. Following service tests, the lowest bidder meeting the standards would be awarded the contract. To Woodring, that was the fair and square way to proceed, and it might be possible to build two planes for the price of one.

Foulois and his staff protested as strongly as they could, and in this they were backed by the War Department. They argued that to get more than one bid on a particular plane, specifications would have to be lowered. Woodring couldn't see it, and in December he announced that open bidding would be used in the purchase of the aircraft to be bought with the $7.5 million of PWA funds. Later, in testifying before the Rogers subcommittee, he would strongly deny that he had done anything to "lower or change aircraft specifications." Nevertheless, directly after Woodring's decision, the engineering division at Wright Field reduced minimum performance requirements on pursuit and attack aircraft to be purchased with the PWA money. Speed at sea level for a new pursuit plane was to be reduced from 235 mph to 176 mph, cruising range from 500 miles to 400 miles, and a service ceiling of 27,800 feet scaled down to 18,700 feet.

It was a bitter pill, and subsequently a great deal of backing and filling was done by the Assistant Secretary on the issue. Foulois also had to retract his statement that the actual change in specifications took place in Woodring's office. But as a result of the switch no new pursuit planes were bought with the PWA money, all of it going for other types of aircraft, principally attack and bombardment.

When Benny Foulois left for his meeting with postal official Harllee Branch, he must have been hopeful that the hearings before Congressman Rogers' subcommittee would put a bridle on Woodring. He would later learn that two of the subcommittee's members, Congressmen W. Frank James, of Michigan, and Paul J. Kvale, of Minnesota, had indeed attacked the Assistant Secretary on the issue, and that Kvale had told Woodring that without the negotiated contract the Air Corps would end up with inferior equipment.

The irony was to become acute, for Foulois's appointment with Branch was the end result of Senator Hugo Black's investigation into a system of negotiated contracts devised by former Postmaster General Walter F. Brown. At exactly the same time, Congressman Rogers' probe was aimed at exploring a similar method of contracting in the purchase of Army aircraft, a method Foulois had adhered to as an accepted matter of course. Shortly, he would be bound within the politics of the two, savaged by the results of congressional and administration failures past and present.

It was a long walk on a cold day from the Munitions Building, on Constitution Avenue, to the Post Office headquarters, at eleventh and Pennsylvania, but Foulois liked the exercise. There was not too much snow on the sidewalks, and he went along briskly, smoking his pipe, knowing by the look of the sky and the feel of the air that more snow was on the way. Dressed in civilian clothes, bundled in a warm overcoat, he would have been taken for just another Washingtonian, hurrying through the cold at noontime.

Harllee Branch, the official he was going to see, had held the position of Second Assistant Postmaster General for just three weeks. A fifty-five-year-old Georgia newsman, Branch was out to make his mark in the ranks of the New Deal, having started out as Jim Farley's Executive Assistant in March 1933. When Foulois was shown into his office, the round-featured, rather solemn-looking postal official circled his way to the point. He began by bringing up a past incident. Two years ago, he pointed out, there had been a threatened strike by the major airlines. At the time, as the General probably recalled, the Post Office Department had drawn up contingency plans in which the Air Corps might be asked to maintain the routes, on a restricted basis, of course. Branch said he'd like to go over that routing now, the scheduling and so forth. Although his manner appeared somewhat offhand, Foulois immediately sensed the direction the conversation was headed in. Through the press and his own contacts, he was fully aware of the airmail-contract battle. "You're speaking in reference to the present situation?" he asked.

"Yes." And now Branch got to it. "If the President should cancel the contracts, do you think the Air Corps could carry the mail to keep the system operating?"

It had been a long time since the Army had carried the mail. Earle Ovington, a civilian, had been airmail pilot number one back in September 1911. He had flown his Queen monoplane five and a half miles from Nassau Boulevard to Hempstead, Long Island, with a sack of mail. Hap Arnold and Captain Paul Beck had been the first Army pilots to double as aerial postmen, during that week-long aviation meet. Seven years later, in May 1918, Army pilots had inaugurated regular government airmail service between New York City and Washington, D.C. Their service was brief, for in August, the Post Office Department took over the job with its own pilots. Again, in 1922–23, Army pilots flying out of Dayton, Ohio, carried mail to New York, Washington, Langley Field and return, and then, in 1927, Arnold had set up his special airmail delivery to vacationing President Calvin Coolidge. But that had been the extent of it.

Now, however, it was not past performance that gripped Benny Foulois. Branch's question had suddenly set before him a God-given opportunity, or so he saw it. If, indeed, the contracts were canceled and the Air Corps took over flying the mail—what better way to show the public what his pilots could do? What better way to test men and materiel in an emergency situation? What better way to get the funds to buy the equipment that had been needed for so long? And what better way to give added impetus to the McSwain hearings on autonomy and expansion?

"Yes, sir." He nodded, poker-faced. "If you want us to carry the mail, we'll do it."

Branch dropped his casual manner. "We can look over the routing. I'll get Steve Cisler to come in, and I'd better call Vidal."

"I'd like to make a call myself. I'll want a couple of my officers to join us." Foulois did not show his excitement, and later he would be roundly attacked for having been so quick to voice his acceptance, but he could not know that in a sense he was being mousetrapped, that events were already in motion that precluded any hesitation on his part, a fact his critics chose to ignore.

At a White House cabinet meeting earlier that morning, the President had acquainted Secretary of War Dern for the first time with the airmail situation. He asked him if he thought the Air Corps could take over the job, and Dern, not having spoken to Foulois or General MacArthur or anyone else on the matter, automatically answered in the affirmative. An emergency was an emergency, and any branch of the military called on must be prepared to act. Following the cabinet meeting, Jim Farley had called his own office and deputized Harllee Branch to get the official word of acceptance from General Foulois. After the call, Roosevelt asked Farley to accompany him to his regular press conference.

It was immediately apparent to the newsmen that FDR was eager to tell all. "Well, I suppose somebody is going to ask me about airmail contracts?" he led off.

"You took the words right out of our mouths," came the response.

Roosevelt then announced that Postmaster General Farley was about to issue an order canceling all airmail contracts. The grounds for doing so, he said, were "sufficient evidence of collusion and fraud." FDR would issue his own Executive Statement directing Farley, Secretary of War Dern and Secretary of Commerce Roper to cooperate in continuing airmail service. Further, the Secretary of War was going to place at the Post Office Department's disposal aircraft, landing fields and pilots to carry the mail during the emergency, "by such air routes and on such schedules as the Postmaster General may prescribe."

Thus, while Benny Foulois was being sounded out by Harllee Branch, the Air Corps had already been committed. It wouldn't have mattered to Foulois, except for the timing. He would have said yes in any case, for beyond all the reasons he could grasp for quickly accepting the challenge, it was at the root of his nature to say yes in the face of almost any odds. He did not pause to consider that, by and large, neither were his men properly trained nor were his aircraft properly equipped to take on the exacting and specialized flying that went with maintaining unfamiliar air routes across the country in all kinds of weather.

Just the day before, in response to testimony given before McSwain's committee, he had sent a note to Tooey Spaatz asking how many planes they had *"fully* equipped to take the field."

At the moment, he answered Branch, he didn't know the answer. But there was no better case history of his willingness to accept a mission that would endanger more than his own life than that of March 19, 1916. Then, with his 1st Aero Squadron, described by Lieutenant Edgar S. Gorrell as being "eight planes in horrible shape, not fit for military service,"

Army Captain Foulois was ordered by General Pershing to fly ninety miles from Columbus, New Mexico, to join Army forces at Casas Grandes, Mexico. Foulois knew that to carry out Pershing's order at once meant a night flight over totally unfamiliar terrain without maps, without navigation or instrument lights and without reliable compasses. Worse than that, only one of the eight pilots had ever flown at night, and Foulois was not that one. The flight could just as well have been made the next morning; Pershing had no use for the planes until then. But Foulois, following a not-to-reason-why attitude, ordered his pilots to take off and follow the leader. As Gorrell said, "if one got lost, *all* got lost."

The result was that, except for Mike Kilner, whose engine quit right after takeoff, they nearly all got lost, some lots worse than others, and none of the planes reached Casas Grandes that night. Only luck or divine intervention kept the attempt from ending in a debacle. Although Gorrell was missing for two days in unfriendly territory and nearly died of thirst, all eight pilots survived, at the cost of two cracked-up planes. Privately, Gorrell, and no doubt others, thought Foulois had been a god-damned fool for taking Pershing's order so literally.

It was shortly after that episode that the plight of the 1st Aero Squadron made headlines in the New York *World*. Foulois was quoted as saying he and his pilots were "risking their lives ten times a day and not being given the equipment needed." Bert Dargue was also quoted as declaring it was "criminal to send men up under such conditions."

On reading this account, Secretary of War Newton D. Baker was enraged. The Inspector General was sent down to Mexico to root out the culprits, but Foulois and his rugged gang pleaded wide-eyed innocence, shocked that anyone could quote them in such a fashion. Nevertheless, the truth of the matter forced Congress to provide a half-million-dollar military air appropriation, part of which was allocated to buy twenty-four new planes.[3]

Now, nearly twenty years later, here was a situation on a much larger and more complicated scale that bore certain similarities to that wild time. This was particularly true in Foulois's mind as to the need for new equipment. And what better way to get it than through the pressure of headlines?

"How long do you think it will take you to get ready?" Branch asked. It was to be the most important question of all, but instead of pinning the reply down to specifics, Foulois's answer was as casual as the postal official's original approach to the matter. "I think we could be ready in about a week or ten days," he said.

Later, he was to underline that he had not thought Branch meant *from that moment on*. But it was his job to know exactly what Branch meant. The failure to be specific was to lie at the root of all that followed. Perhaps his mind was too full at the moment of the potential that lay ahead, but whatever was in his mind, there was little excuse for his unclarified reply.

Foulois reached his office by phone and asked to speak to his Executive Officer, Colonel Chaney. Chaney, who was in the Plans Division office, came on the line and was informed he was wanted over at the Post Office Building right away, and to bring Major Spaatz.

Chaney checked and was told Spaatz was out to lunch. "Bring Captain House, then," said Foulois.*

The pair grabbed a cab, and when they arrived at the designated office a secretary told them that the General was upstairs and would join them shortly. He did, bringing with him ebullient, fast-talking Eugene Vidal as well as Stephen Cisler, General Superintendent of the Department.

"There's a possibility the airmail contracts may be annulled," Foulois told them, "and we will be asked to carry the mail. I want us to get all the available data they have here on routes, schedules, pounds of mail carried and that sort of thing. We'll start some preliminary work on what airplanes and personnel we have available. Mr. Cisler and Mr. Vidal are here to assist. One other thing, the utmost secrecy will be observed. I want nothing said or done that will give any indication that such action is being considered."

For the next two hours they went over the routes, the priority schedules, the number of flights per day, etc. Once they had the picture from Cisler, they planned to return to the Munitions Building and go to work on a tentative program of operations. They had just completed the briefing when Harllee Branch entered the office. "General," he said, "we are going to the White House sometime this afternoon. It would probably be a good thing for the Army to carry the mail all the time, wouldn't it?"

"No." The General knew better than to step into that one. By asking such a question, Branch had revealed his own line of thought. "We are willing to carry it only if there's an emergency," Foulois made clear.

Branch changed the subject, declaring that nothing must be done or said to alert the press. Foulois agreed and, with Chaney and House, carrying the maps, headed for the door. At it he turned, obviously put on his guard by Branch's White House announcement. "Before anything can be done," he said, "I have to see the Chief of Staff or the Secretary of War and report what is going on."

On that note of caution, they returned to the Munitions Building. It was then about half past two. Spaatz had returned from lunch, and Foulois quickly briefed him. An office was selected in the Training and Operations Division, and with several of Tooey's assistants in tow, preliminary work was begun. In general terms, Foulois went over the questions of equipment, loads, aircraft speeds. The planners were to concentrate first on the

* Captain Edwin L. House was Adjutant in the Office of the Chief of Air Corps.

main transcontinental route, from New York to San Francisco via Chicago, but the accent was on *preliminary plan,* and mum was the word. The door of the working office was kept closed. Nothing must leak out.

Foulois kept his overcoat on. He was in and out of the working office, giving additional instructions, in Tooey Spaatz's office discussing the situation, in his own office, calling the War Department for an appointment with Hugh Drum. His restlessness was obvious, his sense of excitement measured by the density of his pipe smoke. It was about three-thirty when he left to see Drum. When he arrived he found that the Deputy Chief of Staff was in conference. As he sat down to wait, he was suddenly approached by a group of newsmen. General, we hear the President has canceled the airmail contracts and the Air Corps is taking over, was the gist of their collective queries. The unexpectedness of it must have nearly put his teeth through his pipe stem. It was not only the meaning of the announcement that shocked him but also the swiftness with which the secrecy of the thing had been blown. He had no comment, figuring the substance of the inquiry was in the nature of a fishing expedition. Fortunately, before he could be pressed further, Drum was ready to see him.

He quickly filled in the Deputy Chief and from what he had to say, Drum assumed that the Air Corps was going to be called on to carry the mails. The only question was when, and while they were discussing it, General MacArthur strode in with the answer. He had just been approached in the hall by Associated Press correspondent Arthur Leverts with a copy of the Presidential Executive Order in hand. "Foulois!" MacArthur seemed to tower about eight feet tall. "A newsman just told me that the President has released an Executive Order giving the Air Corps the job of flying the mail. What do you know about it?"

In his stunned explanation, Foulois had the feeling he was not getting through to his Chief or convincing him that the idea wasn't something the Air Corps had cooked up and somehow planted in the President's mind.

MacArthur interrupted him. "That's all academic now," he said quietly. "The question is, can you do it?"

Foulois replied that he could, provided he got the necessary support.

"You're on your own now," was the response. "Yell when you need help from me and keep me informed. It's your ball game."

What a ball game! Foulois went back to his office in a daze to announce the glad tidings. They had just ten days to get ready, for Postmaster General Farley's shattering order to the airlines stated that all domestic airmail contracts were to be annulled as of midnight, February 19, 1934. In the Office of the Chief of Air Corps, plans for the weekend could be forgotten. An enormous chore of preparation lay ahead.

From the point of view of planes, there were enough P-12 pursuit and open-cockpit observation and attack aircraft to be used, but none of them had been designed to carry cargo.

Instrumentation was something else again. Suddenly, the efforts Andrews had been making without success to encourage the proper instrumenting of his Group and of the entire Air Corps had come home with a vengeance. Of the 274 directional gyros and 460 artificial horizons that the Corps possessed, most were on the shelf at the Materiel Division's depots, being held out for use in hoped-for new aircraft. The Corps simply hadn't had the money to install them in their existing planes, particularly if they were to be used in later models.

Of equal magnitude was the problem of radio equipment. There were not enough radio transmitter-receivers to equip the planes to be used. Worse, the sets on hand were not properly constructed so that they could be easily tuned to the necessary airway frequencies, and they were good only for reception within a thirty-mile radius. As Hap Arnold cracked, "They were great radios, good for straight-down reception only." At the moment of truth, it didn't help to know that the radio equipment of commercial carriers was capable of reception at nearly four hundred miles, whereas the Signal Corps product was both inferior and in short supply.

In command of the Field Services Section at Wright, which encompassed the selection and maintenance of all aircraft to be used in the operation, was Major Hugh Knerr. Unlike most of the officers on the air staff in Washington, whom the New York *Times* quoted as being "jubilant" over the announcement of the Corps's chance "to do a man's job," Knerr was sick at heart. He would write: "This arbitrary assignment to an impracticable task was inexcusable. We had neither the personnel nor the modern aircraft for an operation the Air Mail had taken years to develop. No one paid any heed to our warning of disaster."

No one would ever accuse Knerr of being an optimist. He saw things in absolutes and usually from the dark side. But Tooey Spaatz, as Chief of Training and Operations, knew exactly what the lean, intense Midwesterner was talking about. Aside from half-instrumented planes (the pursuits had nothing but a turn-and-bank indicator) and old Keystone bombers that might cruise at 90 mph with a tail wind, more than half the pilots who would be called on to do the flying were reserve officers with less than two years of service, a good many with less than one.[4] The reason was a combination of no money and the resultant War Department policy of turning over the reserve officers in a squadron at the rate of 25 to 30 percent a year. Most of these officers were second lieutenants. Of the 281 pilots, regular and reserve who would man the mail planes, only thirty-one of them had more than fifty hours of night flying, while the rest had fewer than twenty-five hours of instrument flying to their credit. Spaatz knew the situation as well as anyone could, but he, like everyone else in the Air Corps, was committed and determined to make the best of the challenge.

Aside from the obvious, there was an underlying reason for the general enthusiasm. A page-one story in the New York *Times* the next day reflected it:

> The cancellation of airmail contracts will probably have a far-reaching effect upon the development of commercial and military air service. The proponents of a separate Air Force for defense purposes see for the first time an opportunity to obtain their objective.

In that first wild weekend of preparation, a staff and an operational system were organized. Assistant Chief Brigadier General Oscar Westover was automatically selected to head AACMO—Army Air Corps Mail Operation. He had not been in the office on Friday, so he was Johnny-come-lately to the news.[5] However, as Foulois could not directly command both the Air Corps and AACMO, it was natural that Westover would be in charge of the latter, although the overall responsibility of how the mission was carried out would rest on Foulois's shoulders.

The supporting staff was made up essentially of those holding key positions in the Chief's office: Chaney became the Executive, House the Adjutant and Majors Spaatz, Arnold Krogstad and Asa Duncan doubled as heads of Operations, Personnel and Information, respectively. Hugh Knerr was made G-4, the all-important job of supply, and when Westover telephoned him in Dayton to give him the news, Knerr begged for more time to get the planes ready. Westover was sorry, but ten days was all the time they had.

Foulois, again in conference with Branch and Vidal, established the routes the Air Corps would service. Fourteen of the existing twenty-six were chosen, priority focused on a dozen major commercial banking centers. There would be seventy stops over a course of more than thirteen thousand miles. The system was to be divided into three zones: eastern, central and western. Major B. Q. Jones would command in the East, Lieutenant Colonel Horace Hickam in the Midwest, and Lieutenant Colonel H. H. Arnold in the West.

Byron Quinby Jones, one of the really early birds, who had gotten his wings at Rockwell Field in 1914 and been a pioneer mail pilot in 1918, was now CO of the 8th Pursuit Group, at Langley Field. There was a lot of rugged background to B.Q. He had won the Mackay Trophy in 1915 for breaking the world endurance record, and although the feat deserved acclaim, for another piece of pioneering he should have won a special trophy for pure guts.

At that time, one deadly cause of crashes was the tailspin. Any pilot who fell into a spin became a statistic. B.Q. figured there had to be an answer. In charge of Aircraft Repair, he understood something about aerodynamic forces. He was said to have been the first to loop a plane, and he knew Orville Wright's axiom about the stall. So he studied the problem and decided on a method to solve it. He wore no parachute (for there were none) when he climbed into his Jenny and took it up as high as the plane could climb. Then he purposely stalled it and kicked it into a spin. He had a few thousand feet at an increasing rate of speed to find the answer or become another ugly statistic. The natural reaction was to keep the stick back in order to bring up the nose. He shoved it forward and then applied opposite rudder. The spinning stopped. He was in a dive. He cut down on the power and eased the stick back. The plane responded, and he leveled off at a safe altitude. He climbed some more sky and tried again, and then again. And when he landed he knew he had the answer, but no one will ever know how many lives B. Q. Jones's daring saved.

Now he was swift to brag that his boys were the sharpest pilots in the Corps. In the realm of combat flying, maybe so, although both Andrews and Arnold would have given him an argument on that, but combat flying was as far removed from airmail flying as ducks from elephants, and the ability of such pilots to do the job was questionable.

An indication of how questionable had been pointed up in September, when an element of seven Air Corps pursuit planes trying to land at Mitchel Field were caught in bad weather. The end result was that three crews bailed out, two landed far from the field and only the remaining two made it safely into Mitchel.

There was, of course, no question in the mind of B. Q. Jones. He had taken command of the 8th the previous year, and his arrival at Langley was a well remembered event. The Group's officers and senior noncoms had gathered in one of the hangars to be on hand to greet him. Right on the dot of his ETA, they heard the sound of a P-12 winding up as though it was coming straight in. The assemblage hurried out of the hangar onto the apron in time to see the pursuit level off at treetop level, execute a precision roll and then, grabbing some altitude, go through a series of finely honed acrobatic maneuvers—all against base regulations. The exhibition completed, the pilot made his landing approach and his audience returned to the hangar.

A few minutes later, in strode B.Q., still wearing his leather flying jacket —the hotshot pilot right out of a Hollywood movie. He was escorted to a makeshift rostrum, where he looked over the gathering and was looked over in return.

"Be seated, men," he said, and then, after a sufficient pause, he continued, "my name is B. Q. Jones, and if you wonder what B.Q. means, it means 'be quick.' "

It brought a laugh and, along with his flying performance, it pretty well summed up the Major's manner. He was quick, sometimes too quick, his seniors felt. To his subordinates, as Ira Eaker well knew from Philippines days, he could be mightily brusque. As commander of the Group, his leadership was focused on ways to develop new pursuit tactics, and his approach was, "Let's try it and see if it works." On one issue, he was determinedly dogmatic. Like Oscar Westover, he believed the Air Corps's sole command function was to serve the Army. He did not believe in a separate air force and could not be budged from his position. As he frequently put it, "The day the Air Corps leaves the Army, that is the day I'll leave the Air Corps."[6]

Consequently, a good many separation advocates did not look with great favor on B.Q., combining their dislike for his position with his tough-guy image. Not so Benny Foulois. They were old poker-playing, bourbon-drinking buddies, and B.Q. spoke Benny's kind of language even if B.Q. had a tendency to color the events a bit. On that fateful Friday, he would claim he was up from Langley visiting the Chief, that he was with him in his office when the President called and asked the General if the Army Air Corps could fly the mail. Supposedly, Foulois gulped and referred the question to B.Q., who promptly responded, "Hell, yes, the 8th Pursuit Group can carry the mail!" Whereupon Foulois assured FDR the Air Corps could do the job.

It hadn't happened like that, although B.Q. was visiting on Saturday and undoubtedly he was quick to assure his Chief that the 8th was ready and able. It could be said that the Major didn't mind making a good story better, particularly if he could put himself in the key role. But as Eastern Zone Commander, his first act of preparation was to contact the few officers who were on duty at Langley that weekend and order them to call in everyone to get ready to fly to the eight cities in his zone of command. "Everyone" at the moment consisted of twenty-three regular and twenty-three reserve officers.

"Uncle Horace" Hickam's first knowledge of what was afoot came in a warning order from Foulois by radiotelegraph to Fort Crockett, in Galveston, Texas, where he was CO of the 3rd Attack Group. He was informed to stand by for further instructions on his appointment as Central Zone Commander. His headquarters would be the National Guard hangar at Municipal Airport in Chicago. He was soon to be quoted in the press as saying, "We're not mailmen, but we've got a job and we're going to do it."[7]

When Foulois sent a similar warning order to Hap Arnold that Saturday, Hap was not at March Field. He was up in the mountains with Bee and a half dozen officers and their wives, eating flapjacks and drinking coffee over a campfire, enjoying a real old Army breakfast. It was not an infrequent custom, a dozen or so officers and their ladies rising with the dawn and riding up into the nearby San Bernardino Mountains on horseback. There they were met by an Army chuck wagon to be offered all the comforts of the trail. On that Saturday, however, they were joined by another rider, this one coming with the news that the camping party was over. It was not until the next day, however, that a radiogram was received, giving details of the responsibilities of the Zone Commander. So it was that Captain Ira Eaker, perfecting his golf game at the Victoria Country Club that Sunday morning, was notified by one of the attendants, "Colonel Arnold wants to see you right away."

He hopped in his Ford coupe, and when he arrived at headquarters found a meeting in progress, the plan for the Western Zone already outlined. The territory to be served included almost the entire West Coast, interior desert and Rocky Mountain regions, with landing fields varying from sea level up to seven thousand feet. Some routes would require flying at altitudes of fifteen thousand feet. Zone headquarters was to be established at Salt Lake City, Utah.

"Ira, your squadron will run the route from Los Angeles to Salt Lake via Las Vegas, and from Los Angeles to San Diego," Arnold announced. "Tink, you're going to have the route from San Francisco to Salt Lake, and then to Cheyenne.[8] Castor, Cheyenne to Pueblo,[9] and Charlie, you'll be running things out of Seattle to Salt Lake."[10]

At March, at Fort Crockett, at Langley, they were gripped by mixed emotions, a sense of excitement tempered by a realization of what they were being called on to do, and with what.

The what of it, with regard to equipment, faced them all, but particularly Hugh Knerr. In the entire Air Corps there were only four depots where major servicing was undertaken: Middletown, Pennsylvania; Fairfield, at Dayton, Ohio; Duncan, at San Antonio, Texas; and Rockwell, at North Island, San Diego, California.

When he had become Chief of Field Service, in June 1931, following graduation from the Army War College, Knerr's plan of action was to establish better means of contact with the principal operational bases by better means of transportation. Rail delivery was slow and, to the Air Corps, costly. Knerr's previous duty as CO of the 2nd Bomb Group had convinced him that there was a special need for air transport. His proposal to organize such a group was approved even though there were no cargo aircraft available. By slowly commandeering a flock of worn-out bombers, a couple of trimotors, C-27 Bellancas and anything else that looked as if it might carry a load, he formed the 1st Transportation Group. It consisted of four squadrons of refurbished and modified aircraft purely for cargo operations—a meager forerunner of the Air Transport Command.

Now, with the word from Oscar Westover, they went into round-the-clock service, flying the desperately needed equipment to the units selected by Foulois and his staff, who were to carry the mail. The pace became furious. At Langley Field, a crew of twenty mechanics installed radios in fifty-two aircraft in forty-eight hours. Yet none of the one hundred twenty mechanics at Langley Field was an experienced communications or instrument technician. The same all-out effort applied at every applicable station and depot.

Captain Barney Giles was Chief Engineering Officer at Rockwell, and

he knew exactly how difficult it was to try to place instruments on panels
not properly designed for additional gauges or large enough to hold them.
The result was that under the ground rules of having only a week—from
the twelfth to the nineteenth—to rebuild instrument panels, rip apart cockpits and gun mounts to fashion cargo compartments, and to install radios
and lighting fixtures, a great deal of counterproductive gerrymandering was
done. The problem was further exacerbated by having to fly the properly
instrumented planes to their respective route stations in time for the assigned pilots to get a chance to familiarize themselves with their routes before the gun went off, at midnight of February 19–20. Consequently, gyro
horizons were often placed where they couldn't be seen. Sensitive
compasses were mounted with wire. The Air Corps planes had no shock-
proof instrument panels, as did the commercial airlines. Delicate equipment was mounted on existing panels, where engine vibration quickly
made it inaccurate and potentially dangerous. Knerr, Giles and other
officers engaged in the engineering and supply business knew that under
the circumstances this could not help but happen.

At the outset there were fourteen types of Air Corps planes selected by
Foulois and his staff to fly the mail. The most unlikely was the Boeing
P-12, and within a month it would be dropped from use. It could carry
only fifty pounds of mail, and neither its characteristics as a fighter nor the
smallness of its cockpit made it in any way suitable for the job. Two
Douglas observation biplanes, the O-25C and the O-38B, were modified to
become the principal workhorses in all three zones, their rear cockpits
rebuilt to carry one hundred sixty pounds of mail. Modification of the latter had the tendency to throw the center of gravity off, making the plane
difficult and dangerous to handle in turbulent weather. Further, all O-38's
were equipped with a wing-tip-type antenna that would not reveal the cone
of silence in passing over a radio range station, so necessary for navigation
and landing procedures. Another great drawback of the O-38 and other
Air Corps observation and pursuit planes lay in their exhaust stacks. They
lacked collector rings and, consequently, the exhaust flames were blinding
at night and particularly in bad weather. With cruising speeds of about
130 mph, the O-38 and the O-25 were typical of the gaggle of aircraft—
observation, bomber, attack—drafted for the job.

The Air Corps had only four Ford trimotors, C-4A's, and they were
used principally to ferry equipment and personnel, although they could
carry eleven hundred pounds of mail, and upon occasion did so on the
Newark-to-Chicago run. There were only two Curtiss Condor B-2's available for operations. Vintage 1928 models, they could chug along at about
100 mph.

On Monday, February 12, General MacArthur sent word to Foulois
that he wanted one of the B-2's held in reserve for the Secretary of War's
use, whereas the other could be released for airmail service.

Only a few of the new Curtiss A-12 attack planes had been delivered to Hickam's Group, at Fort Crockett. The baggage compartment and the rear gunner's cockpit were modified to carry approximately four hundred pounds of mail.

When operations commenced, the B-4A and B-6A Keystone bombers were the plodding and poorly qualified heavy load carriers, the 4A in the West and the 6A in the East. They had to be hand flown every minute, and the best that could be said for them was that they could also carry eleven hundred pounds of mail. The Keystones and most of the other planes did not have enclosed cockpits, so the pilots were exposed to the vagaries of the weather.

All these planes, of necessity, would be operating out of municipal airports away from their home bases. It was not difficult to see that maintenance and supply would be critical factors to the success of the undertaking.

If Knerr's major concern was getting the planes properly equipped in the allotted brief time span, Spaatz's concern had to be the qualifications of the men who would fly them. Although the Air Corps, through the pioneering of such pilots as Major William C. Ocker,[11] Captain Albert Hegenberger and Jimmy Doolittle had gone a long way in the development of instrument flying, their procedures, like the equipment necessary to follow them, had remained pretty much on the shelf. This was the condition Andrews had been trying so hard to correct. Now Spaatz suggested to B.Q. that Hegenberger be dispatched from Dayton and the Materiel Division to check on not only the maintenance of aircraft instruments but also the instrument proficiency of the pilots.

In 1933, the Air Corps had issued its first directive on the subject. It required that all tactical pilots with low instrument time take a ten-hour refresher course. Once proficiency had been exhibited, then only five hours of instrument time per year was required. It was a small enough requirement, but just prior to February 9, there were two widely separated communications, emanating from the same source, that indicated why even five or ten hours of extra flying was hard to come by.

On February 1, 1934, Foulois had sent a radiogram to Conger Pratt, at Wright Field. It began:

> Due to a shortage of funds for purchase of gasoline, pilots on administrative duty will be limited to 100 flying hours for the present fiscal year. Pilots who have already accumulated within twenty hours of this total will only be authorized four hours per month for remainder of fiscal year. . . .

On February 5, Hap Arnold wrote to Frank Hitchcock, editor of the *Daily Citizen,* in Tucson, Arizona. He told him that he sincerely regretted that the 1st Wing would be unable to perform at the Tucson air show as planned. He explained:

> Unfortunately, instructions were received which cut down flying time of our Air Corps pilots to such an extent that it is impossible for us to carry out any mission away from our home station during the rest of the fiscal year. This cut in flying time was apparently made due to the shortage of funds for gasoline and oil. . . .

The restrictions, which over the depression years had become almost routine in nature, applied to tactical as well as administrative pilots; in fact, to all Air Corps pilots. And on the very day Benny Foulois was saying yes to Harllee Branch, Hap wrote to Tooey about the matter. He was putting in a recommendation, he said, that all flying time available at March Field be lumped together, with the station commander authorized to dole it out accordingly. Otherwise his group and squadron commanders were going to end up short of time because "the youngsters" were getting in the hours on training flights. He predicted that if his recommendation was rejected, "all Wing, Group and Squadron training in the air must stop and we will revert to the status we had ten years ago when the only training done was for everyone to fly at will—cross country, round the airdrome or any other place. . . ."

The next day, of course, his concerns became academic, but the shortages in flying time—of any kind—were real, and if Foulois could finesse them, Spaatz, as Chief of Training and Operations, could not. The top permitted total, of sixteen hours of flying time a month for squadron and group commanders and ten hours for all other tactical pilots, was pitifully small when compared to the average one hundred hours a month put in by civilian airline crews. This included a great deal of night and instrument flying, on which Air Corps pilots were so woefully short.

On that weekend of galvanized planning, in which it was determined that shortages could be papered over by the use of hammers and pliers and personnel weaknesses could be eliminated by edict, Foulois believed that at least in one area of the undertaking he had no worries.

Saturday morning, with the Chief of Army Finance, he met again with Harllee Branch and Eugene Vidal, and the former agreed to make available eight hundred thousand dollars to get the operation established and airborne for at least a month. The money was to be used to move personnel, military and civilian, to assigned stations, buy fuel and supplies and spare parts, rent office, shop, hangar and field facilities and pay the all-important five dollars a day per diem to the enlisted men and officers who would be working at stations away from their home bases. Foulois also hoped some of the money could be used to pay the salaries of recalled reserve officers.[12]

At the same meeting, Vidal concurred in placing Airways and Weather

Bureau personnel of the Commerce Department under the jurisdiction of Air Corps control officers at the various route stops.

Of the two agreements reached that morning, the second would hold, but the first would come apart at the seams three days later through a ruling by Attorney General Homer Cummings on the legality of the transfer of funds from the Post Office Department to the Air Corps. He advised that Congress would have to enact special legislation to permit the Post Office Department to make the transfer of funds, and Postmaster General Farley immediately requested that it do so. It was generally accepted that in view of the emergency, Congress would move swiftly and comply before the nineteenth.

Important as these decisions were, the most important of all was that the Air Corps gained a degree of independence it had never known theretofore. It came about through a request Foulois had sent to Chief of Staff MacArthur recommending that the Office of the Chief of Air Corps be granted, during the emergency, the authority to control all Air Corps facilities, personnel and equipment; that it be free to issue travel orders, to delegate authority and to issue orders to zone commanders. Corps-area commanders would be so informed but would stay out of the operation.

It is not known whether MacArthur saw in Foulois's recommendation an opportunity for the General Staff—one in which it would step back and see just what this recalcitrant branch of the service could do in an emergency. But in view of the McSwain hearings and the ongoing drive for air autonomy, it is not improbable that part of his reasoning followed such a line, not out of any sense of vindictiveness, for whatever the Air Corps did would reflect on the Army and there wasn't much the War Department could do to help in any event, but because he also saw the emergency as a peacetime challenge. He approved the Air Corps Chief's request with two conditions: "No War Department funds will be used in connection with this project. . . . This authority is to last only as long as the emergency." The Adjutant General would inform the corps-area commanders of the decision.

And there it was, quietly, without fanfare, a form of independence, however briefly granted, in which the Air Corps was set free to prove that it could fly alone—that in the realm of command and staff and operations, it was capable of going solo.

In any other circumstances, no doubt, MacArthur's decision would have brought forth headlines and hosannas. Instead, the headlines that weekend and the week to follow were big and black with reactions to Farley's cancellation message and the President's Executive Order.

It was Charles Lindbergh who led the attack against FDR in particular and the Administration in general. His telegram to Roosevelt was drafted on Sunday and released to the press by his attorney, Colonel Henry Breckinridge, former Assistant Secretary of War. The telegram appeared in the

Monday morning papers. It detailed better than all else the position of the airline industry.

Your actions of yesterday affect fundamentally the industry to which I have devoted the last twelve years of my life. Therefore, I respectfully present to you the following considerations. The personal and business lives of American citizens have been built up around the right to just trial before conviction. Your order of cancellation of all air mail contracts condemns the largest portion of our commercial aviation without just trial. The officers of a number of the organizations affected have not been given the opportunity of a hearing and improper acts by many companies affected have not been established. No one can rightfully object to drastic action being taken provided the guilt implied is first established, but it is the right of any American individual or organization to receive fair trial. Your present action doesn't discriminate between innocence and guilt and places no premium on honest business.

Americans have spent their lives in building in this country the finest commercial airlines in the world. The United States today is far in the lead in almost every branch of commercial aviation. In America we have commercial aircraft, engines, equipment and airlines superior to those of any other country. The greatest part of this progress has been brought about through the air mail.

Certainly most individuals in the industry believe that this development has been carried on in cooperation with existing government and according to law. If this is not the case it seems the right of the industry and in keeping with American tradition that facts to the contrary be definitely established. Unless these facts leave no alternative the condemnation of commercial aviation by cancellation of all mail contracts and the use of the Army on commercial airlines will unnecessarily and greatly damage all American aviation.

The telegram was read by millions before Roosevelt had a chance to see it. Lindbergh was not only a popular hero, he was also a recognized aviation expert, and his words carried great weight. FDR's response came through his Press Secretary, Stephen Early, who commented with more than a little sarcasm, "Except when sending telegrams or other communications out primarily for publicity purposes, the common practice is to allow the President when he is addressed by them, the courtesy of receiving and reading the communications before they are read by others." He also added significantly that he was passing Lindbergh's telegram on to the Postmaster General, since it was *he* and not the President who had cancelled the contracts. It was apparent that when the brickbats started flying, FDR wanted rotund Jim Farley, who doubled as the Democratic Party's National Chairman, out there in front as cover.

Another popular air hero quickly joined the fray. Captain Eddie Rickenbacker, America's highest-scoring World War ace, was a vice-president of North American Aviation Corporation, which as a holding company owned large blocks of stock in five airlines, including Pan Am, TWA, Western Air Corporation and Eastern Air Transport. Rickenbacker had joined the conglomerate in 1933, and since NAA was one of the major airmail carriers through its various airlines, Eddie knew as much as anyone in the business what the business was all about and how FDR's move would affect it.[13] He had voted for Roosevelt in 1932, but by February of 1934 he was no longer a supporter, believing that the President had gone back on all his campaign promises and was leading the country down the path to socialism.

A few days before the cancellation bombshell, Rickenbacker was having dinner with a party at the Central Park Casino, in New York City. Present was Jim Farley's wife, Bess. They had gotten around to discussing the Black Committee's latest revelations and possible Administration action. Rickenbacker was of the opinion that FDR would not go so far as to cancel the contracts, and Bess cautioned, "Eddie, you haven't seen anything yet."

When the "anything" came to pass, he was "shocked and astounded," and like Lindbergh he was as ready to be quoted as the press was anxious to report his comments. It was a dreary and foggy morning in New York when it did so.

Looking out the window, Rickenbacker observed, "The thing that bothers me is what is going to happen to these young Army pilots on a day like this. Their ships are not equipped with blind flying instruments and their training while excellent for military duty is not adapted for flying the airmail. Either they're going to pile up ships all the way across the country or they are going to be able to fly the mail on schedule."

The Washington *Post* quoted Commerce Department spokesmen in Vidal's office along the same lines. A lack of instrument and night flying might prevent the Air Corps from carrying the mail in as efficient and successful a manner as the commercial airlines, they concluded.

In that short blur of a week, airmen were not listening to or reading the comments of the Cassandras. They were going all out to get ready. On the fifteenth, B. Q. Jones departed his temporary headquarters in Washington and led a flight of planes from Langley to Newark in near-record time. He was quoted as saying he would have seventy planes ready at zero hour on the nineteenth, and then he added, typically, "We'll carry the mail, don't worry about that, unless an elephant drops on us. If it does we'll cut it up and ship it out as mail."

Jones's Assistant Operations Officer was First Lieutenant Laurence Kuter, who, since his Panama ferry flight under the command of Lieutenant Colonel Frank Andrews, had served first as Operations Officer for the 49th Bomb Squadron and then in the same capacity for the 2nd Bomb Group. At the time, he was the Assistant Wing Operations Officer. He was the only bomber pilot selected by B.Q. in his original Langley contingent, partly, no doubt, because Kuter doubled as a pursuit pilot.[14] More to the point of the awesome job at hand, the Captain B.Q. had selected as his Operations Officer—one of six of his original staff—turned out to be single-mindedly dedicated to nursing a penchant for strong drink, and it quickly became Kuter's lot to take over the key operational responsibilities.

One of Kuter's jobs on the fifteenth was to close up the temporary headquarters in Washington and fly the B-2 Curtiss Condor released by General MacArthur to Newark. In the rush of last-minute details, Kuter did not arrive at Bolling Field until just after dark. He had never flown the Condor before, but he was impressed with its closed cockpit and its up-to-date instrumentation. He was pleased also to find a warm body sitting in the right-hand seat. Although the copilot was yet to be invented, he took the man to be either another pilot or a crew chief and therefore competent and helpful.

In a hurry and not liking the smell of the weather, Kuter started the engines, taxied to the end of the field, got the green light and poured the coal to the twin Conquerors. The Condor was heavily loaded with reams of stationery, and as the plane moved down the runway, Kuter began to get the idea that it was he who was stationary. He had that stone-cold feeling that the Conquerors were not going to. He was out of field and into weeds before the B-2 staggered off the ground. Then it appeared he was on a collision course with a row of apartment buildings along the Anacostia River.

"Crank the wheels up!" he shouted to his unmoving companion.

"What wheels?" came the startled reply.

Kuter had no time to explain. By some adroit maneuvering and a forgiving aircraft, he managed to avoid the rooftops and angle the plane out over the river. After he began to breathe again, he learned that the man on his right was a clerk and was experiencing his first and almost last ride in a plane.

The weather en route worsened, and they flew north in a mixture of rain, sleet and fog. By the time they were in the vicinity of Newark, Kuter was again toying with rooftops, having trouble locating the field. It was an airport gas station with a distinctively lighted sign that gave him his bearing, and so he brought the Condor safely down to roost, not unmindful of the fact that the plane's instrumentation had been of enormous help. He would not care to compare its comfort and aids with the eighty-two additional planes that were to become available for airmail duty in the Eastern Zone, nearly one third of which were P-12's.

On the same day that B. Q. Jones shifted his headquarters to Newark, genial, unflappable Horace Hickam and a staff of more than twenty breezed into Chicago's Municipal Airport in their A-12 attack and O-38 observation planes. They set up temporary shop in the National Guard hangar, where Uncle Horace, assured of manner but unceasing in getting his zone organized, was soon known as a man whose "hat is his office."

Lieutenant Earle P. Partridge, Operations Officer, and Engineering Officer First Lieutenant Nathan F. Twining were too busy to wear hats, while not far away, at Selfridge Field, Hickam's good friend Andy Andrews was busy also, overseeing the disemboweling of his 1st Pursuit Group as a tactical unit. Thirty-eight of his pilots were to serve at stations in the Central and Eastern zones.

Andrews had just written to Foulois expressing his concern over his pilots' lack of instrument training, declaring that none of them since their arrival at Selfridge had ever seen a directional gyro or an artificial horizon. For seven months he had vainly been trying to improve the situation. He asked Foulois to see that he got just one of each to be utilized by what remained of his tactical unit.

It was also on the morning of the fifteenth that Hap Arnold moved from March Field to Salt Lake City. Three days earlier, thirteen planes, some of them transports, had arrived from March carrying pilots, mechanics and headquarters personnel. Tinker and Monk Hunter, who was Chief of Operations, had set up headquarters in the Newhouse Hotel.

Hap had appointed Major Charles B. "Barney" Oldfield as Regional Commander, and in this capacity Oldfield was given full authority in the control of the movement of all aircraft on the routes from Salt Lake to their first control stop. There the authority passed to the four route commanders. They, in turn, were free to select their personnel from blocks of officers and enlisted men provided by Arnold. At the beginning there would be fifty-seven pilots to fly the routes; twenty-three of them had less than two years' service.

In deciding who would fly, Arnold, like Hickam and Jones, had to take into account that of necessity he would have to draft his most experienced officers for executive and staff duties. As a result, twenty pilots who had just been graduated from the Navigational School were signed on at Rockwell, and an additional sixteen had to be sent off for a two-week course that would be completed on March 3.

In a letter to General Craig, whom he was keeping apprised of his actions, Arnold indicated his wariness about the situation with regard to the press. He had given the reporters an interview at the Newhouse Hotel and been quoted as saying that 90 percent of the civilian airmail pilots had gotten their training in the Air Corps.

> I have stressed upon all the Route Commanders the necessity for doing their utmost to make this thing a success. I have told them that the Army, the Air Corps, and they themselves are "on the spot," and that any slipups would react unfavorably towards the Army at this time. This unfavorable reaction was brought to my attention this date in an interview which I had with the newspapers. I have put them off all week and would not say anything until today when I accorded them an interview. So we cannot afford a slipup as the undertone of their conversation was that the commercial

lines were so much better than the Army that we would show up unfavorably. I personally am of the opinion that they are waiting like a bunch of hungry dogs to grab up any mistake or misfortune which may overtake us and make the most of it.

Thursday the fifteenth was moving day across the country, and Ira Eaker shifted his route headquarters to United Airport, at Burbank, California, a suburb of Los Angeles. Without any funds, but on Arnold's go-ahead, he had rented office space from Western Airlines, the commercial company whose route he would be maintaining.

The headquarters was in "a lean-to (portion) of a hangar in which the planes were stored," reported Eaker. "This made the whole administration personnel closely available to their work, gave operations and engineering a chance for close liaison . . . and otherwise proved very satisfactory. . . ."

Up to the nineteenth, Eaker's whole effort was centered on giving his pilots as much training as possible and in working out staff control for operations, supply and engineering. Toward the former, on Monday the twelfth he had started his men flying route-familiarization checks, using the P-12's of his squadron. They flew from Burbank to Las Vegas and then to Salt Lake. He did likewise in the new Boeing P-26 pursuit, which, if possible, was even less a mail plane than the P-12. His Engineering Officer, Lieutenant Edwin Perrin, flew on his wing. The leg from Los Angeles to Las Vegas was mostly desert, but the run from Las Vegas to Salt Lake required crossing two mountain ranges at a minimum altitude of seven thousand feet; the weather over and between the ranges was highly unpredictable, dangerously unsettled. Eaker knew that airline pilots were required to have an intimate knowledge of the terrain involved before being permitted to fly the course on a scheduled run. On the feeder hop from Burbank to San Diego, the usual Pacific-coast fogs would be the major hazard.

He had a total complement of twenty-five young pilots, almost exclusively reserve officers fresh out of flying school. With 90 percent of the flying to be done at night—so that mail could be delivered during the day—it took no wizard to divine what they were going to be up against.

In establishing his operation, with its multitude of functions outside the Air Corps norm, which included close cooperation with Commerce Department weather and communications people, Eaker moved quickly and with solid deftness. He knew how to delegate authority. He knew also that the success of such delegation depended upon those selected to do the job. In this he was both a quick and a good judge, and he needed to be the former, with so many new faces about. It worked both ways, for it was apparent that he possessed the composite qualities of confidence and command that spelled leadership. Unlike Arnold, he was a man more of rock than of fire.

Not everyone with a responsible role was so structured, and Hap Arnold swiftly had to clear up doubt in one officer's mind as to who was running what. It had been immediately obvious to him that a large supply depot must be set up at Salt Lake City. The Commander of the Rockwell Field Depot, from which most of the necessary supplies would be coming, was Major Shepler Fitzgerald. Fitzgerald had two problems of long duration. He was blindly self-centered and possessed a military mind that was built around the inflexibility of Army regulations. When he received a message from Arnold asking him to come at once to Salt Lake, he refused to do so. Arnold had no right to give him orders, he said. He took his orders from the Materiel Division at Dayton. "I'm not going," he informed his Chief Engineering Officer, Barney Giles.

"Well, I sure recommend you do go, Major," advised Giles. "The authority for Colonel Arnold's command has been in all the newspapers. It said all military personnel are to cooperate with him."

"Well, I'm not going unless I get instructions from Dayton," was Fitzgerald's obdurate position.

The instructions arrived from Hugh Knerr the next day, and Giles flew his CO up to Salt Lake in a Martin bomber, staggering through the air loaded with nearly two tons of radio batteries and assorted equipment. After landing they repaired to the Newhouse, where Arnold was waiting.

"Giles, you stay outside," he snapped, and he wasn't smiling.

The Captain did as he was told, and for the next ten minutes the only sound was that of the Zone Commander's voice. It bit through the office walls like a buzz saw and did not appear to pause for breath.

When the voice ceased, Major Fitzgerald appeared with head bloody but unbowed. "Well," he admitted sheepishly to Giles as they left the building, "you were right, but I didn't get any copies of his orders, and I didn't know anything about it." Giles reminded him that he did, but Fitzgerald was still full of excuses. There should have been a telegram. It was the system that was at fault, not Shepler.

While the zone commanders and their staffs were hard at it struggling to arrange the components necessary to fly the mail, the pilots who were going to do the flying were being drawn in from Air Corps stations.

Lieutenant "Pete" Quesada was attending the Navigation School at Langley Field. The school was under the direction of Harold Gatty, an Australian, who was one of the very few experts on the subject. The two-month course was about two thirds completed when the thirty-eight students got the news that the Air Corps was to fly the mail. The announcement took them completely by surprise.

Quesada's immediate reaction was one of bewilderment, for through his own experience and knowledge he knew that mail flying required a degree of skill most of his fellow pilots lacked. His own experience, on the other hand, made him one of the few who was qualified, for prior to coming to

Langley he had been the Curtiss Condor's principal pilot, ferrying the top brass about the country. This was the same plane Larry Kuter flew to Newark, and Quesada was to use it to fly the mail.

Beam flying, which required a pilot to fly between two points by following a radio signal coming to him—he hoped!—through the earphones in his helmet, demanded considerable practice. The steady signal tone was broadcast on a preset frequency. To either side of it were quadrants, each signaling in Morse code either the letter A or the letter N: *dit dah,* or *dah dit.* The method of following the beam was to fly along the right edge of its signal and blend its steady sound with the sound of the quadrant letter. It was called bracketing, and the closer the pilot flew to the beam source, which also transmitted its call letters, the narrower the beam path became and the louder its volume. Its end was marked by a cone of silence. If a landing was to be made at the station, the pilot had to follow a let-down procedure in which he picked up the outward leg of the connecting beam, crossed over it, and bracketed it inward. At a lower altitude he would again pass over the cone of silence, and from that point continue his descent on a prescribed course, hoping to have the field in view by the time he was down to five hundred feet. The entire landing procedure was dictated by accurate timing, and under instrument conditions timing was crucial to reaching the ground in one piece.

In the course of a mail flight it might be necessary for the pilot to fly down and away from three or four transmitting stations, all sending out their signals on different frequencies. For the mail pilot, the beam was a highway not seen but heard, and bracketing it took skillful coordination, the ear the center piece, the eye the check piece, the mind the translator.

Quesada knew this, knew the technique well, and before he and his fellow officers were dispersed to assigned mail stations he was asked to become a teacher and give instructions to his former classmates in beam flying. There was time for only two, brief sessions, for their orders were to report within twenty-four hours.

Driving up to Newark, Quesada and a carload of his compatriots ran into a full-blown ice and sleet storm. The weather became a topic of conversation. They realized the flying could get rugged. But they were young, full of enthusiasm. None of them lacked confidence. There were no thoughts on the merits of their being given the job, only that it was theirs to do, and neither rain nor snow nor the gloom of night would stay them from their appointed rounds.

At about the same time Quesada and his friends were getting a taste of the weather on the ground, reserve officer Second Lieutenant Bernie Lay, a bomber pilot assigned to the Central Zone, was making his first familiarization night flight in a P-12. His run was from Chicago to Nashville, Tennessee, and soon after departure he found there would be no beam-bracketing practice that night. His radio went dead. Then the compass took up spinning, and he knew he was pretty much on his own, with only flashes of scattered radio beacons and a Rand McNally map as guides. He found Nashville, actually arriving ahead of schedule. But when he climbed out of the cockpit, he announced he had passed from youth to old age.

If Lay made a success out of his practice flight, Lieutenant Jack A. Sutherland made a saga out of his. With a compass giving him a reading some 30 degrees off of true, he mistook Demopolis, Alabama, for Selma, his intended destination, and set his pursuit down in a farmer's newly plowed field. Under the circumstances, it was a nice enough landing, but he realized the field was too rutted and muddy to fly out of. With the aid of the farmer and his neighbors, Sutherland got his P-12 up on the road. To the joy and amazement of the citizenry he taxied it through the town, and finding a proper stretch of straight highway, he once again was airborne. This was not quite the happy end of his story, for although he found Selma, the next day in attempting to depart he nosed the plane over and this time was rescued by some members of a chain gang.

While pilots were busy trying to familiarize themselves with their routes with varying degrees of success, the scene on the ground at various route stops and control points presented a sort of Italian tarantella, the motion of the frenzied dance constant.

When Lieutenant Robert S. Heald stepped into the Cleveland section headquarters, he saw a sight he didn't think he'd ever forget: "a room full of men, each one intent on the job before him, others hurrying in and out, typewriters rattling and the hum of many voices, making it difficult to listen to what someone was saying to you. Two or more men were working at each desk, some were working on boards on chairs. The room was crowded with busy people. None were in uniform. You could not identify officers, men or civilians."

Lieutenant Curtis LeMay would have described activity at the section headquarters in Richmond, Virginia, in somewhat different terms, for it had been established in the ladies' room at the Air Transport hangar.

Captain Melvin Asp's first impression of the Airplane and Engine Section he joined was "of being tangled up in a room full of string. No matter which way you started to untangle, you always met a maze which brought you back to the original starting point or else ran into a knot that required backing up to start all over again."

In the Western Zone the majority of the pilots had reached their station stops by Thursday the fifteenth. The next day, the weather was unsettled over much of the routes north and east, but when word came through that Coalville was reporting two thousand feet, air-reserve pilot Lieutenant George D. Grenier was cleared to make a practice run to Cheyenne. In Grenier's A-12 rode fellow reserve officer Lieutenant Edward D. White, coming along to familiarize himself with the route as well. Coalville was the first checkpoint, no more than thirty miles from Salt Lake, situated in a basin surrounded by the high, rugged country of the Wasatch Mountains.

"No one at that time," reported Arnold, "had sufficient experience to realize that the weather changed with startling rapidity in the Wasatch Mountains or that the commonly accepted theory that '1500 feet was sufficient to get through' meant little or nothing in the Rocky Mountains."

Grenier and White never reached the Rocky Mountains, never got beyond the canyon leading into Coalville. The weather closed down behind them. Grenier was unfamiliar with the route and was not a trained instrument pilot. Spotting a wide canyon that appeared to open to the eastward, he flew into it and was engulfed in a swirling snowstorm. He was at eighty-five hundred feet, the A-12 icing up, unable to climb higher. The canyon had a dead end, its top close to nine thousand feet. Desperately Grenier circled, blinded by snow, trying vainly to find a way out of the trap. For him there was no way out, and the A-12 plowed into a mountainside, exploding in flame.

It was shortly after dark on that same day that reserve officer Lieutenant James Y. Eastham took off in a twin-engine Douglas B-7 on a similar hop, from Salt Lake to Boise, Idaho. Again weather reports indicated a ceiling of two thousand feet. Ten miles east of Jerome, where Eastham planned to land, he reported he had four thousand feet. He had been following the radio beacons. Then, suddenly, the one ahead was blotted out. Like Grenier, he had little instrument experience, and he felt his way downward, groping blindly for the field. The plane made contact with a treetop instead and somersaulted into the ground.

The time difference between East and West made a difference that Benny Foulois would like to have swallowed along with his words. On that grim day, not knowing that he had lost three of his pilots, he testified before the House Post Office Committee. He told its members, among other things:

> We have assigned to this work the most experienced pilots in the Army Air Corps. We have had a great deal of experience in flying at night and in flying in fogs and weather, in blind flying, and in flying under instrument conditions. . . .

It just wasn't so, and later suggestions that Foulois's commanders had misinformed him couldn't stand scrutiny. Andrews' prior efforts alone were proof enough of the situation, if any were needed. Foulois's overselling of Air Corps capabilities, of gilding the lily on instrument competency, was in one sense natural enough. He certainly wasn't going to say his pilots were not capable, but his going overboard before the Committee was probably stimulated by an altogether different problem.

The Attorney General's decision that the Post Office Department could not transfer the agreed-upon eight hundred thousand dollars to the Air Corps until special legislation was passed had left the entire operation without funds. This had been learned on Tuesday the thirteenth, and on the following day, at Jim Farley's request, a bill was introduced in the House to remove the restrictions. It was on this bill that Foulois was testifying when he made his less than accurate statement. Actually, before giving his testimony he had sent a radiogram to his zone commanders stressing safety above all else.

The next day, the headlines blackened the newspapers across the country, announcing that three Air Corps airmail pilots had died in crashes. The news set the pace for what was to follow.

Twelve

Ordeal

Arnold was stricken by the loss of his three pilots, but the fact was there was nothing unusual about these tragedies. In the previous six months, twenty-three Air Corps pilots had lost their lives in flying accidents. It was the bitter pill, the end result that could go with being a pilot in the Army Air Corps.

On news coverage, Arnold's appraisal of the press to General Malin Craig had been essentially correct. It wasn't so much that the newsmen had it in for the Air Corps, but the political opposition, which had been lying low for a year, suddenly saw a way to discredit Roosevelt's vaunted infallibility and attack him on his precipitous action against the airmail carriers.

Popular humorist and air enthusiast Will Rogers had dryly commented on the cancellations, "It's like finding a crooked railroad president and then stopping all the trains." In this case, the planes had not yet begun to carry the mails, but the heat was on and the Air Corps was caught in the middle. Foulois's overselling did nothing to help, but it was Captain Eddie Rickenbacker who got the knife in.

His firm, NAA, had been assisting Donald Douglas and Company in the designing and building of a new air transport, the DC-1. To Rickenbacker it was the best plane in the world, and he had arranged with Jack Frye, vice-president of TWA, to fly it with several other pilots from California to New York in an effort to break the transcontinental transport record. Originally, the plane had been scheduled to make the flight in March, but with the sudden cancellation of the mail contracts, the decision was to capitalize on the situation. They would take off with the last load of West Coast airmail, attempting to wing it to Newark in record time, getting in just before the deadline. Newsmen were invited to come along, and the purpose behind it all was obvious.

Rickenbacker arrived in Los Angeles from the East on Saturday morning and sat down with reporters in the airport coffee shop to have some breakfast. When he saw the headlines and read the story about the downed pilots he remarked angrily, "That's legalized murder!"

"Can we quote you, Eddie?" came the quick query.

"You're damned right you can!" was the immediate reply, and the political opposition to FDR had a phrase they would spread across the nation like snow on the wind.

That Rickenbacker was truly incensed there was no doubt. When NBC asked if he would be willing to speak on a coast-to-coast hookup before takeoff Sunday, he readily agreed, seeing it as an opportunity to have his say. To prepare for the blast, he paid a visit to his old friend Harry Chandler, publisher of the Los Angeles *Times*.

"I'm going to make a speech," he told Chandler, "and I want it to be the most vitriolic attack that ever blistered the airways. I want you to lend me your best editorial writer to help me write it."

Chandler was happy to comply.

"Now, will you object if I go to the *Examiner* and get their best man?" The Los Angeles *Herald-Examiner* was the *Times*'s major rival, but in this case Chandler had no objection. Rickenbacker spent the remainder of Saturday and most of Sunday drafting his speech with the aid of the two newsmen. The result felt as if it were smoking in his hands. That evening, he was preparing to leave for the airport when the phone rang. It was long distance from William B. Skeets Miller, in charge of NBC's Special Events. He informed Rickenbacker that orders had come from Washington to cut him off the air if he said anything controversial. Freedom of the airways was not all that free, and the words that Rickenbacker spoke that evening were moderate and acceptable.[1]

Saturday the seventeenth was also a difficult day for Benny Foulois. It had begun with a meeting in George Dern's office in which Foulois had assured the Secretary of War that yesterday's crashes were not indicative of Air Corps capabilities. Of equal magnitude was the matter of economics. There were no funds to pay for the mounting expenses of preparation and the costs that lay ahead. There was only one place for him to go for money, and in going he lost a degree of his brief independence.

That afternoon, he met with Deputy Chief of Staff Hugh Drum and the Army Chief of Finance to discuss what could be done to make the operation solvent. The decision was that Lewis Douglas, the Director of the Budget, must be prevailed upon to release impounded War Department funds in the amount needed.

It would not be until Monday, the eve of the beginning of operations, that the three generals would meet in the morning with the Assistant Director of the Budget. The amount of $800,000 would be knocked down to $526,000 on the basis that $300,000 of the original amount had been included to pay for depot personnel for one year and therefore wasn't necessary.

That same Monday afternoon, Foulois received a telephone call from Drum informing him that the $526,000 had been whittled down to $300,000. The amount, he said, was to cover expenses from February 10 to March 1. Foulois knew it wasn't nearly enough and protested, but to no avail. Then there was nothing he could do about it except hope that Congress would move quickly on the necessary legislation. Until then, the bedrock of his independence was again tied to War Department largesse.

The previous day, when Eddie Rickenbacker was in Los Angeles preparing for his heralded departure, weather reports indicated that a tremendous storm was developing over the Great Lakes region and was predicted to hit the East Coast by the night of the nineteenth. He and Frye figured if they took off by nine o'clock Sunday evening they could get into Newark just ahead of the weather and in so doing demonstrate to the country the superiority of civilian airmail operations over what would follow.

To the cheers of the assembled, they were airborne on schedule with a party of newsmen on board to record the event. Aided by a tail wind, the sleek DC-1 averaged 230 mph to Albuquerque, New Mexico. Onward they went, flying high, wide and handsome, making only four refueling stops in place of the usual twelve. Not only was the cockpit aglitter with the latest instrumentation, but there was also an autopilot to make the effort less strenuous. When they hit the outriders of the massive winter storm, west of Pittsburgh, they climbed to twenty thousand feet, getting above its fury. Then, with their superior speed, they outdistanced its reach over eastern Pennsylvania, descending into Newark through the leaden morning light of Monday, February 19. It was just ten o'clock, and they had crossed the country in the record-breaking time of a bit more than thirteen hours.

A large welcoming committee was on hand, and Rickenbacker anticipated that he might now get a chance to take a crack at the President. But Skeets Miller was also on hand to inform him sadly that NBC News had received orders not to carry anything Captain Eddie Rickenbacker might have to say. The newspapers could not be so controlled, and they drove Rickenbacker's message home to their readers.

At noon the storm struck, becoming one of the worst in Weather Bureau history to engulf the Northeast. A tempest of snow, rain and fog lay in a churning blanket from Ohio to New England. It was matched in the West by blizzards and icy gales ripping through the Sierras and Rockies.

B. Q. Jones, who had been quoted earlier as saying, "The Army Air Corps does not brag, does not predict, we only ask to be judged by results," had to cancel all flights scheduled out of Newark on the midnight inaugural run. It left him and his eager young pilots frustrated and anxious to be about their business.

In the Midwest, Horace Hickam's boys were off and flying. Lieutenant Tom L. Mosely was the first on a flight from Kansas City to St. Louis. Four other flights left Chicago that night—for Omaha, Kansas City, Nashville and Newark.

At Cleveland Airport there was a goodly crowd on hand to see Lieutenant C. R. Springer become airborne in the near-zero weather. They watched him disappear into the frigid darkness. Fifteen minutes later he was back on the ground shouting to a mechanic, "Get me a flashlight so I can get out of this damned town!" Springer had found that his directional gyro didn't work and he couldn't read the headings on his magnetic compass, because the lighting in his open cockpit was practically nonexistent.

At Salt Lake City, Hap Arnold was forced to cancel the first eastward run to Cheyenne. He was adhering to the warning he had sent his zone commanders: "I cannot say too much about the importance of safety. We have a gigantic job ahead of us, but we will take every precaution."

The mail did go through on the southward route to Los Angeles, but it took some real doing on the part of twenty-two-year-old Lieutenant M. G. Griggs. He fought a grueling battle against a combination of sleet and fog, having to land three times at auxiliary fields to scrape the ice off his windshield. He arrived at Burbank eight hours late, but he arrived.

At Burbank, Ira Eaker had come up against an unexpected problem. The Post Office had estimated the load to San Diego would weigh around four hundred pounds. Instead, due to stamp collectors who were anxious to get a new first edition, the amount was fourteen hundred pounds. All Eaker had to carry the load was a handful of P-12's. He called Arnold and explained the problem. "Go over to March and pick up a B-4," was Hap's suggestion. Eaker did so, but none of his pilots had ever flown a Keystone bomber, so for the next week he practically lived in the plane, checking each man out while flying the route.

General Oscar Westover, as Commander of AACMO, had departed Washington on the morning of the nineteenth to make an inspection trip of the route stations. In his Douglas O-25, he had run smack into the raging weather system that Rickenbacker and his crew had raced to avoid. Westover's plane had a cruising speed of about 115 mph, but against the wind and snow he averaged half that. It took him all day to reach Fort Leavenworth, Kansas, and the next day, after six hours of trying to fight his way through a storm center to Cheyenne, he was forced to return. The next day, he made it on the tag end of a blizzard.

The single most imperative word that struck the Air Corps during the first week and then the first month of its assignment to carry the mail was *weather*—every imaginable kind of weather, not in just one location of the country but in most locations. It was as though a consortium of goblins had concocted a stinking brew of foul elements and spread it over the land from sea to shining sea. Their demonic laughter rode in the gale winds that buffeted and tore at open cockpits, congealing their occupants, glazing over instrument panels with ice, snapping off the delicate lifeline of radio communication.

Lieutenant Durward O. Lowry, of the 17th Pursuit Squadron, at Selfridge, was no novice at the controls. A regular officer, he had been one of those who had flown with Andrews in the welcoming escort of Balbo's mass flight. Assigned to the Central Zone, he had learned the proper code letter for each of the ten flashing beacons along his four-hundred-mile route by memorizing a sentence with the first letter of each word representing the code letter for the particular beacon.

At four in the morning, three days after operations had commenced, he took off from Chicago, heading for Cleveland. The cold was intense. Below, the land lay deep in snow. West of Toledo, his radio antenna froze up. Visibility was poor, and got worse. He dropped down to grope his way over unfamiliar terrain and ran into heavy fog. Experienced as he was, he was not a proficient instrument pilot. Caught in the fog and recognizing he had no way out, Lowry tore open his mail compartment and began dropping mail sacks over the side. It was near 6 A.M. when a farmer at Deshler, Ohio, fifty miles south of the Toledo-Cleveland airway, heard a plane circling overhead. He ran out of his house, heard Lowry's engine cut out, followed a few moments later by the awful splintering sound of the aircraft smashing into trees. He ran toward the sound.

Young Durward Lowry's death was the grim proof of what his CO, Frank Andrews, had been talking about in his letters to Washington and Dayton. *If only.* . . . Investigation would indicate that Lowry had tried to bail out, but even if he had gotten clear he was too low for his chute to have opened. The mail bags he dropped were recovered.

The press, which was covering the Air Corps's every move, even to the extent of reporting inconsequential accidents such as a cracked tail skid in Salt Lake City and a bent propeller in a nose-over at Greenville, South Carolina, made big-headline news not just of Lowry's death but also of a number of other, related incidents in the same twenty-four-hour period.

Lieutenant Howard M. McCoy, flying an O-25 from Newark to Cleveland, had his engine quit and put his plane down in a Pennsylvania cow pasture. With the help of a nearby farmer, he lugged the two hundred pounds of mail to the post office in the town of Woodland and delivered it to the postmistress to be carried on by train.

Lieutenant Charles P. Hollestein, heading east over the infamous "Hell's stretch" of the Alleghenies on a hop from Cleveland to Washington, had his radio go dead. Like Lowry, he ran into fog, but he had some daylight to guide him and managed to crash-land on the side of an ice-coated hill near Uniontown, Pennsylvania. Not as badly damaged as his plane, he found help to move the mail onward.

Caught in the same kind of weather on the Newark-to-Miami run, Lieutenant Howard Dietz also had his radio fail. Aware that there was a private field near Salisbury, Maryland, he sought to locate it, not knowing that its beacon had been shut off in an economy move. He circled the town, looking for a way down. Townspeople got the message and rushed to the airport to turn on the light, but Dietz, figuring he was wasting gas, flew on, searching for another field at Crissfield. There, the citizenry made for the airport and lined up their cars with the headlights on to give him a runway. He spotted the illumination through the fog and made his approach. But he failed to see the tree in his glide path. Contact with it flung his plane into a telephone pole and he was hurled clear of the wreckage. Though badly injured with a fractured skull, he was still able to give orders. "Don't bother anything in the plane. Take care of the mail!"

It was just before dawn the next morning that Lieutenant Norman Burnett, flying from Cleveland to Chicago without mail, plowed into a blizzard. Visibility dropped to zero. He wasn't that good on instruments; he'd already missed a beacon and his radio was so much dead weight. He didn't fool around; he bailed out and got away with a broken leg.

In Texas, Lieutenant Frank I. Patrick wasn't so lucky. An experienced pilot assigned to a supervisory role in the Central Zone, he was brought down by weather near Denison, and he was killed in the crash.

On that same storm-whipped Friday, air reserve Lieutenants James H. Rothrock, William S. Pocock, Jr., and George F. McDermott took off from Floyd Bennett Field, Brooklyn, in a Douglas amphibian. Their destination was Langley, Virginia, where they planned to pick up newly instrumented mail planes. They weren't ten minutes off the ground before they were being tossed about by winds and turbulence that threatened to dismantle the plane. Landing in heavy seas off Rockaway Point wasn't much of an improvement. The tide began to sweep them out into the ocean and the waves to swamp the hull. In the gathering darkness, a destroyer and a Coast Guard cutter moved to the rescue. McDermott, trying to signal while clinging to the plane, was swept overboard, and although his two companions were saved, he was drowned.

It was a freak accident that had to do only indirectly with flying the mail. More directly, what had to do with flying the mail was the radio problem. It was all there in a memo dated February 21 from Hugh Knerr to Foulois:

> We have provided SCR-183 radio sets [two-way radios] for 167 airplanes. There are only two more sets available in the United States. Incidentally, this leaves us without spares, although we have a small supply of maintenance material. . . .

The only solution of our problem is to procure immediately 200 additional transmitters.

It is impossible to obtain any SCR-183 sets within a year's time. . . . There are no aircraft radio sets on the shelves of any manufacturer, nor has any manufacturer, except one, tools and equipment set up to produce in a hurry.

The one exception is the National Electrical Supply Co., in Washington, D.C., who are now producing sets on a Navy contract.

The set they make is ideal for our purpose and known as "Navy Model G.M." Technical data can be supplied if desired. . . .

With approximately two hundred fifty aircraft selected to fly the mail, Knerr's cold figures meant that, at the time, one third of them lacked two-way radios, and those in use were anything but reliable.[2]

The press accent was focused on the negative, and lost in the welter of chilling headlines and newscasts was the fact that in all three zones a great deal of mail was being successfully carried by Air Corps pilots, carried in spite of equipment deficiencies and weather that played no favorites.

Lieutenant Donald Wackwitz had fought his way through from the Midwest with the first load of mail into Newark. En route, his radio had gone dead, the ceiling was hardly above the treetops and the snow on the ground was so deep it looked like fog. Winds over the Alleghenies were brutal, and he was practically frozen when he landed.

Will Rogers commented to the New York *Times:* "After flying all night got in here at eight o'clock this morning, right in the wildest of what even Californians would call unusual weather. Plane stuck in the snow after it landed. An Army flyer came breezing in in an open cockpit who made the same run from Cleveland, these being the only two planes to land in Newark all day. So give a big hand to Lt. Wackwitz of the U. S. Army."

While the Lieutenant had made it into snowbound Newark, General Westover found conditions in Cheyenne even more difficult. Air Corps planes, for lack of hangar space, were staked out in the snow in temperatures of ten below. It took mechanics as much as three hours of "cooking oil" to get a plane started.

On to Salt Lake he went, and there conditions were looking up. "The Post Office people were tickled to death," he said, "that the line up to Seattle was functioning without delay."

In a personal account to Hap Arnold, Ira Eaker gave a position report as of that Friday, the twenty-third:

On my way down Wednesday night I came through all right except for a strong wind which I was bucking until I got over Burbank. I could not get in and returned to Palmdale, landed and trucked the mail in here. . . .

We are now in the throes of revising pilot schedules to break the run into three sections. I will then have six pilots at Salt Lake, six at Milford and six here. . . .

As you probably have noticed, we have been getting the mail over the line with regularity, making fairly good schedules, except for the fact that we have been trucking in and out of Palmdale almost exclusively in the last 48 hours and the San Diego run has been trained. This has been necessary due to impossible flying weather on the coast. . . .

The telegram we got saying that we would report as of 6:00 P.M. all delays in schedule to the Chief of Air Corps brings up one important point. This schedule we now have is probably too fast for the equipment we have. This is going to make our daily telegrams to the chief and to you show departure from initial points on schedule, but late arrivals at intermediate points and terminals. . . . I am in hope you will be able to authorize a change in schedule to be reasonable for the equipment we now have. . . .

And so the mail was moving under the difficult conditions that prevailed, but obviously the loss of Lowry, Patrick and McDermott commanded far greater attention than the operational successes—even more attention than the crash of a United Airlines plane and the death of its eight occupants. It happened on that same fateful Friday, and in the same rugged area where Grenier and White had crashed just a week before. The plane had taken off from Salt Lake at 2:30 P.M. on a flight to Cheyenne, and just east of the city had flown into a heavy snowstorm. The wreckage was located on a ridge in Parley's Canyon, only twenty miles from its point of departure.

An airline loss was one thing, a political pot was something else, and this one was continuing to boil furiously, stimulated by the played-up news of Air Corps casualties. The result was a thunderous amount of congressional sound and fury, which did nothing to help the Air Corps or the blood pressure of the sounders. In a Saturday meeting, the House of Representatives held what was described as "the most turbulent session since the Democratic administration took office." Its purpose was to pass the Brunner bill, authorizing the temporary flying of the mail by the Air Corps, and in so doing bring to the Corps the necessary funds it needed from the Post Office Department to continue operations. But, before passage, a good many members insisted upon voicing and putting in the record savage criticism of President Roosevelt and his Postmaster General "for wrecking the private airmail transportation business."

No one was more vitriolic in his attack than Representative Hamilton Fish, Jr., an implacable and verbose foe of the Roosevelt Administration. A third-generation Republican congressman from New York, Fish laid into the issue with a mixture of political sagacity and overripe exaggeration. He said there was no emergency, that the whole thing had been "manufactured" by Farley, and then declared, "We are now called upon to sanction that situation and send these American boys, young American officers with wives and children, to their deaths without even a chance to defend themselves because they have to obey orders even when they know their planes are not properly equipped and they do not know the mail routes, radio beacons or landing places." He emphasized the Rickenbacker term "legalized murder."

"Legalized murder!" was also the war cry of Representative John C. McLeod, Republican of Michigan. His borrowed *J'accuse* set the tone, and Representative Edith N. Rogers, Republican of Massachusetts, in a voice rising close to the shrieking point, proclaimed that "a vote for this bill would be putting a stamp of approval on murder!"

Applause greeted her outcry, and gathering breath, she pointed to the silent ranks of Democrats and declared, "The story of the airmail will be written in blood on the record of the Roosevelt administration!"

Her colleague Representative Florence Kahn, of California, was somewhat more restrained but no more helpful to the Air Corps when she suggested that Congress would be wise "to suspend the airmail entirely until the Army flyers can be trained and equipped properly for their unaccustomed task or until we learn the truth about the airmail contracts. . . . It seems to me," she continued, "we are gambling with the very flower of American youth in sending these young officers to death in ill-equipped planes."

Representative John Taber, of New York, upped the ante by intoning that "in good conscience" he could not be "a party to murder." His Republican cohort Representative Harold C. McGugin, of Kansas, then tried unsuccessfully to gain approval for an amendment that would have prevented the War Department from placing the Air Corps at the disposal of the Post Office Department. But it took G.O.P. Representative James W. Mott, of Oregon, to throw the House into an uproar, when, after accusing Postmaster General James Farley of an act of "pure hysteria," he stormed, "The deaths of the Army pilots in flying the mail are too great a price to pay—whether the cancellations were justified or not!"

On the Senate side, the rhetoric was just as full-blown and damaging to the Air Corps's image. Farley had been trying rather ineptly to answer some heated questions from opposition members of Senator Hugo Black's committee, one of whom was Senator Warren Austin, of Vermont. It was obvious the red-faced Postmaster General was ill-prepared on the legalistic rationale for the cancellation of the airline contracts, and Black and Austin got in an argument over the former's attempt to protect Farley by blocking Austin's queries. "I want it understood that I object to it," snapped Austin.

Black responded snidely that Austin only wanted questions answered his own way, and the Vermonter shot back, "I've sat here through the taking of six thousand pages of testimony and listening to you ask one leading question after the other in this sad affair."

"What's sad about it?" retorted Black.

"I'll tell you what's sad about it," Austin said angrily; "it is the taking of the innocent lives of these poor boys who have tried to carry the mail without experience since these contracts were annulled!"

To it all, Benny Foulois and his "poor boys" might have shaken their heads and said, *For God's sake, enough!* What they needed from the solons was less political bombast and passage of the Brunner bill.

That the political fireworks ignited by press reports had put Benny Foulois on the defensive with the War Department, there was no doubt. MacArthur might have said, "It's your ball game, Benny," but the Air Corps was a part of the Army, and anything that reflected poorly on it had to be an embarrassment to the General Staff.

On that Monday following the Saturday session of Congress, Foulois, in sending a position report to MacArthur, defensively cited his problems:

> To date, February 26, 1934, the Air Corps has been handicapped (1) by exceptionally bad weather conditions; (2) the necessity for the creation of an emergency organization unfamiliar with the details of airmail procedure and operations; (3) lack of funds together with restrictions on the use of the $300,000 allotted as well as (4) lack of knowledge as to the probable period of future operations.

Foulois also recommended that "the President direct the Director of the Bureau of the Budget to immediately release to the War Department an additional $500,000 of impounded funds that the present airmail operation may be continued without interruption after March 1, 1934."

Whether MacArthur ever saw his report Foulois didn't know, but he didn't think so, for it was Drum who responded to it, and more and more he had come to see the Deputy Chief of Staff as an obstacle in his desire to remain in close touch with the Chief.

In his reply to Foulois, Drum stated that MacArthur had told Foulois to base his plans on a three-month operation and to obligate funds accordingly. At the time, no one knew how long the Air Corps was going to have the job, and although MacArthur might have suggested three months as a ball-park figure there could have been nothing positive about it. Further, in seeking depot funds for a year, Foulois was under the impression, gained no doubt from Harllee Branch, that he'd best figure on a long haul. However, it had been through Drum and the Bureau of the Budget that he'd received only $300,000. Told it must cover expenses until the first of March, he already knew that the cost in downed planes and life insurance would exceed that figure. Drum managed to lecture the Chief of Air Corps on how to run an airline, pontificating "that on dates prior to Feb. 26th the Chief of Staff had cautioned you to conduct operations of the Air Mail with 'safety first' as your guide."

There was nothing in what Hugh Drum had to offer to indicate that his understanding of what the Air Corps was up against had passed beyond the confines of his Sam Browne belt. He was more concerned with correcting by criticism Foulois's position on the forces and problems that applied and in so doing to mount guard against adverse publicity reflecting on the War Department.

Monday also brought a telephone call to Foulois from the powerful Senator Kenneth D. McKellar, Democrat of Tennessee, Chairman of the Senate Post Office Committee. What could be done to quiet the outcry of the opposition with regard to fatal accidents? McKellar wanted to know. Foulois suggested that his office would compile a statistical record and put the matter in proper perspective. This was done, and the record showed that over a five-year span the Air Corps had lost an average of about fifty pilots a year in fatal accidents, and the commercial lines, with one sixth as many pilots, averaged approximately one third less. These were not figures that would make one iota of difference in the mounting political uproar. The Republicans had gotten their teeth into the Roosevelt Administration, and they weren't letting go.

Horace Hickam was quoted as saying a policy of "common sense" would govern flights in his zone. Then he pointed the finger at the press, saying that overzealous attempts by his pilots to get the mail through were the "results of unwarranted criticism and false statements. We will not permit our pilots to show off, and I'll take the hide off of anyone who does. Our primary charge is to safeguard the mail both in the air and on the ground. That will be accomplished and this includes the lives and safety of the pilots."

Uncle Horace was not the kind to pull a punch when one needed to be thrown, and when the War Department began to give him static over the command of a supply depot he had established at the Chicago airport, he fired off what was to become a famous telegram. WHO IS COMMANDING THIS THING ANYWAY? he queried the General Staff. MAKE UP YOUR MINDS. LOVE AND KISSES, HORACE.

It was a time of direct telegrams. On his return from his inspection trip, Oscar Westover, King Canute-like, dispatched an order to his zone commanders: THERE WILL BE NO MORE ACCIDENTS. B. Q. Jones promptly fired back a reply that put the situation in proper perspective: THERE WILL BE NO MORE FLYING. Westover got the message and subsided.

Hap Arnold gathered his route commanders: Tinker, Phillips, Eaker and Castor. Most of the pilots were March Field boys, and he knew them all. They were his boys, *goddammit!* and he wasn't going to have them busting their asses just to satisfy the headline writers! There were certain standard rules to maintain safety that must be adopted! Because weather over the mountains was almost totally unpredictable, all pilots would turn back when they could not see the second beacon ahead of them. Because some radio beams in the mountains had a tendency to feather out and give a raft of signals, pilots who were not sure of the route or the signal would put down at the nearest emergency field. There had not been time to set up the particular standards of safety required for the type of flying, but they would make standards, and they would make them work!

Foulois approached the problem in a different way. He went public. Informed by a friendly reporter that the orders from the top were to dig up everything to prove Air Corps inefficiency in both personnel and equipment, he contacted MacArthur. Naturally the Chief of Staff was worried about what was happening, and with a presidential okay he gave his permission to Foulois to take to the airways in an attempt to explain and clarify what the Air Corps was engaged in doing.

Foulois did so at 10:30 P.M., February 27, broadcasting over the CBS network from station WJSV, in Alexandria, Virginia. Perhaps his most salient point was that with all the "hullabaloo about legalized murder" only one pilot had died on a scheduled airmail run.

"Our military pilots are not weaklings looking for sympathy," he said. "They should be treated like men and not children. If there is anything that the average red-blooded American military pilot resents, it is this recent twaddle about inexperienced, rosy-cheeked boys being sent to their deaths."

In defending his men and himself, Foulois described the entire AACMO structure and operation, emphasizing repeatedly the efforts to assure safety. He could not deny, he said, that accidents were bound to occur in commercial as well as military flying and that the latter was inherently hazardous, particularly on such a large scale of operations.

The editorial writers had a field day in taking apart Foulois's plea for understanding, which stressed duty, honor and the risky business in which his men were engaged.

Typically, Congresswoman Edith Rogers wrote to say she was "amazed! Why the attack on the Party which has always helped the Army and the Air Corps and which has been your only defender on the floor of the House?"

Foulois might have responded that with such defenders the war was sure to be lost. One area that he failed to mention in which his troops were being sorely wounded was in the pocketbook, and it was largely the congressional "defenders" who were responsible. Although the House had passed the Brunner bill on the twenty-fourth, granting the much-needed $5.00-a-day living expenses to the officers and enlisted men—and particularly the latter—of AACMO, the Senate "defenders" sat on it while they exercised their partisanship. Senator Arthur R. Robinson, of Indiana, set the tone of debate by first accusing the President of sending Army pilots to their deaths and then proceeded to chant with feeling, "Theirs not to make reply, theirs not to reason why, theirs but to fly and die!" His poetic syntax produced pandemonium in the chamber but very little else.

At the end of the month, B. Q. Jones sent Foulois a telegram asking for authority to put his enlisted men on per diem. Foulois responded in the negative, stressing that everything possible was being done to hurry the legislation, repeating that he was sure "the Senate will pass the bill this week."

In the meantime, enlisted pay ran from seventeen dollars a month for a private to ninety dollars for a master sergeant. Worse, enlisted men found it much tougher to obtain credit than officers. Without having the base mess and quarters to depend on, they were really up against it, and financial hardship became an increasingly debilitating reality in all three zones.

Near the New Orleans airport, the enlisted men lived in a grocery store with no hot water. There was no bus or trolley into town, and a cab cost a prohibitive dollar fifty. There were no rooms available at the airport, and had there been they would have been too expensive.

At Mobile, Alabama, there were no hangar and no repair facilities, and servicing of the planes had to be done from supply drums. The mechanics bunked in the barn of a nearby farm.

Airmail pilot Lieutenant Curtis E. LeMay, assigned to the Richmond, Virginia–Greenville, North Carolina, run, noted his crew chief and fellow mechanics, on a raw, wet day, cooking homemade mulligan in a corner of the hangar. They were heating their thin gruel in an old stewpot over a plumber's flame. He also noted they were using planks between sawhorses to sleep on, protected from the rain if not from the cold by the hangar roof.

"A sergeant would come in beaming," wrote LeMay, "and tell how that nice old lady at the hot dog stand down on the corner—the one they owed for hot dogs and hamburgers until they were ashamed to ask for more credit—that nice old lady had just given him a pillow. She said she didn't need it . . . and she up and gave it to the sergeant."

There were others who helped. At Birmingham, Alabama, thanks to the Department of Commerce radio operator, enlisted men were accommodated in his home. As Section Chief Second Lieutenant Oliver S. Picker was to report, "I would like to call attention to the wonderful treatment afforded everywhere by employees of the commercial lines. In some places such as Mobile and Macon [Georgia], we could not have operated without their help. This was in contrast to the treatment received by the airport managers, particularly at New Orleans. Commercial employees may have had orders to cooperate but they did so not as a matter of policy but in friendliness."

And so did a great many other people, storekeepers, restaurant owners,

boardinghouse landladies. Times were hard everywhere, and they understood. The boys would pay when they got paid.

Some officers took care of their own. Ira Eaker was one. To keep the
wolf away from his men, he borrowed $750 and doled it out in small sums
as needed. At Burbank and control points on his routes he also arranged
for restaurant and rooming-house credit for all under his command.

In spite of the fact that a second lieutenant earned $170 a month with
flight pay, he wasn't much better off than his crew chief, particularly if he
had a family to support. He was a man on the move, and the weather dictated where he would land. A pilot taking off from Newark for Pittsburgh
with exactly eight cents in his pocket told a news reporter, "If the weather
closes in and I have to stay there, it's going to be just too bad. I haven't
any money for a hotel or even to eat on. I've got to get back here, or else."

Another pilot arrived in Chicago from Newark with all of $3.80 in his
wallet. He got a ride into the city. Three dollars went for a hotel, fifty
cents for food. He arose at two in the morning to walk the seven miles
from his hotel to the airport, at Cicero, then took the mail on to Newark.

Captain Melvin Asp had it even tougher. After a long, grueling night
flight, he went to an inexpensive hotel, but because he was short one dollar
of the price of the room, the clerk refused to give him a room, unwilling to
honor the Captain's identity card.

"No see-um per diem" became the wry motto of the airmail carriers
while the senators who professed to be so concerned about their welfare
busied themselves tearing up the political turf while doing absolutely nothing to alleviate the cause of increasing economic hardship.

At the end of the month, Oscar Westover returned from his inspection
tour of the three zones. He did not have a very rosy picture to present to
his boss. Aside from no money for mechanics and pilots, everyone was
badly overworked. The supply situation was not only a matter of shortages
in everything from engine nuts to landing lights but was also tangled in a
spaghetti of red tape.

Equipment was being parceled out throughout the zones even while the
mail was being flown, logistics trying to catch up with operations without
any written transfer of accountability. Consequently, with no forms, no Air
Corps instructions, no technical orders, supply officers had to attempt to
build an orderly system in an atmosphere that was chaotic. One officer
putting in for an emergency requisition for a needed engine part was informed that there "were no funds for purchase of this item in this fiscal
year" but that it would be taken care of next year. Another learned that it
would be a waste of time and gas to send a plane to a depot to pick up an
emergency part; he would have to make his request through channels.

Out of this confusion, Westover reported, a degree of order was being
forged. The challenge required that those employed rise to it, and they
were battling against incredible odds to do just that. Morale was a bright

ray of sunlight piercing the dark overcast. It was there with the pilots, there on the ground with the crewmen, there in the jammed, makeshift administrative offices. There were no hours. At many stations, enlisted mechanics had no hangar space to work under, having to maintain the aircraft out in the open in wretched weather. Because they had been alerted so quickly and moved from their home bases, many were without tools. Often the only way a man could lay his hand on a necessary wrench was to try to borrow one, and failing that, use his meager funds to buy what he needed. There were no complaints. Their spirits were as high as the pilots', and at least Westover could bring back that observation to his chief.

Throughout the extent of his investigation, Westover had been plagued by the foul weather—out of Chicago, out of Cheyenne, out of Salt Lake City, out of Dallas. "The planes coming down from Wichita and Oklahoma City were all coming in with ice on them," he reported. "So that bogged me down for about four hours. And then I had to go over to Memphis, over that route in which our mail was carried in pursuit planes." In Chicago, checking with Hickam, he found there was room for only three A-12's at a time in the National Guard hangar, where landing lights were being installed. It made for slow work. Hickam's staff of over three hundred fifty officers, enlisted men and civilians were cramped for space as well, but under the Colonel's calm direction, the Central Zone was coping.

Westover's report to Foulois simply proved that the problems confronting AACMO were problems of organization created by Foulois's failure to demand enough time in the setting up of a very large operation in which the Air Corps had had no experience and few guidelines to point the way. Time would also have done much to remove the lack of training. The equipment, unsuited as it was, could have been made to function better. Now operational time would provide the experience in all phases, and time would also eventually change the weather for the better. Foulois's overriding argument, which he had used in his broadcast, was that this was an emergency and a great peacetime test for the Air Corps. As for the hardships, physical and economic, his men were, after all, soldiers, and soldiers were there to take it when the going got rough. Their high morale proved how ready they were to prove the point.

If only they could be let alone by the press, and the politicians would go off and do their fighting elsewhere, then they could damn' well prove that the Air Corps was capable of the job ahead.

But it was not to be so. On the last day of the month, Foulois was called to a White House conference, where he was ordered to investigate a rumored campaign conducted by the airline companies of propaganda against the Air Corps and sabotage of its equipment. This last had been emphasized by the rash of crack-ups. He knew it was nonsense. The propaganda, on the other hand, was obviously there in the press for everyone to read, and it took no sleuthing on anyone's part to know that it was aimed at the White House by the political and ideological opposition.

When the word leaked out that the Air Corps was investigating the possibility of sabotage of its equipment, Secretary of War Dern emphatically denied that he had ever ordered such an inquiry either by his own doing or the President's. He termed the press reports of punctured gas tanks, water in the fuel and controls tampered with, "fantastic."

Indeed they were, but Foulois knew what he'd been ordered to do, and so did his disturbed zone commanders. They had enough problems on their hands without having to take time to prove that their crashes had been caused by foul weather and not by foul play. In his subsequent report to the President, Foulois stated that there were undocumented rumors that some airline officers had threatened reprisals against any of their pilots who, as Air Reserve officers, joined up to fly the mail for the Air Corps, and that was as much as could be said on the subject of sabotage.

On the overall subject, however, a great many things were being said to keep the public aroused. Even while Foulois was at the White House, the Speaker of the House of Representatives, Harold T. Rainey, Democrat of Illinois, didn't do the cause any good by stating for wide publication that Air Corps pilots were not properly trained and did not understand how to fly the beam.

Unexpected and welcome support came from ocean-spanner and noted airman Clarence D. Chamberlin. He wrote to Farley—who made the letter public—that the airlines had tried to "magnify army airplane mishaps and minimize their own deficiencies." He revealed that twenty persons had been killed in recent months on United Airlines flights, including eleven pilots, hostesses and an equal number of passengers. The loss in aircraft, he estimated, was more than half a million dollars.

Billy Mitchell was not to remain on the sidelines during the battle. In an address before the Foreign Policy Association, he tore into the commercial airline companies and the National City Bank, which he maintained controlled the financial purse strings of the aviation industry in the United States. What Roosevelt should have done, said Mitchell, was to declare an emergency, seize the airlines on the basis that they were subsidized, and let Air Corps pilots take over the airline planes, utilizing the industry's radio communication network. He accused the commercial companies of being the instigators of a "great hue and cry through their controlled press to discredit the Air Corps." At the same time, he admitted that the Air Corps equipment was "inferior," a lesson he had preached both in and out of the service for fifteen years.

Inferiority was a matter of degree, and Lieutenant Pete Quesada, who was Section Chief on the Newark-to-Cleveland run, knew he was far more fortunate in his "inferior" B-2 Condor than the dozen pilots he commanded on the route. The Condor, with its enclosed cockpit and modern instrumentation, made his nightly trip to Cleveland far less arduous than those who had to fly confined to an open cockpit. And so, when he received orders that his plane was to be grounded, he was bowled over. "What the hell's this all about!" he exclaimed to anyone in shouting distance.

His routine had been to depart from Newark about 11 P.M. carrying over a thousand pounds of mail, the bulk of it collected from various New York City post offices. Flying with him as his crew chief was Master Sergeant Roy Hooe. Hooe was that special kind of noncom the Air Corps could not have flown without. He knew all there was to know about aircraft and a great deal more besides. A square-jawed pro to his grease-stained fingertips, he was totally unfazed by rank, and he knew how to handle a recalcitrant cam shaft as well as the stuffiest of the brass. General Jim Fechet had a great fondness for Sergeant Hooe, loved his wild and woolly yarns. To Spaatz and Eaker, Hooe was the crew chief they were most happy to have along—as he had been on the *Question Mark* flight. Quesada had gotten to know and depend on him while stationed at Bolling Field, which he viewed as an aerial garage for the Washington VIP's.

Having Hooe with him on the run to Cleveland was like having extra money in the bank. There was one regular landing en route at Bellefonte, Pennsylvania. They would come down out of the frozen night onto the empty field, not a soul around, nothing but the whip of the wind and their boots squeaking on the hard-packed snow. They knew where the key to the fuel pump was. Hooe would go get it while Quesada climbed up on the engines and got the gas caps off. Then, by hand pump, the two of them would refuel the plane, the raw smell of the gas as harsh in the nostrils as the air's cut. On one night it was twenty below zero, and even though they were dressed in wool-lined, leather-covered flying suits, the cold left them numb and aching, worried over whether they could get the bird started. At Cleveland, Quesada would catch a few winks in the hangar while Hooe saw to the loading of the return mail from Cleveland and Chicago. Then they were off again, getting in to Newark about eight in the morning.

In spite of the weather, which seemed unceasing in its cruelty, Quesada and Hooe and the Condor were far better able to maintain the schedule than the other planes bucking the route. Thus, orders for the Condor's grounding sent the irrepressible Lieutenant spinning into his CO's office with the query, "Major, what the hell is going on?"

When the Major found out what Quesada's excitement was all about, he explained that orders had come up from Washington that the Condor was unsafe and should be grounded. The word was, there was no way to bail out of it in a hurry. With a long fuselage, and the only means of exit near its rear, the pilot would have a long hike from the cockpit to daylight, and the chances of reaching it wouldn't be all that good.

"So, what are you bellyaching about, Pete?" B.Q. wanted to know.

And Quesada replied, "Fer Crissake, Major, the B-2's the only plane we've got that you don't have to jump out of!"

"What d'yuh mean by that?"

"It's the only plane we've got in the entire section that's properly instrumented to fly in bad weather. Hell, aside from that, it's got a radio that can receive on almost any frequency. So it's one plane you don't have to worry about. It's just the reverse. It's the other planes that should be grounded!"

The Major rubbed his jaw and wondered, "Why didn't somebody tell me that?" And then he made up his mind. "Okay. Good for you, Pete. Forget the grounding."

And that night, Lieutenant Quesada and Sergeant Hooe, sandwich in back pocket, took the mail to Cleveland in the Condor and continued to do so without any more static over the plane's safety factor.

Static of another kind, however, was continuing to batter the operation. There was no doubt in Lieutenant Larry Kuter's mind that the daily uncomplimentary press reports were beginning to have an effect on the pilots. He tried to convince them by statistics—the tons of mail flown, the miles flown—that they were doing one helluva fine job, but some of them were beginning to believe that they were failures and that the Air Corps was falling down on the job. The attitude was dangerous, because in spite of their commander's stress on safety, the criticism goaded them into taking unnecessary chances. It was not the Air Corps that was failing, but the god-awful weather! The result was that there were some days in all zones when more flights were canceled than flown.

In an effort to get around the unfavorable publicity that was plaguing operations in the Western Zone, Ira Eaker got out his pen and, between flights, wrote an article for *Liberty* magazine giving the Air Corps pilot's view on flying the mail. Arnold, in a letter to Chief of Information Walter Weaver on Eaker's bid, wrote, "I think it will do a lot of good, and I hope you can get to the Chief for permission to publish." Hap wanted a telegram of approval, which he quickly received. Unfortunately, although *Liberty*'s West Coast representative had asked Eaker to do the article, the magazine rejected the piece. But Eaker didn't stop there, and eventually the article was published in the New York *American,* in early April. He also wrote other pieces; one, an eight-page counterattack against the critics, was titled "The Air Corps and the Air Mail" and was signed by his former mentor, retired Major General Jim Fechet. Eaker's piece "Will Mitchell Come Back?" was in keeping with current speculation but was wishful thinking.

In a letter to Weaver, he gave his approach and thinking on the problem:

> Roughly our main effort has been to keep everything derogatory or unfair out of the newspapers. With the exception of the Los Angeles *Times,* which is owned by Chandler, a large stockholder in Western Air, we have

gotten a very decent break. This has been accomplished by personal contact with the aviation editors. I think we should grab all of our people who
have any experience in publicity or public relations and get them on this
work. I have already long since learned that publicity is entirely a matter
of personal contact with the editors and the people who write the stuff.
These aviation companies have high-priced public-relations counsels feeding propaganda against us to the press. The only way we can counter it is
by similar methods.

In the Central Zone a different kind of trouble was firmly and deftly laid
to rest by Horace Hickam. In his nationwide radio broadcast, Foulois had
announced that the Air Corps would be employing interested airline
pilots,[3] particularly those who held reserve commissions. Of the six hundred five rated U.S. airline pilots, in the Central Zone only thirty-eight
held commissions or were otherwise qualified. By the last week in February half of them had applied for flying jobs in the zone.

Hickam wasn't all that anxious to sign them up, because he felt since
they were used to flying modern, properly instrumented aircraft with dependable communications equipment, they would be at a disadvantage
manning Air Corps planes. However, when he learned that Northwest
Airways Chief Pilot Fred W. Whittemore had held a meeting in St. Paul,
Minnesota, with the pilots of the line and had reportedly threatened them
with the loss of seniority and possibly their jobs if they flew the mail for
the Air Corps, he acted swiftly. He fired off a telegram to Croil Hunter,
vice-president and general manager of the company:

HAVE INFORMATION THAT YOUR ORGANIZATION HAS THREATENED RESERVE
OFFICERS APPLYING FOR ACTIVE DUTY WITH DEMOTION OR LOSS OF FUTURE
JOBS IF THEY ACCEPT ACTIVE DUTY. CANNOT BELIEVE THIS TO BE TRUE
SINCE THE CONSEQUENCES WOULD BE SO DISASTROUS TO YOUR FUTURE
PROSPECTS. PLEASE WIRE ME AT ONCE AN UNQUALIFIED DENIAL IN ORDER
THAT I MAY QUIET APPREHENSION OF RESERVE OFFICERS.

The next day, another Northwest senior pilot, Mal B. Freeburg, returned from Washington to St. Paul to find his fellow pilots, whom he
represented in the Capitol, upset over their meeting with Whittemore.
Freeburg then had a conference with Whittemore and Hunter and told
them he had just come from assuring the Air Corps that Northwest would
give its wholehearted cooperation to AACMO. The firm's president was
also in Washington, but Hunter and Whittemore, instead of responding to
Hickam's demand, went over his head and contacted Army Assistant Chief
of Staff for Training and Operations Brigadier General John H. Hughes.
They wanted to know what authority the Zone Commander had and how
dared he send them a threatening telegram? They also sent a similar blast
to Westover.

Before the Northwest officials had a response from Washington, they received a follow-up telegram from Hickam stating that if he did not receive a reply to his original query by that evening, he would assume the information on their attitude was correct.

That turned things around in a hurry, and the reply he got back stated that any Northwest pilot who went with AACMO could return to the company with the same seniority and rank held previously: OUR PILOTS WILL IN NO WAY BE AFFECTED BY ACTIVE DUTY WITH THE ARMY. GENERAL WESTOVER WAS IMMEDIATELY NOTIFIED WHEN AIRMAIL CONTRACTS WERE CANCELED OF OUR FULL COOPERATION WITH THE ARMY IN EVERY WAY. YOUR INFORMATION ENTIRELY ERRONEOUS.

A few days later, as Hickam related, "Whittemore personally called me on the telephone and assured me of the desire of Northwest Airways to cooperate to the fullest extent with me personally and with the Army Air Corps."

The end result was that five Northwest pilots were immediately called to active duty to be placed on their former routes, and the airline went even further in its program to cooperate. One of its officers, Colonel Lewis H. Brittin, called at the Air Corps Chief's office, and Executive Officer Jim Chaney wrote to Hickam that the company had now offered the use of its new hangar at Pembina, North Dakota, to the Central Zone Commander.

And so, through a bit of firmness, Uncle Horace had turned around a situation that in less capable hands might have produced additional friction. The War Department's action through General Hughes was to send an inquiry through the Adjutant General asking for an explanation of the directness of Hickam's telegram to the company. By the time he replied, the matter had been successfully resolved.

Would that Benny Foulois could say the same. On March 8, he had finally learned from Postmaster General Jim Farley that the Air Corps should be prepared to continue flying the mail until June, when it was anticipated that reorganization of the civilian airmail operations would be completed and the job would revert to qualifying airlines.

All well and good. But the bill to pay the men per diem and the mounting costs of operations remained fixed in a senatorial holding pattern.

At the end of February, Drum had informed Acting Executive Lieutenant Colonel Jake Fickel by telephone that the Bureau of the Budget had authorized the use of an additional million and a half dollars, approximately, of impounded funds for AACMO. But none of this money had been forthcoming, and in a move to break loose the funds with War Department assistance, Foulois sent a full report to MacArthur.[4] In it he summed up the growing economic hardships of his men:

> Lack of funds to pay enlisted men the per diem rate of $5.00 per day has created an embarrassing condition which reflects most unfavorably on the Army. Officers and men are quartered at considerable distances from

the fields at which they serve. Temporary credit allowed during the early days of the emergency has been strained to the breaking point.

Foulois was sure his position paper would bring action by MacArthur. Instead he received an infuriating response from Drum, wanting to know "without delay what steps are being taken to relieve the situation."

Foulois's private reaction was that if the damned fool bothered to read the newspapers he'd know that the situation was in the Senate, and since it was Drum who had passed the information to Fickel on additional funds, he should know about that, too! The Air Corps Chief believed that Hugh Drum had purposely blocked his direct report to MacArthur, and he vowed that henceforth he would find ways around the unhelpful Deputy.

That Foulois was sensitive to the economic condition of his troops was reflected in a telegram he sent to B. Q. Jones on March 7. He cited a story in the New York *Daily News* of that date quoting an unnamed pilot at Newark complaining that he had received no per diem and that he and many of his fellow pilots were in a financial bind. "You will investigate immediately," ordered Foulois, "and report name, rank and serial number of all cases of financial distress amongst officers and men of your command."

Thirteen

Political Action

March 9 was a long month away from the crucial meeting Benny Foulois had held with Harllee Branch at the latter's office in the Post Office Building. It was also a Friday. In spite of the continuing steaming political battle between Roosevelt's supporters and his opposition, who were after Jim Farley's bald scalp, and in spite of all the obstacles, natural and manmade, that had chopped away at the Air Corps's credibility, AACMO seemed to be getting a firm grasp on its mission. Its pilots were carrying tons of mail, completing close to 70 percent of their runs without being forced by weather and mechanical failure to ship by train. Fourteen civilian airmail pilots had already been called to active duty and more would follow. The long-awaited Martin B-10 bomber would soon be replacing the worn-out Keystones. The twin-engine B-10, with its enclosed cockpits, retractable landing gear and cruising speed of 170 mph, would really make a difference in load capacity and fast scheduling.

In the three zones, staffs had grown to accommodate the multitude of needs the vast operation demanded. Pilots in the Eastern Zone had doubled to number one hundred four. In addition, there were three instrument instructors, twenty-one control officers on the eight routes, backed by eighty-three administrative officers, four hundred eight enlisted men and a dozen civilians. Nationwide, Air Corps pilots were covering nearly forty-one thousand miles a day. The two most serious problems were the continuing bad weather and the absolutely inexcusable lack of appropriations to pay living expenses. Officers and men were in debt to the tune of an estimated quarter of a million dollars. But, even so, one could look ahead. The weather, moving on toward spring, would moderate. Eventually the Senate or the Bureau of the Budget would have to act. Things were beginning to look up.

And then, on Friday, March 9, the sky fell on Benny Foulois, his zone commanders and the Air Corps in general.

The Curtiss O-39 was a souped-up version of the Curtiss Falcon, with the added distinction of a canopied cockpit. The Air Corps owned ten of these biplanes, all having been formerly assigned to the 9th Observation Group, at Mitchel Field. Modified to carry two hundred fifty pounds of mail in the rear cockpit, and with a cruising speed of 140 mph, the O-39 had become a workhorse on the runs over the Allegheny Mountains.

On that Friday night, First Lieutenant Otto Wienecke, an experienced airman who had nevertheless flown only twenty-four hours in the previous year and a half, took off in his O-39 from a refueling stop at Kylertown, Pennsylvania, bound for Cleveland. Weather reports forecast snow squalls along the route, heavy at times but nothing too serious other than a cold, dark flight. Near Burton, Ohio, Wienecke flew into such a squall. The wreckage of his plane indicated he had spun straight into the ground. There was no sure reason why. Both the artificial horizon and the directional gyro were mounted in the cockpit on a level with his knees and were therefore difficult to read, but all that was really known was that the Lieutenant was dead.

At about the same time Wienecke was heading west out of Kylertown, Lieutenant Walter W. Reid was taking off with the southbound mail from Daytona Beach in a Keystone B-6A. On board with him were Private First Class Ernest B. Sell, the plane's mechanic, and Private A. M. Marshall. Barely airborne, the Keystone began to lose power. Reid fought to keep the lumbering biplane in the air while Sell, in the rear cockpit, struggled with the fuel pump, trying to get the line cleared. At five hundred feet the engines quit and the bomber dropped in a stall. It slammed into a cypress swamp adjoining the field. When the splinters stopped flying, Reid and Marshall were clear of the plane and unhurt, but Private First Class Sell had not been so lucky. His head had been smashed by the impact. He was dead.

Air Reserve First Lieutenant Bernard A. Schriever, who had graduated from Kelly Field in July 1933, had been flying the Salt Lake-to-Cheyenne run in tricky Douglas O-38's since late in February. On March 8, he had landed at Cheyenne. Flying in with him in the same type of aircraft were two former classmates at Kelly, Second Lieutenant Frank L. Howard and Air Reserve Second Lieutenant Arthur R. "Duke" Kerwin. Howard was a graduate of West Point, class of 1932, and a good friend of Schriever's. They had come to Cheyenne to work the radio range to improve their beam- and night-flying proficiency. Howard was pilot of the O-38E, which boasted a sliding canopy over both cockpits, and Kerwin was along to observe.

The three spent Friday in Cheyenne. Schriever, a champion golfer* as well as a poker aficionado—who had previously been admitted into the select Spaatz-Eaker-Tinker March Field table-stakes circle—was anxious to be airborne. It was too cold for golf, and there were no indoor squash or tennis courts available in which to unleash his twenty-three-year-old, six-foot four-inch frame. As for poker, they were all too broke. They could hangar-fly and talk about what they were doing.

When darkness came, the three pooled their resources and got a cab to the airport. It was very cold, the air thin and knifelike at the high altitude, and they had parked their planes between two hangars to shield them from the wind's cut.

After they checked the en-route weather, it was decided Howard and Kerwin would take off first. Schriever said so long, he'd see them in Salt Lake. Then Howard got the O-38's Pratt & Whitney "Hornet" started and taxied out from between the hangars. Instead of turning and going to the far end of the field so that he would have the advantage of its total length for his run, he swung the plane around opposite the hangar. Schriever wondered why he was not utilizing all the room available, thinking that perhaps in the darkness Howard did not know he had a good stretch of field behind him. Otherwise, why attempt such a short takeoff?

Howard must have realized he was in trouble before he ran out of field. Certainly Schriever knew it. As he watched the dark form of the plane, enclosed by the flame of its exhaust stacks, he was suddenly chilled by more than the cold. Beyond the field, there were high-tension wires. Howard tried to haul the plane off in a desperate bid to clear them, but the O-38 simply didn't have the power to generate enough lift at the altitude. Horrified, Schriever saw the silhouette of the plane falling, heard the sickening crash of impact followed by explosion, saw the darkness torn by roiling flames. Others around him were running, car engines were starting, a siren wailing. Sickened and dazed, he turned away. Too shaken to fly, he returned to the hotel in Cheyenne and got drunk.

Benny Foulois may have felt like doing the same thing, but he couldn't afford the luxury. With the time differential in the West, it seemed to him that he received the news of the four deaths in such rapid-fire sequence that all three crashes happened at the same time.

Although in his radio broadcast and subsequent comments to the press he had cautioned that there were bound to be accidents, his warnings meant nothing and were lost in the ensuing whirlwind of press and congressional reaction. The outcry against Farley and FDR reached a new high, and it did not matter a whit that only three days past an American Airlines plane had crashed, killing four, one of whom was Hugh Sexton, the aviation editor of the Chicago *Tribune*.

* He broke the Victoria Country Club course record in 1934.

In another emergency Saturday session, Representative Edith Rogers attempted to get the House to express a consensus opinion that the Air Corps should be ordered to stop carrying the mail forthwith. Speaker Rainey refused to recognize her. She persisted, demanding to speak for five minutes. House Leader Joe Byrnes headed her off at the pass, objecting, and then moved that the House stand adjourned. But the adroit parliamentary maneuvering could do nothing to still the rising demand.

In the Senate, powerful Ohio Republican Simeon D. Fess leveled charges of "inhuman, un-American, indefensible conduct" against President Roosevelt. "The whole country is calling for a stop to this legalized murder!" he fumed.

Benny Foulois was being called on as well. MacArthur telephoned him first thing in the morning and asked him to come over. They were expected at the White House shortly. When he arrived at the Chief of Staff's office he found him pacing the floor.

"Benny," he said, "you're in trouble. I don't know how you're going to explain three fatal accidents in different parts of the country in one night, but I think you'd better try to find the words somehow. The President is extremely upset by the public reaction. . . ."

It was a very low point for the Chief of Air Corps, and it got even lower after he and MacArthur took a silent drive to the White House. They were met by Press Secretary Steve Early, who escorted them upstairs to the President's bedroom. Ensconced in the famous Lincoln bed with its overhead canopy, the morning newspapers with their blistering headlines on either side of him, Roosevelt had anything but a bright greeting for the pair. Foulois was the focus of his wrath.

"General, when are these air killings going to stop!" was his angry opening.

The answer was pure reflex on Foulois's part. "Only when airplanes stop flying, Mr. President," he said flatly.

It was a contingency Roosevelt had already moved to implement. After administering ten minutes of what Foulois later described as a tongue lashing, "the worst I ever received in all my military service," the President informed the two that he had just written a letter to Secretary of War Dern, instructing him to take immediate action.

The letter was illustrative of Roosevelt's political astuteness in a bad spot that the less-than-intelligent impulsiveness of his actions had gotten him into in the first place.

To Dern he wrote:

March 10, 1934

My dear Mr. Secretary:

On February 9 the Army Air Corps was given the temporary assignment of carrying the airmail and commenced the actual carrying on February 20. This action was taken on the definite assurance given me that the Army Air Corps could carry the mail.

Since that time 10 Army flyers have lost their lives. I appreciate that only four of these were actually flying the mail, but the others were training or were proceeding to the mail routes. I appreciate also that almost every part of the country has been visited during this period by fog, snow, and storms, and that serious accidents, taking even more lives, have occurred at the same time in passenger and commercial aviation.

Nevertheless, the continuation of deaths in the Army Air Corps must stop. We all know that flying under the best of conditions is a definite hazard, but the ratio of accidents has been far too high during the past three weeks.

Will you therefore please issue immediate orders to the Army Air Corps stopping all carrying of air mail, except on such routes, under such weather conditions and under such equipment and personnel conditions as will insure, as far as the utmost human care can provide, against constant recurrence of fatal accidents.

This exception includes, of course, full authority to change or modify schedules.

As you know the period of emergency will end as soon as the necessary legislation has been enacted and new contracts can be obtained. I am writing once more to the chairman of the House and Senate committees urging speed in the enactment of legislation. Because military lessons have been taught us during the past few weeks, I request that you consult immediately with the Postmaster General and the Secretary of Commerce in order that additional training may be given to Army air pilots through cooperation with private companies who later on will fly the mails. This should include, of course, training in cross-country flying, in night flying, blind flying and instrument flying.

I am sending a copy of this letter to the Postmaster General in order that he may make arrangements with you. He will, of course, modify the instructions given on February 9 to conform with the Army plans.

Both Generals recognized that the purpose of the letter was not only to shift the blame for what had happened away from the White House and the Post Office Department onto the back of the Air Corps and Foulois but also to calm congressional outrage by maintaining it was up to Congress to act on the necessary legislation to enable the civilian airlines to carry the mail again.

When the pair had returned to the War Department, MacArthur, who had remained silent throughout the ordeal, again reiterated that the ball was in Benny's court. He would back him up as best he could, but there was no doubt that the Chief of Staff was shaken by the Air Corps's seemingly poor performance. Aside from all the perfectly sound reasons Foulois could cite that had produced the grim death toll, the reality was that the Air Corps was the *U. S. Army* Air Corps, and even though MacArthur had granted it a large degree of independence in the emergency, the negative side of its record reflected directly on the War Department. MacArthur could and probably did sympathize with Foulois's difficult position. It took little perception to see that Benny, whom he had known for twenty-five years, was the fall guy. But that didn't change anything. In the eyes of some members of the General Staff, such as Drum and Kilbourne, Foulois and his Air Corps had failed miserably, just as they had suspected they would, and the failure was most damaging to the image of the War Department.

The editorial response to Roosevelt's letter to Dern shifted the point of attack away from the President onto his Postmaster General. It was Jim Farley who now, more than ever, became subject to virulent press and congressional brickbats. FDR did nothing to come to his friend and cabinet member's support. The White House refused to comment further on the matter, and Farley, who had only reluctantly agreed to Roosevelt's precipitous cancellation move, was left to fend for himself.

So, too, Foulois. He could not view FDR's action from any other than his own precarious position, could not realize that the President was subject to political realities too and, like MacArthur, could not ignore public disapproval no matter how exaggerated the claims of the opposition. With MacArthur's concurrence, Foulois made two quick moves. First he met with Stephen Cisler, of the Post Office Department, and reduced route schedules by 40 percent. He then reported to the Secretary of War that he had personally verified the "adequacy" of all equipment and that he had informed his zone commanders to use only their most experienced pilots. He added that in view of the weather he expected there would be more casualties. Dern, in approving Foulois's plan, told him in so many words that he could proceed, providing there were no more accidents, and if there were it would be on the Chief of Air Corps's neck.

Foulois called a meeting of his Washington staff and decided in view of the box he'd been put in he would cancel all operations for ten days while he personally visited with his zone commanders and checked their equipment and procedures. He privately blamed the President for retreating in the face of opposition pressure, but now he was retreating too, buying time.

The decision was a hard blow to the commanders and their men. They didn't understand it. In a letter to Walter Weaver, Ira Eaker expressed the general feeling:

> Our morale is pretty low among Army airmail pilots, as you can well guess, in view of the general official attitude. There may be some reason making it impossible, but from our point of view out here we are wondering why someone in authority has not told the wide world that the Army Air Corps can fly the mail, not only can, but has! Out here we experienced no difficulty in getting it through practically on schedule, except for some delays in starting caused by the Post Office Department.

Lieutenant Benny Schriever was incensed when he was suddenly relieved from flying the mail, because he had been out of flying school for less than a year and was now judged, along with all his fellow classmates, not experienced enough to continue his regular run.

In a letter to Major Asa Duncan, AACMO's G-2, Major Charlie Phillips wrote from his headquarters at Pearson Field, Washington: "During the entire period that Rt. 5 was operating from Seattle to Salt Lake City mail schedules were cancelled only once on account of severe weather conditions. We have been advised by the local post office people that the schedule operated as satisfactorily as any they ever had, so you can imagine our chagrin when we were suddenly ordered to cease operations the other day. . . ."[1]

And Hap Arnold was to state pointedly: "Nothing was gained by stopping the mail for the period March 10th to the 19th in the Western Zone insofar as improving the airmail service was concerned."

That statement was seconded by Hickam and Jones, as well as all the pilots under their command. But they had no voice in the decision. They could only obey orders, and meet with Foulois in Chicago that next week.

Although Oscar Westover was the appointed chief of AACMO, it was Tooey Spaatz who accompanied Foulois to the conference. The accent was on blind-flying instruction, the installation of two-way radio equipment and the replacement of pilots who had less than eighteen months of active duty.[2] Except for the replacements, most of it was window dressing. Blind-flying instruction had been going on full blast since the start of operations, although a blind-flying school at Wright Field was to become an ongoing facility and the Avigation School, at Rockwell Field, would be discontinued so that its instructors could set up shop at the main operating terminals of the Western Zone. Installation of two-way radios was a matter of acquisition and numbers, which all the press releases in the world could not change. At the time of the Chicago meeting there were fifty-three planes in the Eastern Zone with two-way radios, twenty-seven in the Central and thirty-seven in the Western, which added up to fewer than half of the total planes engaged.

Two of Foulois's changes, which were a measure of his state of mind after his bedroom encounter with Roosevelt, were realistically modified at the conference. He had decided there would be no night flights and that all daylight flights must be carried out with a ceiling of at least three thousand feet. Since 90 percent of all mail flying was done at night, the move, if allowed to stand, would have finished the Air Corps as a mail carrier. The need for such a high ceiling would have had much the same effect. The zone commanders got it lowered to a more practicable five hundred feet by day and a thousand feet by night. With Cisler, Foulois and Westover had reduced the routes from sixteen to nine. B. Q. Jones suffered the biggest cut, dropped from eight to three. Hickam lost two, and Arnold kept his four with the spur run to San Diego.[3] The cuts meant that the route miles to be flown were chopped nearly in half and the scheduled miles to be flown daily were reduced from over forty thousand to less than twenty-six thousand.

During the stand-down, Foulois went on to inspect some of the sections and control points in the Eastern Zone. Then, with White House consent and War Department approval, he made a statement in which he defended the Air Corps airmail performance by saying that the Corps "must be completely organized, equipped and trained to meet any national emergency upon twenty-four hours notice."

His theme was that peacetime tests must be carried out under conditions of wartime service, and hazard was ever involved. But the underlying key to his message lay in a single paragraph:

> The opportunity afforded the Army Air Corps to carry the Air Mail, with but approximately ten days in which to prepare, constituted, in my belief, an ideal peace-time test of the Army Air Corps organization, equipment and training, and as Chief of the Army Air Corps, I freely, frankly and without reservation, welcomed this opportunity. [Foulois's underlining.]

No doubt he did, but in view of the public impression, the statement seemed ready made to be torn apart by editorial writers. Ten deaths and more than twenty crashes, either directly or indirectly connected to AACMO, in three weeks did not reflect well on either Air Corps equipment or Air Corps training. Foulois did tie fatal accidents in both military and civilian aviation to the need for "adequate Federal appropriations," but behind his effort to seek understanding, it appears that both the White House and the War Department were not unwilling to let him speak out so as to remove themselves further from being a party to the Air Corps's performance. There had been mounting congressional criticism over the revelation that it had been the Air Corps Chief alone, without War Department consultation, who had committed the Air Corps to flying the mail. Roosevelt was called on to name the persons who had advised him that the Air Corps could safely handle the big assignment. In letters to the Chairmen of the House and Senate Post Office committees, he asked for quick action on new permanent legislation for setting up contracts with commercial carriers. In so doing, he made mention that before he had signed his executive order of cancellation, a member of the General Staff had assured him that the Air Corps could fly the mail successfully.

Former Senator Hiram Bingham, who was president of the National Aeronautic Society, demanded to know what officer on the General Staff had done the assuring, since he was sure it hadn't been General MacArthur. Bingham, a man of many parts, not the least of which had been his rank of Lieutenant Colonel in the Air Service during the World War, had his comments published in an interview with O'Laughlin in the latter's *Army and Navy Journal*. And both the interview and Bingham's question were carried to the White House. Shortly thereafter, the President telephoned his Secretary of War. Coincidentally, present with Dern were Generals MacArthur and Foulois. Roosevelt informed Dern that, indeed, MacArthur had told him that the Air Corps was capable of carrying the mail. Dern repeated this to the General, who then asked to speak to the President. Whereupon the Chief of Staff informed his Commander-in-Chief that he was mistaken: ". . . I never telephoned you. I knew nothing about your plan to have the Air Corps carry the mail," he said.

"But you are mistaken, Douglas," Roosevelt replied. "You phoned me, as I have said."

MacArthur responded in the negative, saying the only time he had discussed the matter was "when you called General Foulois and me to the White House and gave us a spanking."

FDR persisted, maintaining his appointment secretary, Marvin McIntyre, had put the call through. MacArthur asked to speak to McIntyre and the President hung up.

Not long afterward, mild-mannered McIntyre appeared at MacArthur's office and tried to reiterate the President's line. MacArthur would have none of it, growing angrier by the moment. Finally McIntyre, after attempting to shift the burden of the supposed telephone call onto Steve Early's shoulders, was forced to admit there had been no call but that Early had said MacArthur would be a good sport and see that the mails were carried.[4]

It was Benny Foulois's willingness to accept all responsibility that deflected the criticism away from the White House and the General Staff. It also laid to rest any further deterioration of relations between the Chief of Staff and the President. But nowhere did anyone make mention that it had actually been Secretary of War George Dern who had casually committed the Air Corps, at the cabinet meeting of February 9, to the undertaking.

Momentarily, the Air Corps felt as grounded in spirit as in flying the mail. Quoted in the press, an unnamed "high official" in the War Department, who sounded suspiciously like either Drum or Kilbourne, managed to put the worst possible connotation on the situation and do an enormous disservice to the Air Corps and its pilots. He said in self-righteous confession:

> For years we have built up a reputation which has been commonly accepted that the Army air force contained the best flyers in the world. That reputation is now wiped out and we must start at the bottom again. The President's letter is a bitter pill, but I think we all realize there was justification for it. We are soldiers and we will do everything we can to bring the corps to the highest standard at the earliest possible moment.

During the ten days of recouping, Dern moved to form a new investigating board—the sixteenth in fifteen years. He had discussed its formation with Charles Lindbergh, who had come to Washington—as had Rickenbacker and Chamberlin—to testify on the drafting of new airmail legislation. Dern had made a special point of inviting Lindbergh to come to see him and asked the noted flyer to serve on the board, whose purpose would be to report and make recommendations on the Air Corps airmail operation and the "adequacy and efficiency of its technical flying equipment and training for such a mission."

Lindbergh refused the offer, maintaining that the entire affair had been "unwarranted and contrary to American principles." He felt he was being used politically. Orville Wright, contacted by Dern at the same time, turned down the offer for reasons of health. Chamberlin accepted it. So did Jimmy Doolittle, who had left the service in 1930 to fly for the Shell Oil Company. Former Colonel Edgar S. Gorrell, of Mexican border and World War fame, also agreed to serve, as did Dr. Karl T. Compton, president of MIT, and Dr. George W. Lewis, director of research for NACA— the National Advisory Committee for Aeronautics. Dern added to this group the original members of the Drum Board and then asked former Secretary of War Newton D. Baker to chair the committee.[5]

On March 19, Foulois gave to the press a four-page statement on new safety measures that were being instituted. Nevertheless he cautioned, as he had from the beginning, "The Army Air Corps and the public must be prepared for future fatalities."

Two days later word was flashed that Lieutenant Harold G. Richardson, a recalled airline pilot, had crashed near Cheyenne and was dead. He had taken off on a training flight in a Douglas O-38E, had climbed to two thousand feet and the plane had whipped into a spin from which it did not recover. In his report on the accident, Hap Arnold wrote: "Investigation indicated that this particular type of plane was very unstable under certain loading conditions and high altitudes, and the plane was quite prone to go into a spin under these loading conditions and once in a spin it required about 2,000 feet to get out."

On the nineteenth, the Air Corps, its wings somewhat clipped, went back to flying the mail on a reduced scale. Its men were still waiting to be paid their per diem. Both Roosevelt and Dern, particularly the former, were determined that Foulois guarantee no more accidents—or they wished the record to show that they had insisted all crashes cease, and should there be any more they could not be held responsible.

Wrote Roosevelt to Dern on March 18:

> I understand that the War Department has ordered the Army Air Corps to resume carrying the airmail at 12:01 A.M. tomorrow. I cannot approve this order unless you have received definite assurances from the respon-

sible officers of the Air Corps that the mail can be carried with the highest degree of safety.

In this connection, I invite your attention again to my letter of March tenth in which I said, "The continuation of deaths in the Army Air Corps must stop."

I wish you would issue new instructions to the Air Corps. In these instructions, please make it clear that, if on any route, on any day, the conditions of weather, personnel or equipment are such as to give rise to any doubt as to the safety of moving the mails, that is from the standpoint of human safety, the mails shall not and will not be carried.

Finally, if the responsible officers believe that the carrying of any part or all of the mail ought to be stopped at any time, you will be prepared to stop it immediately.

Dern's reply to the President and his instructions to Foulois listed the precautions taken in detail. The exchange illustrated the frustration of not having the political power to order a godlike obeyance of a highly technical activity. The only recourse was to say in writing what one could not control through executive fiat. In short, don't blame me.

It was shortly before the nineteenth that B. Q. Jones shifted his headquarters from Floyd Bennett to Mitchel Field. He, too, had been fighting a political battle, although on a somewhat lower and less serious level. His antagonist was New York's colorful, aviation-minded mayor, Fiorello La Guardia. The wrangle was over rent, for the Little Flower, as he was called, would lease the field and facilities only on a day-to-day basis. B.Q. eventually found the arrangement impracticable.

The two had known each other previously in France during the war, when La Guardia as both a congressman and a flying officer wielded more clout than any captain in any Army air service. The similarities in their feisty temperaments had not produced friendship. Neither was the kind to give way in a dispute and both were individualists. The Major told the Mayor that unless an acceptable agreement was reached by 5 P.M. he was going to move out of his god-damned airport and go over to Mitchel. La Guardia's reply was in the same vein. He wasn't going to be dictated to by any pip-squeak major!

When the Mayor did not arrive at the appointed hour, B.Q. and all but one of his staff departed for Mitchel Field. The one left behind to act as a welcoming committee and handle whatever came next was Lieutenant Kuter. Mayor La Guardia came next, at about five-fifteen. His party arrived in two black sedans, the other occupants councilmen. To Kuter it looked as if the Mafia had moved in. The Mayor said he wanted to see Major Jones right now. Kuter explained that the Major had departed at five o'clock. La Guardia insisted. He had come all the way from the city to see Jones. Kuter explained again; he was all that remained. Fiorello, fuming a bit, realized that the grizzled bird had indeed flown away and there was no point debating the matter further with a first lieutenant. He stomped out of the office, and the two sedans disappeared into the winter's gloom.

Of course, setting up shop at Mitchel Field was a big financial improvement for the officers and enlisted men in B. Q. Jones's headquarters staff, because they were on an Air Corps base, where food and housing were provided.

Out on the West Coast, airmail operations were about to begin again, and Ira Eaker had an idea: institute the use of copilots. He wrote accordingly to Hap Arnold giving his reasons why the thought made sense. It was an added safety measure. It would give the younger pilots "excellent training," and should the Western Zone be opening up new lines they would be in a better position to take them on. But his main reason for the suggestion, he said, was that this was the quickest way to answer the accusation that "our pilots are not trained well." Further, he could see a situation approaching in which they would have more recalled civilian pilots on the runs than Army pilots. "The first day the news got out that we had some Western Air pilots called to active duty," he wrote, "the aviation editor of the Los Angeles *Examiner* called me up and tried to get me to admit that we, having found our pilots were unable to carry the mails, were now going to get the civilian pilots to put it over for us."

Using copilots would certainly be a sound move in every respect, and Arnold was all in favor. With his suggestion, Eaker enclosed a copy of his latest article, one of a number he had written in a kind of collaboration with his boss. It was a collaboration that would grow.

Between March 19 and May 8, when the commercial lines began carrying the mail once again under legislated regulations affording more equitable distribution of contracts, there was a single fatal crash. Horace Hickam termed the cause willful and unnecessary pilot error.[6] However, by the time the zone commanders received word to start cutting back operations, the affairs of AACMO were no longer front-page news or much news at all. The reason was that the Air Corps had the situation under firm and well-organized control, and newspapers do not dwell for long on success stories.

There were other reasons, of course. Politically, the Administration, with congressional approval, was anxious to get the carrying of the mail back where it belonged, and its efforts to do so by breaking up the holding companies and instituting open bidding on the routes calmed the criticism to a manageable level.[7] Meteorologically, the country shed its winter storms and faced up to spring. Economically, on March 27, the Senate finally passed the legislation that made it possible for officers and men to receive their very long awaited and much needed per diem. Aeronautically, the crisis had served the purpose of speeding up production of the Martin

B-10 bomber, and it had also established once and for all the importance of instrument flight training.

During the last two weeks of April, six B-10's were placed on the Newark-to-Chicago run and six more on the route between San Francisco and Cheyenne. With their range, speed and mail capacity of two thousand pounds, the B-10's were able to move the mail as nothing had done before. Their only weakness, which Arnold drew attention to, was their need, at high altitudes, for a very long takeoff run.

At Wright Field, Captain Albert Hegenberger, who was in charge of a new blind-flying school, began training all B-10 pilots in his instrument landing procedure. The accent was strongly focused on instrument flying proficiency. The Link trainer made its debut, and by the end of April, Wright engineers had developed plans to install ground transmitters at Newark, Cleveland and Chicago that would permit the use of the Hegenberger system. Had AACMO continued, plans called for installing the transmitters on a nationwide basis.

With the end of April and the orders to start phasing out, the thought was passed from Arnold's command that it might be a nice public-relations touch to put the B-10 to a real test. Eddie Rickenbacker had won no Air Corps friends by his comments or his widely heralded transcontinental flight in February. Why not see if that record could be beaten? Maybe the attempt would earn a favorable headline or two.

On May 8, a B-10 took off with a full load of mail from Burbank, California, and, throttles to the wall, gave it the old school try. Lieutenant Pete Quesada flew the last leg, from Cleveland to Newark. He was the first in the Eastern Zone to fly the B-10, though when he first saw the plane, he approached it warily. After he and Sergeant Hooe had climbed aboard and looked around, Quesada said, "I'll tell you what, Roy, you start it, and I'll steer it." He didn't add that the sight of the thing scared the hell out of him. But not for long; a real pilot could fly anything.

As newscaster Lowell Thomas said that evening on his daily broadcast: "With throttles wide open the plane flashed from coast to coast. The time: thirteen hours and fifty-three minutes. A little longer than it took Eddie Rickenbacker. But a record just the same, because the Army plane flew two hundred and seventy-nine miles farther and made three more stops. . . ."

The following day, Postmaster General Jim Farley wrote Benny Foulois a two-page letter of congratulations and thanks for a job well done. In it, he made mention of the flight, citing the elapsed time as being a bit more than that recorded by Thomas's stop-watch (fourteen hours eight minutes) and mentioned that the point of departure was Oakland, but pointing out: "This illustrates how well the Army Air Corps was performing its job when it began turning the service back to the commercial carriers."

The most salient feature of his letter, however, stated:

> Undoubtedly, the Army Air Corps and the national defense will benefit from the carrying of the airmail during the emergency. The country and the Congress will, without doubt, give a more adequate support to the Army, will see to it that it has the most modern equipment obtainable, and that sufficient funds are provided for the flyers to have the additional hours of flying which have so long been needed.

That the President for political reasons had been forced to realize the Air Corps's need was also apparent. At his cabinet meeting on March 9, he approved the authorization of the seven and a half million dollars in PWA funds for new aircraft and equipment, and he agreed that the Corps would need additional money in the year ahead.[8]

What had begun with a great deal of discordant fanfare, rising to a howling crescendo on the ninth of March, now faded out quietly in mid-May, with the last Air Corps mail flight taking place on June 1, from Chicago to Fargo, North Dakota. It was Lieutenant Kuter who wryly concocted the tombstone epitaph for the Eastern Zone that would serve for all three. On the title page of the Zone's final report it read:

Conceived (in sin) Feb. 10–
Born (Prematurely) Feb. 19–
Paralyzed (Officially) March 10–
Quartered (By order) May 8–
Died (Unmourned) May 16.
Requiescat in Pace

A great many postmortems were to follow. The word *fiasco* became synonomous with AACMO—even used by some who were a part of it. It is still used.

Statistically, in AACMO's seventy-eight days of being, its pilots flew over one and a half million miles, carried more than 777,000 pounds of mail, completed close to 75 percent of their flights, and could claim, unlike the civilian carriers, that they didn't lose a single letter.

In a covering memo on his zone's final report, Horace Hickam gave the real reason why: "It is believed that no report in Air Mail operations can be complete without commenting on the special efficiency, resourcefulness, fortitude and initiative of Air Corps personnel in all ranks and grades. Adverse weather conditions and inadequate facilities seemed only to spur them to renewed efforts. The manner in which all performed their duties is evidence that the Air Corps can successfully meet any emergency provided the necessary facilities and equipment are available."

Where brevity was Hickam's style, complete detail was B. Q. Jones's, or at least it was Kuter's, who compiled and authored the voluminous final report of the Eastern Zone. It was an extraordinary document for its thoroughness, its almost daily recounting of all operations with the com-

This one, after a forced landing in the sagebrush near Elko, Nevada, needed a new wheel and propeller before returning to duty. (Credit USAF Photo)

Ready to depart from Boise, Idaho. The Air Corps, in its seventy-eight days of flying the mail while plagued by the worst winter weather on record, never lost a letter during its February–May mission. (Credit USAF Photo)

The famous Baker Board, named for its Chairman, former Secretary of War Newton D. Baker, and formed as a result of the political outcry over the Air Corps's inadequate equipment and training. Seated left to right: Major General B. D. Foulois; Dr. Karl T. Compton; Chairman Newton D. Baker; Hon. George H. Dern, Secretary of War; Major General Hugh A. Drum, Vice-Chairman; Dr. George W. Lewis; Major General George S. Simonds; standing left to right: Brigadier General John W. Gulick, Mr. James H. Doolittle, Mr. Edgar S. Gorrell, Brigadier General Charles E. Kilbourne, Mr. Clarence D. Chamberlin. (Credit USAF Photo)

In June 1934, Lieutenant Colonel Hap Arnold was selected to lead a squadron of ten new Martin B-10 bombers from Bolling Field, Washington, D.C., to Fairbanks, Alaska, returning by way of Juneau and Seattle. (Credit USAF Photo)

Getting ready to depart from Bolling Field, July 19, 1934. From the left: Glenn Martin, builder of the B-10, Assistant Secretary of War Henry H. Woodring, Arnold, and Air Corps Chief Benny Foulois, who was anxious to get his hand on the broadcasting microphone. (Credit USAF Photo)

The citizens of Fairbanks, Alaska, turn out to greet the flyers and examine their ships. Flying time from Bolling was twenty-five and a half hours (Credit USAF Photo)

Executive Officer on the flight was Major Hugh J. Knerr, who offers directions to Arnold in the off-loading of supplies for the planes and their crews.

August 20, 1934: The squadron returns to Bolling Field after having flown a total of nearly eighteen thousand miles in an epic undertaking, giving an indication of U.S. bomber capabilities. (Credit USAF Photo)

General Benny Foulois was never happier than when in a cockpit. On one of his final flights before retirement, in 1935, there's that look in his eye. (Credit USAF Photo)

West Point Cadet Frank M. "Andy" Andrews, class of 1906. In the yearbook, it was said of him that he found no greater pleasure than that of being on horseback—a cavalryman to the core. (Credit Jean Andrews Peterson)

In September 1920, Andrews, as a major, took command, at Weissenthurm, of the first U.S. air force in Germany: thirteen officers, eighty-eight men and a dozen aircraft, mostly DH-4Bs, like the one behind him. (Credit Allen Andrews Photo)

Weissenthurm was a busy air post. The Le Pere pursuit-reconnaissance plane behind Andrews and a visiting brigadier general had just been flown in from England by Captain Brogan Arthur. (Credit Moslander Photo)

Who turns to look at the camera with Major Harvey B. S. Burwell? The lady in black is Johnny Andrews, her husband facing in the wrong direction. (Credit Moslander Photo)

When the announcement was made, in December 1934, that Andrews was going to head GHQ Air Force, noted newscaster Lowell Thomas invited him to tell the radio listeners all about it. (Credit Allen Andrews Photo)

Brigadier General F. M. Andrews, with his staff, watch the 2nd Wing pass in review saluting the creation of GHQ Air Force Headquarters, at Langley Field, Virginia, March 1935. From the left, Majors Knerr, Bradley, Burwell, McNarney, unknown, and possibly twins Captains Barney and Benjamin Giles. (Credit USAF Photo)

The Boeing B-299 is rolled out of its Seattle, Washington, hangar, July 1935. The hopes of Andrews and army airmen were pinned on its performance. (Credit USAF Photo)

In the air it handled like a dream; it had speed, range, capacity. The press dubbed it a "flying fortress." It was to be designated the B-17. (Credit USAF Photo)

At Wright Field on maiden test flight, October 30, 1935, air hopes went up in flame and smoke in a tragic but avoidable crash. (Credit USAF Photo)

Proficiency brought progress in other areas. Major Ira Eaker pulls up his canvas hood, setting out from Mitchel Field on the first transcontinental blind flight, June 1936. (Credit UPI Photo)

Andrews arrives at 1st Wing headquarters, to be greeted by Wing CO Hap Arnold. (Credit USAF Photo)

Andrews' indefatigable Chief of Staff, Hugh J. Knerr. He'd fight the Army, the Navy, the White House, anybody, to see that his country had an air force second to none. (Credit K. Le Roy Thiem Photo)

GHQ Air Force received the first of a total of thirteen B-17s March 1, 1937. War Department policy blocked the ordering of more. (Credit USAF Photo)

In a demonstration of what the B-17 could do, six flew from Miami, Florida, to Buenos Aires, Argentina, in February 1938. U. S. Ambassador Alexander W. Weddell greets B-17 pilot Major Harold Lee George and fellow pilots. (Credit USAF Photo)

On return from South America, flight commander Colonel Robert Olds, CO of the 2nd Bomb Group, is greeted on a job well done by Air Force Commander Andrews. (Credit USAF Photo)

The May 1938 GHQ Air Force East Coast maneuvers were the largest on record, utilizing over three hundred planes, flying off of nineteen army and civilian airfields. Commanding General Andrews makes a point to two of his wing commanders, Brigadier General Arnold N. Krogstad, CO of the 2nd Wing, and 1st Wing CO Brigadier General Delos C. Emmons. (Credit USAF Photo)

On May 12, 1938, Andrews' B-17s hit their target—the Italian liner *Rex*—on schedule, 776 stormy miles off the East Coast. Chief navigator was Lieutenant Curtis E. LeMay, photographer Captain George W. Goddard in a third B-17. (Credit USAF Photo)

Warm greetings from GHQ Air Force Chief Major General Andrews to XB-15 command pilot Major Caleb V. Haynes, having returned on February 14, 1939, from delivering over thirty-two hundred pounds of medical supplies to earthquake victims in Chile on a round-trip flight of nearly eight thousand miles. (Credit USAF Photo)

Hail and farewell. Andrews relinquishes command of the air force he built, marching off to Billy Mitchell-type exile in Texas, reduced in rank from major general to colonel (Credit USAF Photo)

When Brigadier General H. H. Arnold became Assistant Chief of the Air Corps, in 1936, the family moved to the nation's capital, although Hap refused to live in the District of Columbia. It was the last year the entire family was together under the same roof. Henry, Jr. (Hank), next to his father, David between Mrs. Arnold (Bee) and Lois, with Bruce at far right. (Credit Wm. Bruce Arnold Photo)

Air Corps Chief Major General Oscar Westover, although a latecomer to flying, was an active pilot, his Curtiss A-17 a familiar sight at Air Corps installations across the country. (Credit USAF Photo)

September 21, 1938, General Westover holds a last-minute conference with two Vultee Aircraft Corporation executives before taking off on a planned flight to Burbank, California. Westover crashed in making his landing, and both he and his crew chief, Technical Sergeant Sam Hymes, were killed. (Credit USAF Photo)

The new Chief of the Air Corps in the office he loved most. (Credit USAF Photo)

President Roosevelt comes to Bolling Field to look over Air Corps equipment. Munich changed his mind on the need for air power. Hap smiles for the camera, as does Assistant Secretary of War Louis A. Johnson. (Credit USAF Photo)

At Wright Field, July 1942, General George C. Marshall looks skyward with the Army Air Force's two most important air leaders, Lieutenant Generals Frank M. Andrews and Henry H. Arnold—joined by Major General Oliver P. Echols, Commanding General of the Materiel Division. (Credit USAF Photo)

ments and recommendations of the section chiefs included on the many phases of their own commands. Hap Arnold's twenty-page summing up was a cogent account that described conditions faced, problems met, operations conducted and conclusions recommended. It illustrated how the inherent weaknesses of supply, communications and operations could be eliminated. Separately, he termed the experience "the greatest peacetime training in Air Corps history" and was so quoted in the New York *Times*.

Essentially, the three commanders had the same things to say. Only the style was different. Everyone knew there had been three major problems: lack of time to prepare, lack of proper equipment and lack of training. The first had been Benny Foulois's decision, and the correctness of it was argued pro and con and still is. The second and third were the result of traditional political failure to appropriate adequate funds. It appeared that the shock of realization would now have positive results.

Fourteen

The Scapegoat

The bruising three-month political conflict was ended. The President, who had been its principal instigator, had emerged relatively unscathed. His Postmaster General, who had taken the brunt of the attack by the opposition, would bear some personal scars as a result of what he considered Roosevelt's failure to support him when the going was rough. But with new airmail legislation passed by the Congress and the airlines back in civilian operation, the press had moved on to other causes. However, by June, when it was all over, it was not all over for Benny Foulois.

When Foulois had been called to testify on McSwain's bills before the full Military Affairs Committee on February 1, McSwain in an introductory preamble assured him he was among friends.

"This is sufficiently executive and confidential, General," said McSwain; "the testimony which is being taken will be for our information only.

"It will not be printed.

"You will have the privilege of revising and extending your remarks, if you wish.

"We want to assure you that so far as lies within the power of this Committee you are absolutely protected in what you have to say.

"We want your honest opinion based on a lifetime of service and we should like to have that opinion free from fear or hope of favor of any sort.

"We have enough confidence in your integrity and honor to believe that you will give it to us.

"We want you to make your statement as if the life of the nation were at stake.

"Tell us what you would do if you had the power of a dictator or a

Czar, to do whatever you thought best to put the air power of the nation on an efficient and adequate basis to serve the national defense."

With such assurances, as previously noted, Benny Foulois held nothing back. But McSwain's assurances were not kept by the subcommittee he had appointed, chaired by Representative William Rogers. As a result, after four months of hearings, most of them held in executive session, the members of the subcommittee in their final report, released in mid-June, concluded:

> We are most firmly convinced from the evidence and records submitted that before any substantial progress in the upbuilding of the morale and materiel of the Army Air Corps can be attained, Maj. Gen. Benjamin D. Foulois must be relieved from his position as chief of the Air Corps. We unanimously recommend that the Secretary of War take such action without delay.

Foulois was also cited in the report for "mismanagement" and "inefficiency" in the handling of the airmail.

The accusation, indictment and recommendation for dismissal had evolved out of the procurement issue—the negotiated contract, as opposed to open competitive bidding. The principal opponents when the hearings had commenced were Foulois and the Assistant Secretary of War, Henry Woodring. Woodring had made competitive bidding not just Air Corps policy but Army policy. While at the beginning of the hearings some members of the full Committee, principally Representative W. Frank James, attacked Woodring for his stand on aircraft procurement, they soon came around to his way of thinking and praised him for the changes he had made.[1] It was then they turned their guns on Foulois.

By the end of February, McSwain, a staunch supporter of competitive bidding, had become upset by some of the statements made by Foulois and other Air Corps witnesses such as Oscar Westover, Conger Pratt and Jake Fickel. He introduced a resolution seeking a broad investigation into "charges of profiteering and irregularities involving the expenditures of public funds for national defense."

The House approved, and voted the Rogers Subcommittee ten thousand dollars with which to proceed. The timing was important in view of the congressional uproar over the airmail situation and the grim reports of the Air Corps's problems in carrying it. Long-time supporters of the Air Corps on the Subcommittee such as Representatives Lister Hill, Paul V. Kvale and Edward Goss felt embarrassed and exposed. What had started out with bright promise for air independence advocates had gotten snarled up by unexpected developments. And since John McSwain was the congressional leader of the move, he was more than a little chagrined by testimony taken within the Committee as well as Air Corps action outside of it.

Further, although General MacArthur had testified before the full Committee prior to McSwain's introduction of his Air Corps-drafted bill, Secretary of War Dern had declined the invitation. Instead, on February 21, he sent McSwain a double-barreled blast that angered the Congressman and gained wide publicity, particularly because relations between the Committee and the War Department had been generally amicable. Dern's letter and scathing comments on McSwain's two bills, coming in the midst of the airmail battle, had considerable effect. The letter strongly implied that the Congressman had pulled a fast one by first indicating support for the War Department's bill proposing a GHQ Air Force under General Staff control, and then submitting his own, controversial proposals that sought air autonomy. Wrote Dern, in part:

> The result of adding these disruptive issues to the simple problem which was originally advanced by the War Department is to destroy any possibility of concrete and constructive action. To these two measures I am unalterably opposed—opposed to such an extent that I will not attempt to advance the constructive thought involved in the simple increase of the Air Corps, if it is your intention to couple it with these other issues. I am most desirous of the increase that I have mentioned to you, but I am equally opposed to having this approval, even by indirection, being interpreted as approving in any way these other issues. I would prefer, therefore, to oppose the entire plan rather than to risk any inference, which might be construed publicly or otherwise, of my approval of the other two proposals. I am so unalterably opposed to these destructive proposals as to feel the necessity of sacrificing entirely the benefits of the War Department suggestion if they are to be coupled therewith.

What one could read into the angry, repetitive, three-page letter was the syntax, style and pipe smoke of Douglas MacArthur. Either our way or no way, was the message; withdraw the air autonomy proposals or the War Department will withdraw its own bill on the creation of the GHQ Air Force and an increase in aircraft.

In his accompanying statement to the Military Affairs Committee, Dern scoffed at the idea of an air force of 4,383 planes. "The first of these two bills," he maintained, "would provide an air force so far beyond any sane estimate of our defensive needs and so costly that its passage could be construed by the world only as an evidence either of ardent militarism or immediate war."

Continuing with a traditional philosophy of focusing on the here and now, he declared the airplane had "too many limitations to be decisive alone," that it was "the fantasy of the dreamer" to believe that the destruction of armies or populations could be brought about by the use of projectiles or gas dropped from aircraft.

And then he gave the Navy a mighty plug, observing that a fleet was not curtailed by weather but airplanes were, and even if the weather was good it was doubtful if planes could destroy a naval task force.

McSwain's low-key public answer to the slashing put-up-or-shut-up attack was to say again, "The Committee has several times requested Secretary Dern to appear before it, but he had never come. We are still ready to have him appear at his convenience."

But Dern had said all he was going to say for the moment, and for the moment the formation of a GHQ Air Force was stalemated. In a March 5 letter to General Pershing, John C. O'Laughlin, of the *Army and Navy Journal,* gave his appraisal of what was going on:

Such an ugly situation has arisen between the War Department and the House Military Committee that I doubt if any legislation will be enacted for the Army which is inspired by Secretary Dern. The tense condition of affairs is undoubtedly due to the irritation of McSwain and [Representative Ross A.] Collins [chairman of the House Appropriations Committee] over the failure of the Department to protect them in connection with the motor truck scandal now being investigated by a Grand Jury here.[2] Ostensibly the Committee is concerned over the action of the air corps in acquiring airplanes by negotiation rather than competitive bid. McSwain alleges that the law has been violated and Collins in his report on the War Department appropriations bill indicates the Appropriations Committee's disapproval of the air corps practice. There is no doubt that the Committee will do everything it possibly can to embarrass both Dern and MacArthur, but it is blaming the former because of its realization that Dern is merely signing whatever the Chief of Staff presents to him.

A different kind of reaction came from Frank Andrews, who had recently returned to Selfridge Field after an official trip that had taken him to Washington and then to Chicago to meet with Westover and Hickam on airmail matters. In Washington he had visited with Walter and Elizabeth Weaver, "and it was like old times." On the fifth of March he wrote to Walter:

I hope Mr. Dern's statement in his letter to Congress . . . does not have too much influence. Do you know how far it represents the administration's views on the subject of our air defense? It shows a total and unfortunate lack of appreciation of what air power is or its place in national defense. Part of Mr. Dern's letter reads like General MacArthur's annual report. I wonder what pains Mr. Dern took to inform himself on air defense, other than to accept the arguments and opinions handed out to him by the General Staff. Has he ever talked with General Foulois or any other of our Air Corps officers on the subject?

How often is that old saying "that the new war begins where the old one left off" true. Our national defense certainly seems headed that way. No real conception seems to exist in our "higher ups" as to what the tremendous strides in the development of the airplane in recent years means, or should mean, in our plans for the next war. It would not be so bad if all other nations were content to make the same mistake, but they don't, and the consequences may easily be disastrous to us.

However, we have got to keep plugging away, making what progress we can in the face of all handicaps, and hoping that we can get somewhere before any real emergency comes up.

We are in a bad fix right now. The First Pursuit Group, and, I imagine, every other tactical organization in the United States, simply does not exist. Even if this airmail job were called off tomorrow it would take us months to get back to where we were when we took over. . . .

The War Department view over the dispute between the Secretary of War and the Military Affairs Committee Chairman was trenchantly put forth by General Kilbourne, who felt passage of the McSwain bill would rob Peter to pay Paul and in so doing gut the strength of the Regular Army. Said the War Plans Chief, "The outstanding feature of this bill, is to appeal to human selfishness. It will be supported, openly or sub rosa, in spite of the disastrous effect upon the national defense as a whole, by some because of its promise of special opportunities for rapid promotion and increased pay privileges. Our aviators have been taken to a high mountain and shown the kingdom of the world—few men can resist such temptation."

The deep-rooted dispute now became intertwined with the highly publicized Air Corps safety record in flying the mail. Both sides were shocked. Congressionally, the Rogers Subcommittee had gone after Foulois, first on the Air Corps's method of letting contracts and then on every level of his command. Part of the members' growing displeasure could be attributed to having previously spoken out strongly in favor of Air Corps wants, proclaiming the Corps's expertise. Now they felt they had been sold a bill of goods. They were out to prove villainy.

On March 13, Rogers moved to attack Foulois through Foulois's own Assistant Chief, Oscar Westover. Oddly, even though Westover had recently returned from his inspection tour of the three Air Corps airmail zones, he was unaware that officers as well as enlisted men were not receiving per diem. He was of the opinion that the War Department had passed funds to the Air Corps to handle this matter. More to the procurement issue at hand, he testified that the decision to lower the performance requirements on the aircraft to be bought with the PWA funds through competitive bidding instead of negotiated contract had been a decision made by the legal officer at the Materiel Division, at Wright Field, and not by Assistant Secretary of War Woodring. Foulois had testified to the latter, meaning that the decision made by Woodring had compelled the legal decision, but the Subcommittee saw this as an outright lie.

In another area, Westover's testimony did his Chief no good. Rogers got down to the nub of things when he asked:

. . . Had you been called into conference on February 9, relative to the handling and delivery of the mail by the Army Air Corps with a week or

10 days preparation, for the best interests of the service, for the safety of those engaged in it, would you or would you not have approved of such a suggestion?

Westover tried to avoid a direct answer, not wishing to use hindsight as a guide, and as Rogers led him, said he would have had to "analyze the proposition, calling in staff officers and looking at the conditions."

That wasn't good enough for the Chairman, nor was Westover's response:

> . . . From one angle, I would say I probably would have been influenced to do it, if I could see my way clear to do it, and that would have been from the standpoint of mobilization and training. From that standpoint it has certainly given us training experience that we never would have gotten without it.

Rogers was out to get a direct answer to his pointed question, and he got it. "Considering the other factors," he pressed, "danger, equipment, landing field accommodations—considering all those other factors in connection with it, do you feel that at the time you would have recommended in favor of carrying the mail by the Air Corps?"

And Westover gave him the answer he wanted, and impugned the judgment of his Chief.

> I doubt whether I would. I certainly would have consulted with the War Department officials with reference to their complications, the question of leasing and doing things like that. I mean there are so many angles to it that it would have taken some time to have reached a decision.

Rogers reiterated, "You would not have felt like doing it until you had considerable time to look into the matter further?" And Westover replied, "Yes, sir."

At that point Congressman James moved that the Chair ask the district attorney for the books of evidence that the morning newspapers reported Foulois had turned over to the grand-jury investigation looking into possible procurement illegalities.

Concurrent with the Rogers hearings, the Subcommittee on Aeronautics of the House Naval Affairs Committee was holding open hearings on exactly the same subject. Having heard Rear Admiral Ernest J. King, Chief of the Navy's Bureau of Aeronautics, as well as other witnesses, both civilian and naval, the Subcommittee reached the logical, commonsense conclusion that "the policy pursued by the Navy Department since the adoption of the Aircraft Procurement Act of 1926, is a practical and prudent one and should be followed until a better plan is proposed."

Said the Subcommittee in its final report:

> Since negotiated contracts are necessary until the aeronautical art becomes more stabilized, and since under existing statutes procurement is

being administered by the Naval Department in an efficient and satis-factory manner, no changes are recommended in the Aircraft Procure-ment Act of 1926.

While the Navy in its design to build up its air arm was making steady headway through calm seas, the Air Corps was caught in severe turbu-lence.

After the less-than-cordial meeting Generals MacArthur and Foulois had had with President Roosevelt at his bedside the morning of March 10, Secretary of War Dern, with presidential approval, made his move to set up a War Department-sponsored board of investigation. There was specu-lation that the idea for the board had come out of the Chief of Staff's office and that MacArthur, "disgruntled at the way the Army Air Corps fell down in the carriage of the mails," wanted to assume control of the inves-tigation himself, relieving Foulois and other unnamed members of his staff.[3] Whether this was so or not, Dern, as noted, appointed the six civil-ian members of the Board, asking Newton D. Baker to serve as its chair-man.

The underlying reason for the Dern/MacArthur move was not difficult to perceive. Aside from obvious concern over Air Corps training, equip-ment and administration, creating a new board was a method by which War Department plans could be put back on the track. MacArthur, far more than those around him, understood the role of aviation. He wanted to give the Air Corps what it needed, short of independence and within monetary reason. He wanted to see the establishment of a GHQ Air Force under General Staff control. McSwain's move, backed by Foulois and others, had been met head on, producing a wrenching stalemate, the thrust further deflected by procurement concerns plus doubt over Air Corps quality. A board headed by Newton Baker and backstopped by Drum, Kilbourne, Simonds and Gulick made good War Department sense —an end run around McSwain and the independence rebels, even though Foulois would be a member.* Such a board's recommendations would be well received by a Congress that was presently thrashing about in the stew of airmail cancellations and Air Corps crack-ups.

Baker at sixty-two could look back on a long career of public service. A small man with a Wilsonian nose and pince-nez to go with a professorial manner, his selection was a clear signal to the perceptive. Even though he had served as Secretary of War during the World War, he was at heart a pacifist. During the war, he had been unalterably opposed to the idea of bombing strategic targets behind enemy lines—whether they were indus-trial, commercial or civilian. In 1919 he had single-handedly scotched the efforts of his Assistant Secretary of War, Benedict Crowell, who, along with an official delegation, recommended after an investigation of Euro-

* With Foulois, the four had made up the Drum Board.

pean aviation capitals that the United States set up a Department of Aeronautics. On all matters pertaining to military aviation, Baker believed the Army-Navy Joint Board was the place to work out policy, tactics and strategy.

The eleven-man Baker Board got down to work on April 17. In its three months of hearings, it questioned one hundred five witnesses—Army, Navy, civilian—took over four thousand pages of testimony, and visited a number of Air Corps installations and aircraft factories. Air Corps officers not called as witnesses were asked by the War Department to submit any "constructive suggestions" direct to the Adjutant General for submission to the Board.

One who complied was Lieutenant Colonel H. H. Arnold. His recommendations carried the concerted opinion of his route commanders and the members of his staff. Together, they called for a separate Air Corps budget free of War Department strings and a separate promotion list with seven conditions involving retirement, rank commensurate with duty, regular commissions, extended duty for reserve officers and a classified cadet status.

They did not call for complete independence from the War Department but recommended instead a reorganization of the Air Corps with a GHQ Air Force to be in charge of all Air Corps training. This was a significant shift on Arnold's part. Although in recent years he had had little to say openly on a separate air force, it was assumed that his position on the matter had not changed. Possibly as realists, he and his staff were opting for what they saw as the remotely obtainable and not the obviously impossible. Also, the influence of the airmail experience had been a powerful lesson. Under the present organization, they recognized, they were not equipped administratively or mechanically to go it alone. Thus the recommendation for reorganization. The thinking was Arnold's, and the ten other signatories concurred, how completely is not known.

Lieutenant Colonel Jack Curry, the Commandant at the Air Corps Tactical School at Maxwell Field, Alabama, had some comments on the submission of his long-time friend "Pewt."

> We were much interested indeed in receiving here the recommendations that your outfit forwarded to the Baker Board. I think your recommendations were fine. We here put in for a separate Air Force, in which we believe most thoroughly. In addition before receiving your letter I submitted one of my own making certain recommendations for reorganization of the Air Corps within the Army provided it was not possible to secure a separate Air Force. Briefly, they were: first, reorganization and control of plans and training; second, personnel legislation for both officers and enlisted men; third, separate budget. So it was really very much along the lines of what you submitted. If we get all of those things we will have an Air Force in substance if not in fact. . . .

We are close enough to Washington to get a number of garbled rumors. The Chief's office is in a flat spin right now and nobody knows what is going to come out of it all. I do feel that we will be much better off as the result of all this investigating. It is most unfortunate that we let ourselves in for it [airmail], but certainly we need so much and have so little that something had to be done about it.

A great many of Curry and Arnold's fellow officers were far more determined to push the cause for separation, either in direct testimony like Tony Frank, who said there was only one officer (Westover) in the Office of the Chief of Air Corps opposed to air autonomy, or by an organized letter-writing campaign. Andrews received his solicitation at Selfridge Field in the form of a telegram signed by an officer student at the Air Corps Tactical School, an airman of long standing.

MAXWELL FIELD OFFICERS ARE JOINTLY SUBMITTING FOLLOWING CONSTRUCTIVE SUGGESTIONS . . . FOR CONSIDERATION BY AIR CORPS COMMITTEE QUOTE FIRST THAT THE AIR CORPS BE REORGANIZED AS A SEPARATE AND INDEPENDENT BRANCH OF OUR NATIONAL DEFENSE CO EQUAL WITH THE OTHER MILITARY SERVICES SECOND THAT IT BE CHARGED WITH THE RESPONSIBILITY OF PROVIDING FOR THE AIR DEFENSE OF THE UNITED STATES THIRD THAT IT MAY OR MAY NOT INCLUDE NAVAL AVIATION FOURTH THAT IT PRESENT ITS REQUIREMENTS TO CONGRESS THROUGH THE MEDIUM OF A SEPARATE BUDGET FIFTH THAT IN EFFECTING THIS SEPARATE AND INDEPENDENT ORGANIZATION IT IS OF NO VITAL MOMENT WHETHER IT BE ACCOMPLISHED THROUGH THE CREATION OF A SEPARATE DEPARTMENT OF AIR OR BY A REOR-GANIZATION OF THE WAR DEPARTMENT PROVIDED THE MILITARY HEAD OF THE AIR FORCE IS MADE RESPONSIBLE DIRECTLY TO THE SEC-RETARY OF WAR IN THE SAME MANNER AS THE CHIEF OF STAFF OF THE ARMY UNQUOTE. BELIEVE UNANIMITY OF OPINION CAN BE OBTAINED ONLY BY CONFINING RECOMMENDATIONS TO MAJOR REQUIRE-MENTS STOP THE VAST LIST OF MINOR OBJECTIONS OCCASIONED BY PRES-ENT ORGANIZATION WILL BE CORRECTED WHEN THE MAJOR RECOM-MENDATION IS ACCOMPLISHED STOP BELIEVE TREMENDOUS WEIGHT WILL BE GIVEN TO RECOMMENDATIONS IF CONCERTED ACTION THROUGHOUT AIR CORPS IS OBTAINED STOP PLEASE SOUND OUT BROTHER OFFICERS AND GIVE CAREFUL CONSIDERATION TO SUBMISSION OF SIMILAR JOINT RECOMMENDATIONS.

JOHN B. PATRICK

In letter form, this same message was hand carried by Major Follett Bradley on his nationwide inspection tour of Air Corps airmail installations. The overall result was that over five hundred air officers either signed the letter or sent in similar letters.

Drum and his General Staff Board members were furious over the manner in which "constructive suggestions" had been handled. At the outset,

Drum had seen to it that Major General Albert E. Brown was named executive vice-chairman of the Board. Brown, a member of the General Staff, was selected to do the questioning and act as official recorder. In this way, through Brown, Drum had fairly close control over the agenda and the line of interrogation. Brown attempted to have any talk of independence, verbal or written, excluded, but in this he was overruled by Baker. Nevertheless, the letter-writing campaign was passed off as a put-up job. And so, in whatever form, the concerted effort by the airmen was shot down before it ever had a chance to fly.

In simplest terms, the Baker Board issued three reports: one that parroted the Drum Board's recommendations for a GHQ Air Force, another reiterating the Drum Board's attack on air autonomy and the potential of strategic air power including a repeated denigration of the Balbo flight, and finally a brief minority report by Board member Jimmy Doolittle.

Doolittle surmised he had been asked to serve on the Board because of his technical knowledge in the aeronautical field. There would be questions as to the kind of tools needed in the future, and his background in instrument and test flying, his work with the Shell Oil Company, were considered valuable. He also believed the War Department did not want to select someone whom they saw as a devout believer in air power "lest there should be less than complete unanimity."

Soon enough Doolittle came to realize that he had been naïve in thinking that an objective investigation was to be conducted. From a military point of view he was the one civilian on the Board who understood the ramifications of what was happening. On May 19 he wrote to Arnold:

> Have been running around in circles endeavoring to sit on the Air Corps Committee five days a week in Washington and still run a rather difficult job in St. Louis.
>
> I am heartily in accord with any plan that will improve the efficiency of our Air Corps. There is no question but that the Air Corps is the greatest single unit in our scheme of national defense and every effort should be brought to bear to make it as effective as possible.

That Drum did not appreciate the potentials of air power was as obvious as was his influence on the Board. And it wasn't that former Secretary of War Baker could be led around by the Deputy Chief of Staff; it was simply that they agreed.

Doolittle didn't, and he showed his individualism by breaking ranks with the other members and filing his own, minority report, a paragraph appended to the seventy-five-page majority opinion:

> I believe in aviation—both civil and military. I believe that the future security of our Nation is dependent upon an adequate air force. This is true at the present time and will become increasingly important as the science of aviation advances and the airplane lends itself more to the art of

warfare. I am convinced that the required air force can be more rapidly organized, equipped and trained if it is completely separated from the Army and developed as an entirely separate arm. If complete separation is not the desire of the committee, I recommend an air force as part of the Army but with a separate budget, a separate promotion list and removed from the control of the General Staff. These are my sincere convictions. Failing either, I feel the Air Corps should be developed and expanded under the direction of the General Staff as recommended above.

Five years before the beginning of World War II and seven years before Pearl Harbor, the Baker Board supported the General Staff view that air power alone could have no decisive effect in war and that those who deemed otherwise were at best visionary. It pointed the finger of blame at the abolished office of Assistant Secretary of War for Air for having psychologically stimulated the idea of air independence, and noted that since the office had been done away with a marked improvement had been noted. It didn't say improvement in what. The report criticized McSwain's proposals for a separate budget and promotion list in his H.R. 7601. In fact, in fulfilling its mandate of "making a constructive study and report upon the operations of the Army Air Corps and the adequacy and efficiency of its technical flying equipment and training for the performance of its mission in peace and war," the Board cut a broad swath. It struck down just about every idea and concept of air autonomy since Billy Mitchell while admitting that under certain circumstances there was a role for strategic air action.

On what could be called the positive side, the Board drew attention to the years of niggardly appropriations, saying this must cease. It recommended greater emphasis on instrument, night and beam flying, exonerated the Air Corps mail pilots from unwarranted criticism, and *supported the use of the negotiated contract*. Its most significant contribution was its firm support of the basic reason for its own creation: recommendation of a GHQ Air Force under General Staff control as a part of the Air Corps.

Through it all, Benny Foulois played a passive role. A signatory to the Drum Board report, he couldn't very well contest its findings. In retrospect, he was to admit regret for not signing the Doolittle minority opinion, but at the time there was no more talk of autonomy on his part. He was four square for a GHQ Air Force.

Because of the Rogers Subcommittee, his back was to the wall. In May, the Congressman had released an interim report accusing the Air Corps Chief of using illegal procurement methods in the purchase of aircraft. It found him "in clear violation of the law." That the accusations were utter nonsense was evidenced by the House Naval Subcommittee on Aeronautics, which had found nothing illegal in the use of the negotiated contract. Foulois, like his naval opposite member, Rear Admiral Ernest J. King, had followed a procedure in use since 1926. It was a procedure that had made good aeronautical sense to the War Department as well as the Air Corps. Woodring had forced the change to open bids, which had necessitated a lowering of aircraft performance requirements. Woodring was praised by the Subcommittee in its interim report, Foulois was damned and seven of his officers, including Westover, Conger Pratt and Jake Fickel, were castigated.

Even before the release of the report, Representative Edward Goss, Republican of Connecticut, signaled the way the wind had shifted. He told the *Army and Navy Journal:* "The military committee, you know, has always been pro Air Corps. Now, however, I believe that any proposal for a separate department, a separate budget, or any other separation would stand no chance. Many of the members sitting on the Rogers Committee have completely turned around in the matter."

Benny Foulois was a beleaguered warrior. His public denial of the accusations did nothing to change the Subcommittee's determination to finish him, and in this they now enlisted the aid of those sitting closest to him on the Baker Board. In late May, Generals Kilbourne, Simonds and Gulick were called to testify in concert before Rogers and his colleagues. The forty-two-year-old chairman, with a strong strain of New England vitriol in his blood, read excerpts from Foulois's wide-and-wild-swinging testimony of February 1, when he had pulled out all the stops on the stultification of War Department control.

This was not only testimony taken out of context, but it had also been given under the impression of strict confidence, for it will be remembered that Foulois had been encouraged by McSwain to lay it on the line without fear or favor. Foulois had also testified before the Subcommittee, and in his appearance on March 7 he had been less than judicious in his claims and manner. Later he was to say he had not been well, that he had been emotionally worn out. With AACMO perched on his shoulder like an overweight vulture, it was probably a fair description of his state of mind.

Reading Foulois's words back to the generals, Rogers got the reaction he wanted, and the generals got the opportunity to respond accordingly. The way Foulois's words were served up could not have brought any other reaction even from those sympathetic to him. Kilbourne summed it up when he said, "For a man to come up here and make such statements as he has made to you, which are easily capable of being refuted, it looks like he's crazy."

But it was Hugh Drum, testifying in early June, who told the Subcommittee what it really wanted to hear:

> My personal opinion is that he is not a fit officer to be Chief of the Air Corps, and I come to that opinion not only in view of these misrepresentations that have been presented to me, but from the state of affairs in the Air Corps. The management of the Air Corps, in my mind, has demonstrated that he is not fit for it.

In mid-June, the Rogers Subcommittee came out with its final report, demanding Foulois's ouster and asking the Secretary of War to take action. Included in the report were extracts of Foulois's confidential testimony. Foulois demanded the right to have the full hearings, and several days later, Rogers, speaking on the floor of the House, said the record of the Subcommittee was available to the General. But when Foulois requested same, Rogers reneged. A lawyer, he wrote in a less than juridical reply: "It was not my purpose to imply that testimony given in strictly executive session with the understanding that it would be held in strict confidence, would be made available to you or any other person at this time."

Benny Foulois had been a soldier almost all of his life, and he knew there was no backing off now in the face of heavy enemy fire. He demanded a public hearing and said he would welcome a court-martial. Powerful forces were lined up against him, none more so than General Pershing, whose influence extended all the way from his retirement home in Arizona to the War Department and Congress. To O'Laughlin, Pershing revealed his position: "The report of the Committee of the H.R. submitted by Mr. Rogers is a most scathing thing and shows him [Foulois] to have acted very indiscreetly both in handling contracts and his statements thereof. I guess there is nothing more for Gen. MacArthur to do, but let him out. The good result of all this is that it places the Air Service back where it belongs under the control of the General Staff."†

MacArthur evidently was not of the same opinion as to his options, nor was Secretary of War Dern. On August 21, in a masterful reply to the Subcommittee, Dern made it clear that under the circumstances he was not about to dismiss the Chief of the Air Corps in the manner prescribed. The heart of his rejection lay in the basic fundamentals of fair play. He wrote:

> Major General Foulois appeared before the committee, but it was in the capacity of a witness and not of an accused. I understand that he was later given an opportunity to appear again before the committee before the report was submitted, but for some reason this did not eventuate. It does not appear that he was informed of the accusations contained in the report, nor was he confronted by the witnesses against him and given an opportunity to hear their testimony and cross-examine them, and then to offer evidence in his own defense. Nor was he permitted to be represented by counsel of his own choosing. All these rights are sacred to every American citizen and are guaranteed by the Constitution. . . .

There is something inherent in the American character that seeks to support the underdog faced by fierce odds, and very shortly Benny Foulois found he was not alone. The press, which had been tough on him during the airmail ordeal, now swung strongly to his defense. On September 1, in

† To General Pershing, apparently, the Air Corps was still the Air Service.

an editorial titled "The Foulois Case," the Washington *Post* laid it on the line. The lead paragraph read:

> Secretary of War Dern has taken a proper stand in refusing to remove Maj. Gen. Benjamin D. Foulois as chief of the Army Air Corps. Because of its dictatorial procedure the House Military Affairs Sub-committee deserves the rebuke administered by the Secretary. At the time the report demanding Maj. Gen. Foulois' dismissal was made public The Post pointed out that the committee had exceeded its power and that the report was in the nature of a judgment rather than an indictment. A similar contention is made by Secretary Dern.

O'Laughlin, in his faithful correspondence with Pershing, commented: ". . . the country had gotten the impression that Foulois is a victim of a Star Chamber proceeding. In the end Foulois will have to go, for it would be foolish for the War Department to promote bad relations with Congress. You know how these fellows on the Hill back each other up."

That was an underlying reality that could not be overcome by editorials, letters to the editor or fan mail. Nor could it be overcome by considerable internal support such as that offered by retired General William Gillmore. Gillmore, who had worked with McSwain on the subcommittee that had drafted the procurement section of the Air Corps Act of 1926, offered to testify to the Congressman's knowledge of the matter. "It seems ridiculous," wrote the General, "for him [McSwain] now to countenance your being persecuted for doing exactly what he knows was necessary in the way of procuring aeronautical equipment."

And within the Corps came offers of assistance as well. Lieutenant Colonel Henry Clagett, at Kelly Field, told his boss, "There is no doubt in my mind but what you have the support of almost the entire Air Corps. It has been suggested to me that I write a letter for the entire field to show that Kelly Field is behind you. . . ."

Andrews, in a letter to his father, summed up his own reaction, which undoubtedly was subscribed to by many of his fellow officers: "The investigation committee gave General Foulois a pretty raw deal. He is no friend of mine, but to charge him and men like Westover, Fickel and Pratt with crookedness is a very unjust thing. They have no way of defending themselves and it casts reflections on the honesty of all of us who have been in responsible positions in the Air Corps."

In the end the support did not help and the injustice did not matter, for Benny Foulois had gained the dedicated ire of a congressional subcommittee. The vindictiveness of its Chairman and some of its members was further aroused by the support for and subsequent exoneration of the General. The major charges against him, following an investigation by the Inspector General's Office, were dropped. Rogers would rant on the floor of the House that the General was a "liar and a perjurer." Other congressmen would come to the General's defense, but it was not until the summer of 1935, when he announced he would retire at the end of the year, that the attack on him let up and Rogers' threats to block increased Air Corps appropriations ceased.[4]

In the meantime, apart from the ordeal of the Air Corps Chief, other movements were afoot within the Air Corps, of a more positive and lasting nature. GHQ Air Force was on its way to becoming a reality, and Hap Arnold was chosen to go to Alaska.

Fifteen

The Movers

When Frank Andrews had written to his father mentioning the "raw deal" Benny Foulois had received from the Rogers Subcommittee, he also told him, "There is soon to be a flight of bombers on a goodwill tour of Canada and up into Alaska. Westover was to have commanded the flight, but he is now held in Washington for possible developments on the Congressional charges. I am trying to get command of the flight. My chances are slim."

He was right for several reasons. One was that, on May 28, he had come down to Washington from Selfridge to head a board whose assignment was, he said, "to make recommendations on organization of the Air Corps for greater mobility," which meant setting up a GHQ Air Force. The board had been appointed to study and submit its recommendations to Foulois's office with intended approval by the General Staff. The plan for such action was to be based on MacArthur's earlier move toward reorganization. The fact that Andrews had been informed of his duty on May 23, while the Baker Board was still in the process of reaching its conclusions, was an indication of how forgone the conclusions were with regard to creating a combat air force.

Working with Andrews were some good friends and old pros: Tooey Spaatz, Horace Hickam, Hugh Knerr, B. Q. Jones and Edwin House. With one interim break, they finished their work on June 16, and Andrews, who had been staying with brother Billum and his wife, Ashlyn, was preparing to start the drive back on Wednesday, the twentieth, with his son Allen.

Allen had been visiting with Scott Royce, whose father, Major Ralph Royce, was attending the War College and would be assuming command at Selfridge in the fall. Royce, too, was hopeful of getting a command position on the Alaskan venture.

Andrews knew that he, in turn, would be taking on a new duty in the fall. But having completed the work of the Andrews Board—as it became known—made him hope he might be able to wangle the top Alaskan slot. There were other factors in the choice, however, that ruled him out and prevented a dedicated angler named Arnold from casting his line in the trout-filled waters around Jackson Hole, Wyoming.

During the airmail operation, Hap had lived exclusively at the Newhouse Hotel, in Salt Lake City, sleeping at his office, as did his staff. In April, at Easter, Bee and Lois had come up to visit and break him free from the daily grind, bringing a bit of female cheer to the staff as well. Their presence helped his morale, and he began thinking of a long-sought leave once the mail operation was over. Then, shortly after wife and daughter had returned to March Field, his spirits went into a steep dive brought about by an announcement from Washington. Bee had already heard the news when he telephoned her from the Newhouse.

"I'm sorry you married such a bum as you did," he said.

"Whatever are you talking about?"

"Did you hear the report that Chaney got the star?"

"Yes, I did, but that doesn't make you any bum," she said.

"Well, I don't know whether it's worth going on or not. I don't know what's ahead for me."

"I'm not surprised at many things. Well, just forget the whole thing and do your job."

Forgetting was hardly that easy. Benny Foulois had assured him some months previously, before they had begun to fly the mail, that when the next vacancy for brigadier general came along he'd recommend H. H. Arnold for it. Naturally, there had to be approval from the Chief of Staff's office. Now Jim Chaney, who was junior to Arnold in time of service, had received the coveted star, jumping two grades, from lieutenant colonel to brigadier general.

It wasn't just ego that made him feel low. The hunger for rank was a gnawing one over the slim pickings of years, and at age forty-eight he had to ask himself again where he was headed. Foulois's promise not being kept, he reasoned, meant somebody back there in Washington still had it in for him. What would that do to his future?

When they closed down the airmail operation, in mid-May, he returned to March Field, exhausted and ready for that leave.

"Have you decided where we'll go?" asked Bee.

"I've never been to Jackson Hole." He had that faraway look in his eye.

"Let's go!" She was all for it.

They went out and bought new camping equipment, having a great time at it, looking forward like a couple of kids to their first adventure in the mountains.

On May 31, he asked the auto club to arrange an itinerary to take them to Yellowstone Park. On June 8, the auto club maps arrived and they pored over them eagerly. They'd be gone for a wonderful thirty days. Son Hank was up at Lone Pine Camp, working in the forests. Bruce was at Big Bear, where there was an Air Corps detachment, and Lois and David would be holding down the fort with Maggie and Pooch. Major Lohman was to take command of the post, and Clarence Tinker the wing.

On the seventeenth, they were off and running, heading for the majestic peaks of the Grand Tetons and the sun-splashed Snake River, where the big ones lay waiting. Near Salt Lake, they camped and wet their lines in the Walker River to get in some practice. The next day, they figured that while passing through the city they'd check at the Newhouse just to see if there might be any messages. To their surprise and not exactly joy, the manager informed them there were a great many messages, including a telegram from March Field headquarters and that from Lois. They opened it first. It read: *Don't open a thing. Head for Jackson Hole.*

Hap looked at Bee. This was their first vacation in longer than either of them wanted to remember. "I've got to open them," he said.

"I know," she sighed.

March Field had been paging him all the way up the west side of the Sierras. The word was that Westover could not lead the flight to Alaska but Hap Arnold would. He was to report to Patterson Field, at Dayton, immediately. Foulois's message informed him that the duty would require approximately two months. He could leave his Sam Browne belt, garrison cap, boots and leggings behind. The matter was considered "confidential," said Benny; "no publicity will be announced."

Someone in Washington didn't grasp this last, and almost as soon as the Arnolds learned the news it was in all the papers. They knew if their vacation and fishing at Jackson Hole had to be put off, it couldn't have been delayed for a more challenging reason. He'd be leading ten of the new B-10's on a mass flight that might just put the Balbo epic out of Hugh Drum's mind. The idea for the mission was said to have come from Foulois's office, and Benny took credit for thinking it up. However, it was Arnold's belief that the flight was conceived by the General Staff at least six months earlier. Whoever thought it up, to MacArthur and the War Department it was an idea that had to make good sense. Secretary of War Dern, along with Assistant Secretary Woodring, were in favor. In a note to McSwain the year before, Woodring had written, "I, too, am impressed by the Balbo flight—it makes one think." Maybe not enough, but enough to know a year later that a mass flight of the Air Corps's most powerful bombers—being touted as the world's best—to Alaska via Canada and return would capture meaningful headlines. The undertaking would do much to restore the Air Corps's tattered image.

The Navy, not to be outdone, was planning to dispatch two squadrons, eleven of its patrol aircraft, under the command of Lieutenant Commander James W. Shoemaker, from San Diego to Fairbanks. It was reasoned the double action would send a signal to both the Japanese and the Russians that Alaska was not so vulnerable to considerations of expansion.

It had been Billy Mitchell, in 1920, who had launched the first long-range flight to the territory. His announced objective for dispatching Captain St. Clair Streett and four DH-4's of the Black Wolf Squadron from Mineola, Long Island, to Nome, Alaska, and return had been to keep his pilots sharp and to give them experience in long-range navigation while gathering map information. The five-year program of expansion as envisioned by the Air Corps Act of 1926 had recommended the stationing of a composite squadron in Alaska consisting of thirteen planes. Congress in 1927 had failed to pass a bill to do so. In 1929 Captain Ross G. Hoyt had established a record solo day-and-night flight to Nome in forty-eight hours. Now, five years later, the planned event indicated the degree of aeronautical progress. Hoyt was again involved in the operation. Only, this time he and Captain Edwin Bobzien, flying in Douglas 0-38's, would act as an advance team, preceding the B-10's along the route.

As the Arnolds reluctantly headed south, he told her he thought Benny Foulois had chosen him for the command to compensate for his having been turned down for promotion. Whatever the reason, she was sure he'd find the fishing in Alaska just as good as at Jackson Hole. The pity was that she could not go with him.

On June 25, he arrived at Patterson Field to take command. Hugh Knerr and Ralph Royce were on hand to greet him. Knerr was his Executive Officer, and Royce was in charge of Operations. They had been assigned the Air Corps's twelve B-10's, a BT-2B equipped for blind-flying practice, as well as the two O-38's to be used by the advance party of Hoyt and Bobzien. Of the dozen bombers, only one was airworthy and available for training. Previously, Knerr had taken six of the planes on a test flight to Dallas.

But the reason only one was ready to go was that all the Martin bombers had been modified for airmail use. Now they had to be remodified back to their original configuration, overhauled and set up to carry aerial cameras. One of the major missions of the flight was to conduct a mass photographic effort, using six of the planes to map on film strategic points and suitable landing areas from north of Fairbanks to south of Seward, covering over twenty thousand square miles of terrain.

Hugh Knerr's job was getting the planes ready and, as Chief of the Field Section at Fairfield Depot, it was one of the reasons he had been selected to be Executive Officer of the flight. He had been hard at it for several weeks by the time Arnold arrived, on the twenty-fifth.

Two days later, Arnold and Royce flew to Washington for a conference with the brass. Hap was not pleased with the setup. As usual, the General

Staff, he said, had waited until the last minute to start preparations, and now everyone was in a rush to have him name a starting date. Well, much as he wanted to get the flight launched, he wasn't going to be rushed. The planes were all in the shops and he didn't know when they'd be ready—he thought maybe by the thirteenth, but the only person he mentioned the date to was Bee.

At a meeting with air and ground officers he aroused a storm when he was asked, if he'd been in charge of planning the flight, whether he would have used B-10's, and he replied, "No. They're service test planes."

"What would you take, then?"

"B-4's."

"What?!"

He was being perverse, yet the outcry allowed him to make his point—that the new plane was an excellent one, but it had to be properly prepared and the installation of equipment and instrumentation could not be hurried. They'd had enough of hurry with the airmail. When he took off he wanted his "ships in such condition that we will have a reasonable chance of completing the flight 100 percent."

He broke some hearts, too. The original plan had been to carry a complement of twenty officers and ten enlisted men. He changed that. He was in favor of having more competent mechanics aboard. It took only one pilot to fly a B-10. The plane was not designed to carry a copilot. The result was that the balance was shifted to fourteen officers and sixteen enlisted men, and five very unhappy officers were sent back to their bases. He felt, in deciding on the shift, he had hurt Westover's and Spaatz's feelings. Maybe it was the merciless heat that was making tempers short.

The sixth officer to be denied the flight was Lieutenant George W. Goddard. Instead, he had been designated Transportation Officer. Goddard, whom Knerr considered to be a genius in the field of aerial photography, was to be in charge of the photographic aspect of the mission, but since a great many supplies would be needed when the flight reached Alaska, he was dispatched to San Francisco by train to bring them by boat and rail to Fairbanks.

Arnold attended yet another meeting before he and Royce flew back to Dayton. He had arranged to visit his old Army-Navy Club neighbor Douglas MacArthur.

After they had dispensed with the small talk, Hap got down to it. "I want to know what you've got against me," he said. "I have to know, because my whole career depends on it."

MacArthur's long features took on a quizzical look. "I don't understand. What are you talking about?"

"Well, I was told I was to have the next air star, that my name would head the list. Someone else got it."

MacArthur made no reply for a moment; then he said quietly, "Your name wasn't even on the list," and to prove the point he had the list brought in.

Hap Arnold never brought to Benny Foulois's attention the promise given, but from then on he held little respect for the Chief. For the moment, he put the revelation out of his mind and went back to Dayton to prepare for the big flight.

Aside from organization and training, the largest chore was getting the bombers ready. Knerr stopped all other work at the depot, but most of the employees were civilians, and that made a difference. As late as July 14 there were only two of the aircraft ready to go. In a long-distance telephone conversation with Westover, Arnold gave some reasons why.

"I can't answer those confidential letters this morning," he said.

"What's the matter?" asked Westover.

"The depot doesn't work today and Wright Field has a picnic. Just those things the NRA is responsible for. I can't help it and nobody else can."[1]

"Chances are, then, you won't be getting away during the middle of the week."

"I don't know. I can't see when we'll be getting away." He was offering much too glum an appraisal of the situation, impatient and anxious to be airborne.

Knerr and his people came through, as it was expected they would, and three days later Arnold was discussing with Major Asa Duncan, Head of Air Corps G-2, the squadron's time of arrival over the nation's capital. "Your arrival and departure," said Duncan, "will be covered coast to coast by a hookup with the Columbia and National Broadcasting companies. Circle the city between nine-thirty and ten the day of your arrival."

On Tuesday morning, July 17, the ten aircraft took off for Washington. They were positioned in three elements, Arnold leading the first, of four planes, Knerr and Royce leading the other two, with three planes each. They came into Bolling Field less than three hours later, met by a delegation whose leader was Assistant Secretary of War Woodring. Woodring was anxious to have a ride in the Martin and Hap complied, taking the Acting Secretary for a ten-minute spin over the capital. Woodring, with shining morning face, might not understand the finer points of air power but he flew every chance he got.

The official account of the arrival of the "gigantic Martin bombers painted yellow and black" stated, "All personnel posed for newsreels."

Unofficially, Hap gave his daughter "Lo" a running description of what it had been like:

> The merrygo round—
> Visitors one after another no end of them—
> Pictures endless—hands up hands down-smiling-frowning—shake hands

with Mr. Woodring—with Gen. Foulois—Now just once more—Say a few words over the mike—Wait a minute I didn't get it—Please stand over here. Wait a minute please I didn't get a shot—Over here by the prop please. They are waiting for you at the Mike.

Gen. Drum wants to see you.

—Hello Bart—Hello Tooey. Yes, I'm staying with you Mildred. What lunch at 12:00—I'll try to get there. Another picture over here.

Hello Gen—Why Mrs. Foulois how are you—Hello Jakie—There's Krog—No, General, you take the car. I must get up to the Younts to wash up. I'll get down as soon as I can.

Hello Muley—I won't be long. Yes one more picture but no more. All right but this is the last. Take me up to the Younts!

Mildred for gods sake give me a glass of Sherry! Never mind, I'm sorry that I asked—Don't you do it. I'll get washed up—Only ten minutes—Well I'll make it.

Why where did you get the Sherry! Here's how, Bart—My God isn't it hot—Look at this shirt. Why how do you do Mrs. Foulois. Where's the General?—He couldn't come, too bad. Why here are all of my old friends. General you take this plate—I'll get another—Mrs. Muhlenburg you have to come out and sit with the general and me. It's awfully nice to give this party for us. Hello Elmer—No I can't sit here Jakie I must sit with the General and Mrs. Muhely—Don't look so sad.

OK General we'll leave at once—It's such a relief to ride over to Washington—Well, call up first and see—Gen. Drum will not be in—too bad. Neither will Gen. MacArthur—also too bad—Well I'll be going back to Bolling. . . .

And so far into the night.

The same this morning with Gen. Drum—Gen. Kilbourne—Elliot Roosevelt etc. Mrs. Drum gave me a goodluck piece but it looked so "catholicy" that I gave it to Mike. He sings over the radio and plays a mouth organ for us. Sounds fine while flying but some of the pilots claim it makes them air sick.

In spite of his being unable to call on General MacArthur, MacArthur telephoned him and offered his best wishes for a safe journey, suggesting Arnold "play it safe."

There were other crew changes before departure. In the original orders it had been determined that a newsman should accompany the flight. So that there was no favoritism, he would send back dispatches to a press pool. And further to ensure accuracy and a knowledge of what was going on, the reporter selected was second Lieutenant Harris Hull, an Air Reserve officer who worked for the Washington *Post*. Hull was the son of former Congressman Harry E. Hull, who had served as Chairman of the House Military Affairs Committee and was a staunch air advocate.[2] Young Hull, who was also a pilot, doubled as a radio news commentator on station WMAL. The Air Corps couldn't have found a more suitable candidate to send back accounts of the flight. Further, it had been Hull who had arranged for Arnold and other dignitaries, such as Glenn Martin, present with his mother, to say a few words to the radio audience. This worked well enough until he turned the microphone over to Benny Foulois. Somehow Benny did not wish to relinquish his hold on it, enchanted by the "splendor and serenity" of the occasion. He went on and on.

Sadly for Hull, he learned the next day he would not be going to Alaska after all. Two rumored reasons for the cancellations were passed around among the newsmen. One was that General MacArthur decided he did not want so much public attention showered on the Air Corps. The other was that Arnold would rather have another mechanic along in the place of a reporter regardless of his credentials. There was some truth in both assumptions. During the week previous, in his telephone conversation with Westover, Arnold had learned that someone in the War Department had decided there would be no newsman on the flight. Such a decision would have required MacArthur's approval if not his instigation. The reason given was that originally the War Department had not envisioned anyone from the press going along. It meant that someone had had a change of mind, and for once, Arnold was willing to forgo the public-relations angle in order to carry an extra mechanic. Only later would Hull's being dropped take on more significance.

Came the morning of the nineteenth and the mission got under way. Present to see Arnold and his men off was a goodly throng of well-wishers, including Generals Foulois, Drum and Kilbourne. Present, too, were aviation writer Elliot Roosevelt and his wife. Foulois again got hold of Hull's microphone and put in a grandiloquent plug for the Air Corps, proclaiming: "Let us show our people in Alaska that while thousands of miles separate them from the parent country only a few days intervene between them and a possible emergency and the arrival of powerful air defenses."

MacArthur was not on the scene to bid the fliers bon voyage. Perhaps it was just as well, because when Arnold, in the lead plane, started his engines and taxied out to swing onto the runway, he managed to blanket the assembled brass in a cloud of dust.

Five days later, on July 24, all of Fairbanks' twenty-five hundred citizens, led by Mayor E. B. Collins, turned out at Weeks Field to give Arnold and his crews a rousing Alaska-style welcome, climaxed that night by a big hoedown in which the somewhat weary airmen grabbed their partners and do-si-doed. They had flown out of Minneapolis, crossing Canada by way of Winnipeg, Regina, Edmonton, Prince George and Whitehorse, and they had done so without incident except for the fact that Hugh Knerr reported Arnold had said over the radio as they were preparing to land, "Where are we!"

Along the more than thirty-five-hundred-mile route the greatest hazard had been the entertainment, for at every Canadian stop they had been greeted and treated royally, the welcoming speeches, the banquets and the toasts seeming endless. In flight, each plane carried a crew of three. Airborne commands were given by radio transmission, Arnold surprised and pleased by the improvement in the newest two-way communication system.

Two days after the squadron's arrival in Fairbanks, Captain Goddard and his trainload of nearly forty thousand pounds of supplies came whistling up the track from Seward. Stormy seas, an old tub of a boat to move the equipment and a longshoremen's strike had delayed him a bit, but not all that seriously. Now it was Goddard's job, with some enlisted assistance, to install and load his Bagley cameras in the clamshell bomb bays of six of the B-10's.

The number was inadvertently reduced to five on August 3, when Captain Ed Bobzien, one of the 0-38 pilots, asked Arnold and Knerr if he could make a flight in a B-10. They had flown down to Anchorage with Ralph Royce on an inspection trip of a new airstrip there, and Bobzien had come along for the ride. Royce put in an encouraging word for Ed, Knerr a discouraging one. The Captain was not checked out in the Martin, and Knerr didn't want to take any chances of not getting all ten aircraft back home. Arnold, in a fit of impulse, overruled his Executive and said, "Oh, hell, let him go."

He went, all right, but very briefly. Not understanding the plane's cross-feed fuel system, Bobzien was hardly off the ground before he was disappearing down behind the trees into Cook Inlet. Neither he nor the sergeant with him was badly hurt, but the B-10 was in the drink. Although the plane was ingeniously salvaged and made flyable by Knerr, using steel drums as pontoons and a railroad wrecking crane, the camera in the B-10 was damaged by saltwater corrosion, and that meant there would be only five aircraft for the mapping. But not right away, either, because foul weather moved in and acted as if it planned to stay for the rest of the short summer.

As Hap looked at it, if one couldn't fly, one could fish. Major Mike Grow, the squadron's medico, was of an equally enthusiastic mind. So, too, was Mayor Collins, who quickly made the necessary arrangements, and shortly thereafter they were battling graylings in the Tanana River, thirty miles from nowhere.

In their party were two young sports writers, hunting and fishing aficionados Corey Ford and Alastair McBane. Around the campfire and over the nightly card game they came to know and like Hap Arnold. The friendship would ripen and later be reflected in their writing about the Air Corps. Arnold did not wire Bee about the good fishing, but he did telegraph to ask her what kind of fox furs she and Lois would like to have him bring home.

On a far more serious note, Arnold had an unexpected visitor at the small wood frame hotel at which he was staying in Fairbanks. The man was a German who said he had been a pilot during the World War. He had recently left his homeland to come to Alaska to start his own airline.

Having read the press accounts of the superiority of the Martin B-10, he had looked the planes over carefully at the field. He informed Arnold that already the new Nazi regime had bombers that were more advanced than the Martin. Arnold was not inclined to believe him, but the visitor suggested that the U.S. military attachés in Germany be assigned carefully to look over and check out what was going on in the Dornier, Heinkel and Junkers plants. Arnold found it a shocking bit of intelligence to receive in faraway Alaska.[3] Although at the time that the mysterious German informant came to Arnold's hotel room and offered his disturbing news German aircraft manufacturers had not yet produced anything that would measure up to the Martin's range, speed and bomb load, they were well on their way to doing so, for Goering had secretly given them the green light on January 30, 1933.

After nearly two weeks of gray skies, the weather broke. Goddard was off and running. In three days of intense flying, he and the five crews carried out the photographic phase of the operation. It was a unique piece of aerial mapping, and Arnold congratulated the inventive and indefatigable newly promoted Captain on its success.

On August 16, once again the entire population of Fairbanks turned out, this time to say farewell. Into the planes were loaded bear skins, moose antlers, scrimshaw and a husky pup one of the men had acquired. And then, as Knerr put it, "we joined the geese on the long flight southward . . . over the glacier-clad St. Elias Range to Juneau." There, since Arnold's plane had already been christened *Fairbanks,* Royce's ship was dubbed *Juneau.* Royce, who had already had a trick played on him while panning for gold when someone had slyly slipped a five-dollar gold piece into his pan, got even. He and Mayor Oscar Gill expropriated the bottle of champagne that was to be used for the ceremony and enjoyed the contents behind the plane when no one was looking. In its place they substituted a bottle of soda. Knerr protested the desecration, but Hap laughed at him. In Juneau they were also presented with a colorful winged totem pole as a gift to the Air Corps.

By the time they were ready to depart for home, individual flights to look over possible landing sites had been made to Fort Yukon, above the Arctic Circle, to Ruby and to Nome as well as the aforementioned stop at Anchorage. But it was on August 17 that Arnold led his squadron on the most significant and important segment of their passage. From Juneau they winged nonstop for nearly a thousand miles, most of it over the Pacific, to Seattle, Washington. They did it in six hours, with no landing fields in between. It was an aerial feat that had never been attempted before, and its significance was implicit if not heralded.

In Seattle when they arrived were some other visitors of note: Jimmy Doolittle, who was full of praise for the accomplishment, and Rear Admiral Ernest J. King, who did not like the Air Corps flying over water anywhere because, to Ernie's way of thinking, all water anywhere belonged to the Navy. The fact was that the Admiral's eleven patrol planes out to show that they could grab some of the Air Corps's thunder had failed to do so, their equipment hardly on a par with the Martins.

Three days later, on August 20, Arnold led his men into Bolling Field. They had been away just a month, had flown safely over eight thousand miles, a great many of them without anywhere to land in case of trouble except at point of departure and arrival.

Although Secretary of War Dern and Assistant Secretary Woodring were on hand with a large throng to greet the returning fliers, again the hierarchy of the General Staff was not. Perhaps the absence of MacArthur, Drum, Kilbourne et al. was of no real moment, but simply indicative.

General Foulois accepted the totem pole from the people of Alaska with cheers and clapping, and all the crews autographed it.

It was obvious that Secretary Dern had two things on his mind when he addressed the fliers and the assembled. "You have forged a new link between the people of the United States and Alaska," he said. "You have demonstrated anew the skill and daring of our Army and its fliers and the thoroughness of their training." And then a bit later, "I will use every effort to put into effect the recommendations of the Baker Board with the purpose of placing an Air Corps where it belongs in the first rank of world air power."

Assistant Secretary of War Woodring's major contribution was to permit the use of his private plane to fly the enlisted men who had participated in the Alaskan flight to Langley Field to spend the night.

The next day, Arnold and his officers gathered in Tooey Spaatz's office to prepare a report on the details of their aerial adventure, an exercise in debriefing. But who could write a report with fourteen experts telling how it was all at the same time? It would not be difficult to picture Spaatz just sitting silently at his desk, smoking a cigarette or two, listening to it all pour out. They had been invited that night to a stag dinner at the Carlton Hotel, their hosts the Air Corps personnel stationed in Washington, and the report was not completed by the time they adjourned for an evening of many speeches.

In a letter to Walter Weaver, vacationing in Pauling, New York, Spaatz described the climax of the affair:

> I don't know whether any one has written you about the dinner which was given for the Alaskan flyers at which were present the Secretary of War, General Drum and of course all the Air Corps personnel in and about Washington plus the Alaskan flyers. The Secretary of War made quite a lengthy talk stressing particularly his keen satisfaction with the findings of the Baker Board as being a means of developing the Air Corps. At the conclusion of his speech our Chief gave a talk which ended by saying he hoped he was addressing officers of the Army. At the conclusion of

his talk he called for three rousing cheers for the Secretary of War. Whether the call was rather unexpected or for other reasons, there was a long awkward pause followed by a rather feeble attempt at cheering.

Which may more or less have summed up the general Air Corps attitude toward the recommendations of the Baker Board.

The report of the trip was written soon thereafter. It was weighted with conclusions and, of course, recommendations. The most salient with regard to air power dealt with the need to establish an Air Corps base in the Territory primarily as an advance supply depot where tactical units "essential to the air defense of Alaska may be based," and "that special projects such as this flight be given to a tactical unit to execute."

This last was no sooner suggested than it was tried out in an Army General Headquarters Command Post exercise, featured as the first operational application of a GHQ Air Force. The exercise was to run for a week from September 2 to September 8, inclusive. Its announced purpose was to train commanders and their staffs in a combat situation that foresaw all tactical Air Corps units engaged on the West Coast while an aggressor suddenly attacked in the East. This would "require the immediate concentration of the entire Air Force to the Eastern Coast." With the defending Infantry there would be both "horsed" and "Mechanized Cavalry."

Since this was a theoretical staff exercise, in all but one instance actual aircraft would not be called on to assemble, nor would ground forces. The exception was Hap Arnold's squadron of B-10's, which were now at March Field. On September 2, Arnold received a secret order from Army General Headquarters, temporarily established at Raritan (New Jersey) Arsenal, to move his planes eastward with all dispatch.

Before taking off, he had meteorologist Dean Blake brought up from San Diego to give his forecast on the weather the squadron could anticipate en route as well as the best route to fly to Mitchel Field. Originally, Arnold had intended to go by way of Chanute Field, Illinois, but Blake had advised taking a more southerly route.

Time was of the essense, and before departure, on the evening of the third, Arnold signaled his itinerary and ETA to the Commanding General of the Armies, at Raritan. The fact of the matter was that he was angry over the unexpected orders to move his B-10's back across the country again when they were in need of overhaul. The unpredictable weather did nothing to soften his mood. Just past Amarillo, Texas, they plowed into an extremely nasty storm system that dogged them all the way to Shreveport, Louisiana. There, bad gasoline added to their problems. Arnold had hoped to reach Mitchel Field by nightfall, but he called it a day at Atlanta, having been at it for nearly fifteen hours.

At Langley the next afternoon, he received orders for his squadron to conduct simulated bombing attacks on four targets, including New York

City. This they did, short one plane because Lieutenant Ralph A. Snavely had been delayed by engine trouble. Swede Larson had gotten stuck in the mud at Langley but had finally caught up with the squadron at Wilmington, Delaware. The bombing runs were single passes over the designated targets before the final landing at Mitchel Field. What was most important was the manner in which the bombers were used, pointing up the General Staff strategic thinking.

Commanding General of the defending Blue forces was General Kilbourne. Oscar Westover headed GHQ Air Force. Under him, Andrews commanded the 1st Pursuit Group; Major Earl L. Naiden, Hickam's 3rd Attack Wing, from Fort Crockett; and Major Willis H. Hale, the 2nd Bombardment Wing, out of Langley. There were also two observation groups, Majors Tony Frank commanding the 9th, from Mitchel, and Barney Oldfield the 12th, from Brooks Field, Texas. Each of these officers had a small staff to assist in the war gaming. All came under the command of the Commanding General of the Armies.

In a detailed description of what the exercise was all about, the War Department official release stated: "First of all, the primary mission of an air force is to attack enemy objectives beyond the range of other arms; to prevent enemy air operations; to assist the ground forces directly by attacking enemy ground troops and installations; and to conduct reconnaissance flights in connection with each of these various missions. . . ."

In the first and last of these uses it would have seemed obvious that the purpose in winging Arnold's bombers from coast to coast would be to strike at the invading forces before they could get a foothold on the shore. Not so. Instead, they were utilized in the exercise as part of the enemy attacking force, and attention was focused on the technical problems of long-range movement and not on the strategical use of long-range bombardment. This was a General Staff decision under MacArthur's guidance in setting up the parameters of the exercise.

The release went on to explain:

> During the initial phase of warfare, when the opposing ground armies have not yet come in contact, the Commanding General will outline a certain plan of strategic employment for the Air Force and will direct the Commanding General of the Air Force to operate in conformity therewith except when the Commanding General of the Armies himself finds it desirable or necessary to direct the performance of a specific strategic mission. *From this it will be seen that the organization of the GHQ Air Force is such as to lend itself to operations of an independent nature.* [Italics added.]

Oscar Westover and his staff found that last piece of contradiction difficult to swallow. It was Kilbourne who detailed how the defending air units would be employed, and there was nothing independent or strategic about any of it. The entire concept of his defense was cemented in the past and was further reflected in the public description of the military situation as seen by the War Department. An analogy was drawn of how Generals Grant and Sheridan had trapped General Lee and his Confederate forces with the coordinated and combined use of Infantry and Cavalry. There was no consideration of what Lee might have done, as Arnold once observed, had there been an air force at his disposal. In this case, Kilbourne drove Westover and his staff up the wall in his failure properly to use the air available to him in his defenses against the Black invaders, who had been permitted to make an uncontested amphibious landing on the East Coast. When Westover prepared his attack wing to hit the enemy on the beaches, Kilbourne said no. Instead he pulled back thirty miles and dug in. Airmen were incensed, and a bitter argument ensued between ground and air staffs. Westover threw up his hands and said he was done butting his head against a stone wall. Arnold had not cooled off over what he considered the misuse of his aircraft in bringing them cross-country and, red-faced and fuming, he told MacArthur exactly what he thought of the idea.

The exercise, in which all the major officers of the General Staff including General MacArthur either participated or observed, showed if nothing else that although a GHQ Air Force might be on its way to actual formation, the concepts of its utilization left in the minds of the participating airmen a great deal to be desired. To many of them the exercise illustrated just how far a cry a GHQ Air Force would be from the real independence they sought, anyway. That the basic conflict in the respective positions had not been resolved and still remained in doubt lay thinly beneath the surface of the War Department's concerted public-relations push to sell the findings of the Baker Board.[4] The move was stimulated by the somewhat ironic fact that there was yet a new investigating board in being, and the War Department was worried that this new board's all-civilian makeup threatened its policies as McSwain and other air-autonomy advocates could not.

When the Air Mail Act of 1934, which put the airlines back in business, was approved by FDR, on June 12, it authorized the President among other things to appoint a commission of five members—the Federal Aviation Commission—"for the purpose of making an immediate study and survey and to report to Congress not later than February 1, 1935, its recommendations of a broad policy, covering all phases of aviation and the relations of the United States thereto." All aviation included, of course, military aviation. In this case, there would be no Drum quintet to set the theme and orchestrate the score.

A month later, on July 11, the Federal Aviation Commission held its first meeting. Its Chairman was Clark Howell, editor-in-chief of the Atlanta *Constitution,* who had also served as a member of the National Transportation Committee.[5] When FDR had contacted him to ask if he'd chair the Commission, Howell had responded that although he was interested in aviation he was no expert, particularly on military aviation.

"That makes you just the man I want for the job," said the President, not adding that he was weary of experts.

When Howell announced that he had a completely open mind on the development of air power and that he and his Commission planned to review not only the Baker Board report but also all the previous investigations into the role of military aviation, the General Staff went into a fast huddle. When it came out, Kilbourne, as Chief of War Plans, was announced as the sole War Department liaison with the Commission. Drum let it be known that all Air Corps witnesses either called by Howell or offered by the War Department would clear their statements through Kilbourne. MacArthur then had the General Staff prepare a position paper on air power, which was submitted to the Commission. It was a reiteration of the Drum-Baker boards conclusions, containing a veiled warning that these recommendations had to be followed or the national defense might be imperiled. MacArthur then issued orders that all Air Corps officers called to testify would be required to read the eighty-six-page document, and that he did not want any of them volunteering information opposed to the War Department views.

The question came back from Foulois's office about what to do if personal opinions were asked. Kilbourne responded:

It appears that there is still some misunderstanding of the freedom of witnesses to express personal opinions. No expression of opinion contrary to War Department policy should appear in the presentation of the War Department representatives, but it appears inevitable that, in questions by the Commission, personal views will be asked and must be given. The only requirement in case such opinion is at variance with the approved policy of the War Department is that the witness call attention to the fact, as well as to the fact that these policies have all been formulated after consideration of testimony from all available sources. I must depend upon you to make this clear to the witnesses.

Assurances had been given by the Navy that it would follow a unified line with the Army in all its testimony. Its position was: "Any other arrangements than the present one of having the Naval air component an integral part of the Navy would be fatal to the efficiency of the fleet and jeopardize national security."

Since the General Staff held firm control over questions of strategy, requirements and use, Foulois was asked to prepare a submission to the Commission that covered the nuts and bolts of training, procurement and mobilization. He was no longer at his fighting best, worn down by the ruinous attacks suffered during the airmail mission and still facing the unresolved accusations and demand for resignation by the Rogers Subcommittee. Nevertheless, his staff, which somehow had missed the message from the War Department, handed him a paper that thoroughly upset Kilbourne. It concluded that with the exception of the United States and Japan, all major powers had their military aviation under a single department and that "the trend is definitely toward a unified Air Corps under a minister of air."

Take it back! ordered Kilbourne. Benny's subordinates had no right to include such a controversial issue.

Foulois did as he was told; his revised statement to the Commission was in line with War Department thinking.

The Chief of War Plans also took Oscar Westover to task for indicating in testimony that long-range bombardment missions, crossing seas, would be feasible in the near future. Westover, never an independence advocate, ever loyal to the official line, was informed on and fully aware of the underlying meaning of Arnold's Alaskan flight. After all, having been originally selected to head it, he had been in on its planning and formation. More, he had just had the experience of the General Staff Post Command exercise. As head of GHQ Air Force, he had been party to the misuse of air power and was undoubtedly concerned about it. In his corrected fifty-five-page statement to the Commission he hedged his controversial prediction because of Kilbourne's insistence that he follow the party line, and stressed the overwhelming problems of long-range missions. But under questioning by the Commission, Westover reiterated it. "There seems to be no doubt," he said, "that it will only be a matter of a few years before operating ranges of three to four thousand miles will be entirely feasible."

Howell and his group had gotten some special insight into Westover's claim, because they happened to be in Seattle gathering information at the Boeing plant when Hap Arnold and his squadron flew in from Juneau. That night, Arnold and his officers had been invited to attend a dinner at which the guests were the Howell Commission members.

In his brief daily diary entry Arnold noted with regard to the dinner, "talked to the guests."

In so doing, he was one of the very few Air Corps officers who spoke to Howell and the members of the Commission free from General Staff restraint. Later, Howell quoted him as saying that he believed that a GHQ Air Force should be given a two-year trial before insisting on Air Corps separation from the Army. No doubt Arnold also discussed the capabilities of his aircraft and therefore aircraft to come. This was the same dinner at which Admiral King had raised objections to Air Corps planes flying even a few miles offshore.

The Howell Commission, continuing its investigation, traveled abroad in a wide-ranging exploration that kept the General Staff worried and gave hope once again to the air-independence champions. The agitated reaction of the former lent encouragement to the latter. Kilbourne went on examining the drafts of all military statements, letting each witness know what was expected of him. Then either he or one of his assistants sat in on the hearings.

That Arnold had spoken to the Commission off the record without prior briefing by Kilbourne convinced the members that they wished to question him on the record. At the beginning of November he received orders to report to Washington at the request of the Commission but with no further explanation. He flew his B-10 from March Field to Midland, Texas, and then on to Wright Field, where he had a brief reunion with Conger and Sadie Pratt and reported to Bee that he had visited with Andy Andrews, B. Q. Jones, Ralph Royce and others, who were locked in on a series of conferences. It appeared to him that all the rest of Wright Field was on the move to Washington, summoned to appear before the Commission under War Department sponsorship. "Conger, [Major Oliver] Echols, Knerr, Hegenberger and I don't know 10 others. Just like the Sinclairites streaming into California," he wrote to Bee.

He recognized that his own summons must be tied to theirs, but he wasn't sure. His plane needed some minor repairs, so he borrowed a Douglas O-25 and arrived at the Chief's office by way of Bolling Field, to find out the reason for his trip, but "no one seems to know what it is all about." No one knew why he had been ordered in, and it appeared to him that the War Department had his colleagues "buffaloed and apparently scared to death," or so he wrote to Bee. Seeking clarification, he called first on Commission members Edward Warner and Jerome Hunsaker. They told him they had some questions they wanted to ask as a result of their conversation in Seattle.

Were there any strings attached to anything he might say? Arnold wanted to know. There were none, they informed him. They simply wished to draw on the benefit of his long experience, that was all.

He paid a call on General Kilbourne. Did the General have any instructions? He did not. He had no idea what the Committee wanted with Arnold, but since Arnold's written views to the Baker Board had not been inimical to War Department policy, he was not all that concerned.

Arnold was. No longer the fiery rebel of the twenties, he'd come a long way in nine years—from eviction to a major command. At forty-eight, he was one of a handful of Air Corps officers who could look forward someday to the rank of general, in spite of Benny Foulois's oversight. He had become a good political boxer. He knew how to bob and weave, when to hold firm and when to give. He recognized the impulsive quality in his nature and, with Bee's helpful steadiness, had it under fairly tight rein. But the Howell Commission was bound up in the gut issue—an issue he had risked his career over more than once. Where did he stand on it now? If he said what he thought, whose hopes would he betray, what friendships would he endanger, what would he do to his own future? The War Department did not scare him to death, but he knew its power, knew what it could do to him; and yet, it was not within him to hedge his bets on beliefs and knowledge acquired over a quarter of a century of experience.

He knew he would have to temper his testimony with care and delicacy, and he did just that. He was on the stand for three hours, and then Howell threw him a curve.

"Now, don't blame me for my testimony," Arnold wrote Bee, knowing perfectly well she wouldn't. "I was extremely guarded and careful, but the last question asked by the Committee was. . . ."

"Colonel Arnold, I'm going to ask you a question which you may consider embarrassing. Could you straighten out the Air Corps were you given the opportunity?"

My gawd what could a poor man do! he later asked Bee. He hesitated and then replied, "If I were given the authority, I'm sure that I could."

That was that, and when the meeting adjourned a few minutes later, Howell invited him to lunch and Warner asked him to dinner. The last he had to decline, because Tooey and Ruth had already asked him to be their guest.

FDR wanted to congratulate him on the Alaskan flight, and this honor tied in with Arnold's plan to drop by the War Department and find out what was delaying the promised awards for the airmen who had participated in the flight.

The meeting in the oval office lasted only ten minutes. During it, Roosevelt asked questions principally about the Alaskan landscape and not about the aircraft that had flown over it, yet Arnold was buoyed up by the command performance. He knew that the Distinguished Flying Cross had been proposed for his crews, and he left the White House believing that the paper work had simply gotten stuck somewhere in the bureaucratic pipeline. What he didn't know was that, back in September, at the same time the General Staff was reviewing the Baker Board recommendations, the Adjutant General, Major James F. McKinley, an officer of considerable influence, had decided the Alaskan flight did not warrant such a blanket award. A letter of commendation signed by Secretary of War Dern in the 201 file of each member of the flight would be adequate, he thought. General MacArthur decided even this was too much. A single letter to Arnold, written in the name of the Secretary of War, was sufficient unto the day. McKinley drafted the letter, using a number of praiseworthy adjectives. MacArthur edited them down to a rather perfunctory nothing. It can be surmised that his action was predicated on two integral considerations: One was that he did not wish to bring undue attention to the Alaskan flight during the Howell Commission hearings, in which the flight's strategic meaning might be stressed by War Department recognition of the maneuver. The other was that he was engaged with the Navy in working out a Joint-Board-GHQ Air Force agreement, and he did not wish to antagonize the admirals by drawing attention, through the awarding of medals, to the overwater capabilities Arnold and his bombers had demonstrated.

So it was that when Arnold called on the Deputy Chief of Staff that November day, the reception was not what he had expected. Drum informed him in so many words that the Alaskan venture hadn't been all that great. It didn't really warrant handing out medals en masse. Arnold, coldly fed up, told him that Air Corps people were going to find it "awfully hard to be loyal to the General Staff when they pulled things like that." Then he left to pay a call on Kilbourne in the hope that he could enlist the War Plans Chief's support. It was two hours of frustration, out of which he concluded that, for Kilbourne, "the straightening out of a Post Exchange is the most important event in his career."

When he arrived back at March Field, flying through "mile after mile of rain and low visibility from Muskogee to El Paso," he fired off a memo to Drum comparing the Alaskan flight with others that had garnered the coveted DFC for thirty airmen, beginning with the six pilots who had made the first Alaskan flight, in 1920. Unknowingly, unthinkingly, he made two points that must have galled the Deputy Chief and indicated exactly what MacArthur was anxious not to have played up. Wrote Arnold:

Balbo's flight covered 12,000 miles and had months of preparation, with landing fields available to his flying boats at all times.

The Alaskan flight from organization to demobilization covered 18,000 miles of which about 6,000 was covered without any landing fields except those at points of departure and landing.

The Alaskan flight by covering the distance of 980 miles from Juneau to Seattle non-stop performed a feat never before accomplished by individual planes, much less a group of ten.

And at the end of his brief, Arnold said he hoped Drum could find a way to reopen the case, "in order that the youngsters who made the flight such an unqualified success will know that their efforts were appreciated."

It was a vain hope. The War Department sent no such signal to the youngsters or anyone else.[6]

If the month was one of personal frustration for Arnold, it was also one of great personal loss for the entire Air Corps. On the winds of November came again the sad reminder that in the air, hazard knows no rank. Lieutenant Colonel Horace Hickam made his last landing.

The field at Fort Crockett, Texas, home of the 3rd Attack Group, was too short. Because of its smallness and the roughness of its southern end, planes landing to the south, even against a light wind, made it a point to touch down between its boundary lights—the field's only lights—just beyond the shallow embankment of its northern threshold.

On the evening of November 5, Air Reserve Second Lieutenants Harry N. Renshaw and Andrew N. Wynne were standing on the porch of

Group Operations talking to Captain Charles C. Chauncey, the Operations Officer, watching Uncle Horace Hickam shooting night landings in his Curtiss A-12. It was close to eight o'clock as they observed the Colonel coming in for his second touchdown. They realized he was low and was going to undershoot. So did Hickam. He applied power to correct the error and then chopped it off too soon. The watchers saw the A-12's wheels hit the embankment just below its top, saw the plane flipped on its nose, skidding along the ground, the weight of its engine tearing up the turf, and then saw it snap over on its back with a thunderous crash, slewing completely around.

The three men were running toward the aircraft before the sound had died. Wynne arrived first, yelling, "Colonel, are you hurt? Can you hear me?" There was no answer. The cockpit rim was flat on the ground.

A group of enlisted men came charging up, followed by the crash truck and an ambulance. Even after Renshaw had driven the cab of the ambulance under the broken tail fin, with the men holding up the fuselage, they could not get Hickam free of the cockpit. It was necessary to dig a trench to do that. By the time Renshaw and Wynne had managed to get the Colonel out of his parachute and onto a litter, Captain Byrnes, the base doctor, had arrived. While the ambulance raced to the Marine Hospital, Byrnes did what he could, but it was too late. Renshaw believed his CO was dead before they had managed to free him from the cockpit.

It was Spaatz who had noted in his diary, back in 1925, that after age thirty a pilot's reflexes are not as sharp as they should be. Lieutenant Colonel Horace Hickam was forty-nine, determined to keep up his proficiency and flight time in a plane he had fought to have produced but which, in the interests of economy, lacked a fuselage projection to protect the pilot in case of a flip-over.

Captain George S. Warren, who had been on Frank Andrews' staff at Selfridge, summed up the feeling throughout the Corps when he wrote his former CO two days after the accident, "We are all broken up over Colonel Hickam. He was a true officer and a gentleman and his loss creates a vacancy in the Air Corps which will be hard to fill." Arnold called the death "an almost irreparable loss . . . at a time we need strong men so badly."

Andrews agreed, for personal as well as military reasons. He and Johnny had known Horace and Helen Hickam since the early days, and it was he and Johnny who saw to Helen, who, with her children Martha and John, came to Washington for the funeral and burial at Arlington Cemetery. Uncle Horace's popularity was such that every officer who could make it was there to give him the last good-bye.

Taps, the heart-wrenching notes of the bugle sounding the soldier's requiem. The bark of rifles, the final salute echoing down the hill toward the river, and overhead, a flight of pursuit in sharp formation, with one position empty, winging through the autumn sky, signaling farewell to a fallen comrade.

It had been a fall season in which principal Air Corps officers felt themselves to be circling, flying in heavy weather, not sure of course or destination. Hickam's death was the tragic end of the scale, but perched on the other end was the weight of uncertainty. They could not perceive, indeed no one could at the time, that this had been the most galvanic year in peacetime Air Corps history, that the forces that had beset them internally, politically, operationally were without parallel. And the year's course was not run.

Boards and commissions could ask questions, pile up thousands of pages of testimony, make recommendations for future benefit, but as Tooey Spaatz, heading Training and Operations, knew, the shortage in aircraft was so pronounced that required training could not be properly carried out, and as things stood, procurement for 1935 would not equal losses for the year.

Yes, there were more Martin bombers coming off the production line, joined by the B-12 version of the same aircraft. But they were finding bugs in them of a serious nature. Arnold informed Spaatz and Hugh Knerr on the sorry state of affairs, reporting to the latter: "In addition to the airscoop trouble, we have now developed propeller trouble, and it looks as if the propellers are all beginning to crack under ten hours."

Two of his had cracked, he said, in nine hours. The wing was just about grounded and they were in "one hell of a fix." Worse, there had been a fatal crash in which Captain Bob Selff's B-10 had developed such severe vibrations that an aileron had torn in half and then ripped away from the wing. A cadet, riding in the rear cockpit, bailed out safely, but inexplicably Selff and the other two crew members had gone down with the plane.

Aside from bomber trouble, there was trouble with the Boeing P-26. Only three of them at March Field were flyable when Arnold wrote Knerr explaining that the others were awaiting fuselage repairs, their bulkheads working loose. "I'm morally certain that the P-26, as is, will never be a satisfactory plane for tactical purposes," he advised. "I flew it with flaps the other day, and while it's a great improvement over the standard model, any such change will merely be a means to permit us to continue the use of the P-26 in service until they wear out. Such changes will not make it a satisfactory combat plane."

It was against this background that Andrews had been instructed to reconvene his Board at Wright Field, and that was where Arnold had met with him on his way to Washington in October. Although the Board had taken up the problems of engineering and supply, its primary purpose was to set up a tentative framework on which a GHQ Air Force could be constructed. This had been done hastily in June, but later, with further input from Captain George C. Kenney from the Plans Division of the Chief's

office as well as from the Materiel Division, it was hoped real progress
could be made. It was the seventeenth of October when the official announcement came that Andrews had been assigned to duty in the G-3 Section of the General Staff to work out the well-entangled details of organizing the new combat arm.

Immediately he took up his duties, he "found the instructions for the
creation of a GHQ Air Force had been drawn up and approved by the
Chief of Staff." He also found there were several roadblocks to its implementation. One of the biggest was the tables of organization, which, by
their very nature, created a number of potential promotions for those who
would command the Air Force. There was strong resistance in G-3 to the
idea, and there were those in the Section who were not all that interested
in making Andrews' work easy. His office was in the Barr Building, looking out on Farragut Square, across from the Army Navy Club, not in the
massive, tiered and pillared keep of the State War and Navy Building. It
was an easier atmosphere to work in, but what concerned him most was
not G-3's attitude, which he was capable of handling, but the future of
General MacArthur.

"It was at that particular time," Andrews related, "that there was a
good deal of uncertainty whether General MacArthur was to be retained
as Chief of Staff and activity in the General Staff on the Baker Board recommendations was waning."

That MacArthur was the motivating force for a GHQ Air Force, in both
the War Department and the office of the Secretary of War, was obvious.
He wanted, needed and was determined that the Army was going to have a
combat air arm worthy of the name. The doubts raised by Roosevelt over
MacArthur's continuance as Chief of Staff brought about within the War
Department a sudden loss of interest in its creation. To add to the speculation, if not confusion, an official document over General MacArthur's signature, detailing the Joint Army-Navy Board's acceptance of a GHQ Air
Force, was circulated within the War Department. It spelt out, often in
contradictory and ambiguous terms, the command responsibilities between
the two services with regard to coastal defense in case of war. Supposedly
an update of the Pratt-MacArthur agreement of 1931, which the Navy no
longer considered valid, its approach restricted the GHQ Air Force to
coastal defense in such a way that under certain conditions it was difficult
to know who would have commanded what, and just how far offshore air
force elements would be permitted to fly under varying circumstances.

Major Follett Bradley, on duty in the General Staff War Plans Section,
sent "Pewt" a copy of the doctrine, pointing out: "There are a few vague
expressions which are susceptible of interpretation either way. Most unfortunate, the War Department sincerely believed that this document cinched
the matter for the Army in Coast defense and are now chagrined to realize

that it is being interpreted as having surrendered to the Navy. Steps are being taken to correct the impression."

Arnold responded that he sure as hell hoped Bradley's surmise was correct! Meantime, the Howell Commission had gone on with its inquiry. It was late in November when it requested by name the testimony of six Air Corps officers. The names were selected through the suggestion of Florida Congressman J. Mark Wilcox, a newly elected air enthusiast. They were Captains Harold Lee George, Claire Chennault and Robert Webster, Air Corps Tactical School instructors at Maxwell Field; and Major Donald Wilson, Captain Robert Olds and Lieutenant Kenneth Walker, students at the Command and General Staff School, at Fort Leavenworth. With the exception of Chennault, the group was firmly wedded to the theory of strategic air power through the independent employment of long-range bombardment. All six, Chennault included, were staunch believers in air force autonomy and considered the Baker Board findings a pile of hash.

Shortly after Howell contacted them, they received personal letters from Kilbourne saying they were free to testify but there was no money to pay their expenses. The six discussed Kilbourne's letters by phone and saw the lack of funds barrier as a not too subtle signal from the War Department that their presence before the Commission was not being encouraged. They were not willing to knuckle the forehead, and Hal George, Chief of the Bombardment Section at the Tactical School, telephoned Howell and explained the problem. Two days later the proposed witnesses received personal telegrams from Kilbourne saying they were free to travel by military aircraft and that the Commission would pay them eight dollars a day for expenses. Before arriving in Washington they arranged, with the exception of Chennault, to meet secretly in Louisville, Kentucky, for the weekend. There they planned their strategy, recognizing they were probably laying their careers on the line, that what they were going to say ran counter to the official policy. They also knew that many of their colleagues were now willing to accept a GHQ Air Force under Drum-Baker guidelines, figuring it was the best that could be hoped for.

All six had certain traits of character in common, in that they were as intractable in their beliefs as the General Staff was in its. George was the most dynamic intellect of the group, Wilson the calmest, Olds and Walker the most intense, and Chennault, with that chin stuck out, was as fiercely partisan over the importance of pursuit aviation as the others were over the primacy of bombardment. It was five against one on that issue, but Chennault liked the odds, and battling against them was the elixir of his life.

When Ken Walker had been an instructor in the Bombardment Section of the Tactical School, he and Chennault, who was Chief of the Pursuit Section, would go down to the mat in their differences over the respective roles of the aircraft they espoused. Walker could make a typewriter smoke with his ideas, Chennault a pursuit plane. Far more emotional than Chennault, his intensity reflected in chain smoking, Walker worried about the internal rivalries being generated in the Air Corps by "prejudices we have built up in considering ourselves as bombardiers or pursuiters." He wrote Spaatz accordingly while Spaatz was serving on the Andrews Board, suggesting that training at Kelly Field should be changed so that every officer could fly any type of plane and know at least the minor tactics and techniques of each branch. "I feel that any step that we may take to eliminate internal prejudice is worthwhile and practical," he wrote, saying that they must develop "a type of officer who is thinking in terms of air force as well as in one particular branch. . . .

"Take your own case. I know damned well that you are confident in your ability to command a bombardment, observation or attack group as well as a pursuit group. Where then, is there to be found a place for this fetish of overspecialization?"

Spaatz showed Walker's letter to Andrews and the other members of the Board, all of whom agreed with his position, but as Spaatz said in his reply:

> We are still in the midst of considerable controversy and until the various committees and commissions have completed their findings it is difficult to foresee exactly what is going to happen. However, one thing is apparent and that is the people in authority are beginning to realize the importance of combat aviation which is the one bright spot on the horizon at present.

To those gathered in a hotel room in Kentucky, the realization was neither all that apparent nor sufficient; on one thing they were agreed: that they would lay it on the line.

A year prior to his appearance before the Howell Commission, Hal George had given a lecture at the Tactical School entitled "An Inquiry into the Subject of War." His audience, of some seventy officers, was held by his presentation, pinpointed to the clarity of his reasoning into the causes, objectives and effects of war as well as his thoughts on the question of things to come. "We are attempting to peer down the path of future warfare," he said. "We are not discussing the past."

In doing so, he examined the role of air power in future wars, questioning whether it had brought into being a new method of defeating an adversary independent of armies or navies. The answer, he noted, was of paramount importance, for its resolution could determine victory or defeat. With intellectual skill he laid before his audience this central question that must occupy their thinking. He did not answer it directly, adroitly staying clear of what War Department planning would consider heresy. "Whether air power can by and of itself accomplish the whole object of war is certainly academic," he said, "but that the air phase of future wars between major powers will be the decisive phase seems to be accepted as more and more plausible as each year passes."

Now, a year later before the Commission, he opened up, using his previous lecture as the basis for his testimony. No longer did he ask questions. Following the standard disclaimer—"Before presenting my testimony, I desire to state the opinions which I give are my own and not the official view of the War Department"—he gave answers.

He described the object of war as breaking the enemy's will to resist, and said that before the advent of air power there had been no method of doing so short of a superior invading land force. But, he said, "a new method of waging war can only be realized when its employment is understood, and when it is given an opportunity to develop itself primarily for the waging of independent warfare instead of as an auxiliary of the other armed forces."

To do this, he proposed an air force headed by its own chief of staff who would share common administrative services with the Army. In short, freedom from the General Staff.

Bob Olds's theme was that a strong, independent air force was a better guarantee of peace than treaties and conventions or armies. "A determined air armada," he said, "loaded with modern agencies of destruction, in readiness within range of great centers of population and industry, may eventually prove to be a more convincing argument against war than all the Hague and Geneva conventions put together."

Wilson warned that world conditions were leading toward war and stressed the vulnerability of the United States to air attack, thus the need to have an air force capable of both repelling and striking back at an aggressor. "We have no such air force now," he said, "nor will we have until its creation is definitely fixed as the responsibility of an agency whose primary purpose is that of providing for air defense for the nation."

Walker and Webster followed the same lines of thought, stressing the need for independence, observing, as Wilson had pointed out, that how the air force was brought into being was not all that important "provided it has representation coequal with the Army and Navy on all questions concerning national defense before the President, the Bureau of the Budget, the Congress and such joint boards as have been or may be created."

Throughout the testimony of these five, General Kilbourne sat silently in the back of the room with two secretaries working to get it all down. What he had heard were arguments advanced by intelligent, forward-looking junior officers, each of whom had close to twenty years of service and all of whom were extremely zealous in their beliefs. What they had presented was a concise and detailed refutation of a very determined War Department policy—exactly what Kilbourne and the ranking members of the General Staff had sought to avoid. The Chief of War Plans rose and asked Chairman Howell if he could make a statement.

Howell replied that since the hearing had been reserved for the Air Corps officers, it was up to them to decide whether he should speak. Chennault had yet to be heard, and the five who had testified took a fast vote. They figured Kilbourne would blast them, but better he do so in the open, where they could counterattack. They gave him the floor and then nearly fell through it, for Kilbourne, brass hat that he was, proved that he was not a blind man, not so fixed in his thinking that he couldn't listen and admit to learning something contrary to a held point of view. He said he wished to congratulate the witnesses, that he had gained a great deal of insight from their testimony. Their position, he confessed, had not been clear to him before. Now it was. Their thinking on national defense from now on would have a bearing on his own opinion.

That this new degree of enlightenment on Kilbourne's part would bring no changes was no surprise—he was due for a new assignment anyway—but the fact that he had stood up and offered his congratulations to a handful of juniors who had presented a case that was the antithesis of his own was a measure of his attempt to be open-minded.

Perhaps he spoke too soon, for Claire Chennault testified next, and where his colleagues had discussed the theory and practice of strategic air power under air force direction, Chennault ripped into the Raritan Command Post exercise and the waste of air power under War Department direction. The exercise, he said, had degenerated into trench warfare, because the air forces of the Blue defenders had not been properly utilized, and even the bombers had been reduced to attacking trenches. He used the case as the latest illustration to show how inept the General Staff was in its understanding and employment of aircraft.

Again Kilbourne asked to be heard, and Chennault had no objection. The General offered the explanation that the tactics used were the only method to bring the two forces together. Thereupon Chennault replied, "General, if that's the best you can do in planning for future wars perhaps it's time for the Air Corps to take over."

The General in this instance did not congratulate the Captain for his outspokenness, and Chennault was to maintain that as a result of his bluntness his name was removed from the list of those slated to go to the Command and General Staff School, which was so essential to promotion and advancement.

In any case, the Howell Commission thanked the witnesses for their informative testimony and continued their deliberations. The six officers returned to their respective stations, hopeful that what they had had to say would have some meaningful result. They knew, just as Kilbourne had said earlier, that "the Federal Aviation Commission will be the last board or committee reporting on the subject of aviation for some time, and its recommendations will have great weight with the President and Congress."

This might be so, but MacArthur, now assured by Roosevelt he was to remain as Chief of Staff for at least another year, gave orders to move full speed ahead, believing that a GHQ Air Force in hand would be solid protection against the recommendation of a Howell Commission still out in the bush. MacArthur knew that Secretary Dern had sent a letter to Roosevelt with the Baker Board recommendations and his own. They were that the Air Corps remain a branch of the Army, that its strength be built up to an agreed-upon 2,320 planes and that a program be instituted to rearm and reequip the entire Army. FDR was in favor, provided the money could be found. It was obvious that the President was keeping his options open, waiting to see what the Howell Commission would have to say sometime shortly, after the first of the year.

November ended with Billy Mitchell telling the Howell Commission that a proposed GHQ Air Force was "nothing but a subterfuge which merely divides aviation into more parts." Howell recommended that he pass his views on to the President, which Mitchell did.

December began with a number of questions, the biggest of which was, who would command the new combat arm? In 1933 Brigadier General Oscar Westover had been named the titular head of a GHQ Air Force that existed in name only. Prior to that, Benny Foulois thought that he might command the force as well as the Air Corps, and when that was obviously not acceptable to the General Staff, he may have believed that when his tour as Chief was ended he could move into the combat slot. That such a possibility was no longer in the realm of reality was indicated by the failure of General McKinley, the Adjutant General, or anyone on the General Staff, to inform him in October that it had been decided that headquarters for a GHQ Air Force would be at Langley Field, and not Patterson, which he had strongly recommended.

Also, Westover, as head of AACMO plus his tangle over the procurement problem with the Rogers Subcommittee, had raised doubts regarding his own fitness to command a tactical air force. Equally, it was realized that Oscar Westover, conscientious and dedicated as he was, was not a popular officer with the Corps. True or not, one observer's impression was that when Westover arrived at a field, the feeling he generated was that it was his field and everyone on it was there to serve him. Popularity was a factor MacArthur and his staff had to take into account. It was MacArthur who would make the final decision, and it was logical to hope that, given an air force leader who was widely acceptable, the independence boosters would be mollified.

When it came down to the selection, more than seventy officers were considered for the job, which MacArthur felt should be held by a major general. There were, however, few candidates with a permanent rank above lieutenant colonel, and the Air Corps Act of 1926 did not permit any more than a two-grade jump. The answer was that should the selected officer be a lieutenant colonel, he would be promoted to brigadier general and soon thereafter to major general. MacArthur's reasoning was that the duty warranted the rank, and since the Chief of the Air Corps was a major general, he felt the air force commander should be of equal stature.

Early in October, the New York *Times* had come out with a story that Colonel Charles H. Danforth, CO at Langley Field, had been "tentatively selected" to head up the command. Danforth was nearly fifty-nine, and his career had not been all that impressive. He had not learned to fly until 1921, had finally made the switch to the Air Service from the Infantry with the rank of colonel in 1924, and had served largely in the training command. In 1930 he had gained a temporary brigadier generalship, serving as an Assistant Chief of the Air Corps. But in June 1934 he had reverted back to colonel, assigned to Langley. There were not many Air Corps bets made on his selection to head the new Air Force.

As it was, the rumor mill in Washington cranked out the likely names, although in the minds of many there remained a very large question whether the General Staff would ever let the air force get off the ground. Still, everyone knew that when and if the announcement was made, it would make a memorable Christmas for the officer chosen.

Lieutenant Colonel Frank Andrews was to remark later, "Under pressure of haste and admitted insufficient consideration of many important points," tables of organization were submitted to Drum, and "it was just at this point that I was informed I was to be the GHQ Air Force commander."

The date was the eleventh of December, and when on that same day he came to the Chief of Air Corps's office for a conference with Lieutenant Colonel Arnold N. "Krog" Krogstad and his staff to discuss the tables of organization, he said nothing to Krog or anyone else about his appointment. Krogstad, as head of Air Corps personnel, was closely concerned with the problem and not at all happy about what was or was not going on.

In a letter to Hap Arnold the next day, in which he said that Drum wanted the Air Corps to submit its "64th study, including a program to set up the 2,320 airplane units and where located," Krogstad wrote:

> Of course with all these additional units we have to have more personnel and without knowing what if anything has been approved no such plan can be made out. There is simply a stalemate now in procedure and no one can make a move. We think the War Department doesn't even want to make a move just now. The question is so confused we don't know even what we are dealing with or shooting at. . . .

He was selling gloom. If the question was all that confused on the eleventh, it was cleared up before the end of the week, when Andrews came over from the War Department and sat down with him again. Present also were Jim Chaney, Chief of Plans, and his assistant, Captain George

Kenney. Tooey Spaatz was not on hand, because he was on sick leave, waiting, he said, for the Howell Commission's conclusions. While he was waiting, he would have been surprised to learn, as were those who were present, that Andrews brought to them an unexpected conclusion of a different sort. Drum had given approval for the reorganization plan and the immediate creation of a GHQ Air Force. They still did not know they were discussing its formation with its commander. Nor is there any indication that they comprehended the reason for the sudden haste to get the thing done was caused by what Spaatz and a good many other airmen were waiting for from Howell.

Although the Andrews family celebrated the appointment at Christmas, the official word was not released until the day after. Then the response to the news was immediate; a blizzard of congratulatory messages began swirling into their Washington home. They came not just from fellow officers but also from a wide range of well-wishers: government officials, congressmen, aircraft makers, friends and relations. Within the Air Corps, Tony Frank's letter was indicative of a great outpouring of relief:

> Like Little Orphan Annie, I was "glad all over" when I read of your appointment in the New York *Times*. It certainly looks as if we are going to get out of the slough of despond and really have something with hope in it to which we can look forward. After all the heartaches and disappointments and depression in the Air Corps during the last two or three years, your appointment to me is the first ray of sunshine that has appeared on the horizon. You certainly have my very best wishes and support and I know you will knock them for a loop. Congratulations a thousand times over. I am not so crazy about firing this Reveille Gun out here, but it would certainly do my heart good to waste a little powder on you if you would only come up and have lunch with us some day so that we may congratulate you personally.

Even the Navy piped Andrews aboard. Captain George C. Logan spoke for the future when he wired:

> As the old saying goes, I'm proud to know you. It was a real pleasure to learn of your selection to head GHQAF. There was also a feeling of added strength to the Navy for when that much talked of and carefully considered moment comes for the U.S.N. to attack the enemy from its advanced base, it will be supported by a striking force under the command of one who understands our needs and whom we learned to admire and respect in our War College contacts.

General Kilbourne was not all that effusive, but he was frank enough when he told Andrews, "While I had hoped that Westover would get it as a reward for the burden he has been carrying the past two years, nonetheless, I know the service as a whole will welcome you in that position."

Hugh Drum, in a letter to Newton Baker, explained why the General Staff had chosen Andrews.

> We all feel he will be able to meet the situation and develop the Force along the lines contemplated. Furthermore, in addition to being an efficient flyer, he has been in harmony with all the War Department has been trying to do. Consequently, we hope the future of the GHQ Air Force will be in good hands.

That was the broad-gauge reaction from the high and the mighty and from those who knew him best, but the feeling that swept the Air Corps from officers and men who didn't know him but simply knew about him was equally enthusiastic. Second Lieutenant Tom Darcy, West Point class of 1932 and graduate of Kelly Field 1933, was serving as Adjutant to B. Q. Jones's 8th Pursuit Group when the word came that Lieutenant Colonel—now Brigadier General—Frank M. Andrews was to command the air force. Darcy had never met Andrews, the distance between a twenty-three-year-old second lieutenant and a fifty-year-old colonel being one of great magnitude, but his reaction and that of his fellow pilots was automatically one of soaring approval. Everybody knew that Andy Andrews would make it fly, and everybody wanted to be a part of his command.

Hap Arnold summed it all up with his message from the Coast when he wired Andrews: "I know of no one in the Air Corps I would rather serve under."

Part III

Sixteen

The Maker

The congratulations had come from far and wide. The press coverage had been broad, *Time* magazine pointing out incorrectly that the three-grade promotion (it was two) of Frank Maxwell Andrews from lieutenant colonel to general—though temporary—was the first of its kind since 1906, when Teddy Roosevelt had jumped John J. Pershing from captain to brigadier. Radio newscaster Lowell Thomas had interviewed the newly appointed air commander, General and Mrs. Foulois had given the Andrews a tea at the Shoreham, and the Woodrings had given a reception at the Mayflower. But for all the pomp and ceremony, Andrews, referred to in one press account as "a heretofore obscure field officer," began his duty as a commanding general with a one-man staff and no office space.

The one man was First Lieutenant Pete Quesada, who had been serving as Operations Officer at Bolling Field, and also held down the same title for the nonexistent job in the nonexistent headquarters squadron of GHQ Air Force. Andrews was pleased to have the personable, fast-moving Lieutenant to assist him in getting started. Aside from his flying ability, Quesada was a diplomat and an organizer, with a broad background and a facility for knowing how to get things done. He proved the point immediately by obtaining office space with a special significance. At the close of Secretary Trubee Davison's tenure of office, Quesada had served as his Executive, and he learned that the former Air Secretary's offices, in the northwest corner on the second floor of the State War and Navy Building, had remained vacant. He quickly made the necessary arrangements, and Andrews walked down Seventeenth Street from the Barr Building and took up his new residence on Pennsylvania Avenue, just a stone's throw from the White House.

Quesada soon realized that beneath the surface of what appeared to be an atmosphere of close cooperation between the General Staff and Andrews there ran an ugly undercurrent. It emanated out of the G-3 Section of the War Department, where Andrews had been employed since October in working out tables of organization for the Air Force. The problem was not with General Hughes, the Chief of Plans and Operations; Andrews got along with him well, he and Johnny dining with the General and his wife at the Army Navy Club at the end of January in celebration of the promotion. The roadblocking and the foot-dragging lay within the next echelon, headed by several disgruntled colonels.

To Quesada, the attitude was a product of pure jealousy, stimulated by the knowledge that setting up the combat arm was going to require a good many Air Corps promotions. Andrews' promotion was a case in point to the naysayers, and Quesada observed they did everything they could to slow the organizational plans and make life uncomfortable. He saw "there was no desire to help Andrews to do the job to which he had been assigned," either before his appointment or directly after it.

However, the unruffled commander was far more than a match for them, not by table pounding or complaint to Hughes but by calm adroitness and the ability to outthink them, outflank them and get the job done. He was not to be deterred, deflected or slowed by intraservice rivalry. He said nothing to anyone about the effort to make his job difficult, and only Quesada was aware of it.

On January 10, Andrews received a letter of instructions from Secretary of War Dern directing him to have GHQ Air Force organized by March 1. Initially, he was informed he was to have a staff of four officers: a chief of staff and three assistants. He was free to make his own selections and could expect Air Corps and War Department cooperation.

His choice could be viewed as an insight into his thinking. He selected a backfield of individualists, yet all were driven by the same qualities of energy, dedication and determination. Hugh Knerr was to be his Chief of Staff; Follett Bradley, G-2; George Kenney, G-3; and Joe McNarney, G-4. All but Kenney were majors, he still a captain. All were in their forties, with approximately twenty years of service apiece, old pros on the ground as well as in the air. All but McNarney were known among their colleagues to be strongly on the side of air independence, and all but Knerr were combat veterans.

There were other things they had in common. Knerr and Bradley were Annapolis graduates, Knerr in 1908 and Bradley two years later.

Knerr had switched services in 1911, a result of his first major encounter with hard-core brass. As Officer of the Deck on board the U.S.S. *Lamson,* he caught two sailors coming aboard after a night of heavy weather in Norfolk. Ensign Knerr noted they were carrying a goodly amount of grog, inside and out, and he suggested they drop the latter over the side. The

next day, he was informed by the Captain that he was under arrest for neglect of duty, having failed to report the pair.

Hugh Knerr, at twenty-four or forty-four, was not a man to await the
inevitable. He departed the ship and journeyed to Washington, where he
made a call at the office of the Naval Judge Advocate to ask for advice
concerning his rights. He found he didn't have many and that his daring to
ask the question while under arrest was a grave act of insubordination. His
Captain would be so informed in order to prefer charges against the hap-
less Ensign. The frying-pan-into-the-fire act was not for Hugh. He was at
the State War and Navy Building, and he decided right then and there he'd
resign from the Navy and see what the Army had to offer. He inquired at
the War Department and learned that the Army had to offer an immediate
commission in the Coast Artillery for anyone properly qualified. He was
so qualified and signed the necessary papers.

Three years earlier, directly after his naval graduation, on his way home
for leave, he had stopped in Washington to see if he could be assigned to
the new Naval Aviation Section. He was told no. He was on the Navy
Rifle Team and that was considered a far more important duty. Now, joining the Coast Artillery, he made the same request and again was told no. If
he was good enough to be on the Navy Rifle Team, he could certainly try
for the Army's. For a man whose interest in aviation had literally begun
under the feet of the Wright brothers and whose Aunt Aida was a good
friend of Orville's, he was having a hard time getting into the air.

A year before leaving the Navy, he had married Hazel Dow. They had
met on visitors day aboard the U.S.S. *Albany* in San Francisco. She was
slim and trim, with honey-blond hair, and he was attracted to her by her
quick, intelligent manner. She had to be both to adapt to his fiercely independent nature.

It wasn't until July 1917 that the War Department inquired whether
Captain Hugh Johnston Knerr was still interested in aviation. Two months
later, he reported to North Island's Rockwell Field. His visions of a vast
military airdrome disappeared in a cloud of wind-blown dust. He saw a
scattering of wooden hangars, a handful of Jennies lined up in front of a
crude operations office and a small group of men dressed in flying gear
lounging in what shade they could find. Knerr became a member of a class
of thirteen. His instructor was Ira Biffle, a case-hardened old pro.

To Knerr, as he approached his first Jenny, it was a "magic moment."
The last plane he had seen was with Aunt Aida and Hazel when, seven
years earlier, they had gone out to Simms Station to watch Orville
Wright. Now Knerr climbed into the rear cockpit of the JN-4 wearing the
standard apparel. Biffle strapped him in and told him to make damn' sure
to keep his hands off the stick.

To Knerr, nothing would ever match the thrill of that first flight. He'd
had a haunting premonition of it at the turn of the century, before

powered flight was more than a dream. Then a boy of thirteen, he had climbed up on the roof of the family home in Atchison, Kansas, and gazed away beyond the Missouri River, "strangely disturbed" with a yearning to be airborne. He wasn't disturbed in the air again until Biffle turned him loose to solo. He'd had ten hours of instruction. The takeoff was smooth enough, and so were the climbing and the flying, but he circled the field several times, so excited that he'd forgotten how to throttle back to come down to land.

Hugh Knerr had wanted to be an engineer. His father, a professor of sciences, might have had something to do with the desire and the talent. Perhaps his mother, tall and fair, noted for her wit and ability to play the guitar, was responsible for the winglike flight of his mind, which in working with the mechanics of aviation could see developments and use beyond what was at hand. Andrews could perceive the shape of the future in the air by the sweep of his intellect, Knerr through the nuts and bolts of the art.

In 1920 he was assigned to McCook Field, under the command of Colonel Thurman Bane. There Knerr worked with Chief Engineer George Hallet on the development of the air-cooled engine, then in its infancy. Hallet, a small, quiet man who had begun his aviation career with Glenn Curtiss, was in the forefront of testing new power plants even if there was no money to produce them. Knerr was in his element until, once again, he ran into another brass wall and nearly switched back to the Navy.

It was learned in the office of the Chief of the Air Service that he had permitted the medical officer at McCook to use aviation gas for the field ambulance after the ambulance had been unable to answer a call because it had run out of its own, restricted supply.

Knerr was called to Washington to face General Menoher on the charges of misuse of government property. Menoher implied without saying so that he didn't believe the lean, hard-jawed Captain's story, and Knerr was about to tell him to go to hell when he saw Billy Mitchell standing in the doorway behind the General. Silently, Billy shook his head, cautioning Hugh to keep his temper. The advice was only partially heeded, for Knerr ended by telling the Air Service Chief that, should the occasion of a gasless ambulance needed in a hurry arise again, he would do exactly the same thing. Menoher dismissed him not only from his office but from the Air Service as well, seeing that Knerr was transferred back to the Coast Artillery. Fed up, Knerr was on the verge of resigning from the service altogether when he learned that his duty station, at Pensacola, boasted a Navy HS 2L flying boat. The plane was used for artillery spotting, and he became its pilot, rejoined the rifle team and, two years later, in February 1922, through Mitchell's efforts and Menoher's departure, was returned to duty with the Air Service.

A year later, following training at Carlstrom, Brooks and Kelly fields, Mitchell made Knerr CO of the 88th Observation Squadron, at Wright Field. There, he and Hazel and their two sons, Hugh, Jr., and Barclay, settled into their beaverboard quarters, the heating supplied by a potbellied stove.

The planes of his squadron were DH-4's, with all that that meant.[1] It meant two close calls for him in a short period of time, the first on a fair and balmy day in April 1923.

He and Sergeant John McKenna were returning from a reconnaissance flight, Knerr enjoying the noble sport of hedgehopping—skimming treetops, fence posts, barn roofs and anything that jutted up above ten or so feet. It was dandy fun until a valve in the old Liberty got stuck. At that moment, he had forty-two thousand volts of high-tension wires ahead of him. He could try to go either over or under them, and he had to decide right now! He went under, and planked the plane down into a rough-hewn, sloping pasture. When the sound and fury ended, the DH-4 had taken down about one hundred feet of wire fence and sideswiped a budding cherry tree. In so doing the plane had lost its lower wings and the landing gear and was, alas, "damaged beyond repair." Its occupants were less damaged, Knerr having strained his neck and McKenna temporarily acquiring a pale greenish hue. Farmer Puthoff, in whose field they had come to rest, later sent a bill for one hundred dollars, citing the loss of ten fence posts, one hundred feet of wiring and his pet cherry tree. Colonel Warnen Robins, the field CO, presented the bill to Major Hugh Knerr, even though investigation showed "the accident was unavoidable and through no fault of the pilot."

Shortly thereafter, Knerr was acting as a mail pilot, carrying an important load from Wright Field to a congressional committee engaged in investigating the scandal surrounding the lack of production of DH aircraft during the war. It was a foul day, the sky full of gray clouds, heavy rain close to the freezing mark. Over the mountains of West Virginia, with no real place to land, Knerr's engine caught fire in the V of the cylinder block. He had two choices: he could bail out or ride her down. He chose the latter because he figured the mail was important. Spotting an area on the side of a mountain that had been logged, he put the aircraft into a sideslip and purposely came in nose high, wings down to take the impact. He was not wearing a seat belt, and when he picked himself up from the rain-soaked earth, he found he was still in one piece and the DH-4 was folded up on one side like an accordion. A crowd gathered, and a schoolteacher in it offered to drive Knerr to Moundsville, where the Air Service had a depot. There he picked up another DH-4 and all in the same day continued on to Washington. On his return to Dayton he had his chute drop-tested. It failed to open. All of which proved that a pilot had to have his luck riding with him if he was going to ride far.

In 1925, Knerr was assigned to take the course at the Air Corps Tactical School at Langley. Three of his fellow students in the five-man class were Major Horace Hickam, Captain George Kenney and Lieutenant St. Clair Billy Streett. Captain Earl L. Naiden was the Director of Instruction, and Major Oscar Westover the Commandant. At the time, pursuit was considered the principal weapon of air war. Knerr disagreed. Maybe it was now, but it wasn't going to remain so. His position was: "When you take away an enemy's bullets and beans from him, his air power becomes useless." Bombardment was the key. His minority belief was, of course, a variation on the Douhet, Trenchard and Mitchell theme. But Knerr added a further proviso to his thinking. It was one that Hap Arnold would come to appreciate and subscribe to five years later, when he served as Foulois's G-4 during the 1930 maneuvers. Knerr insisted that the supply-and-maintenance function was far more effective in deciding the issue of air combat than any other single factor.

It was over the issue of logistics plus a couple of more close calls due to defective aircraft that Knerr got in a brass jam again. He began complaining about the poor quality of maintenance, and when no one listened to him, he made his own investigation and filed a report direct to the Air Chief's office. For his pains, once again he was placed under arrest, pending the outcome of an inquiry into his independent action. He called for an investigation by the Inspector General, and the action against him was quickly dropped.

Mitchell's court-martial was also taking place at about this time, and Knerr offered to come and testify on Billy's behalf, but Billy said he had plenty of witnesses and thanks just the same.

In spite of his reputation as a very free thinker and stubborn-minded individualist, Knerr's intelligence was recognized, and following the year at Langley he was approved to attend the Command and General Staff School. At Christmastime, 1926, the Knerrs suffered a devastating blow. Ten-year-old Barclay was killed while riding his bike, hit by a truck.

It was two years later that Andy Andrews and Hugh Knerr got to know each other well.[2] Andrews had come from duty with the Training Command in Texas to attend the Tactical School at Langley Field, and Knerr had arrived earlier to assume command of the 2nd Bomb Group—one of the three full combat groups in the United States.

The 2nd's former CO was Major Lewis H. Brereton. Brereton, a former World War combat leader and Mitchell aide, was noted for being a debonair type and, at the time, not too serious a practitioner of pushing air power anywhere but up in the blue to fly here and there without a great deal more in mind than the joy of plugging along in a Keystone or a Martin B-2.[3]

Knerr quickly sized up the situation and put his serious mind to work. He made wiry, energized Lieutenant Ken Walker, CO of the 11th Bomb

Squadron, his Operations Officer, and they went to work. The Group was streamlined, with thirteen planes to a squadron. Serious practice was begun on developing bomber formations that afforded mutual fire support and a method of cross fire to drive off attacking fighters. Technically, there were two major problems. Aside from having bombers that could plod through the sky for only a short distance before heating up, there were no bombsights worthy of the name. To attempt to solve the latter difficulty, a do-it-yourself method was devised. Strings were rigged in the bomb bay, over which bombardier Knerr could draw a bead on the target. The strings were tied on one end to the pilot's arms and Knerr could guide him by putting pressure on one string or the other. It was no Norden bombsight. Rather, it was an example of ingenuity, an indication of what was needed to make the aircraft effective in a time when aerial inventiveness soared without benefit of economic support.

It soon became apparent, following the maneuvers of 1927, that Knerr had converted the 2nd Bomb Group from a playboy command into a premier unit on a par with the 1st Pursuit and the 3rd Attack groups. The sharpness of the Group's ability was never more apparent to Benny Foulois than the night Knerr brought six of his bombers to Wright Field and, with landing lights gleaming from their wings, signaled them down to roost. Knerr was to say that Foulois nearly fainted at the sight. Bombers at the time were not equipped for night missions, and their pilots were not thought to be qualified for flight duty after dark.

Knerr changed all that, but he wished others, somewhat far distant and more powerful, to take note. Fuel supplies were meager, and therefore training was restricted. He decided to call attention to the situation by using up the Group's entire gas supply for a year on a single mission. The 2nd would take off at daylight and fly until sunset the following day, refueling wherever necessary. They had no navigation aids. They flew by compass to Kansas City and attracted a great deal of attention, so much so, in fact, that Knerr, instead of being hauled up on charges for misuse of government supplies, received a congratulatory telegram from Chief of Staff Major General Charles P. Summerall for having drawn congressional attention to the need for additional appropriations.

In 1927 Knerr had submitted a letter to the Materiel Division urging the development of a bomber that could carry a 1,000-pound bomb load and be capable of cruising at 150 mph at 10,000 feet. It was pie in the sky at the time, but Chief Engineer Jan Howard had been thinking along the same lines. He asked Knerr to come up, and they gathered with some others in Benny Foulois's office and hashed the concept over, Knerr being called upon to present his ideas.

At Langley, where the National Advisory Committee on Aeronautics had its laboratory, aircraft engineers attending a meeting there paid a visit to Knerr's Operations office and, with Knerr, Walker and others in attendance, they were given a thorough briefing on bomber operations. Knerr believed that these meetings laid the groundwork for the design competitions that were eventually authorized to construct a new bomber. During this brief period, Andrews, having finished the Tactical School course, served on the 2nd Bomb Group's headquarters staff and became closely associated with Knerr. The appreciation was mutual, and the Andrews and the Knerrs became associated socially as well as aeronautically.

In 1930 Knerr father and son went to school, Hugh Senior to the War College and his son, following in his father's original footsteps, to Annapolis. It was then that Benny Foulois, as Assistant Chief of the Air Corps, informed Knerr that he was going to take over the duties of Chief of the Field Service Section at the Engineering Division at Wright Field. There, Foulois said, Knerr would be able to do something about all his previous complaints on maintenance and logistics.

It was during maneuvers in 1930, however, that Knerr demonstrated the importance of mobility, a requisite deeply ingrained in Andrews' thinking. If you did not have mobility, aircraft would be tied to ground transportation for support. The answer was to develop aircraft that could supply air units in the field. To illustrate the point, an exercise was carried out across the bay at Virginia Beach, which was inaccessible to ground transportation. The unit operated out of a base in the sand dunes for ten days. Everything that was needed was brought in by bombers. As Knerr later said, "My report on the exercise stressed the need for specialized military transport; eventually such transports were developed." What he did not say was that out of his concept was eventually to come the Air Transport Command.

When Andrews requested Hugh Knerr for his Chief of Staff, he well knew the kind of officer and gentleman he was getting, his strengths and weaknesses. In the latter view, the lean, rather handsome Midwesterner had a black-and-white way of looking at things, saw issues in absolutes, believing he was absolutely right on questions of aeronautical substance. Admirably, he had no hesitation in laying his career on the line to prove a point, but his dogmatism often antagonized his opponents, whom he saw lurking in the War and Navy departments plotting dark deeds against those who worked to foster air power and independence. On this last there was no more determined officer, a zealot who could foresee that bright goal and was determined to reach it. He had a Scotsman's ingrained wariness, and this combined with the stubbornness in his nature was inclined to lead him to erroneous conclusions about those who did not see eye to eye with him.[4]

Andrews fully understood his determined, sharp-minded Chief of Staff, and the two would work closely together, the bond between them deep and enduring.

Major Follett Bradley, as Andrews' G-2—Information and Intelligence Chief—was the first of the small staff to report while the preliminaries were still being ironed out in Washington. Bradley, since his job as Air Corps Inspector General of the airmail operation, had been assigned to the War Department, working in the War Plans Division under Kilbourne, and in such capacity he had participated in the early GHQ Air Force conferences.

Andrews had never served with Bradley, but he knew his reputation as a thinking airman, solid and circumspect. An independence believer, Bradley was cagey in his dealings with the General Staff, polite and cooperative. Unlike Knerr, he moved quietly and kept his own counsel. Similarly, however, he had gone almost the same route in his switch from Navy to Army to Air.

The son of Brigadier General Alfred E. Bradley, Surgeon General of the A.E.F., he had made the change from the Navy to the Ordnance Section of the Field Artillery in 1912. His reason for leaving the former was acute seasickness. The problem didn't carry over into the air, for in 1912 he had been Hap Arnold's observer at Fort Riley before that fateful flight in November when Arnold was nearly killed. Bradley had wanted to fly then, but it took him a while to earn his wings although in December 1917 it looked for a moment as though he would get to do so. The Adjutant General sent him greetings with notification that he was to be detailed for aeronautical duty "to serve on such duty for a period of two years after having fulfilled the requirements of a military aviator unless sooner relieved therefrom by proper authority." The "proper authority" was not about to relieve him for such duty in the first place, and he went off to France as the CO of Battery E of the 17th Field Artillery. There, near Vaux, he was cited for gallantry in action. He and two subordinates—a second lieutenant and a private—occupied an advance observation post for twelve hours while under intense enemy fire, directing the shelling of German positions. Bradley's helmet saved his life when a piece of shrapnel made contact. For his diligence he received the Silver Star and the Croix de Guerre. His two companions received the Distinguished Service Cross, but when Bradley later made application to acquire the same award, the Adjutant General rejected his claim, maintaining that "extraordinary heroism had not been displayed."

In the Army reorganization of 1920, Bradley was able to arrange the transfer he sought, and two years later, as a major in the Air Service, he was a pilot making a name for himself in more ways than one. In October 1922 he received a commendation from Secretary of War Weeks for taking second place in the Liberty Engine Builders Trophy Race. This was a time of air races and a time of controversy among Army airmen over the use of the parachute. Some pilots thought parachutes should be banned, that an airman must be prepared to stick with his plane. Bradley was one who agreed, and he said so in a letter published in the monthly *Air Service News*.

He was serving then as the Assistant Commandant at Chanute Field, and he claimed his view was shared "by a majority if not all, the older pilots at this station." There were only three occasions, he declared, when it was justifiable to use a parachute: fire, collapse of the airplane in flight or collision. He felt making the wearing of a chute mandatory would "encourage faintheartedness."

Major Eddie Hoffman, a pioneer in the development of parachutes at McCook Field, took Follett Bradley to task, pointing out that no plane was worth a pilot's life and that the wearing of chutes at McCook had not produced any faintheartedness but instead had saved a good many lives.

Bradley's attitude was more an indication of his resolve than a hard-pants view of the need to save life and limb. On the latter score, he had no hesitation in risking his own. During the airmail undertaking, having stopped to inspect Hap Arnold's headquarters at Salt Lake City, he had flown to Burbank, California, to have a look at Ira Eaker's Route 4 operation.

Eaker had met the Major in 1926, when Bradley was serving as CO of the Air Corps Observation unit at France Field, Canal Zone, and he knew him to be a very able type. Two incidents at Burbank were illustrative of his quiet way of doing business. He flew on a regular mail run to Las Vegas, riding along in the crew chief's position. On the ground, he was mistaken by a mechanic for an enlisted man, the mechanic singing out, "Hey buddy, where shall we put the gas?" Bradley told him and said nothing to correct the mistaken impression. When the mechanic learned of his error, he feared he might end up in the guardhouse on bread and water, for many officers were known to take umbrage over a failure to be properly recognized and addressed. It was not the sort of simple mistake that occupied Bradley's thinking.

As noted, during the airmail crises, when he was Air Inspector, what did occupy his thinking, aside from his mission, was the independence-now letter he was carrying that had been jointly drafted at the Air Corps Tactical School for submission to the Baker Board. During the three days that he and Eaker had adjoining rooms at the Crystal Hotel, in Hollywood, they discussed the letter, and Bradley did some quiet missionary work among the pilots. The result was that Eaker later wrote him to say that "ten of our reserve officers have guaranteed to get 10 telegrams each in on it."

Eaker also sent Bradley photographs, newsclips and a story on how the Hearst papers were handling AACMO, for, like every involved Air Corps officer, he was acutely aware of the importance of public relations, particularly because so much of it of late had been adverse.

As Andrews' Chief of Information, balding Follett Bradley at forty-five was a pro who had tested the air above early on, and in the ensuing years had come to know well the rugged shape of the political topography below. He was elated to have been chosen to serve on Andy Andrews' staff of the new air force. So, too, was Captain George C. Kenney.

An airman of many parts, Kenney had begun his flying career in 1911, while a student at MIT. He and an adventurous pal put together what they believed was an improvement over the plane Louis Blériot had used to cross the English Channel two years previously. The latter was the same plane that had captured the fancy of Second Lieutenant Henry Harley Arnold when passing through Paris. A photograph of the flying machine had inspired Kenney. The result was an aeronautical wonder that turned out to be more a hopper than a flier. Kenney, on the short side, with a full-lipped mouth and a go-to-hell look in his eye, would get the contraption fixed up, run it across the field and climb to an altitude of about four feet. Airborne for all of fifty or a hundred yards, he'd then come down with a crunch, smashing the landing gear. They'd repair the damage, drag their creation back across the field and then it was his pal's turn to take wing. They literally flew their bird into the ground until it was "busted" beyond recall, whereupon the more essential need to tend to their studies took over.

The war brought twenty-eight-year-old George Kenney into the cockpit for good and all. He began his flight training at Mineola, Long Island, in the summer of 1917, with Bert Acosta as his instructor.* When he continued his training at Issoudun, France, he had all of seventeen and a half hours to his credit. Halfway through the course, in February 1918, he and eighteen others in his group were sent to the front to fly two-seater French Salmsons. They formed the 91st Squadron, which Billy Mitchell made into a long-range reconnaissance outfit. They had very little combat training. None of them had been given gunnery or acrobatic instruction or shown how to fly in formation. It was strictly on-the-job training, and the first time Lieutenant Kenney was in combat he figured the only thing that saved him was the fact that his opponent was just as green as he was. They flew around in circles, chasing each other, shooting holes in the air. For a time thereafter when George saw trouble coming he'd duck into a cloud and then fall out of it in a spin because he didn't know anything about instrument flying.

On one of his first missions behind the lines, he had Lieutenant Asa Duncan in the rear cockpit with a camera, their job being to come back with photographs of a supply dump. But the weather was foul, and they were headed for home at treetop level when Kenney spotted a column of German troops marching along the road. He banked fast to give Dunc a shot at them with more than a camera, but Dunc, a friendly soul if there ever was one, waved at the troops and they waved back.

* The site became Mitchel Field.

"Shoot! Shoot!" yelled Kenney.

"No!" shouted Asa. "They're Americans!" Then he recognized the coal-scuttle German helmets.

Low clouds put an end to the case of mistaken identity. In his report, Duncan sheepishly made note of marching German soldiers, and that night the beer was on him.

In Kenney's way of thinking, it was a good thing they had the rugged Salmson, because flying deeply into enemy territory, they had to fight their way in and out. To him the DH-4 in similar circumstances was indeed a "flaming coffin" and was a measure of U.S. aeronautical expertise. In one encounter, he saw seven DH-4's shot down and burning in a single pass by attacking Fokkers. The cause was the installation in the Liberty engine of a pressure-feed fuel system instead of a gravity feed. A hit anywhere in the system meant gasoline spraying and fire. Following a mission, Kenney, having been chased and low on gas, landed at the 11th Squadron's field to refuel. The 11th flew DH-4's, and Kenney found there was only one sur-viving pilot in the Squadron. "That makes me the commanding officer," said the pilot, "and I'm putting myself on extended leave." He took off for Paris that afternoon, and according to George, when the war ended he was still there sampling the wine.

As for Kenney, at the Armistice he was CO of the 91st. He had shot down two German planes, been shot down himself, flown seventy-five mis-sions and received the DSC and the Silver Star. A real professional, he de-veloped a hard-nosed attitude regarding the Regular Army officer. Al-though reserve officers of low rank, he and his buddies such as Monk Hunter, Hal George and a good many others had become combat-wise pilots. They would take the newly trained, half-trained kids into their squadrons and teach them how to stay alive until they, too, were battle tested. But all too often when a new squadron was formed, a regular officer from home, usually a major with no combat experience, would ar-rive on the scene to take command and the losses would be awful. "It hap-pened time and again," said Kenney. "The average guy hated the guts of the Regular Army guy—he had no use for regulars."

There is little doubt that emotions produced by the life-and-death out-look of young combat pilots lay at the root of strong feelings carried over into peacetime, where they were rallied around the banner of Billy Mitchell.

After the shooting stopped, Kenney became a captain, remaining in Germany until June 1919. From the first, he was a Mitchell man and an air independence advocate second to none. He was also an airman with three years of MIT behind him and a bent for aeronautical engineering. After duty on the Mexican border and with the Artillery, he was assigned to the Engineering School at Wright Field, where Major Edwin E. Aldrin was his instructor.[5]

It was during this period and later, when Kenney was Chief of the Factory Section at Wright, that a great many aeronautical ideas were being explored at nearby McCook Field, which had been established in 1917. McCook's purpose was to design and test aircraft, their engines and their ordnance. At Wright Field, experimental equipment was further tested with the thought, if not the monetary wherewithal, for eventual purchase and production by an aircraft manufacturer. Kenney had little regard for what was going on at McCook, believing it was staffed by a bunch of "pseudo engineers" and that the place where aeronautical design should be pursued was with the civilian aircraft companies.

Billy Mitchell wouldn't go so far as to say that, but after he had test-flown a McCook creation full of experimental equipment, he was not impressed. He summoned the section chiefs from both fields and in a wide-ranging, typical Mitchell exploration of ideas, Kenney first heard about the jet engine.

Said Mitchell to his engineering chieftains, "I've been down to the power-plant lab, and I've seen the work you're doing on the new engine that they say is going to turn out six hundred horsepower. You know what's going to happen to all the work that new engine does? It's going to turn crankshafts and push pistons up and down and wear out bearings and wear out the sides of cylinders and spend a whole lot of energy internally, doing no useful work; then it'll push a lot of stuff out into the open through the exhaust pipes. What's left will turn a silly little fan out front, and what's that fan doing? All it's doing is putting a thrust to the rear, and the reaction to the thrust is what makes the airplane go forward.

"Now, the most important part of all that power that's being used there, you're throwing away," he lectured. "It's going out into the air through the exhaust pipes."

His followers looked at him; they looked at each other. They worshiped old Billy and anything he said was gospel, but they didn't know what in hell he was talking about.

Mitchell observed their reaction and said, "You're a dumb bunch. I suppose I've got to tell you how to build that engine. Well, go get yourself a big piece of pipe about ten feet long and three feet in diameter. Bore a hole in the top, pour gasoline into it, touch a match to it. Then all this stuff will blast out through the rear and the reaction will shove this thing forward if you'll just tie it into an airplane."

They still didn't know what he was talking about, but later Kenney would realize he had given them the basic idea for the mechanics of a jet engine. The year was 1924.

Philosophically, Mitchell's pungent aphorism "To entrust the development of aviation to either the Army or the Navy is just as sensible as entrusting the development of the electric light to a candle factory" was Captain George Kenney's maxim.

Like Andrews, Kenney attended the Air Corps Tactical School, the Command and General Staff School and the War College. He returned to the Tactical School as an instructor in 1928, and it was there that he and Andrews came to know each other, in the classroom, on the golf course, over cards and in the air. Andrews recognized that the voluble Kenney was an energetic and enthusiastic airman whose mind crackled with ideas and opinions, most of them strongly held. His tactical experience marched along with his technical thinking. A prime example of his mind's reach lay in his pioneering the use of fixed machine guns on the wings instead of the cowling of a plane. He had developed the idea in 1924, and against a welter of engineering criticism, he mounted two .30-caliber machine guns on the wings of a DH-4 and demonstrated how firepower could be increased. "Putting guns in the wings of airplanes—my God, how they fought me on that one!" he was later to reflect. And well he might, for when the United States went to war in December 1941, its most noteworthy fighter, the Curtiss P-40 carried just two .50-caliber machine guns, mounted on the cowling.[6]

In a more strategic vein, it was Kenney who was involved in obtaining and correcting Dorothy Benedict's translation from French to English of Douhet's writings. Finally, while serving in the Air Corps Plans Division under Jim Chaney during the tumultuous year of 1934, if George Kenney was not the principal drafter of the McSwain bill on air autonomy, his writings formed the basis of what was fed into it.

There was nothing dour or reserved in Kenney's Irish-New England outlook. His crew-cut pate was as much a trademark as his good-luck dice, given to him by a priest while Kenney was going through training at Issoudun. His personality was unlike that of Knerr's or Bradley's. But beneath the flair and friendliness, the sharp wit and good humor, lay identical beliefs and concepts on air power and its proper utilization. He'd been a captain for seventeen years when Andrews chose him as his Assistant Chief of Staff for Operations.

Major Joe McNarney had departed Hap Arnold's command at March Field in August 1933 to become an instructor at the Army War College, a measure of his caliber as seen through the eyes of the General Staff, not to mention the Office of the Chief of Air Corps. Andrews' selection of him for his G-4, Chief of Logistics and Materiel, came not through close previous association but undoubtedly from a knowledge and a recognition of McNarney's long record of solid dependability in everything from aerial combat to intelligence operations. His reputation for thoroughness, his ability to think and maneuver his way through thickets of military obfuscation to levels of clear reasoning, were known to his associates. Lieutenant John Whiteley,[7] who was to serve as one of McNarney's half dozen staff assistants, observed: "For sheer calculated thinking and decision, Joe

McNarney had it. . . . His unadorned personality was disconcerting until one got to know him, when he could display a submerged warmth."

Joe McNarney was not a cold man, just a quiet and very thoughtful one, a good man to have holding down a tough spot, and Andrews was happy to have him as a top member of his small team of assistant chiefs.

There was still another member of the team to be named: G-1, Chief of Personnel. The choice must have brought some quizzical expressions. Major Henry B. S. Burwell somehow did not seem to fit the slot, either in looks or in temperament. Some called him Spider Burwell, for he was a man who moved along quickly, almost scuttling, his eyes usually on the ground. It wasn't that he was searching for something he'd lost. More likely, he was deep in thought, for Harvey Burwell was a man full of intellectual curiosity, and one pictured him amid the halls of ivy rather than in a cockpit circling the cloud tops. But looks are deceiving.

Harvey, a Nutmegger, from Windsor, Connecticut, and a graduate in 1913 from Norwich University, had been an "early bird," taking his flight training in 1916 and serving with the 1st Aero Squadron on the Mexican border. During the days at Rockwell Field in 1919, as CO, he had been plagued by the wild aerial antics of one Lieutenant Jimmy Doolittle. One of the few remaining instructors at Ream Field, a North Island pursuit and gunnery station adjacent to the Mexican border, Doolittle's behavior was the talk of Rockwell. It was not considered good military form to sit on the crosspiece between a Jenny's wheels in flight or in landing, even if there was a five-dollar bet riding on the success or failure of the stunt. And even though the government was bent on self-destructing its air strength to practically zero, a second lieutenant chasing a duck up a dead-end ravine and wiping out a ten-thousand-dollar aircraft in the process was not recognized as an acceptable method of reduction. Further, any very junior officer who made a game out of using his motorcycle to cut off his CO not once but several times while the latter was attempting to land, deserved to be grounded and was . . . at least temporarily. To his fellow pilots, Jimmy Doolittle was something special, but none of them would have put a plugged nickel on his chances for the future and that included Harvey Burwell.

In the early twenties, for a time Burwell had been a proud member of General Allen's Occupation Army on the Rhine. He became closely aligned with Andy and Johnny Andrews in that highly cosmopolitan, polo-playing, fox-hunting scene. As Operations Officer at Weissenthurm, the Air Service's major flying station in Germany, his was the official signature on all flight logs, attesting to "passenger carrying, cross country flights and practice flights."

When Burwell returned to the United States for duty that in the next decade would involve him in both staff and tactical commands, including command of the 3rd Attack Group, he looked upon Andy Andrews as a special person, a friend without parallel. Andrews, for his part, enjoyed Harvey's unorthodox manner and appreciated the wide range of his intellectual pursuits. Burwell was a little like Socrates in that if he had an idea, no matter the subject, he would corner the unwary and lose all track of time in explaining his theories. Andrews once remarked jokingly that he didn't mind Harvey's using such big words, if he'd only use them in the right places. More seriously, he knew that Burwell had much to contribute and had already proved the point by collaborating on the standard form for the maintenance and inspection of aircraft.

Prior to 1925, there were no standardized flight records kept on Army planes. A pilot climbed aboard, a sergeant noted the time of takeoff and return or destination elsewhere, and that was that; the system was slipshod, inefficient, and often hazardous. More important than the hours flown was the plane's condition. It, along with the hours flown, should be recorded after every flight, so that maintenance was maintained and was not a matter of labor employed only after trouble developed. Working together, Burwell and Second Lieutenant Donald B. Phillips produced the form 1-A, known thereafter to every military pilot and mechanic. The pilot signed the form before he took off, indicating he knew the condition of the aircraft he was flying, and when he landed, he signed it again and made note of time flown and the aircraft's condition. Like most brilliant concepts, this form was a simple and obvious way to improve an existing condition, but no one had thought how to do it until Phillips and Burwell put their minds to it.

When Andrews summoned Burwell to join his staff as head of G-1, Personnel, he had been serving as commanding officer of a small Air Corps unit at the University of California. He departed the academic backwater, excited by the prospect of serving the new Air Force on the staff of his old friend Andy.

From the mail that poured in to Andrews, it appeared that almost everyone in the Air Corps was anxious to do that. The congratulations, the praise, the *thank God it's you!* were more often than not tied to *Andy, can you use me? I'll take any job you've got to offer.*

Wrote Bob Olds from the Command and General Staff School: "Naturally I'm hopeful of being able to be of some direct service to you."

Major George Brett, anxious to bid farewell to the classroom at Fort Leavenworth, was more plaintive. "I'm not crying, but if you could just rig me up a job, I know these people will let me go."

Major Lewis Brereton, CO at France Field, wrote: "Needless to say, I would be delighted to serve under you in any capacity wherever, and I would more than welcome the opportunity it offers."

And so forth and so on.

To everyone who wrote wanting a job in his command, Andrews responded with a specific answer. Olds would come to serve. Brett was slated to go to the War College.[8] Brereton was slated to take Brett's place at the Command and General Staff School. Etc. There was nothing easy in the selection of his staff or that of his wing commanders and their staffs. It was a matter of too few candidates to fill too many jobs. It was a matter of the special place of seniority, of previous performance and finally of General Staff approval, although at the outset all his recommendations were honored.

To B. Q. Jones he wrote a letter of regret that B.Q. was leaving Langley to teach at the War College. In response to Bert Dargue's opinion that the commanding general of GHQ Air Force should carry the rank of major general, he sent a copy of the Tables of Organization as a guide and informed Bert, who was the Assistant Commandant at the Air Corps Tactical School, that his temporary headquarters staff would number only twenty-three officers and eighty-five enlisted men. It was not going to be a very large air force. In response to Jakie Fickel's happy salute, Andrews wrote: "There is apparently an attitude of sympathy and understanding in Air Corps problems more prevalent in the General Staff, certainly at the top starting with the Chief of Staff, and I feel that we are really going to get somewhere."

And in that regard as to the matter of numbers and selection, Andrews had a trait for which he was to be criticized. It was said his loyalty to his close friends made him blind to their shortcomings. He would have disagreed, for in seeing strength he could overlook weakness, and that was how his gaze was focused. He would get the best out of those who worked for him, because they wanted to give him their best. He was fully aware that within the exclusive structure of military life prejudice, jealousy and personality often played a part in who got what or went where. And although there was no doubt he was helpful and protective of his friends and would try to aid them if he could, his actions were always within the proper framework of the structure.

To Harry "Sue" Clagett, West Point classmate and one of his oldest friends in the service, Andrews wrote in confidence that the three wing commanders of the air force were to be Conger Pratt in the East, Jerry Brant in the Midwest and Hap Arnold on the West Coast. Then he gave Harry "a little friendly tip. Get a detail to the Industrial War College and from there go to the War College." If Clagett did that, Andrews felt sure—"if I stay in this job"—that Harry would get the appointment to command a wing on the next vacancy.

"I am a great believer in command going by seniority," he informed his friend, "and have told the Chief of Staff that that was my policy and would govern my recommendations unless there were particularly exceptional circumstances. Of course, this is all on the assumption that the senior man is qualified for the job." Andrews suggested that if Clagett was interested in the idea, to let him know as soon as possible. Clagett, who was CO at Kelly Field, fired back a telegram that indeed he was!

With old friend Walter Weaver the situation was quite different. Weaver had written his congratulations before year's end, saying, "Thank heavens someone is seeing the light! Gee, Andy, I'm glad. I wish I could be there and laugh with you. It would do me a helluva lot of good."

Weaver had been transferred as head of Air Corps G-2 at the height of AACMO to duty in New York to serve as a representative on procurement with the Materiel Division. He felt that he had been exiled by the General Staff a la Billy Mitchell because of his outspoken position on air power. Andrews took it upon himself to check with General Kilbourne and others, all of whom assured him that they had nothing to do with Weaver's removal from the Washington scene, "but in spite of that," he admitted to Walter, "I find considerable reluctance to do anything about it."

Weaver had been hoping to attend the War College, but his name was not on the final list prepared by the Office of the Chief of Air Corps and sent to the War Department for approval. Previously, through Pa Watson's White House influence, Weaver knew that his name had been number two on the list, beneath that of Mike Kilner. However, the first submission was rejected because of too many names, and when it was sent back from the Chief's office, Weaver's name had been deleted.

"So you see," he wrote Andrews, "the same old crowd in the Chief's office apparently did me dirt. Apparently there is either a group or an individual there who feels that I have in some way done them a personal injury. . . ." On whom the knife wielder might be he did not offer an opinion, but he did specify that whatever his next duty assignment might be, he was not interested in "serving under any activity over which my good friend Benny has any control."

In the end Andrews was able to work the problem out to Weaver's great satisfaction. He would join the GHQ Air Force headquarters staff at Langley as Air Inspector. "Rank will be that of Lieutenant Colonel," Andy informed him, "and will mean no advance in rank for you, but at the same time we request orders we will institute steps to get you on flying status so that you can make inspections by air."

A good and loyal friend, Weaver recognized the hazards of the chase, and although he and his wife, Elizabeth, were only too anxious to join the special few at Langley Field, he cautioned: "I cannot impress upon you too much, Andy, that you have just got to make good on this thing. Your life's work, of course, is wrapped up in it all, and you just must not let friendship or feelings jeopardize the success of your project. And please do not let our mutual friendship in any way operate to hurt you officially."

That Andrews was careful to draw the line on his recommendations was apparent in his brother Billum's case. He told their father, "I'm sorry Billum cannot be on my staff, but even if he had been through Leavenworth it would be an unwise move. I have never seen it work. I might be meaner to him and require more from him than any member on my staff, yet there would always be suspicion that he was favored."

He showed, too, that better than Weaver or anyone else he was fully aware of what he was up against. When informing Lieutenant Colonel Jerry Brant of Brant's appointment to command the 3rd Wing of the Air Force, Andrews made his position patently clear. "I believe the whole future of the Air Force in the Army rests upon what we do with the GHQ Air Force. If we don't make a success of it, we are going to lose it as a part of the War Department. I believe it would be passed over to the Navy."

While on the surface and in the press reports it appeared that the GHQ Air Force was to be a formidable combat arm numbering a thousand fighting planes, its three tactical wings ready to move on its commanding general's order, Andrews was faced by problems and conditions that projected a somewhat different reality. He described "the fundamental mission" of his command as "to so organize and train as to operate efficiently in any weather and against any proper objective within range of the equipment available. . . . If we can deny a potential enemy air bases, fixed or floating, from which he can operate by air against the Continental U.S., then, and only then, are we in a position to accomplish our mission."9

The Drum and Baker boards, with MacArthur's approval, had recommended a force of 980 planes with a 12 percent margin for overhaul. Andrews had 446 aircraft, less than one half the officially press-reported figure, and only 176 of them were considered modern and suitable.

The same boards had recommended, and the General Staff had approved, a complement of 1,245 pilots and 3,267 enlisted men to fly and maintain the aircraft and the bases where they were stationed. Andrews had 555 pilots and nearly 4,400 enlisted men; only in this last was the number greater than specified, but not really enough. Those were the numbers, and it was hoped that where they were so lacking they would be built up in time through administration support and congressional approval.

In the organization of the command, not only was Langley Field designated headquarters for the Air Force—Andrews had opted for Mitchel Field—but it was also the home of the 2nd Wing, headed by Brigadier General H. Conger Pratt, and the 8th Pursuit Group. The wing consisted of two bomb groups and two pursuit groups, one of each at Langley with the 1st Pursuit at Selfridge and the 9th Bomb Group at Mitchel.

There were also four observation squadrons. Two, using amphibians, were to be organized at a later date. A balloon squadron was also attached to the 2nd Wing.

Hap Arnold, whose long-awaited promotion to Brigadier General was a well-celebrated event at March Field, commanded the 1st Wing. It consisted of two bomb and one attack groups, principally stationed at March and Hamilton fields. There were to be three observation squadrons as well, but like the 2nd Wing, two of them were on paper.

The 3rd Wing, under Colonel Jerry Brant, with headquarters at Barksdale Field, in Shreveport, Louisiana, consisted of the 3rd Attack Group, at Fort Crockett, and the 20th Pursuit Group, at Barksdale. Included were a scattering of independent bomb, attack and pursuit squadrons as well as a Scott Field Airship Group. There were also ten service squadrons stationed on the nine bases the GHQ Air Force made use of.

And so, on paper, the tactical components of the Air Force appeared formidable, and could they be brought up to strength they would be. But between what was meant to be and what was, there were vast discrepancies. The press reports stressed the former, capturing attention at home and abroad.

In Berlin, U. S. Military Attaché Colonel Jacob W. S. Wuest sent word to Andrews in February that "the change we have made in our Air Corps organization is being watched with great interest here. Questions of the formation of an Air Force now occupy a prominent place."

Indeed they did, and by coincidence, at almost the same date Andrews assumed command at Langley Field, Hitler let it be known openly what had been going on secretly for a year, that the German Air Force, the Luftwaffe, was now an independent arm of the Reichswehr.

Air Minister Hermann Goering was to head the force, with the title of General der Flieger (General of the Airmen). The rotund World War ace informed Europe and the world that Germany demanded "equality" in the air, while his subordinates in the huge, newly built Air Ministry were engaged in a crash program to attain superiority.

The French reaction to this latest violation of the much-violated Versailles Treaty was to lodge a protest with the League of Nations. At the same time, a British delegation meeting with Hitler in Berlin assured him they wanted peace.

In his reply to Wuest, Andrews wrote: "We have a year of service test and though I feel there will be many changes during the year, I am confident the GHQ Air Force is with us to stay."

There were two changes he would like to have made at the outset. One concerned temporary promotions and the other the clumsy division of authority and command the General Staff had put together in establishing the Air Force. Andrews had gone along with both conditions because one was a compromise that had the agreement of Benny Foulois and his staff, and the other offered no immediate choice but to accept it and try to work for a change.

On the matter of temporary promotions, the device had been a sticking point since its inclusion in the Air Corps Act of 1926. The provision provided for promotion for duty commensurate with the rank required, the promotion not to exceed two grades above the officer's permanent rank. Airmen had long opposed the inherent inequities in the idea. The promotions were exclusive, because they applied to some commands and not to others. Only a few could benefit, to the exclusion of the many. If the temporary grade did not become permanent while the officer held it, he could be rotated to another duty and revert back to his permanent rank with the attendant economic and psychological losses.

In 1926, Assistant Secretary for Air Trubee Davison had made it clear to congressional leaders that the Air Corps was not about to adopt the provision. Before that and afterward, airmen from General Patrick onward had agitated for a separate promotion list, a stipulation that had been included in the McSwain bill.

The War Department and the other service branches were unalterably opposed to the idea for reasons previously cited, but the very nature of the new force demanded an increase in rank for those who were to command it. The Baker Board had made a point of noting the low rank of Air Corps officers holding down important commands, and suggested that the War Department do something about it by granting temporary promotions but on a far broader scale than indicated in the Air Corps Act. A ruling by the Judge Advocate General gave an interpretation that supplied the needed breadth, and Benny Foulois agreed to give it a try. In setting up the duty assignments that would bring temporary advancement, Andrews had worked with Foulois's staff as well as G-3 in an effort to establish an equitable system. Andrews wanted as many regular officers as possible to benefit from the move. Foulois's office wished to limit the promotions to the top command and staff officers. Assistant Chief of Staff for Operations General Hughes wanted the list made even smaller. MacArthur supported Andrews. At the end of January, the War Department in issuing its Circular 7 recommended expanding the opportunity for temporary promotion. The end result was that nearly half of the Air Corps regular officers received temporary promotions, 616 out of 1,333.

GHQ Air Force benefited the most. In a talk he gave to his staff on March 12, Andrews made note that although the system of temporary promotion was not desired, since it had been established they must submerge personal feelings and make the best of it for the good of the Air Corps. He did not believe the method could last, and felt that at a future date it must be replaced by a permanent system. He also felt that the General Staff in utilizing the present method was tacitly admitting that there was something wrong with Air Corps grades with reference to the rest of the Army. Optimistically, he thought there was an excellent chance that a separate promotion list for the Air Corps would result if they all made the experiment that lay ahead of them work. He believed, he said, that the War Department had made a 180-degree turn in its attitude. It genuinely wanted to see the Air Force succeed. The misunderstanding and hostility were gone, and the General Staff appreciated the necessity of keeping Coast Defense in the Army.

The matter of promotion was a personal question affecting the individual officer, his morale, his bank account and his wife's social standing in the military pecking order. The matter of the new Air Corps/GHQ Air Force command structure was something else again. Approximately 40 percent of the Air Corps strength made up GHQ Air Force. Foulois and Westover and those Air Corps officers concerned with the problem naturally enough were anxious to have the Air Force come under the command of the Chief of Air Corps. Foulois had originally hoped to command it himself. MacArthur and Drum and the General Staff disagreed. On the level of military strategy, their thinking, particularly MacArthur's, dealt with unity of command and the Four Army Plan. GHQ Air Force was a mobile strike force that was trained and equipped to repel an aggressor until such time as the ground forces were ready to act.

While the surface reasoning sounded logical enough, as had the War Department press release at the Raritan post command exercises in September, the ulterior reasoning was glued to the fundamental issue of Air Corps independence. It could be argued by the General Staff that to give the Air Corps a unified air force was to give it clout and prestige that could only encourage and aid those who were determined to lead it out from under War Department control. There would be those, such as Hugh Knerr, who would maintain that the General Staff in establishing two air commands was out to divide and conquer, because in setting lines of authority the two air staffs were bound to compete and become antagonistic, which in turn would help to block moves toward autonomy.

It is more likely that the planners in the G-3 Section of the General Staff were influenced by the Drum and Baker board recommendations and were motivated instead in their decisions by the complexities of national defense in a nondefense-minded atmosphere. If this was their first concern, then maintaining tight control was their second, to aid in insuring the first. Certainly the architects of GHQ Air Force were not bent on building a convoluted system of command simply to produce friction within the Air Corps. That the system was convoluted was more a measure of the planning capabilities within the General Staff, influenced no doubt by the political conditions of the moment, than of any Machiavellian intent. The structure was experimental; it had to be tried out. But it didn't take a prophet to see that a three-way split in authority would hardly produce harmony.

As the Commanding General of the combat arm, Andrews was charged with the tactical training and employment of his force. But, immediately, his line of authority was broken by the inclusion of corps-area commanders in the structure. These ground generals, just as before, were to retain administrative jurisdiction, including courts-martial, over the personnel manning the GHQ Air Force bases. This meant that Andrews' degree of internal command was limited. Foulois had protested the retention of the standard system, maintaining that the Baker Board had been opposed to it as well. The decision to keep it intact was made by Drum.

Under the new arrangement, the Chief of Air Corps was responsible for Supply, Training, Schools and the control of personnel. With GHQ Air Force concentrating on combat readiness, the division of effort seemed equitable enough. Instead, it was to become a dividing wedge.

Some months after the GHQ Air Force came into being, Lieutenant Colonel Ralph Royce, commanding at Selfridge Field, detailed how the divided chain of command was affecting and complicating his efforts. He was answerable, he said, to four commanders: the Chief of Air Corps for personnel and technical inspection; the Chief of the Materiel Division on matters relating to maintenance; the Commanding General of the GHQ Air Force for tactical training and development of group and service squadrons; and the Commanding General VI Corps Area for matters relating to courts-martial and the administration of non-Air Corps troops on his station.

It was not a system that engendered smoothness of operation, and Andrews was determined that he would bring changes to it. These, then, were two major internal problems confronting him. A lack of equipment and personnel could only be corrected by additional appropriations. Divided authority could only be corrected through his powers of reason and persuasion.

Most Air Corps officers saw GHQ Air Force as a compromise, but it is doubtful if any of them saw in its creation a parallel to events of nearly a decade earlier. Then, to head off Billy Mitchell's rambunctious drive for air autonomy, President Coolidge had appointed the Morrow Board, whose recommendations produced a compromise through the establishment of the Air Corps. Through Secretary of War Dern, with President Roosevelt's approval, the Baker Board had headed off the independence drive again with the compromise of GHQ Air Force. In both battles the War Department had kept its control over what it had first considered an auxiliary branch but was now coming to recognize as a vital component in the scheme of national defense. However, in this latest conflict, with the President's appointment of the Howell Commission, a new element of hope/doubt had been added, at least until January 22, 1935, when the Commission handed in its report to FDR.

The Commission made no recommendations for a separate air force, unwilling to buck the Baker Board findings. Instead, it recommended that the GHQ Air Force be given a try. The Commission's failure to come to grips with the issue was a great disappointment to those who felt it was their last, best hope of attaining air independence.[10]

No doubt, in the War Department there was a great sigh of relief, and following the Commission's release, General MacArthur made clear his own views for the present and the future. He felt that for at least a five-year period there should be no change in the status quo of the Air Force. Its establishment, he declared, was a far more economical one than would have been the formation of a separate air arm. This new Air Force was equipped to perform "every mission that could be carried out by an air force organized separately from the Army." He believed that now there would be no occasion for the submission of what he considered "false and irritating issues that not only impeded progress but for the moment at least are inconsequential."

It had been just one year since he had gone before Chairman McSwain and announced the War Department's oft-repeated plan actually to bring into being an air force. Now he had done so.

And Frank Andrews—how did he view the chances for independence now that he had been assigned the most important air command in the Army, the greatest challenge of his career?

To his officers at Langley in those first exciting and hectic days of moving in and getting organized, he offered his thinking on the past and the future:

Now just a word about the past. Some of us perhaps believe in an independent Air Force. Some thought perhaps that an Air Defense could be best developed as a separate part of the War Department not under the General Staff, and others perhaps had still other plans, but now that the decision has been made, and by the President himself, to develop our air power as an integral part of the Army, it is up to us to get behind the plan and push it loyally to success. Gentlemen, I give it to you as my sincere belief that a separate Air Corps is a dead issue for many years to come. The GHQ Air Force is a part of the Army and it is our interest and duty to keep the fact constantly in mind, for therein for many years, at least I believe, lies the best chance of developing Air Power and the best interest of National Defense.

Seventeen

Transition

The year 1935 entailed far more than General Frank M. Andrews' and his staff's preparing GHQ Air Force for the service test mandated by the War Department. It was also a year in which there was continuing congressional inquiry into matters aerial, a year in which strategic theory at the Air Corps Tactical School became more fixed, and a year that saw a changing of the guard at the top.

In this last, Major General Hugh A. Drum and Brigadier General Charles A. Kilbourne were shifted in February to Pacific commands, Drum to head the Hawaiian Department and Kilbourne to take charge of the Harbor Defense of Manila and Subic bays. Drum devoutly anticipated being tapped for Chief of Staff when MacArthur stepped down. His replacement as Deputy Chief was Major General George S. Simonds. Although five years older than Drum, Simonds was more flexible in his thinking and less of a self-assured pedant, an easier officer with whom to reason.[1]

About GHQ Air Force, it was Simonds' opinion that regardless of future progress made in the realm of weapons, tactics and organization, setting up the Air Force was the proper move at the proper time. Any other step, he felt, could be "compared to stepping off a precipice in order to take the shortest line, whereas by developing a road as we go along we are more sure of reaching the proper destination." He didn't say what that destination was, but the conclusion was obvious.

Kilbourne's replacement was Major General Stanley D. Embick, who, since 1932, had held the command Kilbourne was assuming. An Artillery officer with a considerable amount of General Staff duty, which had included serving as Chief of Staff of the American Section of the Supreme War Council, Embick's attitude toward air power was on a par with

Drum's. He did not have a high regard for its potential strategic uses, nor for those who espoused them.

Other changes of command were in the offing as well.

Congressionally, in spite of MacArthur's desire no longer to be bothered by "false and irritating issues," Congressman John J. McSwain had taken the field once again. He remained determined to try to breach the walls of the War Department and free the airmen from their long captivity. It was a new year, a new Congress, a time for the Military Affairs Chairman to make a new attempt—despite the towering opposition, which now included a goodly number of those selfsame airmen. McSwain did not seem to have gotten the message, or if he had, he was out to have it repealed. He brought up for hearings before his Committee a bill to "Promote National Defense by increasing the efficiency of the Air Corps." It was submitted with five additional bills, but HR 7041 was for all intents and purposes a bill to promote national defense by creating a department of air. The department would include a joint national defense board to supersede the existing Joint Army Navy Board. It would have fifteen members equally divided among Army, Navy and Air.

McSwain's hearings began late in March and continued through the first half of April. They were significant hearings not so much because of their familiar intent but because of what some of the witnesses had to say concerning the matter at hand.

At the outset, McSwain quoted the Baker Board's opposition to his proposals of the previous year(s), to wit: that such a plan "would be a serious error, jeopardize the security of the nation in an emergency and be an unnecessary burden to the taxpayer." The Congressman and members of his Committee wished to explore the Board's contention and, if possible, through the testimony of those appearing before the Committee, disprove it.

All of this was far removed from March Field and newly promoted Brigadier General Henry H. Arnold. When he received orders from the Adjutant General to report to his office in Washington on April 3, as in November he made no connection between the summons and a request to give testimony. General McKinley's wire gave no reason for the appointment, but since it did state Arnold was to appear in uniform, Hap concluded the matter had to do with his being awarded the Mackay Trophy for commanding the Alaska flight.

He arrived at Union Station with a half hour to get checked into the Army Navy Club and then to rush over to the War Department for his nine o'clock date with Adjutant General McKinley. He made it with ten minutes to spare and no one present to bid him good morning. A short while later, in strolled Captain Corley McDermott. He was wearing civilian clothes. McDermott was followed by Bert Dargue and Follett Bradley, similarly dressed. In short order, Arnold realized that his being on the scene had nothing to do with the Mackay Trophy. His friends further enlightened him, saying the rumor was they were all to appear before McSwain's committee at nine-thirty that morning. Next on the scene was the Adjutant General's Executive Officer, whose only contribution to their joint inquiry was that General McKinley would not be in before ten. Arnold asked him if he knew why they were there, and the Executive said he did not. Realizing that he could not possibly appear before the House Military Affairs Committee and keep an official appointment with McKinley, Hap raced out of the building and returned to the Army Navy Club, where he changed into civilian clothes and, as he put it, "arrived back in the Adjutant General's office at 9:58½ in time to learn that the dignitary would not arrive until 10:15."

When the Executive Officer appeared once again it was to inform the assembled that they were to be present in the hearing room of the House Military Affairs Committee the next morning at ten. Arnold wanted to know if there were any specific instructions and was told there were not. Now, more than in the fall, when he'd testified before the Howell Commission, he did not wish to offer testimony at variance with War Department policies on doctrine "if I could help it."

When he arrived in the hearing room the next morning, he sat down and wrote Bee a running description, giving time, location, date and the cast of players, starting with Chairman McSwain and Representative Sam Collins. He noted that Generals Embick, Moses and Westover were present accompanied by "scores of brass hats from the General Staff." There were seven Air Corps witnesses including himself, plus Jimmy Doolittle, veteran pilot Casey Jones, Dr. Warner of NACA and a great many reserve officers. As Hap put it, "Everyone has an axe to grind," and then observed that the purpose of the gathering was "to consider about 8 or 10 bills concerning National Defense. Perhaps something out—Perhaps not."

There in the hearing room, writing to Bee, he listened to the ex-Governor of Arkansas "making the eagle scream and lauding the advantages of a Department of National Defense and talking in glittering figures of what should and should not be done."

Retired General William Rivers followed the ex-Governor, and he, too, came out for a department of national defense. McSwain appeared to be calling on his big hitters, for Jimmy Doolittle was up next. "Many cameras clicking and flash lights flashing, Jimmy, as usual very short and snappy in his answers," commented Arnold, describing the scene but making no mention of Doolittle's opening statement.

Said Doolittle:

> I feel that the Air Force is destined to be the first line of defense and the first line of offense. I feel that the duty of the Army as an Army will eventually be to prepare the bases from which the Air Force will operate and to hold positions that they take. I feel that the duty of the Navy will

be to patrol the high seas and protect our maritime commerce and to a certain extent our coast line.

I feel there should be a supreme commander; and under that supreme commander, who will perhaps be a Secretary of National Defense, three secretaries—one of Navy—one of Army—and one of Air.

There will be a defense council which will be made up of the Chief of Staff, Chief of the Air Corps, the Chief of Naval Operations.

I feel that commercial aviation is the background of military aviation; that first it furnishes a training ground for military personnel. Commercial aviation constitutes a proving ground, facilities for production, factories.

I feel that only through a separate air force, entirely divorced from the Army and Navy, can we efficiently promote and develop a defense machine that will provide adequate security in case of a national emergency.

As Arnold listened, writing to Bee at the same time, he asked, "Is it any wonder that my heartbeat has jumped and I look forward with some considerable dread as to what it's all about and what I can say and still maintain my self-respect and not offend anyone?" It was November again, only worse because of the change that had not only given him a long-awaited star but also made him one of the three principal commanders of GHQ Air Force. Even so, his orders without explanation, his unexpected appearance before the Committee, indicated just how closely informed a senior Air Corps officer was kept by the General Staff on Washington political moves affecting the military.

He became occupied in telling Bee about his inability to reach Bart and Mildred Yount and that Tooey was expected in today, when suddenly he heard his name being called as the next witness. *"Now what the h--l!"* he scribbled, *"they are calling on me to give my expert testimony."*

He had anticipated that the War Department generals would precede him as a matter of prestige and rank. To be put on the stand unprepared as he was, he considered a dirty trick, and he could see "the ever-present studious watchful note-taking General Staff officers" with their pencils poised.

He was there to testify at McSwain's request and undoubtedly the Congressman expected to hear considerable support for the Committee's bills. What Jimmy Doolittle had said was certainly familiar enough, but Hap Arnold's thinking on an independent air force, if not patently clear before the Howell Commission—whose testimony was confidential—was clear enough now.

In answering the questions of Chairman McSwain and his colleagues, Arnold did not sound or appear unprepared. He pointed out the aeronau·tical development of the bomber—five years ago a plane could not carry a two-thousand-pound bomb one thousand miles. Now we have such a plane. Around the corner there is a plane that will carry it three thousand miles and beyond that, a five-thousand-mile plane coming through. "Six months ago I would have said that that pointed directly toward a Department of National Defense. Right now I am not so sure that we are ready for a Department of National Defense. In order to get a properly trained and effective Department of National Defense in our opinion you have got to have three arms—land, sea and air, equally self-supporting, and the Air at this time is not self-supporting for many reasons and that is the main reason I say now we are not ready for a Department of National Defense. . . .

"We have to go through a series of training, equipment equipping and operations before the Air can stand on its feet. We have to train first our individual pilots and get enough of them. We are short now in pilots. We are short now in planes. And we are short in equipment that goes in planes. . . .

"So there is no use talking about air power and air force until you have something to talk about. We have not got it now. We are on the way, however. I am of the opinion that the GHQ Air Force is heading toward a bigger and better fighting force, so we are on our way."

He went on speaking about the move to standardize aircraft and to equip them properly as warplanes, and then touched on what had long been a War Department criticism. "We have got to train commanders of air units that know how to handle their units. We have got to train staffs that can handle those units under commanding officers and until we get these things we will not have an efficient fighting unit air force. . . ."

And finally, "I am of the opinion that sometime in the future this thing called a Department of National Defense is bound to come. I do not see how it can stop."

And that was all he had to say, and while he was still on the stand McSwain called for adjournment, making no mention of having Arnold return for more questions. *Thank God!* thought the witness, feeling that his testimony had not set well with the Committee.

There was at least one exception, for California Republican Representative Sam Collins, whom Arnold had been acquainted with for some time, invited him to lunch. They picked up fellow Californian Representative George Burnham and, joined by Follett Bradley, sat down to break bread in the Capitol dining room. It was a relaxing interlude after a long morning, and when it was over they said good-bye to Burnham and went down to the War Department to talk with officers there on the continuing problems of acquiring more of the Muroc territory for a bombing and gunnery range. Arnold was not sure they made any progress in convincing the ground types of their need.

When he returned to the Munitions Building, he had a long session with Oscar Westover and several others. All of them told him his testimony had been sound. It made him feel better. It was not what he would have said less than a year ago, and his position had to have surprised more than Chairman McSwain.

He finally met up with Tooey, and later, at the Army Navy Club, he joined Charlie Chandler, now long retired from the service. In the war, Charlie had headed up the Balloon Section in France, and his enthusiasm for ballooning and aviation had continued. He had been and was a great supporter and friend of Billy Mitchell.

Although Hap made no mention in his letter to Bee of what he and Charlie talked about for a couple of hours over a glass or two of sherry, it would be safe to surmise that he spoke of his appearance before the Committee and the rationale for his position. With Charlie that would have been easy enough, but not so easy later on, when he had a reunion of sorts with Billy Mitchell at the Carlton Hotel. Jimmy Doolittle and his wife, Jo, were there with Casey Jones, as well as Tooey and his pal aircraft designer Henry Berliner. They repaired for dinner down on the docks and had a gay old time, telling bright stories of yesteryear when the air was young and their flimsy wings had dared the imperious blue. It is unlikely that Mitchell was aware of his former principal supporter's testimony. Recently, Billy had damned the Howell Commission report as a "sellout," was "fed up" with FDR and was going to go after him. Only fifty-six, he had less than a year to live, and the stridency of his accusations and claims no longer carried weight or real meaning. It would take war to prove the correctness of many of his predictions. But better now to light a cigar, pass the brandy and remember when.

The next morning, in a complete change of pace, Arnold had breakfast with the noted sculptor Gutzon Borglum, who told him all about his work going forward at Mount Rushmore, in the Black Hills of South Dakota. Hap was genuinely interested in Borglum's epic carvings on the mountain's face, and it was only the realization of where his presence was commanded that took him down from the heights and back up onto the Hill.

There he observed: "Same people—same bored Congressmen—same hopeful and expectant reserve officers—same flashing picture-taking cameramen." They had taken his picture yesterday. He was very anxious to be off for Langley, by tomorrow at latest. In reflecting on yesterday, he still thought it had been dirty poker to put him on the stand unprepared and perhaps he had "surprised them by announcing [his] stand against a Department of National Defense at this time."

When the hearings began again, he watched as reserve officers, one after another, took the stand "openly spreading themselves as to why they needed a chief of air reserves with the rank of major general, of how it cost each one 300 dollars a year to go to airdromes to get in their flying. How they fly obsolete and antiquated equipment. How they are the first line of defense, etc. etc."

He sounded weary of it all, a bit sarcastic, citing their claims which he knew so well. He brightened up when Professor Edward Warner declared that General Arnold's testimony had been sound and he felt much better when the hearings adjourned that day, for Chairman McSwain thanked him and said he would no longer be needed. Arnold thanked the Congressman in return and said he was leaving for Langley in the morning.

One result of his departure was that he did not hear what his fellow officers had to say about the immediate need for a department of national defense. He knew, of course, that Oscar Westover was not in favor. In 1920 Oscar had predicted that the thought of a separate air force would not be worthy of consideration in less than fifteen years. Now, fifteen years later, before McSwain, Westover repeated what he had told the Baker Board: not for another ten years at least. He added the thought that he believed the creation of a separate air force might be unconstitutional. He saw a lack of money as one root of the problem and that any moves in the direction entertained would complicate the omnipotent watchwords *unity of command.*

Bert Dargue, the Assistant Commandant at the Air Corps Tactical School, was direct and brief: "I think in years to come we are bound to have a Department of National Defense in which our air forces are more or less coordinated with land and sea forces. I do not believe that time is here yet."

Follett Bradley didn't think it was either, principally because of what was lacking in supply and administration. There was a severe shortage of personnel. The Air Corps, with approximately one seventh of the Army's regular officers, had no permanent generals of the line. As to the key question: "If you think there should be a department of national defense, I think rather than creating a separate Air Corps or separate Air Department, that you should do it all at one time." He was against further hearings. "We have thousands of pages of testimony from various—sixteen, seventeen commissions," he said. "What we need is a synthesis of the facts."

The facts as Captain Harold L. George saw them were somewhat different from those seen by his fellow officers. He was "heartily in accord" with the bill to create a department of air and the office of secretary of air right now. But yes, the creation of GHQ Air Force was "a great initial step forward."

On that score all the Air Corps witnesses were agreed. To John McSwain and his fellow committeemen the message must have been clear enough. A step had been taken, if not directly forward, obliquely. In time the next step might be more direct, particularly if in the interim the necessary funds were forthcoming.

GHQ Air Force had been in being little more than a month and a sense of newness and anticipation still prevailed, when Johnny Andrews noted in her diary: "Another month gone. Andy says I must like it here because I

haven't been away yet." It would be safe to say that her liking was reflective of the general atmosphere. Her husband's presence, his actions and
those of his staff generated a feeling of cohesion and of things going forward. "My days and nights continue full," he wrote to his father. "We
have many visitors, official and social." The most important of the former
had been the new Deputy Chief of Staff, George Simonds. He had stopped
in on his way to Texas to talk things over and have a quick look around.

Socially, the pace of activities along "Gold Coast Row," where the
Andrews lived in a rambling Tudor-style house backing on the river,
seemed unrelenting. "Had supper for 25—then dance at the Officers Club,"
noted Johnny. "Dinner at the Knerrs', played poker . . . Jean's birthday
[her eleventh] 17 for supper . . . Burwells and Bradleys here for supper,
went over to Pratt later," etc.

"I have to entertain all these people," Andrews informed his father,
protesting mildly, "and although we enjoy it very much it is quite a drain
on the endurance. I will have to find an aide soon to take some of the burden of showing people around."

He added, "Tonight the Headquarters GHQ Air Force were guests at a
dinner given by the Rotary Club of Newport News. I, of course, had to
make a speech, a thing I don't believe I shall ever enjoy but at any rate I
am becoming more accustomed to it."

It had been his first public address as GHQ Air Force Commanding
General. The text was possibly drafted by George Kenney or Hugh Knerr
or both with Andrews' editorial modifications. It was a speech that deserved a larger audience, for while it adroitly tread on no official toes,
Army or Navy, it neatly made clear the overall importance of air power.
Andrews outlined the role of a *defensive* air force attacking an aggressor's
vital centers of operations, his air bases, his navy, his transportation and
communication facilities. In citing strategic targets such as factories,
refineries and power plants, he said these would be the points in the
United States an enemy would seek to strike. Thus while he kept the emphasis on the need for a strong defensive air arm, he left with the listener
an image of what long-range strategic bombardment was all about.

General MacArthur came in for genuine praise for his Army reorganization plan, which had foreseen the need for a GHQ Air Force, a force that,
like the Navy, must be in a constant state of readiness. The War Department General Staff was given overall credit for recognition of the independent and perhaps decisive role of the Air Force in phase one of an aggressor's attack. Had Billy Mitchell been present, there might have been
the sound of raucous laughter, but Andrews' purpose was to be positive
and there were those in the War Department who had changed their outlook.

There was only one sentence in his speech that Andrews crossed out.

"These independent operations," the original text read, "are, of course, beyond the sphere of influence of the surface forces ground or sea." He saw no value in misunderstanding. He changed it to read that independent air operations were "always in furtherance of the general strategic plan."

His address was a concise picture of what strategic air power was all about, and although bombardment had been the key, the parts played by pursuit, attack and reconnaissance aircraft were described. The balance was there, a factor that was becoming more difficult to maintain at the Air Corps Tactical School. Balance was also there in his broader declaration that national defense required an Army and Navy as well as an Air Force. Then, in concluding, he got down to the nub of it all: "The Air Force is not ready now," he said, "and cannot be made ready in less than two years. It cannot be ready then without a full realization on the part of the public and the Congress of the necessity for its preparedness, and that re-alization must be translated into money for airdromes, for airplanes, for accessory equipment and for pilots. The sums involved are small with re-spect to the results to be accomplished."[2]

In enlisting support to carry the message to the public and the Congress, Andrews invited a third type of visitor to Langley, the interested newsman, and in this case, the widely listened-to broadcaster and columnist Boake Carter. Carter was heard nightly offering his views of the news over Phila-delphia's Radio Station WCAU. He had given Andrews a strong plug at the time of his GHQ appointment, and Andrews was quick to tell him, "I made up my mind at that time that on the first opportunity I wanted to have you visit me at Langley where I could have a talk with you and let you see the GHQ organization. . . ."

Carter had been a pilot in the RAF and was a keen aviation enthusiast. He was only too happy to get to know Andy Andrews and his staff, and to "secure firsthand knowledge of the existing situation." Boake Carter was to become a frequently valuable nationwide spokesman for the cause of GHQ Air Force.

As for the existing situation, when Hap Arnold appeared at Langley fresh from the congressional hearing room, there were a number of matters for him and Andrews to discuss in private before the meeting with Conger Pratt and Jerry Brant. Andrews knew that Hap had been before McSwain's committee. In February, he had been before it too, testifying in executive session on a bill drawn up by Representative J. Mark Wilcox on the need to construct additional air bases in strategic areas including Alaska and Ha-waii. Previous to his own testimony, Captain Hal George had stated to the Committee, "The best defense against an attack is an offense against places from which the air attack originates." And then, "The offensive avi-ation is aviation that carries bombs and goes out to destroy."

Representative Paul Kvale, of Minnesota, cautioned sharply, "I am not interested in offense one bit. Predicate the building of your air bases on defense and not offense." It was a key point.

McSwain questioned Andrews on the subject the next day, and Andrews assured him and the Committee that the Air Force viewed the heavy bomber as a defensive weapon. The B-10 and the B-12 did not have the range to be considered as offensive weapons. As Andrews had later stressed in his talk before the Rotary Club and would continue to stress, defense—*frontier defense*—was the doctrine, the umbrella, under which the strategy of long-range bombardment must be couched. The seeming ambivalence, the tightrope walking, was dictated not just by the War Department but also by a political attitude that demanded a totally defensive outlook and a public reliance on geography that seemed to assure it.

Behind closed doors in executive session both Andrews and General Kilbourne had discussed with the Military Affairs Committee the outline of the General Staff's contingency Rainbow War Plan in the eventuality of an aggressor getting a foothold in Canada or British- and French-held islands in the Atlantic and West Indies. It was an eventuality that included the outside chance these countries might be aligned with a hostile coalition against the United States. Andrews spoke of the proximity of such possible enemy bases to our cities and industrial centers and said they would need to be kept under surveillance and bombed if necessary.

He informed McSwain he was in favor of the War Department's compromise bill on the building of six new bases for the use of GHQ Air Force, and he had every reason to believe that his remarks on any remote need to bomb targets in Canada or elsewhere were highly confidential, his own statement marked SECRET.

That had been two months before, and the hope that the bill would pass and the funds would be forthcoming to build the specified air bases was but one major item on a long shopping list replete with major items necessary to build a modern air force.

As one possible indication of the necessity, Andrews pointed out to Hap, in passing, the Sunday Rotogravure Section of the New York *Times*. On its front cover was a full-page spread of Hitler and Goering inspecting the newly emerged Luftwaffe. Conversely, on the back page was an aerial shot of March Field with a concentration of one hundred planes made up of the 17th Attack and two bombardment groups, the 19th, from Rockwell, and the 7th, from Hamilton Field. From the air and to the public eye, the array of aircraft appeared formidable and possibly had been included to offset the impact of page one. But, as both men knew, there were few planes among those assembled that could fit the designation *modern*.

The 31st Squadron of the 7th Bomb Group did fit the designation, for it was equipped with B-10's. Even so, Andrews told Arnold he had felt compelled to cancel the squadron's proposed flight to the Canal Zone.

The plan for the flight had been announced in the press on the day Andrews had taken command of the Air Force. Said the *Times:*

> As a demonstration of the mobility of the land planes the General Headquarters first objective was designed to show that it can go over water with the same precision that flying boats have recorded. Ten fast bombers were groomed today for a one-stop flight from Washington to France Field, Panama, a distance of more than 2,100 miles. The first leg will be to Miami and the second straight to the isthmus, 900 miles of which will be over the Caribbean Sea from the southern coast of Cuba. It will be the most strenuous undertaking of its kind in the history of the Army Air Corps.

This last wasn't quite correct, the newspaper apparently forgetting Arnold's Alaska flight. Nevertheless, when Andrews quietly called off the project there was not only great disappointment among the crews of the 31st Squadron but also in the Office of the Chief of Air Corps. To many officers there and elsewhere, the flight seemed tailor made to exhibit the Air Corps's capability of long-range reconnaissance over water and coastal defense. Tooey Spaatz was one who believed the flight was very essential "in order to bolster up the Air Corps."

At the moment, Benny Foulois was being investigated by the Inspector General as a result of the charges made against him in June of 1934 by the Rogers Subcommittee, and it was Assistant Air Corps Chief Westover with whom Andrews met to discuss the mission. Andrews pointed out that although the flight had been planned in the Chief's office, its timing was such that the authority and responsibility for it now came under his own command, and not that of the Air Corps. Westover didn't think so, and it marked the first such division between the two headquarters. The question went to MacArthur for decision, and he upheld Andrews.

Since the squadron selected to make the flight was a unit of Hap Arnold's wing, the decision to cancel affected him directly, and he was anxious to know how come. Andrews explained confidentially and in detail. He said he had sent George Kenney up to Wright Field, where the squadron's equipment was being checked. At the same time, he had contacted Brigadier General Augustine W. "Robbie" Robins, who had taken over as Chief of the Materiel Division, where he had formerly served as Executive Officer. Robbie was an old and valued friend, and Andrews told him what he was after. Based on the evaluation made by Kenney and Robins as to the state of the aircraft and the proficiency of the flight crews, Andrews decided to cancel the operation temporarily. He would, he told Hap, keep the proposed mission as a line project, rescheduling it after the rainy season had ended in Panama.

In explaining his reasoning, he stressed that they must always keep in mind the responsibilities of the Army and Navy within the disputed sphere of coast defense, and that in so doing it was essential that every precaution be taken to ensure such a flight's success. As he put it, "The vital necessity of meeting all occasions as they arise before they affect the operation must be firmly established. Nothing that can be foreseen must be left to chance." In this case, some things had been left to chance.

He gave an illustration of what he meant. He had learned that the B-10's of the 31st had no segregators on board, a segregator being a device to remove water from the fuel. The attitude of the squadron engineering and supply officers was that they did not foresee the use of segregators, and should they be needed, it was up to the base where they landed to supply them. Andrews considered the attitude sloppy.

Another reason for delaying the operation was that Wright Field was about to set up an intensive course in instrument training, and Andrews wanted all his pilots either to attend the school or have their proficiency ratings updated through January 1, 1936.

Still another question to be explored was whether the B-10 or the B-12 should be selected to make the flight. It was a matter of engine reliability against flotation gear. Also he believed, and Hap agreed, that the flight should be made by a group and not a squadron, and the personnel for it must be chosen from a wider selection of pilots. After all, the 19th Bombardment had operated the Avigation School, and he didn't think its personnel should be excluded just because the Group was below strength.

Further, two weeks prior to the mission he wanted the status of the equipment tested with a fifteen-hundred-mile flight. He wanted the pilot training for the undertaking to include flying through fog without loss of contact or control. One lesson to be learned from the preparation and the operation itself was that it would determine the limit to which the refinement of the equipment should be carried. "All too often," he pointed out, "the requirements of commercial operations are allowed to exercise too great an influence over the equipment provided for military operations."

The care and planning that he foresaw for the flight and described to Arnold were a far cry from his own journey to the Canal Zone but a short three years previously. Perhaps the difference proved that if the pace of aeronautical development was to be properly utilized, it demanded an equal pace of thought on the part of the air planner.

Andrews had made his own personal observation of the crews, having ordered the squadron to stop in at Langley Field. They had been informed of the cancellation, and the reaction was one of laxity. He knew they were disappointed, but nevertheless, flight discipline should have been maintained. In the final analysis, he told Arnold, "The political effects of anything unfavorable happening now, not on our personal futures but in the future role of GHQ Air Force in national defense, are such that no chances must be taken on its happening."

And he was taking none. Hap agreed with his reasoning; it matched his own. The plan for the flight, he said, had started prematurely.

A few days after Arnold had returned to March Field, birds of another feather descended on Langley; the entire thirty-seven members of the House Military Affairs Committee dropped in to see how the new Air Force was getting along. Johnny was the hostess and Hazel Knerr had the whole kit and kaboodle in for cocktails. Just a month later they returned for a second visit, accompanied by Assistant Secretary of War Woodring.

This time McSwain's purpose for coming, if not the whole Committee's, should have been to apologize profusely to Andrews, for in the interim, through a so-called "clerical error," the transcript of the secret testimony given previously by both Andrews and Kilbourne on the Wilcox bill had been made public. A hue and cry rent the peace of springtime. It was not just the howling of the pacifists, some of whom had recently tried to prevent naval maneuvers off the West Coast for fear of antagonizing the Japanese. It was Franklin Roosevelt, the Canadian Government, the State Department and much of the Congress, and particularly the press. The headlines made startling reading, as did verbatim quotes from the transcript by Andrews and Kilbourne in which Canadian, British and French cities and territories were mentioned as possible targets for bombing attacks.

Immediate dismissal from the service was demanded for both officers by a conglomeration of organizations. A delegation of clergymen came swirling down from Baltimore, calling on the President to present a papal bull that indicated the headsman's block was none too good for the pair. McSwain, not wishing to get caught in the storm of righteous wrath, was quick to say that neither he nor the committee members had been at fault, placing the blame for the gaffe on the back of an overworked clerk.

This did nothing to repair the leak, and Roosevelt, caught in the deluge, reacted snappily. He wrote McSwain a stinging letter telling him in so many words that if he and his Committee were not capable of keeping testimony taken in executive session out of the public domain, then the President, as Commander-in-Chief, would require henceforth that all such testimony be passed to him for approval before it was heard by the Committee. Roosevelt sent a copy of his letter to Secretary of War Dern, who responded with a letter of his own, saying that he had been considering taking similar action and he was grateful for FDR's having anticipated him. "In all fairness to the officers concerned," he wrote, "it was their understanding that the testimony was entirely secret and was not to be made public under any circumstance." And then he added, "It is needless to say that their views on the points you mention were individual and had not been submitted to the Chief of Staff or the Secretary of War," a statement which was not entirely correct.

However, the letters were for public consumption in an effort to quiet the criticism on both sides of the border and to reassure and mollify the Canadian Government. McSwain proclaimed, "The President cannot believe any more strongly than I in preserving our undefended frontier with Canada and our peace with all the world."

Copies of the official letters were sent to Andrews by Adjutant General McKinley with an accompanying brief memo that stated in part that the Secretary of War called to Andrews' attention "that you should not have expressed personal views of the character contained in your testimony without previously having submitted testimony for approval of higher authority."

Andrews in his nine-line response for public viewing admitted his testimony "represented his own views on *an abstract military study* with no concrete political thought or reference." (Italics added.)

All the official exchanges were in the nature of a military cover-up over a political embarrassment. The abstract military study to which Andrews referred was one of the War Department's contingency plans in its Rainbow color file. Each plan posed a set of circumstances in which the United States was attacked by one or more nations and a color and number were given to the particular attackers. There were variations of these contingencies, and Andrews and Kilbourne had testified in general terms on one such plan. As previously detailed, the Air Corps in 1933 had been asked to base its own coastal defense ideas on such a color scheme.

Had either officer been testifying in the irresponsible manner the letters seemed to indicate, they could have been in serious difficulty—Andrews possibly being relieved of his command. As it was, he came out of the affair with a bad taste in his mouth. When the news had broken, on April 28, which happened to be Johnny's birthday, he had just returned from Mitchel Field and a visit with Tony Frank. He had nothing to say on the matter, and he was not given to bringing problems home, particularly on his wife's birthday. He appeared more interested in the news that Jeannie, the hard-riding apple of his eye, had become a girl scout and that her mother was teaching her how to jump Baby Boy.

The same calm probably did not prevail in the Secretary of War's office or at the State Department. There was a great scurrying around. The President was understandably upset. Of all our potential enemies, Canada ranked about last on the list, and although both the British and French governments had more or less ignored the scenario that made them possible enemies, faces were red and action was demanded. The moves made to still the criticism and rebuke the offenders were political backfill, as was the rush to get off the hook. The letters were quickly drafted and the proper diplomatic apologies made.

There was one letter, however, that was not made public. It was sent to Andrews from MacArthur and it contained a severe admonishment, castigating the Air Force Commander for his testimony.

Andrews stuck the letter in his pocket, climbed aboard his Douglas DC-2 and flew directly to Bolling Field. The glint in his eye had little to do with the sun's glare. Neither was there anything genial or deferential in his approach to the Chief of Staff. This was not what he had bargained for, he told MacArthur. Why had such a letter been written to him, he asked, when his testimony *had* previously been approved by the War Department? MacArthur offered no satisfactory explanation as far as Andrews was concerned. He simply shrugged off the complaint, suggesting that Andrews might just as well forget about it.

Andrews did not forget. One of the primary laws in his code of conduct was loyalty downward. Senior officers protected their men. To Andrews, MacArthur's act was one of rank injustice done to protect himself and the War Department. The price would not affect Douglas MacArthur, but never again would Frank Andrews hold him in respect.[3] It was too bad, for, after all, it was MacArthur who had made the final selection choosing Andrews to be Commander of the GHQ Air Force.

Fortunately, the press could not long belabor the release of secret testimony. Once it was out, it was all out, and after the editorial writers and special-interest groups had taken a whack at the culprits, everyone could get back to business. Beyond his brief official response, Andrews had absolutely nothing more to say on the subject, but when Woodring and McSwain came to call on May 11, he got down to business with them of a far more pressing and lasting nature.

The badly needed additional appropriations for the Air Corps that Mac-Arthur had requested in February, although once rejected, were nevertheless still hanging fire in May. Just prior to Woodring's visit to Langley, he had met with Benny Foulois and Jan Howard, now assigned to the Chief's office. Following the meeting, Woodring had gone before Congressman Rogers' subcommittee to make a demand for a $30-million appropriation right now! He told the legislators he had just made a survey of Air Corps installations and that he had found conditions "deplorable." He stated that if the United States had to go to war within three years, before the buildup to the Baker Board level, "we would be caught with our pants down." The Assistant Secretary, his bullish features set beneath his balding pate, was described as being greatly exercised over the situation. The thirty million a year must be in addition to the regular appropriation. They must start to catch up at once!

The next day while in conference with Jan Howard, who was supplying him with additional data on the proposed increases for the next three years, Rogers telephoned to say that he was going before the full Military Affairs Committee to recommend that the necessary funds be approved.

Woodring offered to come up and support him on ten minutes' notice. They agreed that if the full Committee approved, the two of them would go see the President and get his okay for unexpended PWA funds to start the procurement ball rolling at once. Howard was impressed with Wood-ring's apparent determination. So was Andrews when the Assistant Secre-tary arrived at Langley with McSwain and company several days later to fill him in on the plan to approach Roosevelt. Woodring was equally impressed by the Air Force Commander and later wrote to say so, speak-ing for all concerned. He said they now had a much better idea of the problems and requirements, and he wished to thank Johnny for enter-taining them all so royally.

For the moment it appeared that the optimism in both Foulois's office and at Langley Field was justified. But only for the moment. Woodring and Rogers did not call on the President, the typical explanation being that the Committee needed to hear more testimony. Once again, Dern, Wood-ring, MacArthur and Westover went before the legislators to repeat what had already been said repeatedly. The only difference was that the magic figure was raised to 40 million. But, by now, the elusive unexpended PWA funds had been expended elsewhere and it was necessary to obtain a sup-plemental appropriation. Finally, late in July, Congressman Rogers wrote Roosevelt with the request. FDR responded by refusing to do anything about it.

If Roosevelt's refusal signaled hopes not to be realized down the road a piece, the immediate situation was essentially the same as it had always been. In a letter to George Brett, who was vacationing with his family at Cranberry Lake, in the Adirondacks, where Johnny and Jeannie were soon to go, Andrews confided that considerable progress in the standard of training was being made. But he was not getting very much of anywhere on receiving new equipment or on increasing the number of experienced officers in the Air Force. "The latter thing is what is holding us back more than anything else," he wrote,

> because in our units today we have three or four Regular officers, and the rest untrained youngsters. Our units are far from measuring up to the standard of efficiency to be rated an M-day force. I am also afraid this shortage of experienced officers is affecting our accident rate, particularly with the pressure of additional training which we are facing in all our wings.

Hap Arnold was more vocal and specific about the problem. "Airplane accidents have me at my wits' end," he reported to Andrews. He thought he had either the dumbest pilots in the world or they weren't properly trained or that his group commanders didn't know how to train them and therefore they must be the worst in the Army!

> "As far as I can see, the only way to stop accidents," he moaned, "is to keep all airplanes on the ground. Just take the series we have had within the last few weeks as an example. Starting with the cadet who, with the whole of March Field to land in, strikes a wing tip against a boundary

light. Ken Walker, supposed to be one of our best pilots, apparently cuts out completely, uses up 4,000 feet and finally hits a concrete block and spoils a perfectly good airplane when he normally would have given her the gun and gone around again. A B-12 comes in and makes a landing at Medford, Oregon. Apparently everything is okay when the landing gear gives way. Captain Malone, returning from a night flight, reports having engine trouble out over the desert, drops his flares to make a landing at Palm Springs, hits a clump of vegetation, rolls over on his back, and ruins a P-12. Captain Morrow, landing at Modesto, either neglects or forgets or can't get his wheels down and lands on his belly in a B-12.

Andrews knew the problem existed in all three wings and had discussed the safety measures to be taken with both Conger Pratt and Jerry Brant. He informed Hap that he thought one of the most important factors in reducing accidents was the correct attitude on the part of the pilot. He could remember the day when it was considered a feather in the cap to crack up and walk away from the wreck. That day was past, and he believed pilots were gaining a greater sense of responsibility. However, he cryptically noted, "I was very surprised in reading the report of the Inspector on the taxiing accident at Hamilton Field at the attitude assumed by as fine an officer as Eaker on his responsibility for operations in connection with his group."

When Andrews had been in command at Selfridge he had made a study of the methods to reduce accidents. He had developed a system that had produced results, and he passed it to Arnold. He would have his inspectors classify aircraft defects under two headings: those that might cause a forced landing or crash and those that would not. Every month, he'd post a chart showing the class of defects in each squadron and the results. This in turn gave an indication of the quality of maintenance in the squadron, and when there was a large disparity between squadrons he moved to find out why and acted accordingly.

In the realm of inspection, Walter Weaver had come to Langley in early April to take over as Chief Inspector. For assistants he had two hard-driving pros, Major Ennis C. Whitehead and Captain Bob Olds. Like Olds, Whitehead was a World War combat veteran who had been a test pilot and flight instructor at Issoudun under Tooey Spaatz. He had broken his nose somewhere along the course of a very active flying career, including his bail out on the Pan American Goodwill Flight of 1926, and the slant of it somehow went with his leathery manner. He had the well-earned reputation of being a tough hombre in the air or out of it.

Possibly the criticism of Ira Eaker later mentioned in the letter from Andrews to Hap Arnold had been leveled by Whitehead. Certainly it had not been leveled by Weaver, for after he returned to Langley he wrote Ira to say how sorry he had been not to see him before he'd had to leave for Rockwell Field to inspect that station. "I appreciate so much your many kindnesses to me while I was at March," wrote Weaver, "and as of old, you were always running around doing something for someone. You do not know how much I enjoyed the few minutes conversation I had with you. I am sorry that could not be stretched out into days."

Among other things revealed by the summer wing inspections, in which Andrews himself participated, was the problem of working out the clumsy and ridiculous command split between Air Force and corps area. In August, Andrews approached General Simonds on the matter, asking that corps-area control be eliminated—that all Air Corps stations where GHQ Air Force units were based be exempted from such control. Instead, he proposed that the stations come under the command of the senior Air Corps flying officer assigned to the particular post. As it was, the lines of demarcation between who commanded what were not at all clear. In fact, said Andrews, the archaic system was interfering with training and operations. Under the circumstances it was difficult to establish a proper organization based on sound operating principles.

Langley was a good case in point. The station CO was Colonel Charles Danforth, who had missed out on getting Andrews' job. Danny was an old compatriot, genuinely liked for his rough, rawhide manner. He had commanded at Langley back in the twenties and everybody knew the story about how he had turned over an SE-5 when landing and tried to blame the mishap on a hole in the ground, which the inspecting officer claimed Danny had made himself when he was thrown from the plane. Friendships and anecdotes aside, Danforth was responsible for maintaining the post and answerable to the Corps Area Commander for the administration thereof.

Conger Pratt, CO of the 2nd Wing, was the tactical air commander at Langley, answerable to Andrews. Between the station and wing commanders there were three types of units: tactical, service and station. The first two complements were attached to the wing, but the administration of their affairs on the base came under Danforth. Not only that, if Danforth's staff found it was shorthanded for station duties, enlisted men would be assigned such duties. The result was obvious, and what was true at Langley was true at all the other Air Force installations.

Simonds took Andrews' request to MacArthur, who rejected it. MacArthur was opposed to any change in the structure at the present time that he considered radical. Behind his caution quite probably were his own immediate personal concerns. For some months he had been engaged in a complex and delicate political maneuver aimed toward being named the Philippine Islands High Commissioner when that territory was granted commonwealth status, in the fall. Even if he did not receive the appointment he knew he would be stepping down as Chief of Staff, and he did not wish to do anything that might disturb the acceptance of his new defense plan. There was the possibility that the change Andrews recommended could have political repercussions, and MacArthur wanted none of that. He planned to remain on the agreeable side of Roosevelt, who would make the Philippines appointment.[4]

On another front there was a change in August that did have major significance. Benny Foulois announced he was retiring as Chief of the Air Corps as of the end of the year. It was not a move he was really anxious to make, weary as he was, but having been so savagely attacked and mauled by Representative William Rogers and his band of experts, Foulois recognized that whatever his effectiveness, it had been permanently damaged within the War Department and the Congress. Even though the Inspector General had found him not guilty of the major charges leveled by his accusers, he was reprimanded for his intemperate statements. This last was of no consequence to him. But he saw there was no point in continuing, not if the Air Corps was ever going to get what it needed in the way of appropriations.

Although he was only fifty-five and retirement was not mandatory until he reached sixty-five, his four-year tour as Chief was up, and where else could he go? He had worked and fought for the Air Force now in being. He had stuck his neck out and risked his career for it time and again. He and his staff had been given a primary role in working with Andrews and the General Staff in bringing the combat arm to a reality. Now he was through, finished. He kept his bitterness to himself.

His life had been a deep-rooted, all-consuming love affair with flight. He had never grown weary of it; his loyalty to the Air Service, the Air Corps and the establishing of a combat air arm had been steadfast, tireless. His ambivalence on air autonomy was his way of seeking a means to the best possible end. Whatever his weaknesses, they had nothing to do with his dedication and courage and faith—all in the face of high odds. In many ways, there was more strength of character in Benny Foulois's pipe smoke than there was in the collective cabal of congressional pontiffs who, in the self-righteousness of their vaunted power, brought him down.

The question following the announcement was, Who was going to replace him? The last two chiefs had been succeeded by their assistants, which in this case would mean Oscar Westover. But nothing was all that certain. Input would come from various quarters on the selection. Westover, gentleman that he was, was not all that popular throughout the Corps. Further, there were others, such as Conger Pratt and Danny Danforth, who had also served as assistant chiefs and by date of service were senior to Oscar. On that score, word got around that Benny had recommended Hap Arnold succeed him, since Hap was the Air Corps's senior active pilot. The suggestion might well have been an effort to make up for the broken promise of the previous year. As for Hap, through the rumor mill he got wind that Andy was going to be named as Foulois's successor. "Has that any foundation?" he asked.

"I have heard no rumors that I am to be the next Chief of the Air Corps," responded Andrews. "Conger inquired very carefully from me about that point, and I assured him that I had not taken, and did not intend to take, any steps along those lines. Conger and Danny both have their hats in the ring, and I suggested to Jerry Brant that he throw his in the ring also. I said that the GHQ Air Force was backing no candidate. We might find ourselves on the wrong horse and that would be very embarrassing."

Privately, he confided to Johnny that he hoped Conger would get the job. With Conger in Washington and Robbie Robins in Dayton and himself here at Langley, he believed real progress could be made in spite of bollixed-up command divisions. There was also the question of who would be replacing MacArthur when and if the President asked him to step down. The belief was that Simonds was MacArthur's choice, and that would be all right with Andrews, too. But mostly the speculation centered around who would take over as Chief of Air Corps, and the general feeling stifting down from Washington was that Conger Pratt and Danny Danforth had the inside track.

Other changes of lesser political impact were also in the wind. Ira Eaker, with a temporary promotion to major, had been accepted for the next class at the Air Corps Tactical School. He and Arnold had become collaborators on a book they jointly carved out of mutual feelings concerned with what the Air Corps was all about. During the airmail mission, Arnold had become impressed with Eaker's writing ability and the clarity of his manner in turning a phrase on matters aerial. Their venture, published by Funk & Wagnells in 1936, was titled *This Flying Game* and was aimed at the young man considering the career of military pilot.

The Arnolds would miss Ruth and Ira Eaker, of whom they had become extremely fond. The officers would particularly miss Ruth, her charm and beauty helping to soften the blow of her cigar-smoking husband sitting there at the poker table raking in their money. Still, the path upward was found not only in cockpits. The classroom was a sure step in the right direction, and when Arnold received a request to recommend a half dozen candidates for the Tactical School, he had recommended Eaker for that reason. He knew that if and when the opportunity arose, he would want Ira in his command again.

Another candidate wending his way a bit more unwillingly to school was Tooey Spaatz. He had written to Hap on the matter: "I am going to Leavenworth not because I expect it will do me any good, but primarily because I am ordered there and secondarily to get away from here." What he really wanted was to be assigned to the 1st Wing and he hoped someday to join up again. Like Hap, he was in full agreement with the command selection for GHQ Air Force. But early on he had foreseen exactly what Andrews was now protesting. ". . . there are going to be a number of very difficult problems in connection with its organization," he wrote, "particularly with the position of the post commander, working under the three or four bosses." Also he felt that temporary promotions would be demoralizing and, of course, the lack of sufficient officers and equipment was going to make it rough.

In agreeing to take the course at the Command and General Staff School, Spaatz was influenced by another consideration. The course had been cut from two years to one. His plan was to take some leave and drive Ruth and the children to California, where they would live for the year with Ruth's mother while he took up bachelorhood and the Civil War. All he would need for himself was a couple of rooms and a bath. He wrote to George Brett to ask if there were any squash courts. If not, perhaps he could be of assistance in finagling some funds to build a court or two in the hangar. "After resisting this detail for a long time," he told Brett, "I am fully reconciled and am looking forward to getting there. It will certainly be a relief to get out of this melee here." His buddy Strat Stratemeyer, who had been trying to prevail on him for a long time to come take the course, was so overjoyed by the news that he was willing to stay on and spend another year as an instructor. Actually, as Tooey confided to Mike Kilner, he would be leaving the turmoil of Washington "with great glee."

The turmoil that had swirled around and through the office of the Chief of Air Corps during his posting there had not bruised him all that much. He was too experienced and wily a campaigner to expose himself needlessly in the running battle with the General Staff or to get caught in the political cross fire. He ran the Training and Operations Section without fanfare or outcry, knowing better than anyone the realities of the equipment shortages and how they affected training. He gave his support to Benny Foulois, helping him as best he could with the ammunition of facts free of all hyperbole.

Captain Harry Halverson, of the *Question Mark* flight, had served Spaatz as his Chief of Training. He, too, was structured on a low key, hefty of chest but retiring of manner. They did what they could, but quietly.

When Spaatz drove away from Washington's summer heat with his family, he was happy to be gone. In September he was promoted to the temporary rank of lieutenant colonel, his first promotion in seventeen years. At Leavenworth his quarters faced the polo ground, from which the summer breezes blew. It was said that no more irreverent student passed through the school, and that frequently he made his instructors, particularly those teaching courses steeped in the care and use of military nomenclature relegated to the glories of past cavalry charges, blanch at his pungent remarks and sardonic queries. His dedication to the game of poker was no less, nor were his losses. He had to be playing for reasons that had nothing to do with financial liberation. It could have been the good fellowship of the game or that no one had to say much, or maybe the distant dream of winning a pot. For the moment, he was out of the main-stream of Air Corps activity, which was restful, but things were going on of which he would have liked to be a part.

In mid-August, Andrews confided to Arnold, "I am having a number of battles with the War Department and although we are not making big gains we are making some progress. The GHQ Air Force and the Wings are definitely on the map as an important element in our defense forces. Training regulations coming out soon will give us definite status and the Wilcox bill recognized the GHQ Air Force in legislation."[5]

The first six months had been in the nature of a shakedown, which would continue. In sessions with his staff chiefs, Andrews prepared for the planned December service test, its purpose to prove the mobility of the Air Force. Were they ready to move swiftly to any danger point and to take the skies to stop an aggressor? Tangentially, the originally proposed Canal Zone flight was canceled altogether on Hap Arnold's recommendation. The staff agreed that in the face of shortages, all their efforts should be centered on making the December maneuvers a success.

Andrews' staff sessions had an electric quality in which suggestions for provocative action from such as Kenney, Bradley, Whitehead and Olds made the air crackle. Knerr, McNarney, Weaver and Burwell were less demonstrative in their considerations but no less determined in offering ideas to get what was needed. And what was needed first was to break the hold of the corps-area commanders. Pete Quesada, now a Captain and Operations Officer of the Headquarters Squadron, was present at some of these conferences. The contrast impressed him, for in the months previous to GHQ's formation he had observed how Andrews had fought almost sin-gle-handed the roadblocking efforts of the Organization Section of Army G-3. Then the pressure had been to get movement by General Staff colo-nels who were damned if they were going to assist Air Corps lieutenant colonels get promoted to brigadier general, even if the rank was tempo-rary.

Now it was all turned around, and here was Andrews being pressed by a staff that often sounded like a conglomerate Billy Mitchell. To Quesada, it seemed that Kenney and Whitehead formed a chorus urging extremes. *Why don't you do this?! Why don't you do that?! Don't let them do that to you!*

Quesada saw that Andrews did not seem to mind the extremes of their suggestions. He liked the ideas they stimulated, but he was not about to be unduly influenced or directed by them. He would extract the wheat from the chaff and use it accordingly. One example in which Quesada became involved was the end result of the decision to cancel the mass flight to the Canal Zone. At Langley there was a single B-12 on pontoons, and Quesada was one of the very few pilots who flew it. Andrews came down to the dock of an afternoon and said, "Pete, how about checking me out in your B-12?"

And Quesada, happy to comply, said, "Yes, sir. Climb right in."

The B-12 did not have dual controls, so it was necessary for the Captain to give oral instructions. They went out into the bay and taxied back and forth while Andrews got the feel of it. He had flown every type of land plane available and had close to three thousand flying hours. Still, it took some doing for him to learn how to lift the plane out of the water quickly. Then the trick was in landing, and again it required lots of practice and fast talking by Quesada before Andrews got the feel of it. It was not at all like setting down on terra firma. After the lesson was over, Andrews turned with a grin to his instructor and said, "Gee, Pete, this isn't as easy as I thought it was."

But, like his becoming an instrument pilot, he kept at it, for there was more behind his learning to fly the floatplane than gaining proficiency. A plan was afoot, born of a staff meeting, to bring the over-water capability of GHQ Air Force to the public eye without having to fly very far out to sea. And what better pilot to bring attention to the fact than the Air Force Commander himself, an air commander in far more than name and title? The announced purpose of the flight was to make an attempt at breaking some existing plane records. The whole endeavor, however, was a sagacious substitute for the Canal Zone mission. It was a subtle plan, coming from a staff that knew how to put the right elements together in such a way as to make a point or two—namely, that the Air Force was better equipped to fly over water than the Navy, that it could do so at record-breaking speed, and that its premier pilot was its Commander.

On August 24, with Crew Chief Staff Sergeant Joseph Moran and Radio Operator Harold ·O. Johnson, Andrews lifted the B-12 off from Willoughby Spit, at Hampton Roads, and headed for Floyd Bennett Field, Long Island. The plane carried a payload of two 1,100-pound bombs. By the time Andrews had completed the 2,000-kilometer course—over 1,200 miles—he had broken three seaplane records. They were records set the previous year by Charles Lindbergh, Edwin Musick and Boris Sergievsky, flying a four-engine Sikorsky Pan American Clipper. Andrews' speeds were for 1,000 kilometers, even though his average over the entire course was faster than the previous record.[6] The reason given to the press for the failure to break the 2,000-kilometer score was head winds. This was not so. It was a matter of spark plugs.

In relating to Robbie Robins the details of his widely acclaimed effort, Andrews cited how a shortage of funds was forcing unfortunate economies, one of which was the reconditioning of spark plugs. "I lost eight spark plugs on my attempted records with the B-12," he reported. The single sentence spoke volumes. All of the meaningful plans that evolved out of his own thinking and that of his staff could be made meaningless by such unfortunate economies.

They were economies that could kill and very nearly did a few weeks later. Andrews was taking off in his DC-2 from Bolling Field headed for Selfridge and a visit with Ralph Royce when the left engine cut out. He was barely airborne, in the worst possible position. He ordered his copilot, Second Lieutenant Hiette Williams, to get the gear up, and then on one engine, as he put it without descriptive adjectives, "we had sufficient power to remain in the air and make a turn and get back to the field." What he didn't say was that he and his crew were very close to a crash. The problem again was spark plugs. All the forward bank in the left engine had to be replaced.

Along with the shortage of parts, the shortage of personnel continued. In August, although Congress took action to bring the Air Corps up to its authorized strength of 1,650 officers and to give regular commissions to fifty air-cadet graduates a year, it failed to provide the necessary funds. This meant the War Department would have to wait to include the increases in its proposed program for 1937.

Linked to all these actions was the central question of who was going to be the next Army Chief of Staff. His selection would have a direct bearing on the problems confronting Andrews. The answer came on October 2 when Roosevelt announced that he had named General Malin Craig to succeed MacArthur as of that date. The announcement was a surprise to the War Department in general and to MacArthur specifically. He was on a train traveling to the West Coast, preparatory to leaving for the Philippines, when he received the news in a telegram from Woodring. The announcement infuriated him, and as his aide, Major Eisenhower, reported, he let go with "an explosive denunciation of politics, bad manners, bad judgment, broken promises, unconstitutionality, insensitivity and the way the world had gone to hell." He recovered enough to send the proper telegrams of congratulations to both FDR and Craig.[7]

Hugh Drum's supporters, one of whom was Postmaster General Jim Farley, were also taken by surprise, if less explosively so. Drum was no doubt greatly disappointed, but he could reconcile himself that in four years he would be in an even better position to assume the leadership.

At GHQ Air Force headquarters, the fear was that Craig's appointment would mean Danforth would probably be named Chief of Air Corps and not Conger Pratt. But that was in the future, and before the end of the month Andrews had approached the new Chief of Staff on three important matters that concerned the present. He appealed again to have his Air Force bases exempted from corps-area connection. General John H. Hughes, Assistant Chief of Staff for G-3, brought his weight to bear against the change, maintaining the problem was being exaggerated. Craig, having so recently served as a corps-area commander, his relations with Hap Arnold apparently without complaint, accepted Hughes's recommendation.

Andrews' second point dealt with conditions here and now. At the outset he had told Hap Arnold, "As I see it, we have two main missions to accomplish this year. Test our tables of organization, but primarily and above all is to increase our combat efficiency."

Toward that end he gave Craig the cold facts. "We started out in March 1935 with approximately 446 planes. . . . Today [October 28] we have 346 planes of which 168 are standard types and the remainder are of small combat value."

Current contracts would by January 1937 add an additional 335 combat planes, but attrition between now and then meant they "are not expected to do more than provide replacements in the Air Corps, if they do that." The only way the number could be built to that authorized was through congressional action.

In spite of decreasing air strength, Andrews reported that in emphasizing navigation, instrument flying, bombing and gunnery, flight time had been more than doubled, and that whereas when he had assumed command only 9 percent of his officers were qualified instrument pilots, now 77 percent were, and by the end of the fiscal year that number would reach 100 percent. "In some units blind flying equipment was flown 150 hours per plane per month," he informed Craig.

And so Andrews had brought to fruition the plan for instrument proficiency he had had little more than two years before. It is unlikely that Craig could have any real grasp or understanding of the accomplishment. Through Hap Arnold, he had certainly come to appreciate the importance of the airplane, but the details were of less interest.

The third item Andrews came to discuss with the new Chief of Staff in early November dealt with the future of an aircraft that was now in doubt. Andrews first went over the situation with Oscar Westover, who in Foulois's absence was the acting Chief of Air Corps. The plane in question bore the designation B-299. It would come to be known as the Boeing B-17.

Eighteen

The Bomber Advocates

The B-17's genesis lay in the undeveloped application of long-range bombardment instituted by the major combatants in the World War. Research and development in the decade following the war led to the aeronautical feasibility of a plane that had a range of over two thousand miles, could fly as high and nearly as fast as interceptors, and could carry and accurately drop two thousand pounds of bombs on a given target. The theorists at the Air Corps Tactical School, spurred on by Billy Mitchell and their own thoughts, had since 1930 been openly preaching the ascendancy of the long-range bomber as the ultimate aerial weapon. There wasn't anything all that profound in their conclusions. *If* you could build a large enough plane, heavily enough armed so that, in concert with others like it, you could fly to a given target and with an accurate bombsight destroy the target below and return home safely, you could destroy your enemy's capability to wage war.

There were three deterrents to the belief. Pursuit had come out of the war as king. Pursuit was style and glamour and the wild blue yonder and, until the advent of the Boeing B-9 and then the Martin B-10, its advocates, such as Claire Chennault, could argue that lumbering MB's and Keystones would, without ample fighter protection, be decimated by interceptors. Chennault claimed to have proved his point in the 1933 maneuvers at Fort Knox.

In West Coast maneuvers commanded by Oscar Westover later that year, in which Hap Arnold was Chief of Staff, the new Boeing P-26 pursuits were placed against attacking B-10's. It appeared that the B-10's were able to get through time and again without the defenders making contact. In his report on the maneuvers, according to Chennault, Arnold concluded that fighters would be ineffective in wartime. He sent his report to the Air Corps Tactical School for comment, and Chennault fired back a smoking, eight-page critique. His criticism was on the methods used to intercept the bombers; for example, fighters taking off, forming over the field into squadrons, waiting for their group commander to take off and lead them, etc. Back to ACTS came Claire's blast with a query scrawled by Arnold, "Who is this damned fellow Chennault?"

Chennault could have told him he was a voice crying in the tactical wilderness, for by 1933 the deterrent of pursuit to the emerging doctrine of strategic bombardment was being shot out of the sky by both the theorists and Air Corps planners. As early as 1930, Tony Frank, principal referee at a pursuit-bomber maneuver, had stated the day of the pursuit plane was ended, and he was being seconded by a growing chorus.

A more formidable deterrent, one that made the teaching of strategy, tactics and the techniques of air power at the Air Corps Tactical School somewhat exhilarating, was the War Department's attitude. Thinking there remained cemented in the Clausewitzian principle that "the ultimate objective of all military operations is the destruction of the enemy's armed forces by battle. Decisive defeat in battle breaks the enemy's will to war and forces him to sue for peace."

What the bomber proponents were artfully theorizing was that if you could control the air, you could destroy your enemy's industrial base and his will to wage war, and there would be no need for armies or navies, except for policing and transportation. There was nothing new in it, only that it was being more boldly advanced, stimulated and supported by aeronautical progress. To the General Staff during the twenties, the concept was a form of heresy, and since the General Staff controlled the school's curriculum, such ideas were advanced with considerable care. By the mid-thirties, the school's Department of Air Tactics and Strategy, headed by Major Hal L. George, was advancing the concept of long-range-bomber daylight operations against critical pinpoint targets.

When Major Ira Eaker arrived at Maxwell Field, in September 1935, to take the nine-month tactical course, the faculty numbered twenty-two, seventeen of whom were graduates of the school. Of these, aside from George, the principal bomber theorists numbered no more than a half dozen. Their influence, however, far outweighed their numbers, partly because they were the spokesmen for old beliefs being brought to a boiling point by aeronautical advances, and partly because almost all the air officers attending the school were familiar with the beliefs, whether they accepted them or not.

From 1923 onward, Billy Mitchell's manual *Notes on the Multi-Motored Bombardment Group, Day and Night,* which he had privately printed and distributed, had been the cornerstone on which ACTS bombardment advocates structured their theories. By the mid-thirties it was not the Italian Douhet, or Trenchard and the British, or the French, or any foreign air strategist who influenced the school nearly as much as it was Mitchell.[1] The bomber advocates evolved their own thinking and passed it on. They couldn't make doctrine, but they could and did stimulate thought. Within the classroom, they were able to visualize and project what they liked, regardless of the reality outside. Although only a little more than 50 percent of the school curriculum was allocated to air subjects, and the supposed mission of the school was to train air officers for higher staff duties before going on to the next school, its underlying and actually surreptitious value was that it permitted lecturers and students alike to expound their ideas on the use of air power, free of restraint.

Rank went out the window in the classroom. Two of Eaker's instructors were first lieutenants, Laurence Kuter and Haywood S. "Possum" Hansell, Jr. Both had been students in the previous class. They had so impressed the faculty board as well as the school's Commandant, Colonel John F. Curry, and Bert Dargue, who was Assistant Commandant and ran the school, that they had been asked to stay on as instructors.

Kuter had been assigned to the school following his service with B. Q. Jones and the Eastern Zone airmail operation. Hansell, whose father had been an Army surgeon, had come with the reputation of a premier pursuit pilot. He had entered the Air Corps as a flying cadet in 1928. Three years later, he and Ira Eaker found they had a common bond.

As a pursuit instructor at Langley Field's Air Corps Tactical School, Major Clayton Bissell was anxious to test the P-12's capacity to carry the Signal Corps's seventy-pound radio, and he instructed Hansell to load seventy pounds of sand in the plane's baggage compartment and then take the aircraft up and test it. Hansell did so and found himself in exactly the same predicament as had Eaker: a flat spin and no way out but to bail out. Due to centrifugal force, he had trouble doing so. Whereas Eaker had managed to make his landing on a back stoop, Hansell came down in four feet of water and very nearly drowned. It was the middle of the winter, he was heavily suited, and when he hit he couldn't get the straps of his chute loose. His seat pack was a life preserver, which managed to keep his feet out of the water but not his head. The rescue boat got stuck in the mud, and it was an oyster fisherman who finally saved him. A few of his friends, including Larry Kuter, had been playing golf nearby and had thoroughly enjoyed old Possum's splashing around in the water, not realizing his plight. When the flight surgeon gave him a restorative drink of whiskey and he threw it up, they knew what a close call it had been. Eaker had sent him a note saying that makes two of us.

Hansell arrived at the school a strong supporter of pursuit aviation, for he had flown under the care and tutelage of Claire Chennault. When the school was still based at Langley Field, Commandant John Curry had suggested to Chennault that a precision pursuit team be formed to match the Navy's Hell Divers. The Navy fliers had been thrilling crowds at air meets with their acrobatic skill. Chennault selected for his team two Air Reserve second lieutenants who were serving as sergeants, John H. "Luke" Williamson and William MacDonald.[2] The fourth member was Possum Hansell. For a year, he flew the number-three position off Chennault's left wing, the quartet widely known as "The Men on the Flying Trapeze." Although Curry's idea for the team was to show off the Air Corps's flying ability in competition with the Navy, to Chennault, who was a dedicated student of tactics, the demonstrations were a chance to try out his ideas while thrilling the crowds at the National Air Races.

To Hansell, Chennault's primary characteristics were aggressiveness and self-confidence. Chennault, in Hansell's view, figured there were only two kinds of people—those who agreed with him and those who didn't—and while his flying skill and dedication were admired at the school, his personality was not.

At forty-four, Chennault had been in the Army since 1917, although he had first applied for flight training in 1909 while a student at Louisiana State University. The examining captain had noted, "Applicant does not possess necessary qualifications for a successful aviator." When Chennault applied again, in April 1917, he was rejected because he was married and had three children.[3] He became a ninety-day wonder in the Infantry instead, and then by hook and crook and iron determination, which included learning to fly on the sly, he eventually got to Kelly Field as a flying student and graduated in the spring of 1919, a pursuit pilot. In 1921 he served in the 1st Pursuit Group, commanded by Tooey Spaatz. Monk Hunter had the 94th Squadron, and Chennault was one of his boys, flying Spads and SE-5's. Very shortly he realized, as did Spaatz and Hunter and others, that World War fighter tactics were all wrong. They lacked "the calculated massing of overwhelming force," he said.

In seeking to overcome the problem, he brought into being the three-man element. Since the firepower of opposing forces is not measured by numbers but by the square of the difference in the numbers—two planes against one is not two-to-one odds but actually four to one—he reasoned that two 3-plane elements attacking a bomber force could deliver far more of a punch than mere numbers would indicate.[4] Over the years, he had labored to drive home his unyielding thesis that the fighter in the sky would decide the issue of control both defensively and offensively. He acknowledged, and claimed, that the fighter must work in conjunction with an effective ground observation system and that when that was done the bomber would not get through.

When Possum Hansell came to take the Tactical School course at Maxwell Field, he was better informed on pursuit tactics than anyone there aside from Chennault himself. Yet by the time he had completed the course, he had become a thorough convert to the zealous belief in the efficacy of long-range strategic bombardment over all else.

Larry Kuter didn't need such conversion. He had long been imbued with the theory, and flying the B-9 and the B-10 simply confirmed his deep-rooted tactical opinions. As instructors, Kuter and Hansell were the most recent spokesmen emanating from the long line of Billy Mitchell. Two of the torchbearers in the early thirties had been Ken Walker and Bob Olds, and of them Kuter was to say: "Between the two, Mitchell's work was continued, expanded, augmented, separated into its several components, including tactics and technique of bombardment aviation, and the employment of air forces." In that employment, however, the bomber champions rode high and hard both at the school and beyond it.

Nevertheless, Chennault was not entirely alone in his stand on the importance of pursuit over bombardment. Captain Hoyt S. Vandenberg, who would replace him as the head of pursuit instruction, was on his side. So were Captains Earle Partridge and Lotha A. Smith, but they were outnumbered on the ground and outmaneuvered in the air. Further, Chennault was not an articulate speaker, which did not help foster his theories.

In Kuter's view, Chennault saw the fighter as both an offensive and defensive weapon and refused to concede that bombardment and attack aviation could do more damage to ground targets than pursuit. Later, Chennault would claim he had been in the forefront fighting to get a long-range pursuit aircraft as a protective escort for bombers. Kuter said no, Claire wouldn't even discuss the idea, because such escorts would be placed in a secondary role to the bombers. Apparently, the contradiction lay in *how* the escort was used: as close protection for the bombers, or with the fighters ranging ahead of the bombers seeking out enemy interceptors. That the bomber supporters were as dogmatic in their rejection of the need for fighter escort in any form as Chennault was in refusing to see the need for bombers, made for some hot and heavy classroom sessions. However, down the road a few years, when reality caught up with the supremacy of the bomber theory, the price paid for the error would be bitterly high.

The independent employment of the long-range bomber was still essentially contrary to the views of the General Staff. Most Army planners refused to see the bomber cast in a strategic mode beyond the purpose of supporting the ground and naval forces. Key targets, in their thinking, did not shift away from the scene of action: enemy forces, land and sea; airfields, ashore or afloat; supply depots; etc. This entrenched point of view had forced air leaders such as Foulois and Arnold, who had not attended the Tactical School, and Andrews and Spaatz, who had, to frame their requests and public statements always on the basis of needing the bomber for defensive purposes and for long-range reconnaissance. Fortunately, there were other factors that helped to modify General Staff resistance to the development of a long-range bomber. Aeronautical development could not be ignored, either in the United States or abroad. Most, if not all, of the nearly one hundred officers then serving on the General Staff had seen duty in the Philippines, Hawaii, and the Canal Zone. They knew how far distant were these possessions, as well as Alaska, and if planes could be built to reach them, then, on the merits of defense, it made good military sense to have such a plane. This did not mean, as attested by the Drum and Baker board findings, that the General Staff believed for a minute such a plane could either win or play a decisive part in battle. It could only be a helpful adjunct.

The third deterrent, which had a major bearing on the other two, was money. At the time, both the War Department and a preponderance of Air Corps planners felt that the very limited procurement dollar should be spent on building up a balanced air force. They did not wish to see the lion's share of that money spent on a long-range bomber whose use, they felt, would be limited and whose cost made it possible to buy practically two smaller bombers for the price of one large one. The ground forces were partial to attack and observation aviation, because the first was designed to be there to hit the enemy they faced, and the second could act as the eyes, visually and photographically, for Artillery and Infantry.

With the Air Corps's mission of coastal defense, the War Plans Division did recognize the need for over-water reconnaissance within limits, but this was a sore point with the Navy, and in early 1933, when General Kilbourne rejected Benny Foulois's request for a long-range amphibian, it was believed by Foulois and his staff that Kilbourne was acting in deference to the Navy. If he was, he was also rejecting the proposal because he didn't believe the Air Corps should use its money for such aircraft. In so doing, fortuitously and ironically, he set forces in motion that were to produce the B-17.

There were a considerable number of Army bomber advocates, from Billy Mitchell on, who were willing to take credit for actions that ultimately brought forth the four-engine bomber, overcoming the deterrents cited. Benny Foulois, Hugh Knerr, Conger Pratt, Hal George, Jan Howard, and fellow engineers Jake Harman, Don Putt—the list is long, and all of them were correct in some measure. All supplied input, either through advanced engineering thought, administrative planning or political finesse.

It was this last, however, that launched the movement. In July 1933, the Materiel Division, under Conger Pratt, and with Jan Howard as the principal engineering officer, conducted a feasibility study on the question of the maximum range for a four-engine bomber carrying up to two thousand pounds of bombs. The answer they came up with was five thousand miles at 200 mph. At the end of the year, Foulois submitted a proposal to the War Department to build such an experimental plane. He termed the secret undertaking Project A. His argument for building the plane stressed the need for protection of far-distant possessions as well as the capability of being able swiftly to guard either coast. The money was available, he said, to construct the plane as the result of recently released procurement

funds, and since the proposal for the long-range amphibian had been rejected by Kilbourne, the $610,000 needed for Project A was attainable.

Kilbourne was opposed. So was Drum. They saw no need for the aircraft. It would be more practicable to buy more B-10's and station them
on our possessions as well as both coasts. Others, with less-fixed ideas, notably MacArthur, were in favor of going ahead.[5] At the time, congressional pressure from McSwain, Collins and others had to be a consideration, and in May 1934, Secretary of War Dern gave jubilant airmen the
go-ahead to make the purchase and build the experimental giant.

On May 14, Claire Egtvedt, the Boeing Aircraft Company's new president, was ushered into Conger Pratt's carpeted office at Wright Field. Present with Pratt were Captain Al Lyon, Jan Howard's assistant; Major
Junius Taylor, Chief of the Aircraft Branch; Project Officer Captain Leonard "Jake" Harman; and C. A. Van Dusen, the Martin Company representative. Originally, it was Harman who had sold Taylor on the idea of
working up the requirements of a plane with five times the range of a
B-10. "Why not put down a plan of what we need instead of what we
think we can get?" Harman had suggested. "We're getting the speed; now
let's go after the range."

Taylor had agreed, and when the proposal was completed, Conger Pratt
okayed it and sent it on to Foulois, who knew all about it and was waiting.
Pratt was requesting funds only for the engineering components at an anticipated cost of $600,000. The engineering staff at Wright Field was
willing to put all its experimental money into this one project.

Benny had talked it over with Westover and Spaatz, and Tooey had
suggested that Boeing and Martin be asked to submit proposals. If what
they came up with looked good, a design contract could be let.

Now Pratt got down to cases: "The purpose of this meeting," he said,
"is to discuss a procedure under which the Air Corps will consider proposals for construction of a long-range airplane suitable for military purposes." He then gave the salient requirements and asked before going further whether "you gentlemen are interested in discussing the project?"

They were, and those present filled the two men in with the necessary
data, Pratt saying that a cost estimate should be in his hands no later than
mid-June. Egtvedt assured him it would be.

Of all who contributed to the advent of the B-17, none was more important to its arrival than soft-spoken Claire Egtvedt. His fellow workers said
of him that he dreamed and did. Advancing powered flight was his dream,
and the design and development of new aircraft is what he did. His interest
in bombers went back to September 1923. He'd been an observer on board
the transport *St. Mihiel* when Billy Mitchell's airmen, in a repeat of their
1921 performance, bombed and sank the Navy's ships *Virginia* and *New
Jersey*. Egtvedt had thought of getting into the bomber business then, possibly intrigued by a naval officer's remark that obviously such planes

wouldn't have the range to touch a fleet at sea. But he had backed off when the War Department had canceled all bomber contracts, declaring manufacturers were unable to come up with a satisfactory design. Because pursuit was still king and what money the airmen had was going into its production, Egtvedt sold Boeing's chairman, Ed Gott, and president, Phil Johnson, on a reverse approach. They would design and construct their own pursuit and then offer it to the Army. Although the plane was rejected for the Curtiss PW-8, Boeing was given the go-ahead to build three experimental models of the ship, designated the PW-9.

At the time—1924—Tooey Spaatz commanded at Selfridge Field, and he saw to it that his pilots flew both planes in mock aerial combat. All liked the PW-9 over the Curtiss, and in the next decade two of the Air Corps's line of pursuits that followed were Boeing-built.

Early on, Egtvedt had predicted that at some point Boeing would get into the bomber business, and it did in 1931 with the production of its radically designed B-9. The B-9 was the first monoplane bomber, and the first with a crank-up retractable landing gear. Its long, thin monocoque fuselage with three open cockpits gave it an odd appearance. It was in the nature of an experimental breakthrough aircraft that could fly higher and faster on its twin engines than anything around. The Air Corps bought only eleven of the planes, nine of them going to the 49th Squadron, of the 2nd Bomb Group.

To those who flew it and then went on to take the course at the Tactical School, the plane's performance was solid support for the concepts the bomber instructors were teaching there. Regardless of what Claire Chennault maintained his interceptors had done to the bombers during the 1933 Fort Knox maneuvers, the pilots of the 49th would claim that once they got up to about twenty thousand feet, nothing could touch them, and that they won all the fighter competitions in those maneuvers. A great deal of flying at high altitude was done under conditions of anoxia. Liquid oxygen was being used, and the breathing tube would freeze up. Still, to bomber proponents, Claire Egtvedt's B-9 was helping to prove theories and assure greater emphasis on bomber development. In fact, it was a long step forward in the long climb to Project A.

On the negative side, the B-9 was a killer. Six of the 49th's complement crashed, with heavy fatalities. The problem appeared to be in the long fuselage. Severe turbulence or violent control action broke it apart. A pilot who jammed down the rudder could actually see the fuselage twist. It was a design problem Egtvedt and the Boeing engineers rectified in their follow-on, all-metal airline version of the plane, the 247, termed the first modern passenger aircraft.

When Egtvedt brought the details of Project A back from Dayton after his meeting with Pratt and showed them to Monty Monteith, Boeing's chief engineer, Monteith looked the specifications over and said, "A year or two ago I would have said this was ridiculous." Now he couldn't wait to get started.

Little more than a month later, on June 28, Egtvedt learned Boeing had been awarded the design contract for Project A—an experimental long-range bomber designated XB-15. The purpose of the exercise was to learn how to build an aircraft of maximum size, although it was understood it might be years before a production contract could be awarded by the Air Corps. Two days later, Boeing began its own design plans for a four-engine transport designated No. 300. Then, in early August, Egtvedt received a circular from Wright Field setting down specifications for the next bomber to be produced in numbers for the Air Corps. The plane must be able to deliver 2,000 pounds of bombs and have a required speed of 200 mph with a desired speed of 250 mph. It would carry a crew of four to six, possessing a desired range of 2,200 miles. It was anticipated that 220 of the planes would be produced. Interested companies could submit their bids. A prototype should be ready for testing by August 1935.

In 1928, Egtvedt had been on board ship with Admiral Joseph M. Reeves, then in command of a number of naval air squadrons. A remark Reeves made at the time on the merits of aircraft had stayed with Egtvedt. "The airplane," said the Admiral, "isn't a dreadnought. It just isn't as effective a weapon as those we already have." Perhaps, thought Egtvedt, that wasn't so any more. With the aeronautical possibilities before them, Boeing just might put a dreadnought in the sky. He went east to Dayton to have a talk with Jan Howard.

"Would a four-engine plane qualify?" was the key question.

A grin spread over Howard's face. "Say, now. . . . The word is multi-engine, isn't it?"

No one had attempted more than a twin-engine bomber since the ill-fated Barling behemoth. Egtvedt had a lot to think about as he journeyed back to Seattle. Employment at the company was down to six hundred workers, as opposed to seventeen hundred at the first of the year. United Aircraft, to which Boeing had been joined before the new airmail legislation had forced separation, had given no assurances that it would buy the new four-engine transport, No. 300. Further, profits earned from the sale of the 247 had faded away. Worse, the P-26, with cost increases and design changes, was being produced at a loss.

The company was operating in the red. Egtvedt knew that to take on the competitive risk of designing and building a new Air Corps bomber would require all Boeing's capital and manpower. He talked it over with Bill Allen, the company's lawyer. Would he be putting the company in jeopardy if they went ahead? The answer came in a board of directors meeting soon after. A maximum effort would be launched; it would be the plant's one job; $275,000 was voted to design and construct what was designated Model 299. The work was to be kept secret. Another $150,000 was necessary before the plane was ready to be shipped to Boeing Field for its trial runs, on July 1, 1935.

By the end of July the press was referring to the graceful, four-engine Model 299 as a "veritable flying fortress." Not a dreadnought, but just as meaningful. Chief test pilot was Leslie Tower, a lanky Montanan who had gravitated from engineering and draftsmanship to the flight line. His copilot was Louis Wait; and Henry Igo, of Pratt & Whitney, was the flight engineer.

On August 20, Les Tower pointed the plane's nose eastward on a nonstop, 2,000-mile flight from Seattle to Dayton. Nine hours later he brought the B-299 in for a landing, having averaged nearly 235 mph. On hand to greet Tower and his crew were Claire Egtvedt and Boeing project engineer Ed Wells. No one else was around. How come? asked Tower.

"You're not supposed to be here," Wells said. "Claire and I figured you'd be due an hour from now. The Field expects you in an hour or two."[6]

The early arrival was a most pleasant surprise to the Air Corps. General Robbie Robins, who looked a little like a prizefighter, his face having been injured in a serious crash long ago, couldn't hide a smile. Lieutenant Jake Harman didn't try to. Major Oliver Echols, who had replaced Jan Howard as Chief of Engineering, was anxious to have a closer look at the bird.

For the next two months, Tower and his crew worked with an Air Corps team headed by Major Ployer "Pete" Hill, chief of the flight test branch, and Lieutenant Don L. Putt, who was the project test pilot. The coming tests would determine which of three aircraft in the competition would be selected for production. The Martin entry was an updated version of the B-12. The Douglas submission was also a twin-engine model, and although all metal, it was simply not in the same league as the 40,000-pound Boeing with its four 750-hp Pratt & Whitney engines.

What was going on at Wright Field was, of course, known throughout the Air Corps. Andrews had visited the Boeing plant during his West Coast inspection during the summer. Arnold had been worried over the changes in training the new bomber would bring. After his visit to Langley in the spring, he had written a long memo to Andrews, citing them. And Andrews had answered him in detail, pointing out:

> Well-trained permanent crews are necessary to the full efficiency of our present bombers. The new bombers only accentuate the viciousness of the present high turnover rate in personnel. Bombardment squadrons and groups will continue as the organization to operate larger types. The idea that these larger airplanes will be manned by special crews, under command of high-ranking officers more or less independent of the bombardment, is not favorably considered. Rather, the present organization should retrained and the present flight commanders with but one airplane in their flight.

Andrews and his staff, Knerr and Kenney and McNarney in particular, were right on top of the game. The bomber enthusiasts at the Tactical School were excited by the news of the new plane. Having adopted the theory that air power, if used correctly, could itself defeat the enemy, in the coming of the four-engine bomber they visualized the proof of their assumption.

The morning of October 30, 1935, was clear and sharp. Present on the flight line a goodly crowd was gathered, including representatives of the three competing companies. Claire Egtvedt was not one of them. He was in Chicago, talking to United about buying the new four-engine transport.

The mixed flight crew climbed aboard the B-299, as it was still called. Pete Hill would do the flying, and Don Putt was copilot. Les Tower was to be an observer, and he took his place standing in the aisle in back of the control pedestal. Igo and the crew chief made up the remainder of the crew.

Hill started the engines and taxied out for the takeoff. The plane made a proud sight, and workers stopped what they were doing to watch. The takeoff sounded thunderous and looked smooth, but suddenly the watchers knew something was wrong. The plane was in too steep a climb. The nose was not coming down! The propellers of the four Pratt & Whitneys were clawing to gain a foothold on the air, their sound a frantic roar. Then a wing dropped, the nose fell, and the stricken plane plummeted. The crowd stood transfixed.

Jake Harman was in conference with his boss, Oliver Echols, when the sirens sounded. He sensed their meaning and raced out of the building, hailing a passing car. On duty not far away, Lieutenant Robert K. Giovannoli reacted also. He came around the end of a hangar and saw flame and smoke rising from the wreckage. "Come on!" he shouted to Cecil Dean, a civilian employee, and together they ran toward the burning plane. When they reached it, Dean boosted Giovannoli up on a wing and, in spite of the fierce heat, he reached the copilot's window. Forcing it open, he thrust head and shoulders into the smoke-filled cockpit. He got his hands on Les Tower, and as he struggled to work Tower's body through the window, Harman and others reached the scene.

Civilian worker Joe Harriman had backed his flatbed truck up to the wing so there would be a platform to work from. Bill Hunt, of Douglas Aircraft, and Bernie Winter were on it, and Hunt rushed to aid Giovannoli in pulling Tower clear of the cockpit. While they were getting him to the truck, Jake Harman bulled his way through the cockpit window, coat over his head to protect himself from the flames. He found that Pete Hill was trapped, his foot caught under the rudder.

"Get a knife!" he shouted as Giovannoli came crawling through the window. Together they struggled, choking in the smoke, fighting to free the unconscious pilot, the flames spreading, the heat becoming swiftly unbearable. They knew the whole thing could blow up at any instant.

"Here's a knife!" Herbert Wentz called from outside the window. Giovannoli grabbed the skimpy penknife, went down on his knees, and by feel cut the shoe off Hill's foot. Somehow they lifted him out of the seat and eased him through the window to the waiting hands of the others. Then, badly burned, they made their own exit.

In the crash Don Putt had managed to make his escape through the nose of the plane. He was gashed and burned. Dazed, he kept repeating, "Look at the control stand!" Igo and his crew chief had made it out the side of the fuselage. They were all rushed to the hospital. The daring and courageous rescue had taken no more than five minutes, but the crowd, for the most part, stood immobile, stunned, watching a few brave men risking their lives to save pilots Pete Hill and Les Tower—as all the bright dreams of the moment were consumed in ugly smoke and flame.

Hill never recovered consciousness and died that same day. Tower, who was not considered to be seriously injured, blamed himself for the disaster and died a few days later because he no longer wished to live.

Even before an investigation could determine the cause of the crash, the word reported from the War Department was that the four-engine Boeing was too much airplane for one man to fly. Aside from not being true, the negative opinion added to the pall cast by the accident. If nothing else, it indicated a rejection of the plane.

The Douglas twin-engine bomber, designated B-18, now had no competition in the trials. The Douglas was more to the liking of the General Staff and Henry Woodring, anyway. Its cost would be far less, as would its performance, and therefore it would be not so offensive to the Navy. Woodring felt he had been largely responsible for the go-ahead on the Boeing Fortress in the first place. He well remembered the day in 1933 when Benny Foulois, Conger Pratt and two or three other officers paid him a call. They had with them the first development order for the Boeing. They needed Woodring's signature on it, and he admitted the thought of signing scared him to death, for it meant the spending of millions in the final development. He had hedged, but Conger Pratt, whom he liked and trusted, came to see him privately and, through gentle persuasion, talked him into nervously signing. On that October afternoon, he must have received the news with mixed emotions.

Andrews' emotions over the news were anything but mixed. He was in conference with Westover when the word came. He was doubly shocked because he had known Pete Hill very well. Hill had served for three years as his Engineering Officer in the Coblenz days at Weissenthurm, and Andrews considered him "an excellent officer." But, emotions aside, the question he and Westover faced was how they could still get a production order on the plane when it was no longer in competition. The answer was crucial.

They talked it over, and Andrews came up with a suggestion that he put in writing a recommendation that he and Westover could present to Woodring and Craig and the General Staff. His thought was that if the bomber could not be procured regularly, they could still push for its production by means of an experimental order for service tests. He pointed out that a provision in the Air Corps Act made such a move possible. He reasoned that insurance on the downed plane would pay for some of the development costs, and that obtaining an experimental order for "say thirteen of these planes would not necessarily run the cost to a prohibitive figure." He also figured that if they were then to go ahead and procure the B-18 or additional B-12's, they might end up with more bombers than had they bought the Boeing alone.

"The big disadvantage," he wrote to Robins in Dayton, "is the service will not be equipped with the latest type of bombing planes for a good many years."

As Chief of the Materiel Division, Robins had come up with exactly the same idea, and whereas Westover had discussed it informally with both Woodring and Craig, Robins had telephoned them saying, "With the money we have for total procurement, including experimental funds we can save up, we can buy 80 Douglases and 13 Boeings." Westover in his approach was far more conservative in his figures, mentioning from three to five of the Boeings, but whatever the number, the decision was up to the War Department.

In a follow-up letter to Robins, Andrews queried, "I am very much interested to know just what happened to that B-299."

What had happened was that either through pilot error or mechanical failure, the control that unlocked the elevators on the tail had not been released. The reason for the lock was that the tail surfaces were so large that if the elevators were not secured while the plane was on the ground a strong wind could damage them. The unlocking lever was in the cockpit. Putt believed that Hill had activated it before the takeoff. Les Tower had tried to reach it as the plane fell, and he died believing that he was to blame for not making sure the elevators were free and movable. This, of course, was Hill's responsibility.

Bob Giovannoli and Jake Harman, along with those who assisted them in the rescue attempt, were cited for their bravery. But tragedy was never far away, for Giovannoli received the Soldier's Medal posthumously, killed in a crash before the President could make the award.

Reaction to the loss of the Boeing 299 was varied. Claire Egtvedt had lost a plane and a very old and fine friend. His company was deeply in the red. It was vitally important to know whether the Air Corps would be giving Boeing a purchase order. The answer was needed soon.

At the Air Corps Tactical School, the bomber advocates were dismayed and worried. Knowing the paucity of funds available, knowing the War Department's thinking, they were afraid that reaction to the crash would jeopardize the thrust of the bomber program. They, too, had their ears to the wind, waiting for a signal from the North.

To the West, Hap Arnold's thinking during the summer and fall had been concentrated on bombers but not on trials to be conducted at Dayton. Using B-10's and B-12's, the 7th and 19th Bomb groups had begun tests with a new secret bombsight called the M1. It had originally been developed for the Navy; it became commonly known as the Norden bombsight.[7] In spite of news reports that the development of the instrument would permit bombardiers to drop their eggs down the pickle barrel, Arnold's crews found that from altitudes of twelve and fifteen thousand feet, the barrel would have to be half the size of San Joaquin Valley.

In studying the results, Andrews figured, "It would take about forty bombers loaded with 2,000-pound bombs to get a hit on a single battleship at fifteen thousand feet." He wouldn't guess what the possibilities would be at eighteen thousand feet, which was considered to be out of range of anti-aircraft fire.

"This is not pay dirt," he told Hap, "and must be bettered. Our gunnery is also weak." And then he added the thought that maybe high-altitude bombing was impractical, but he'd have to see a lot more before he would accept that possibility.

So would Arnold. But he agreed fully that the results were lousy. He had some valid reasons and a few side excuses, such as vibration of the bombsight mount regardless of its position in the plane, cross hairs that were much too large for accurate aiming, and the sight's calibration in knots instead of mph. Practice would go a long way toward eliminating these, and certainly did. Andrews passed on later results to George Kenney, who was greatly impressed. Kenney told Andrews that eleven direct hits out of twenty-six bombs dropped at twelve thousand feet at a speed of 160 mph was "superb bombing. He [Arnold] has done well and I believe you should tell him so and to keep up the good work. Maybe in another six months we will find that we can even better the radial error of 144 feet."

Kenney's evaluation cheered Hap, but even without it he was greatly pleased with the improvement shown by the 7th Bomb Group. Tinker and his boys had been working on the theory that practice makes perfect. Hap was hopeful that a demonstration from eighteen thousand feet might be put on for the brass the following May. He suggested every effort be made to have representatives present from Congress, the War College, and the General Staff.

On the day before the fatal flight of the Boeing B-299, Arnold was engaged in giving John McSwain and five of his colleagues from the House Military Affairs and Appropriations committees the grand tour of March Field as well as a firepower demonstration. In this last, he ran head on into the ridiculousness of corps-area control. He wished to show the visitors the effect of various-caliber ammunition on the metal skins of aircraft. To do so it was necessary to borrow a 37-mm cannon from the National Guard. The Congressmen were sufficiently impressed by the devastating power of 37-mm shells over the machine guns. Arnold's report to Malin Craig, who was also interested in the result of the experiment, never got past corps-area headquarters. By what authority, Arnold was summarily asked, had he used the 37-mm ammunition? Whose ammunition was it? Who was going to pay for it? Headquarters Supply Services wanted to know the answers right now. Arnold later went up to San Francisco to explain the situation, and since it was the Chief of Staff who was involved, the matter was dropped. But the interference had been a waste of time and effort and money, an example of what Andrews had been complaining to Craig about.

Directly after McSwain's visit, Andrews flew out to California to observe personally the accuracy of Hap's bomber pilots. He was heartened by the results, but there was another disturbing problem: oxygen, or lack of it, at eighteen thousand feet.

"Why can't we get some data on bombing at eighteen thousand feet right now?" Andrews asked.

"Not enough liquid oxygen, Andy. We can't get it, and when we do get it, it leaks so badly the crews can't stay up there long enough to get the bombing done."

Some months earlier, Andrews had discussed the oxygen situation with Robins. The feeling on the part of Andrews and his staff was that liquid oxygen wasn't reliable and they should go back to gas oxygen. Upon his return from visiting Arnold, around mid-November, he wrote to Robins asking him the status of the problem. Its resolution was an integral and vital part of the grand equation of *if*.

If they could get the Boeing. If the bombsight would deliver the necessary accuracy. If crews could operate at high altitudes with dependable oxygen supply. Then the bomber champions, in whatever branch of the Air Corps they served, would have the total instrument of flight that they had sought for so long and in which they devoutly believed.

Nineteen

The Contenders

When Malin Craig had taken over as Chief of Staff, a considerable amount of his knowledge concerning air power had been the result of his relationship with Hap Arnold. Arnold's ebullience, the impact of his personality and his driving methods of getting things done had appealed to his former corps-area commander. Through Arnold, Craig had developed some understanding of how airmen looked upon the use of aircraft. But Craig was a traditionalist, an organizer, and a Chief of Staff whose attention was now focused on two major areas: his relationships with those with whom he had to deal in the Congress and the government, and in developing a workable, realistic mobilization plan in case of war.[1] The two focuses were interlocked, and in the latter Craig had a strong supporter in Henry Woodring. Both were men whose thinking was attuned more to the logistics of war plans than to their strategical use—particularly when the use concerned aircraft.

The crash of the Boeing prototype had been an unnerving tragedy. It had sent shock waves through Craig and Woodring as well as rattling the hopes of the bomber advocates. Almost as unsettling, it had brought out into the open the growing schism between the Air Corps and the GHQ Air Force staffs, with critical comments crackling back and forth between the camps.

On the very day the Boeing had crashed, Andrews was in Washington discussing with Westover among other things "the lack of cooperation" between the two commands. Colonel Alfred H. Hobley, Westover's Executive, had given his boss "quite a few sidelights on the attitude of your [Andrews'] staff officers."

One sticking point had to do with War Department restrictions on permitting the Air Force staff being given the estimates for 1937 expenditures.

The General Staff believed these estimates should remain secret between itself and Westover's office until they had finally been approved by the Director of the Budget. Three of Andrews' G-4 officers, led by Joe McNarney, had insisted that they should be privy to the information. The repartee evidently got hot and heavy and Hobley told the Air Force boys where they could head. He informed Westover that McNarney and his people wanted the figures "for the purposes of juggling the estimates in case they found it to their advantage to do so."

The advantage they were seeking was for the benefit of the air arm as a whole, more money allocated for the procurement of planes such as the Boeing. But procurement was an Air Corps prerogative, and personalities got in the way of cooperation. To Westover, the implication of a juggling act had steamed up Hobley, who had become "impatient" and then had contacted the War Department to make sure he was on firm ground in his refusal. Assured that he was, he reported the encounter to the Chief of Air Corps, who went over the disagreement with Andrews. Although West Point classmates, Westover was one of the few who addressed the Air Force Commander as Frank, and in return was addressed as Oscar, not Tubby.

Oscar suggested that Frank might come up with a list of questions covering the items of interest, and then the two of them would go over them together. He would not be able to give Frank a copy of the estimates. They were taboo, but he would personally try to give him the information he needed for planning purposes, and in future considerations, having Andrews' requirements for the following year, he would personally see that they were considered along with the others of the Air Corps.

He was trying to be cooperative, but he would follow the letter of the law to its sometimes inflexible conclusion, regardless of the result. By the very nature of a gerrymandered command structure, it was apparent there had to be a greater degree of flexibility in carrying out the mandates of the War Department if there was going to be any degree of unity between Westover's office and Langley Field.

Andrews' suggestion was, "I wish you'd come down and spend the night with me sometime between now and the first of December so we can talk this over."

Instead, in an effort to put a lid on the contentious atmosphere and to gain control over it, Westover, on a Saturday, called on the Chief of Staff and gave him his view of the problem and what should be done about it. Craig said, let me have your thoughts in writing. Westover complied with confidential drafts of memos to be sent to himself, to Andrews, and to all Air Corps commanders, over the Chief of Staff's signature.

In his covering memo to Craig on the policy to be followed, Westover made it clear that it was his "definite belief that the sooner GHQ Air Force was again placed under the control of the Chief of the Air Corps in time of peace" the sooner would there be harmony and the overcoming of divergent viewpoints that had always existed with reference to future trends, plans and Air Corps policies. Westover believed the whole thing could be handled informally with a meeting in Craig's office with him and Andrews present. He wrote a draft of the words Craig might speak, and it was apparent from them that he felt insecure as Acting Chief, so he wanted it made patently clear that until a new Chief and Assistant Chief were appointed, he was in charge and his policies were to be carried out. More, he wanted Andrews to "operate through and under [his] control."

Craig's response was to send out letters essentially in the form Westover's office had drafted but not specifically giving the Acting Chief the domination he wanted. Craig wrote that Westover was to have the "sole responsibility for leadership and direction in Air Corps matters in general." He wanted internal discussion and criticism kept within the family and a united front presented "when we appeal to Congress for legislation or supporting appropriations."

Andrews, who had been on the West Coast with Hap Arnold when Oscar had taken action, replied that he was in complete agreement with regard to cooperation. He had discussed the problem with his wing commanders, he said, and the policy would be henceforth "scrupulously followed." But, of course, the fact was that by War Department directive as GHQ Air Force Commander he reported to the Chief of Staff in time of peace, and while Westover was running the Air Corps he was not running the Air Force. Under the guidelines of divided command, harmony would not be all that easy to find.

Shortly thereafter, Westover sent out an edict to all his principal commanders which was to be read to each and every officer in the command. "And you will render a report to this office not later than November 15 to the effect that you have done so," the order wound up.

Westover's policy statement was more in the nature of a lecture than anything edifying. His theme was loyal obedience to superior officers. Although he felt loyalty should be unnecessary to refer to, he did so at great length. The direction of his scatter shot was aimed principally at all those anywhere in the air arm who had voiced complaint outside of channels at War Department or Air Corps actions.

It was pendantry to men who knew the form as well as Oscar, but he wanted it clearly understood that as long as he was in command, that's the way it was going to be:

> It should be clearly understood, therefore, as a basic policy from now on, that all steps looking toward the improvement and development of the Air Corps, or in connection with the proper carrying out of its operation, will be properly represented and considered through the normal military channels of communication to the point where decision of higher authority results either in approval or disapproval. In the event of disapproval, such

decision must be carried out loyally and cheerfully by all persons concerned who in so doing, are charged with refraining from discussing such
matters with outsiders not in the military service, or in public. . . .

His position was based on convictions, arrived at in 1920, that had not
changed one whit over the years. They had been further reinforced by the
difficult events of the preceding year and during Benny Foulois's tour,
when Mitchellism had again raised its rebellious banner. Westover was not
blind. He fully understood the needs of the Air Corps, understood them
because of his position and experience perhaps better than anyone in the
air arm, but his rigidity did not connote the kind of leadership Andrews
and Arnold and Pratt and Brant and others could inspire in their commands. As the year wound down, the hope, at least in GHQ Air Force,
was that he would not continue as Chief of Air Corps.

On the matter of salvaging the four-engine bomber from complete rejection, Westover went all out to save it. He and Andrews had done their best
with Malin Craig, Simonds and Embick. Then he took Claire Egtvedt to
see Henry Woodring. Woodring sat and listened to all their reasons supporting acquisition of the plane but was very noncommittal. Possibly a
small number of planes might be purchased but he doubted it would be
enough to equip a squadron.

At the time, Robbie Robins had been down from the Materiel Division
and Oscar had told him of the indecisive meeting. Robins immediately
contacted Andy and suggested that he drop everything and try to push
their idea for procurement of thirteen of the bombers through the General
Staff as strongly as possible. Robins' information was that Woodring
would not "buck" the Staff's recommendations.

His information was correct, for although the Douglas B-18 won the
bomber competition by default and not by superior performance and was
to be procured in large numbers, Andrews' suggestion of obtaining thirteen
of the Boeings through the expenditure of experimental funds was approved. A great sigh of relief went up, particularly from the banks of the
Alabama River, where the bomber advocates of the Tactical School took a
firmer hold on their theories.

Evaluation and selection of a new bomber was a decision of great moment, and there was general uniformity of opinion within the Air Corps on
the need for and the inclusion of the choice to obtain at least a handful of
the newly named B-17's. The same uniformity did not apply in the evaluation and selection of a new pursuit ship, and the choice brought a three-
way disagreement among Andrews, Westover and Robins.

There were two planes in the competition, a Seversky model and a
Curtiss. The evaluation board as usual was overlarge, top-heavy with sen-
ior officers from the Air Corps and had but one representative from GHQ
Air Force. The board chose the Seversky, the P-35, and learning of its

method of selection and getting an earful from Curtiss vice-president Burdette Wright, Andrews contacted Westover. "I understand that the Curtiss
Company is kicking up a row about the supposed selection of the Seversky
pursuit plane for procurement," he wrote. "That is your pigeon and I'm
going to keep completely out of it—but it does bring up a point about
which I feel very strongly. . . ." The point was that Westover's evaluation
boards did not take the time to do a proper job, and the pursuit board in
particular. "It did, I understand, very little flying, dropped no bombs, fired
no guns and tested no radio. In other words, they made no test on any of
the functions which represent the purpose for which the plane was built."

Andrews closed by saying that he strongly recommended that in the future, boards put the aircraft in question through every conceivable test before submitting an evaluation figure.

He received no response from Oscar, and a week later he sent an even
stronger blast, suggesting that the board that evaluated the plane for
suitability as to type meet again and run the tests properly, and that this
should be done before any choice was made on a go-ahead.

"If this is not feasible," he added, "could the Division put guns, radio
and bomb racks on the selected plane and send it to Langley Field in order
that we can put it through its paces . . . ?"

When no response was forthcoming, he sent a radiogram to Westover
and followed it with a personal visit, but found the Acting Chief was out
of town, so he called on Assistant Secretary of War Henry Woodring and
made his position on the matter clear to him. The upshot was that Oscar's
nose was put out of joint. As Robins reported to Andrews, "Oscar was
very much upset and seemed to have his feelings hurt a little bit. . . ." As
a result, Westover misinterpreted Andrews' criticism, possibly on purpose,
and informed Robins that Andrews' complaint was over the selection of
the Seversky, not the sloppy method of selection. This necessitated Andrews' putting the Chief of the Materiel Division straight. Since Burdette
Wright had been a long-time friend of Andrews', there was an implication
of favoritism in the information sent to Robins, and Andrews responded:
"Such is not the case. I haven't the slightest interest in who gets the pursuit
contract. Our only interest is that when the planes come to us that they
will function for the purpose for which they are intended. . . ."

In offering an explanation to Andrews of how the decision had been
made at Wright Field, Robins made it clear that when tactical boards arrived at the Materiel Division to make their evaluations, he gave his people strict instructions to stay out of the deliberations. But he suggested that
in the future the GHQ Air Force member be given proper instructions at
Langley to prevent his going off "half cocked." The reason the Pursuit
Board had "cut its testing short was that they were comparing a ship
which would eventually make nearly 300 miles an hour with a ship that
could not possibly be expected to go better than 280, their decision being

that a 280 mile ship would be absolutely worthless as a pursuit plane and that no matter how bad the 300 mile ship was it would be the only satisfactory equipment for pursuit work."

To Andrews it was a very sloppy way to reach an important conclusion, and since Robins considered the Bombardment Board "the best Board that I have ever seen operate at the Materiel Division," it may have been a reflection of the degree of interest and emphasis on pursuit as opposed to bombers. However, Robins agreed with Andrews that the Pursuit Board had been much too large and that too many of the senior members were not all that anxious to throw pursuit ships all over the sky, and that the younger members who would have been glad to do so were held back by the seniors. In large measure, both Andrews' and Robins' feelings on the subject ran parallel to Hap Arnold's, Hap having previously voiced his complaint in 1933, after Monk Hunter and Elmie Elmendorf had spun in. Obviously, no change had been made since that date.

Robins then made a recommendation to which Andrews took quick and firm exception. He suggested purchasing cheaper planes for training purposes. He said he was motivated by the constant need to do something about getting some sort of planes for the Air Force. He felt that all their current troubles were due to the attempt "to develop the most advanced types too rapidly," that it would be better to settle for more conservative aircraft that would get through the factories without fear of crashes or the failure of manufacturers to fulfill the required specifications. Then they'd at least "have a completely equipped air force instead of having the situation we have today where it is very difficult to get any planes whatsoever."

The shattering effect of the Boeing's crash was not the only consideration that influenced Robins' thoughts on settling for second best. He was having a very difficult time with the new B-10's. They had developed tail flutter. It had taken two months for Martin to correct it, and then the plane could not meet its minimum capabilities. Martin blamed the engine manufacturer, and after a delay of another two months, Robins put his foot down and hit the company for a return of more than thirteen hundred dollars on each plane.

He moaned bitterly to Andrews, "You cannot imagine what a job it is to deal with a bunch of crooks, especially when the responsibility lies on me so heavily to get some equipment for GHQ Air Force. Otherwise, I would have kicked him [Glenn Martin] out of here and let him worry for several months."

The fracas with Martin was not over, for when Robins sent the agreement to Woodring, the Assistant Secretary of War disapproved it and slapped an additional $250-per-plane charge on the firm. Martin balked, and there the matter stood: no production; no new planes.

The choice of the Seversky pursuit brought still another battle for Robins. He told Andrews that Burdette Wright and his people were putting in a four-page protest against the award. "This new procurement system may be grand from some points of view," he commented, "but the Chief of the Materiel Division certainly has his troubles for every competition. The loser immediately goes into a huddle with a bunch of lawyers to find out just how much trouble he can make to prevent the award being given to the winner."

And so Robbie Robins had been having a rough few months, and he thought maybe the way out was to step back on quality and go for quantity. Since Woodring had previously encouraged the same line with Burdette Wright's prompting, Andrews brought his foot down hard. "I can't imagine the Navy advocating wooden battleships because they cost less than steel ones and because they could buy more of them," he put it to Robins, and then told him what he thought it was all about. "The role of the Air Force in national defense is becoming more and more important and I think someday the airplane will be the most important weapon we have. It is going to cost now. It is our job to educate the War Department, the Congress and the public to this fact."

Robins' recommendation traveled to Washington, where evidently it received a kinder reception, for Andy Andrews reported to Hap that considerable pressure was being brought to bear on his headquarters to recommend cheaper types of planes "in order that a larger showing may be made on the 2320-plane program." He would have none of it, and he made sure that Malin Craig understood his position.

He also made sure the Chief of Staff understood his position on an equally pressing matter. The month previous, Hap had sent him a short, explicit note on news that the Air Corps had provided the Navy with a ramp for its planes at Hickam Field, in Hawaii.

"Really, I marvel at the childlike innocence of the Air Corps," Hap protested, "and at the many ways in which we stick our chins out waiting for someone to take a swat at it. Realizing that the Navy is like a camel—first it sticks its nose in, then its head, then its neck, and then its whole body—we continually lay ourselves open to them. I predict in another ten years Hickam Field will go to the Navy the same as Ford Island and North Island." In this last, Arnold was referring to the Navy's acquisition through congressional and administration pressure, with War Department compliance, of Rockwell Field.[2]

Andrews was in full agreement with his position. He brought the matter to Westover's attention and then decided that it would be a good idea to inform Craig on the situation as he saw it vis-à-vis the Navy.

Craig's expression was bland and set, and it was difficult to know how much of an impression was being made. He listened, made a few comments, and then kept his own counsel until he had reached a judgment. In this instance, there wasn't much he could say, for Andrews was seeking to fill him in on a long-standing Air Corps fear. He spoke of the efforts of the

Navy to develop its own air force, which would pose a threat not so much to possible enemies but to GHQ Air Force and its mission of direct defense of the coast.

Andrews cited, chapter and verse, recent comments by top naval officers on the issue. When George Kenney had been present during the Howell Commission hearings, Chief of Naval Operations Admiral Ernie King had gotten his oar in by stating that should Army aircraft be called on to assist the Navy in offshore operations, the Navy would send an aircraft to lead the Army planes out to the target, and then, if necessary, show them the way back home. King saw no reason for Army pilots to learn how to navigate over the water. The Navy's policy was that it was responsible for all air operations out of sight of land and that for the Air Corps to carry out such operations was an infringement on Navy prerogatives.[3]

More recently, Andrews told Craig that Bert Dargue had temporarily left his duties as Assistant Commandant at the Air Corps Tactical School to go as an observer on summer maneuvers with the U. S. Fleet. Bert had gotten an earful at a critique given by Admiral William H. Standley. Standley, Dargue stated, said that "the Army had considerably embarrassed the Navy by certain publicity with regard to responsibilities in coast defense but added that they [the Navy] had overcome the handicap of the so called 'Gentleman's Agreement' between Admiral Pratt and Gen. MacArthur when the doctrines for the GHQ Air Force were drawn up because in these doctrines, they preserved for the Navy its proper function in coast defense and they were going ahead with a striking force of, at least, one thousand shore-based airplanes."

Andrews informed Craig that it was apparent that the Navy was out to usurp some of the functions charged to his command. He believed that they must be on the alert at all times to guard against Navy encroachment. The point he was leading up to he later put in writing to Craig: "As far as the Air Force is concerned I am convinced that our only defense [against the Navy] is an adequate Air Force, actually in being, properly equipped, thoroughly trained and with sufficient bases to operate in any possible theater."

He admitted to Craig that his fear of the Navy's actions was deepseated, a fear ingrained among most senior Air Corps officers. He suggested that Craig might wish to discuss the subject with "Arnold, Pratt, Knerr, Dargue, Harold George or any of our other leaders."

Craig took it all in, but underlying Andrews' fear of the Navy moving in was the upcoming Service Test, which was to prove the untested capability of the Air Force. Preparation for the test had been going on since Andrews had assumed command of the combat arm.

Back in May, Colonel John Curry, the Tactical School Commandant, had written to Andy with a suggestion. He said that at the March Field maneuvers two years past he had gotten the thought that in the future it might be a good idea to have the Tactical School draw up the general and special plan for the next maneuver and then hold it at the school.

Andrews had responded that he thought Jack's idea was an excellent one. With his own headquarters understaffed as it was, Curry's aid would be greatly appreciated and the cooperation would benefit all. At the time, however, Andrews said they were concentrating on organizing, training and inspection. He saw little point in "trying to put over an extensive air maneuver until we are welded into some sort of organization that had proven itself capable of carrying out its wartime missions."

Now he felt that time had come. The first two weeks in December was the period chosen for the exercise, the Air Force to be concentrated along the Florida Peninsula. In late October, he took the Tactical School up on its earlier offer. The major purpose of the maneuver was, he informed the school, to develop and test communication methods and procedures. Because of a shortage of personnel it could not be an around-the-clock operation; eight hours out of every twenty-four would have to be eliminated for rest and maintenance. The exercise was to be controlled by Air Force Headquarters and the problem must be worked out between December 2 and 10. The operation would begin with a previously prepared message. The situation foresaw a coalition of forces committed to an attack against the United States and the Panama Canal through the Caribbean area.[4] It would begin with the Air Force having just arrived in southern Florida and the enemy having reached the eastern Caribbean. There was one additional condition. Even though this was to be a problem concentrating on communications, there was only enough money available for a single teletype loop from Air Force Headquarters to the three wings.

The Tactical School went to work drawing up a plan to cover all phases of the exercise. It was the kind of cooperation many officers, along with Jack Curry, had long sought. Bert Dargue had told Andy that it was his purpose to stay in step with tactical progress and to provide a stabilizing influence through the school's graduates for actual operations. He, like Curry, was anxious to assist the Air Force and in turn to benefit by the practical application of GHQ Air Force maneuvers.

Widespread press reports indicated that the December exercise was a full-fledged success. Combat units got the word to move out on December 1, and in less than 23 hours, all were assembled on seven selected bases in Florida. Hap Arnold's 7th Bombardment Group made it from the West Coast in twenty-one hours. This led Andrews to inform Craig that, given the proper advance notice and if bad weather didn't prevent, he could place his force anywhere in the United States within twenty-four hours.

The force, whose squadrons were at full strength for the purpose of the maneuvers, numbered thirty-one bombers and twenty-eight attack, sixty-one pursuit and five observation planes, for a total of 125 aircraft.

Aside from testing mobility and communications, the Tactical School's war plan was focused on simulated combat missions and staff work in the field. It proposed that war had broken out between the United States and a coalition of European powers and that the Fleet was in the Pacific. The decision had been made to move the Fleet to the Atlantic, and the Air Force was to assist by keeping enemy air power away from the Panama Canal. Map reconnaissance had determined where the enemy would set up bases in the Caribbean. Observation aircraft on assumed missions would report these locations to both attack and bombardment units, and they in turn would hit them. However, neither the press nor the public knew what the game plan was, and the aircraft in question struck at simulated targets in the United States, with one exception.

An observation plane on a mission off the coast spotted an approaching steamship. Immediately, the ship was designated as a hostile aircraft carrier. Its position and course were given, and a squadron of B-10's took off to intercept it. In his later report to Craig, Andrews noted in passing that the ship was about fifty miles off the coast. In a more comprehensive roundup of the Air Force, Hugh Knerr gave the distance as approximately one hundred miles off the coast, wishing to draw attention to the over-water interception capabilities of the Air Force. Knerr stressed that "long-range missions requiring expert aerial navigation were performed by bombardment and aviation units," some of them covering between nine hundred and one thousand miles. He went into detail on a record-breaking over-water flight made by the Douglas OA-5 amphibian *The Big Duck,* the only one of its kind in the Air Force. The plane flew to Puerto Rico and then returned nonstop to Miami, a flight of over one thousand miles, in less than seven and a half hours, which broke the world's record for aircraft of its class.

Andrews' use of his pursuit offered food for thought and should have brought close attention by the contesting theorists at the Tactical School. He put to the test a new method of interception, worked out with George Kenney and others of his staff. All bombardment and attack missions were routed through an assumed antiaircraft intelligence net. The net's center was Miami, with reporting stations fanning out to around one hundred miles. There was really no such network, and bomber and attack planes when flying through the area were required to give position reports from time to time. Under these ground rules the interceptors had no problem in locating the invaders. But as the maneuvers went on, less and less information was transmitted, and that purposely delayed. Even so, the fighters were generally successful in making contact.

The method for scrambling was ingenious. As soon as a report of approaching attack was received, a siren was sounded. Pilots manned their aircraft and started their engines. Large signs mounted on the side of a truck gave magnetic course, altitude and estimated distance from the point of attack. The truck raced around the field with the information, and the interceptors were off. It took an average of six minutes from the sound of the siren for thirty-six planes to be airborne. A predetermined air speed and rate of climb were utilized so that Andrews and his staff could accurately plot and direct action. This was five years before the Battle of Britain; yet, with certain modification and without the benefit of radar, Andrews was employing techniques of interception to be used then. He reported to Craig that "although the pursuit made a very high percentage of interception, it must be realized that this could probably not have been done without the establishment of an extensive and complicated antiaircraft warning service."

Once the contact had been made, there had been no attempt to determine how well the pursuit would have done against the bombers. All that could be concluded was that the differential in speed between the pursuit and bombardment craft did not make it possible for the former to execute diving attacks. Aeronautical advancements down the road would prove the observation to be totally incorrect. "Hasty conclusions should not be drawn from these operations," Andrews cautioned.

In the key area of communications, very decided conclusions could be drawn. There was an extreme shortage of trained personnel and no provision in the GHQ Air Force table of organization to obtain them. Equally there was an "almost total absence of field radio equipment." It had been necessary in every instance to improvise ground radio stations from aircraft transmitters or "unsatisfactory commercial receivers inherited from the airmail. Not one unit was provided with ground radio equipment as provided in the tables of equipment."

The press and public knew nothing of the shortages. Newspapers, newsreels and radio broadcasters were out in full cry, reporting on the success of the operations. Before the exercises had begun it had been determined by the War Department and the Air Force that no mention of foreign targets or places would be made, and that even the existence of a war situation would be denied. Consequently, "No hint of these forbidden subjects appeared in print." All that the residents and vacationers in Florida knew was that the Air Force was out in force, and wasn't it grand to see all the planes flying about in formation? The pilots were all handsome, their planes sleek and formidable. The observers didn't know and most wouldn't have been too shocked to learn that the 125 planes collected for the maneuvers made up the entire modern component of the United States Army Air Force.

To further enhance the impression of Army air strength, at the conclusion of the exercise the 8th Annual All American Air Maneuvers were held at Miami. Andrews' combat squadrons were augmented by units from Maxwell Field and the National Guard, swelling the numbers to approximately 350 aircraft, but if what was observed in the sky was impressive, the underlying reality that prevailed was not. At the conclusion of the

Service Test, Andrews summed it all up in a single sentence to the Chief of Staff: "There are not sufficient commissioned personnel in the GHQ Air Force for active operations against an enemy."

Hugh Knerr in his overall report on the first exercise involving all three wings added, "It was evident throughout that, due to lack of well-organized base and shortage of maintenance and supply personnel and transportation, actual tactical operations could not have been carried out."

In short, against a real enemy the minuscule Air Force would have been hard put to carry out its mission. But who was a real enemy? Certainly not Mussolini, whose forces had gotten away with invading helpless Ethiopia in October. His planes had bombed and gassed defenseless civilians to a polite gavotte of protest by the British and the French. In spite of Hitler, not Germany either. Even if he had pulled out of the moribund League of Nations, so had the Soviet Union and Japan. The Saar had voted by more than 90 percent to be reincorporated in the German state, and Hitler himself was on record as having told the French Ambassador that he had nothing but peace in his heart for all mankind. As for Japan, at the moment of the Florida maneuvers a conference was being held in London with representatives from the United States, Britain, France, Italy and Japan to limit once and for all the size of each of the respective fleets.[5] The President's attitude was reflected in a letter to Jesse Straus, the U. S. Ambassador to France. To Roosevelt, the growing indication of an armaments race meant "either bankruptcy or war." But then he went on to maintain:

> Heaven only knows I do not want to spend more money on our Army and Navy. I am initiating nothing new unless and until increases by other nations make increases by us absolutely essential to national defense. I wish England could understand that—and incidentally, I wish Japan could understand that also.

The difference between Roosevelt's thinking and that of his Air Force Commander was that, to Andrews, that time had already come. Tactically, he was planning to conduct his next exercise, a winter test in February, to check on the ability of his men and the capability of his equipment operating in frigid New England weather. However, before year's end, important and far-reaching personnel changes were in the wind.

They began on December 21, when Andrews learned before the official announcement that he was being promoted to the rank of major general.

With the glad tidings came the news of other promotions and appointments. The day before Christmas, Secretary of War Dern announced that Oscar Westover had been appointed to succeed Benny Foulois, who had officially retired on December 21. The news came as a surprise and a disappointment to many airmen, who had been rooting for Conger Pratt, and if not Pratt, Danforth. The word around the War Department was that Westover was not Malin Craig's choice but Henry Woodring's. Secretary

Dern had been on a naval inspection tour of the Pacific when Roosevelt had appointed Malin Craig to succeed Douglas MacArthur, and since Dern did not know Craig that well, he was more inclined to make his decision with Woodring's strong recommendation. He believed, correctly, that Westover's thinking was in line with their own, that he would never buck the tide and was a stickler for following War Department policy whatever it turned out to be. Dern and Craig had conducted several days of meetings prior to the announcement, and it was obvious that the Chief of Staff had another choice in mind.

Reportedly, Craig had recommended Andrews' promotion, for it had been understood at the outset the commander of the Air Force should hold the rank of major general and would have in the beginning but for the law. Since Westover was also being raised to the same rank, Craig, having recognized the growing adversary relations between the staffs of the Air Corps and the Air Force, may have felt that with both men equal in rank and seniority a better balance could be maintained.

In his statement to the press, Secretary Dern added another reason for Andrews' advance. He said it was in keeping with the growing size and importance of the Air Force, "especially in view of the reinforcements which will be given it during the coming year by the addition of over 500 new modern combat planes under the army's air purchasing program." The statement was more pie in the sky than aircraft, and if the Secretary of War didn't know how completely incorrect his figure was, he could have asked anyone from Woodring on down to give him a more realistic appraisal.[6]

As for his promotion, Andrews had not known whether it would be forthcoming or not, for in the many private meetings he had held with Craig, the Chief of Staff had never given any indication of his intentions. The day after Christmas, Craig called Andrews to offer his congratulations, but, as "Maxwell" informed his father, "I am, of course, very much pleased and only hope that I can hang on to it, but it is as temporary as an April shower and at any time I may have a different point of view from the powers that be, I will be out."

There were other changes, some not all that well received and one that caught the recipient 'twixt heaven and earth. Hap Arnold learned that he was to be the next Assistant Chief of Air Corps. The choice made excellent sense, for as Jerry Brant put it, "I believe we have a very fine combination in Tubby Westover and Happy Arnold. Westover can do all the detail work and Arnold can sell our ideas to the General Staff, Bureau of the Budget and Congress." Brant's opinion indicated the differences in personality that tied into considerations that had to have been a factor in Hap's selection. Among pilots, there had long been three categories of airmen: those who were active fliers, those who had once been or were balloonists and those who had been ground officers of high rank who had transferred to the air arm during the war. Fortunately, most in this last category were gone by the end of 1935, but there were a good many officers in the second group, and although Westover continued to do a considerable amount of flying, he was placed in it.[7]

Hap, on the other hand, was recognized as an extremely active pilot and known to be an excellent one. This, again, was a reflection of his personality, and the announcement of his move to the second-ranking slot in the Air Corps was greeted with wide enthusiasm . . . except for one officer and one family. The Arnolds, from Pop on down, were stricken. He did not want to give up his command of the 1st Wing. It was the best command he'd ever had; it was right in the midst of what he loved most: operations. The family did not want to give up March Field and California, which they all loved dearly. The hell with the advancement! Arnold protested to Westover, and perhaps Tubby reminded him of the time Hap had written to say how much he'd like to serve under him. Well, now he was going to get his chance. Hap did the unheard of: he offered to be reduced in rank to lieutenant colonel if he could just keep his job at March!

Ten years ago the powers had literally thrown him out of Washington. Now they were demanding his return. He would go, but he would not like it, nor would anyone in the family!

There was another reason for Arnold's great reluctance in the shift. He would be leaving the Air Force for the Air Corps, crossing the barrier that now existed between the two camps. He had written to Andy: "I have to tell you again how sorry I am that I have been taken out of your organization. I hate to leave the command here and the Air Force."

For Andy's part, while he, too, was sorry to lose Hap, he wasn't all that sorry, because having him in Washington might do a great deal to smooth relations between Langley Field and the second floor of the Munitions Building. At least he hoped so.

On the basis of change, directly after Christmas, Andrews contacted Malin Craig and recommended that War Department Circular 7, which had established the tables of organization of GHQ Air Force, be amended so that Jerry Brant could be given his temporary star as CO of the 3rd Wing. He had also recommended the same promotion for old friend Harry Clagett, who, through Andrews' strong support, would be taking over the 1st Wing from Arnold.

Clagett's appointment was not appreciated by a number of his colleagues. Arnold didn't like him. Neither did Jerry Brant, and Andrews told Jerry he hoped he would put his personal differences aside and cooperate with Harry. Brant, who was known for his easygoing nature, was willing to give it a try. Others would not be so charitable. Clagett had a bullhorn voice and a hard-line manner with subordinates and equals alike that didn't go down well. But, Andrews saw in his friend qualities others missed, either because he genuinely believed they were there or, through his loyalty and the bonds of an old friendship, he secretly hoped they were. He described Clagett to Malin Craig as "a soldier with a flying heart." He was sure Harry would turn out to be an outstanding wing commander.

Andrews' own promotion brought, as it had the year before, a great outpouring of congratulatory messages. The telegrams and letters and phone calls came from all branches of the service and the Navy as well. Former classmate at the War College Colonel Charles Grunert cabled from the Philippines, "See that it [promotion] doesn't improve your golf game."

Trubee Davison, who in November had lost out on an election campaign and since his departure from Washington had been Director of the American Museum of Natural History, sent his heartiest. Hans Adamson, still with Trubee, informed Andrews that the former Air Secretary seemed "to thrive mightily on a diet of dinosaurs and habitat groups."

On the aeronautical manufacturing front the praise was sometimes coupled with business. Burdette Wright was looking for a way to get a service-test order on the Curtiss A-14. But he did manage to observe, "It is a strange world when we think of you and Johnnie and Elizabeth and me sweltering in little wooden houses at Kelly Field over 10 years ago and where our destinies have led us since."

Larry Bell did not reflect on the past; he was looking to the future. "Signed papers in Dayton yesterday," he wired, "to start program on multiseater fighter—proceeding full blast—make presentation to Materiel Division with another competitor 15 March."

But Claire Egtvedt summed it all up with a hopeful New Year's message. "It is very gratifying to know that you and GHQ Air Force, which you command, are being recognized not only by this significant promotion but by leaders of importance throughout the country who feel that your command means something real in national defense."

Broadcaster Lowell Thomas sent an autographed photograph of himself and an invitation for Andrews to share the microphone with him on the evening of January 13. Andrews accepted.

Although the interview with Lowell Thomas was brief, Andrews, through the broadcaster's leading questions, made some New Year's pronouncements that held U.S. air defenses up to the cold light of January. Uncle Sam was woefully unprepared. Of the 980 planes authorized for the Air Force, maybe by year's end they would have a third of that number. The recent concentration in Florida proved that the Air Force could assemble promptly, but, said Andrews, "we need landing fields. Without a base of organization to support operations a big concentration of planes is useless." He believed that the Congress and the public were beginning to recognize the necessity for an adequate air defense, and Lowell Thomas said that he certainly hoped so.

Amid all the changes at the end of the year there was one departure, somewhat sad and touched with bitterness. Since mid-August, Benny Fou-

lois had been on terminal leave, he and Elizabeth adjusting to life in Ventnor City, New Jersey. Oscar Westover had kept him posted on developments and had planned a farewell dinner in his honor at the Army-Navy
Country Club. It was to be a big send-off, but Benny Foulois said no. He
wanted no farewells, no eulogies sincere or otherwise, no gold-plated
watch. He simply wished to fly away. Accordingly, on the official date of
his step-down, he went out to Bolling Field and took his final flight. . . .

The old year went out on the winds of change and the new one came in
on the wings of hope. There was a similarity to the previous year's beginning, but whereas then everything was centered on establishing the new air
arm, now the external battle was in trying to foster its growth, while internally the conflict between the two staffs went through a series of smolder-
ing gyrations.

On growth, Chief of Staff Hugh Knerr's report of the first year of GHQ
Air Force operations offered a stark picture:

> The GHQ Air Force Headquarters is inadequately staffed.
> Hard-surfaced runways are essential to operations of modern military
> aircraft, and the lack of such runways is causing excessive deterioration
> and loss of valuable equipment.
> There is insufficient commissioned and enlisted personnel in the GHQ
> Air Force to care for any increase in the numbers of aircraft assigned.
> There are insufficient regular officers in the GHQ Air Force to properly
> staff, command, and train the units even at existing strength.
> The term of active duty of the reserve officer is insufficient in length.
> Etc.

As to the conflict between Air Force headquarters and the office of the
Chief of Air Corps, one of the root problems lay in the assignment and rotation of personnel. As Andy had put it to Hap before the latter's transfer,
"The basic trouble is that all personnel is handled by the Chief of the Air
Corps' Office and priority setup which places almost every requirement for
personnel above those of GHQ Air Force, particularly as to qualifications
of the personnel."

Andrews, working with Harvey Burwell and others on his staff, had
come up with a plan of lateral coordination between his office and that of
Westover's, a system essentially of cooperation in exchanging information
on personnel changes. Where such cooperation was forthcoming, the results had been satisfactory, but the instances had been all too few, and
Andrews hoped that Hap, knowing the Air Force side of the coin, was
going to be of help.

On the issue of corps-area control, which Andrews considered the most
serious internal handicap the Air Force faced, he finally won, with the assistance of Deputy Chief George Simonds, a solid victory. In September
1935, following his two previous attempts to get Air Force stations put on
an exempted status, plus a number of conferences at the War Department,

a special board was set up whose titled purpose was to make a Survey of Personnel Situation of the Air Corps. It was headed by Colonel William S. Browning, of the Inspector General's Department, and included Lieutenant Colonel Follett Bradley and Major Rosenham Beam, of the Air Corps. It became known as the Browning Board. When its chief suffered a heart attack, Bradley's influence became paramount. Appointed to study matters involving the personnel problems that confronted the Air Corps and the Air Force, the Board injected itself into the corps-area question and recommended in January the Air Force be freed from the arrangement.

Prior to the two-week winter maneuvers in northern New England in mid-February, Andrews sat down with his staff and they came up with their third study on the need for the Air Force to rid itself of corps-area entanglement. This recommendation came on the heels of the Browning Board submission. But the G-3 Section of the General Staff, still headed by Gen. John Hughes with his chief planner Colonel W. E. Cocheau, remained locked in its rigid mold. And as it had previously done, it rejected both reports, saying there was no need for a change, that air stations should continue under the command of field-force Army commanders.

Twenty

A Lack of Unity

Andrews refused to accept the turndown. He, Kenney and Bradley climbed on board his DC-2 and flew up to Washington to battle for acceptance of their proposals. It was a day-long session, with the three of them arguing "to have the base commander in the Air Force chain of command and to have the Wing commander not in command of any particular station but to have jurisdiction over all stations occupied by the troops of his command."

After all, Andrews pointed out, "the success of the Air Force is going to be judged by what we can do in the air and not how we command our stations. Of course, we realize the two go hand in hand, but the station is only the means to an end."

A point that may have had no bearing on the final outcome was that temporary as his rank might be, Major General Frank Andrews now outranked Brigadier General John Hughes, the Assistant Chief of Staff for G-3. Certainly the recognition must have appealed to George Kenney's sense of humor if not to Follett Bradley's.

When the meetings had ended, the three were greatly encouraged, for George Simonds, unlike his predecessor, Hugh Drum, made it apparent that he saw the soundness of their reasoning. There was no real point for the existing control. Corps-Area commanders, instead of being a help to the Air Force, could only continue as an unnecessary and divisive hindrance. Simonds' support was to gain Malin Craig's approval, and the convoluted connection, with the exception of court-martial jurisdiction, was at long last severed officially in May 1936.

When the round of sessions at the War Department was ended, the three dropped in at the Air Corps offices. There they learned that the advent of spring appeared to have stimulated the Congress. Seemingly successful hearings were being held before McSwain's committee on three Air Corps bills, the principal one concerning the increase of reserve officers. To Andrews this was vital, because he anticipated that by July 1, 1937, he would need about twelve hundred pilots and eleven thousand enlisted men to handle the proposed strength of his force. At the time, he had about half that number, and he knew the one practical way to get pilots was to call reserve officers back to active duty.

He stopped by Hap's office and found that his former wing commander was not all that happy. In fact, he had not been in a sweet-tempered frame of mind since the family had bid a tearful farewell to March Field, in early January. He had driven across the country like a man who believed that just because the family chariot didn't have wings was no reason it couldn't fly. Bee was sure he was trying to kill them all! Nonsense! He just wanted one thing clearly understood. They were not going to live in Washington. He hadn't forgotten one penny of the money owed him by those utility snakes.

"I don't want you to sign any leases. I want you to get a place to rent from month to month. I won't last in that city three months. Don't get tied up in anything. I will have no official social life. I refuse to buy tails and white tie. I will not go to parties where I have to wear white tie. I don't like the life there!" And so on.

"All right." Bee was in full agreement. "We'll follow through on that." But she wished he'd realize he couldn't pass the truck ahead by driving over it.

They had finally arrived—safely! As soon as they'd checked in at the Brighton Hotel, Oscar had come to greet them, so happy they were there, and to bring them some news. Mrs. Westover was not well. She could not act as hostess at the many official functions. Their daughter Patsy had been helping out, but now that the Arnolds were here, he would ask them to take on the duty. And as a matter of fact, tomorrow evening. . . .

Now, three months later, they were about to leave the Brighton, and no one had indicated that Hap Arnold was going to be ordered out of the city as in days of yore. They were in the process of moving to a six-bedroom house on Bradley Lane, in Bethesda, Maryland, renting a home owned by Admiral Van Kuren. The expenses and social obligations were damned awful! And Bee was doing it all, and with no funds for entertainment they were going broke!

Andrews could only smile and commiserate. Johnny was a great one for social functions and club dances, which he did his best to avoid.

Before returning to Langley they all sat down with Westover and discussed the day's progress. Oscar had read Andrews' February report, and obviously Simonds was going to want his comments on it. Westover's thought was that in order to clarify further the chain of command, he was going to recommend—as he had previously done privately to Craig—that the GHQ Air Force be placed under the Chief of Air Corps. This had been Foulois's and his aim since before the Drum Board. Andrews carefully said he agreed with the idea in principle, but in reporting to the Chief of Staff, he ran his own show as to training and tactics. Privately, he was not anxious to give up his prerogative of being able to go directly to Craig and Simonds in matters concerning the Air Force. Still, integration with the Air Corps under a single head should bring about a stronger organization in the long run. But, as he wrote to Harry Clagett, "I am afraid that the plan will not get to first base because it puts too much power in the hands of one branch chief, and may be regarded by many as a step towards a separate Air Corps."

Even so, Andrews had no sooner returned to his headquarters than rumors began floating in that the Air Force was going to be put under Air Corps command. Bob Olds, on a round of inspection trips to Mitchel, Barksdale and March fields, picked the rumors up from Tony Frank, Jerry Brant and Clagett. Andrews put them all straight that no such thing had happened. To Knerr, Kenney, Bradley, Olds and most of the others on Andrews' staff, there must have been a certain amount of ironic contradiction in the Westover proposal. He, of all of them, was not seeking a separate air arm. They could certainly see how his proposition would be interpreted by the General Staff, even if he couldn't. It is probable they were also not anxious to have it accepted because of the nature of the man, and they were in doubt that such a change would eliminate the polarization that had developed between the two staffs.

Conversely, at Maxwell Field, where theory at the Tactical School often took the form of reality, the belief was that keeping the Air Force apart from the Air Corps was not the way to build the road to independence.

In April there were two more conferences with the General Staff on exempted status for the Air Force and on Westover's recommendation. Then it was left to Malin Craig to make up his mind. To aid him, he asked Andrews to give his opinion on the Air Corps Chief's plan and to defend his own position on both proposals. The key sentence in Andrews' response to the first was: "A continuation of present divided control is an unjustifiable and unnecessary violation of one of the basic principles of war—Unity of Command."

The heart of his critique on Westover's bid was also contained in a single sentence: "With reference to the recommendation of the Chief of the Air Corps to place the GHQ Air Force under his command, and Air Corps bases on exempted station status also under his command, it is believed that this is an unsound organization in that it is an organization for peace and not for war. . . ." To support his contention, he quoted both the Secretary of War's annual report for the previous year which described the combat arm as a war-ready force, and a basic War Department

Training Regulation, stating essentially the same thing: That in war the GHQ Air Force would be under the command of the Chief of Staff and not the Air Corps.

Westover was very upset over Andrews' position of nonsupport, and there is every indication that some harsh words were spoken. When Andrews had submitted his February proposal and the Browning Board report, they had been received first by Hap Arnold, who had given orders that substantiating material be prepared to accompany their transmittal to Simonds, at the War Department. In other words, strong support had been given by the Air Corps staff. As Westover saw it, the reverse kind of cooperation had been forthcoming when he had attempted to take the matter a step further. But, to Andrews, it was not a step in the right direction, and thus the disagreement. That it was a real one was noted by a cryptic reference in Johnny's diary for April 26: "Andy worried about fight with Westover."

It was not the sort of worry he would be inclined to discuss with others. He sought its resolution through reasonableness and not acrimony. Personalities aside, he told Oscar what was wrong with his plan. It did not take into account that in time of war, the Office of the Chief of Air Corps would be fully occupied with Zone of the Interior duties. If Westover expected also to command the Air Force, a top-heavy dual staff organization would have to be established on top of the already functioning Air Force staff. Aside from all the logical reasons against such a move, where would the personnel come from, not to mention the finances?

In the end, Andrews and his staff had won a double victory, but at some cost to harmony.

How much that lack of harmony was reflected in the hard blow that hit Andrews early in May is not measurable, but in view of it, an obvious conclusion is apparent. Westover's office requested that orders be cut detailing Lieutenant Colonel George C. Kenney as an instructor to the Infantry School, at Fort Benning, Georgia. For those who believed that quick-thinking, fast-talking George exerted too much of an influence on his boss, there wasn't any more pointed duty to separate them. Andrews sent an immediate protest to Craig, pointing out that G-3 was the "heart of the GHQ Air Force staff organization at this time, where there are still so many problems of organization, training and operations to be worked out."

Kenney, he said, had demonstrated his ability to head the section. Then he pointed out the turnover in G-3 section in less than a year and a half: Major Eugene "Gene" Eubank had been relieved due to eye trouble. His replacement, Major John E. Upton, had been ordered to Fort Leavenworth. The remaining officer, Major Clements McMullen, was at Walter Reed with a heart problem. Kenney was about all that was left. As for

Kenney, he had worked long hours and many nights to keep the work of his section up. Wrote Andrews:

> His training and experience are an invaluable asset to the GHQ Air Force, and his relief at this time would impose an unnecessary handicap at a most critical period in our existence. He is a loyal, efficient, well-educated officer with war experience that is a particular asset to this headquarters. He does not desire the detail to the Infantry School.
>
> I feel I cannot too urgently recommend his retention in his present position, not only in justice to the officer himself but in justice to this headquarters, and for the best interests of the service.

Craig's reply was a sympathy note, so sorry but the Chief of Infantry had made a specific request for Kenney and the Chief of Air Corps could find no one else eligible for the job.

When it came down to how the Chief of Staff viewed the importance of the two commands, his indication was clear enough when he wrote: "I realize this is a serious blow for you to lose him at this time, however, I believe he can be more easily replaced on your staff than at the School, consequently I do not feel justified in making any change in the assignment." And then in his own hand, he scribbled, "Ask Westover for a competent replacement. If the replacement does not fit the bill, I will give you another."

Andrews went up to Washington to see Craig and got the Chief of Staff to agree that he could have Kenney back at the end of the year, when the course at Fort Benning was over. In the meantime, McMullen had returned from Walter Reed with the news that his heart was still beating and Andrews made him his G-3 and asked for no replacement.

Directly after receiving word that Kenny was going to be removed from his staff, he sat down and got off a memo on the continuing bugaboo of personnel policies. He recommended that machinery be established to ensure that GHQ Air Force have the opportunity to present its views and requirements on personnel to the Air Corps *before* the actual assignments were announced or ordered. He referred to the previous recommendation in the Service Test report submitted to the Adjutant General, that a permanent board of officers be appointed in which both camps would have equal representation.

For the moment, there was no response. In the latter part of May he flew out to March Field to see how Clagett was getting along. Harry didn't seem to have gotten the message straight on the new chain of command and had brought the Wing Headquarters into the Post Headquarters, which was exactly the opposite of what Andrews wanted. It had taken Conger Pratt and the 2nd Wing almost a year to get the kind of organization desired.

He also went over the rising possibility that one of Clagett's bomb groups would be called on to bomb a naval target at sea. The consideration had been rumbling around for some time, the test idea said to have been FDR's. The latest rumor was that a couple of destroyers rigged with radio control would be the targets. As far as Andrews was concerned, they were not the kind of targets that were meaningful. Destroyers would not play much part in attacking a coast. Battleships and carriers would. He knew it was unlikely the Navy would agree to risking evidence of its vulnerability from the air. Still, he told Clagett to get with it in practicing bombardment from all altitudes. Information on the maneuver was to be kept from the press; both sides were in agreement on that.

Upon his return to Langley, Hugh Knerr informed him that Congressman J. Mark Wilcox had won a hard-fought primary in Florida, and Andrews immediately sent him a note of congratulations saying, "The air defense of this country could not afford to lose in Congress such an able friend as you have proven to be."

Wilcox responded that he had told Knerr in Miami his chief desire in being reelected was to be able to render more effective service in building up an adequate air force, and he looked forward to seeing Andrews on his next trip to Washington. The connection was to grow in importance.

It was during this time that the writing team of Hap Arnold and Ira Eaker broke into print with their coauthored book *This Flying Game*. It was noted that while the authors' prose waved no red flags, on the subject of strategic bombardment they established the Douhetian doctrine by attributing it to European theory and tactics: "The outlined missions of attacking bombers all tend toward achieving that one end desired in any war —breaking down the will of the people." On the critical tactic of bomber escort, they took the position, being advanced so glowingly at the Tactical School, that bombers could handle attacking pursuit with ease.

A familiar and much less publicly noted document was issued in June by the War Department. It followed the same old Army theme. Secretary Dern had appointed yet another board to look into the idea of standardizing airplane design. Three of its five members were ground officers, and its two Air Corps members, Lieutenant Colonel B. Q. Jones and Captain Lowell H. Smith, had never been accused of being strategic thinkers. It was called the Bryden Board, named for its Chairman, Colonel William Bryden. The collective findings were another put-down of the "romantic appeal of aviation" that foresaw superbombers eliminating an enemy's industrial resources, his will to fight, etc. The members felt that more attention should be given to the ground forces and to "smaller, more economical and limited-performance types of aircraft" to assist them. They could see no reason whatever why a policy of standardization similar to the automotive industry could not be instituted for each type of aircraft on a two-year basis.

If nothing else, the report indicated the tugging forces of contradiction at work. The ground forces were essentially beggared and desperately needed modernization, but trying to put a fixed stamp of standardization on an industry that was figuratively soaring in its development was hardly a solution to be accepted by air leaders. In the two to five years it took for a military aircraft to reach the flight line, it could already be obsolete. That there was recognition of aviation's expanding reach by some in the War Department was evidenced not alone by the acceptance of Project A (XB-15) and then the go-ahead to procure a handful of B-17's, but also by its previous approval of Project D. This was a proposal for the construction of a still larger experimental ultralong-range bomber.

In October 1935, the War Department had signed a contract with the Douglas Aircraft Company to build the model but made clear this did not indicate a change in thinking or policy on long-range bombardment.[1]

In the end, it would be the personalities in power who would make the decisions. The contradictions were but manifestations of a tug-of-war in which the airmen were outranked, outnumbered and outpulled, but through keen intellect and determination fought to adjust the imbalance. For the moment, they seemed to be making some progress, and although *This Flying Game* would change no minds in the War Department, it did have public appeal and went into a second printing. Originally, Amelia Earhart had been approached on writing the foreword. She had tried, but as she had told the authors, she was so unalterably opposed to war she didn't think she was the right person to do it. They reluctantly agreed after reading her submission, and Donald Douglas was enlisted.

In June, while still at Maxwell Field, Ira Eaker gained some publicity of a different sort. In May, he had with the blessing of Bert Dargue and John Curry approached Arnold and thus Westover on the idea of making a transcontinental blind flight in a P-12. The purpose was to demonstrate flight training in extended navigation using radio and instrument aids. Two planes would execute the flight, Eaker under the hood and fellow student at the Tactical School Major William E. "Bill" Kepner in an accompanying chase plane. Approval was given, and on June 3 the two took off from Mitchel Field.

From Mitchel to March fields "and just over the Pacific Ocean," there were nine landings, beginning with Bolling. Eaker was under the hood on each leg from directly after becoming airborne until they were over the landing stop. With refueling and rest, it was a four-day venture. On one leg over Texas they ran into real instrument weather, and Eaker instructed Kepner to stay in close and he'd fly him through the mess. Ira's capabilities as an instrument pilot had become well known at March Field when, in the same fashion, like a mother duck he would guide the young and newly winged in neat formation up through the clouds. Now he made a new record for himself and in so doing demonstrated how far the Air Corps had advanced in its ability to outfly the weather. And if anybody wanted to think about it, what could be done in a pursuit could certainly be done in a bomber.

Andrews was thinking about it. So was George Kenney, even though he was slated to leave for Fort Benning at the end of the month. Captain John F. Whiteley, on Joe McNarney's staff, observed that Kenney was helpful in building up the image of flying leadership by organizing flights to break existing world records. He wanted to keep Andrews in the limelight. Whitely believed it had been Kenney's idea that had engineered Andrews' previous record breaker, and now he had an even better idea. Whether it was inspired as a result of Eaker's transcontinental hop is not known.

The idea was to make a nonstop flight from San Juan, Puerto Rico, to Newark, New Jersey. The plane to be used was *The Big Duck,* the Air Force's twin-engine Douglas amphibian. In a previous nonstop record flight it had flown from San Juan to Miami; the newly projected course would double the distance.

Andrews signed on John Whiteley as his copilot. Lieutenants Hugh F. McCaffery and Joseph A. Miller would do the navigating. Corporal John McKenna was Crew Chief, assisted by Private First Class Ralph A. Miner, and Private Charles J. Archer was Radio Operator.

The official purpose of the flight was to test long-range observation and navigation equipment and to look over landing and servicing facilities in Puerto Rico and possibly St. Thomas. The day before departure, Major General Frank R. McCoy, II Corps Area Commander, arrived at Langley as Andrews' guest. He had been invited to come along as a passenger, ostensibly to inspect an Infantry unit in Puerto Rico that was under II Corps jurisdiction but also to show him how the Air Force could manage over-water flights.

On Wednesday, June 24, Andrews lifted *The Big Duck* off from Langley in a rainstorm that stayed with them to Jacksonville, Florida. After an overnight in Miami, they went on to San Juan to be greeted upon arrival by the island's Governor, Blanton Winship, who years before had been one of the judges on the Mitchell court-martial. Andrews spent the next few days looking over possible field sites. Then he prepared for the record attempt. The early morning takeoff in a choppy sea was long and difficult, the *Duck* reluctant to get up and go.

Once airborne, over the wave tops, celestial navigation was the order of the day. Since the *Duck* flew with a pendulum motion, McCaffery and Miller had their work cut out. About four hundred miles from San Juan they hit a line squall, and since there was no getting around it, Andrews bore into it. At five hundred feet they had ten minutes of battering turbulence and blinding rain. Then they were through, and for the next twelve hundred miles, although Kenney radioed foul weather ahead, they had nothing but bright sunshine and salt spray on the windshield. "We did not see a sail or a bird—nothing but ocean," observed Andrews.

There was one problem: strong head winds. McCaffery reported their ground speed was only ninety-five knots. There would be no making Newark Airport with the fuel aboard. Andrews informed McCoy, who expected to be dropped there, they would be landing at Langley instead and he would have his DC-2 standing by to take McCoy the rest of the way. Radio Operator Archer informed the War Department accordingly.

It was obvious in spite of the head winds that they were going to break a distance record if nothing else, and Conger Pratt decided there should be a fitting reception. The press was alerted, and Lieutenant Colonel Arthur "Gilkie" Gilkeson took off with twenty-five planes to fly out and escort the *Duck* in from Cape Hatteras. Somehow in the vastness of the big blue sky, Gilkeson and his pilots failed to make contact. Andrews' navigators had estimated their time of arrival at 5 P.M., but they arrived a half hour earlier, having been in the air for just over eleven hours.

Along with Pratt, Weaver, Kenney, Knerr and a good-sized reception committee, Andrews' DC-2 was waiting. General McCoy's baggage was off-loaded, and he just had time to pose for a news photo before he was on his way home, arriving in daylight, thoroughly impressed with how far and how straight one could travel by air.

In a letter to Blanton Winship, Andrews indicated the underlying purpose of the entire effort:

> With the delivery of our new long-range equipment, beginning early next winter, I am planning to use this route over which we pioneered yesterday as a navigation test for our bombardment and long-range observation personnel stationed on the Atlantic Coast, and it will be valuable to have our GHQ Air Force personnel acquainted with operating conditions in the important Caribbean area.

Boake Carter gave the flight a big play, and Andrews in a letter of thanks told him: "This is the forerunner of flights which will become more or less routine when we get our new long-range equipment."

If June was a month of new aerial records with an aim of looking ahead, July was a month of more earthly change, some of which was sudden, unexpected and divisive. In a personnel shift, Harry Clagett lost his job. The news from Washington took both Andrews and his Wing Commander completely by surprise, and Andrews went directly to Craig to protest the axing.

The rumor was that in March, Clagett had received a poor efficiency rating from the IX Corps Area Commander, under whom he had been serving. This was not so, for at the time Andrews had called the report to Craig's attention, saying he hoped the Chief of Staff would find time to glance at it, and then adding, "I am still confident that Clagett is going to prove one of our outstanding Wing Commanders. . . ." Someone had thought otherwise, and the rumor was that Hap Arnold had been behind the move.

In taking command of the 1st Wing, Clagett had complained about the laxity of the whole wing in submitting correspondence reports in their proper form. Many reports had to be corrected, which to Clagett showed either "indifference or ignorance." This was a reflection on the previous wing commander. The complaint also offered insight, for while the problem was one of detail and not all that horrendous, Clagett had built it up. Such an officer could not have been all that popular with the troops, and no doubt the word got back to Washington and to Arnold, who had little regard for him in the first place.

However, when Andrews sat down with Craig over the sudden uprooting, the Chief of Staff informed him that Clagett was going to be replaced for having disobeyed a War Department edict. He had permitted retired General Malone to ride as a passenger on a military flight to the East. Without prior approval from the War Department, this was in violation of regulations and had been used as the excuse for the dumping and demotion of Harry Clagett. While not saying so directly, Craig implied that the move had been politically motivated. From whence the political thrust had come was not explained, but Clagett at least had no doubt that it was Arnold who had orchestrated his dismissal.

Whether the politics of Clagett's relief were centered solely on his own personality or the same political desire to replace him with Lieutenant Colonel Delos C. "Lucky" Emmons is not known. But Emmons, who was in command of the 19th Composite Group, in the Hawaiian Islands, received the unexpected word by cable and wrote to Andrews on July 1: "My appointment came as a great surprise to me, and I am delighted to have the 1st Wing. I would choose that above any job in the U. S. Army." He and his wife, Billie, were anxious to get started.

Emmons, at forty-eight, was a vintage airman who had graduated from West Point in 1910. He had come to flying via the Infantry, earning his JMA rating in 1917. During the war, he had gotten to know Hap Arnold well and had served with him both in Washington and then on the West Coast, while in command at Mather Field. He had also been in on the testing of the Kettering Bug in Arizona and later spent five years in the Engineering Division at McCook and Wright fields. Somewhere along the way, he had picked up the nickname "Lucky." Maybe it was the night Billy Mitchell decided they should work on formation driving. A whiskey front had passed through the Dayton Country Club when Emmons, Mitchell and another officer decided it was time to go home. We will fly in formation, announced Billy. Emmons would be on his left, the other officer on his right, and he would hold to the streetcar tracks. It seemed like a grand idea. The country club was a goodly distance from town, and it would give them time to perfect their technique. Off they went, barreling through the night, accelerators to the floor, hitting at least sixty. Fortune smiled, for at that hour there were no other cars on the road and the streetcars were all safely tucked in their barn for the night. Later, Emmons was to reflect, "I've often thought what damned fools we were!"

In spite of the episode, Delos was not really one of Billy's boys. It was not his nature to become involved in causes. He had followed in Ira Eaker's footsteps, serving as Executive to Air Service Chief General Mason Patrick and then as Executive to Trubee Davison. At the time, he had gotten Mitchell's goat by sending back to him some badly written reports for rewriting, but politically he was careful to keep clear of the rising conflict. This was not so much a matter of luck as it was of personality; a year at Harvard Business School was suited to the orderliness of his thinking.

That he could think and act fast was proved the day Frank Lahn had inadvertently come within a prop spin of running him down in a DH-4. At the last second, Delos glimpsed the approaching juggernaut, its propeller about ready to turn him into hamburger. Then he made his leap. The mud was soft anyway, and perhaps that was the point at which his friends began to call him Lucky.

Or maybe it was the episode with Mason Patrick. The General paid a visit to McCook Field and then wished to have his pilot, Billy Streett, fly him over to Wright. He had taken off his helmet and was talking with Emmons and Colonel Thurman Bane, the field CO, when Streett fired up the DH-4's Liberty. As neatly as an Indian's scalping knife, the prop blast unhinged Patrick's toupee, and it went flying away in the dust like a tumbleweed. "Get it! Get it!" shouted the distraught Air Service Chief, his hands clamped over his naked skull. Those who were not convulsed with laughter at the unfrocking sought to obey. Major Delos Emmons had joined in the hunt, but he had not laughed even when Patrick had quickly pulled on his helmet. It took an hour to locate the lost pelt, and in that time Emmons' solicitousness had made an impression on the General that later stood Lucky in good stead.

Over the years, there had been a kind of ease and smoothness to his career not found in those of his contemporaries who were out to battle for both the tactical and strategic aspects of air power and were willing to be counted in the front ranks in the fight for independence from the Army. Efficient, well liked, an officer with all the attributes of a long career served in many capacities, he did his job well and made no waves.

It was Mary Clagett, Harry's wife, who was out to make waves, arriving at Langley on July 3, ready to single-handedly take on the War Department, protesting her husband's relief. Andrews was able to calm her down, no easy chore. After she had departed, Harry flew in, understandably upset. He was not only losing his job but his star as well, the meaning of temporary promotion hitting home with a vengeance. As he had with

Mary, Andrews stressed patience. Maybe a battle had been lost but not the war. There would be new vacancies coming up. In the meantime, Ralph Royce was slated for a new command, and possibly Andrews could see that Clagett would be assigned to take over the 1st Pursuit. Later Andy would write Harry, who had left his glasses behind, telling him, "Everywhere I go I hear expressions of regret that you were not reappointed," which didn't change the situation one bit.[2]

Aside from all else, Clagett's precipitous removal illustrated how little control Andrews had in the final selection of his personnel. He had managed to put Clagett in as wing commander against considerable opposition, but the same opposition had managed to get rid of him.

Fortunately, there were other changes of a more enjoyable sort in which he thoroughly concurred. Tooey Spaatz and the Command and General Staff School had parted company, mutually anxious to be rid of each other. It was rumored that Tooey's irreverence would prevent him from receiving his graduation certificate, but reason prevailed and he happily departed for Langley Field to become Conger Pratt's Executive Officer, of the 2nd Wing.

There were other arrivals at Langley that summer, although of a far more temporary nature. A gathering of old hands to play a few hands and cuff the ball around the fairways included Hap Arnold, George Brett, Jim Chaney and Krog Krogstad. Johnny took rueful note that Andy and the men left her to play poker. She would like to have been dealt in. Another welcome guest was Helen Hickam, who had been spending considerable time with Tony and Hazel Frank at Mitchel Field. She came to Langley for an extended stay, and throughout the summer and fall there were the ebb and flow of other visitors, official and otherwise.[3]

The family highlight was when Andrews' father, his sister Josie and her husband, Gillespie, and children arrived for a few days. This was a big moment for the elder Andrews, a chance to see his son's command and to meet those who staffed it. It was a big thing for Gillespie, too, because he insisted on a poker game, possibly having heard of Tooey Spaatz's expertise. He went away seventy-two dollars the poorer but said he'd had a fine time losing his money to some real sharks.

That was the social level, in which on the surface at least a style of ease and gentility and good-fellowship prevailed. It didn't mean that everybody was all that fond of everybody else at Langley. Knerr's taciturn absoluteness, Bradley's air of condescension, McNarney's withdrawn manner, Weaver's pompous tendencies, all were quirks of character that at times produced their own frictions and disagreements. They were representative, essentially minor and relatively unimportant in that under Andrews' leadership the bumps and warts of personality were not permitted to intrude on the larger issues.

They were issues that in the second half of the year began to take on a sharper edge, a new momentum and a renewed tone of hostility. For over a year, Andrews had been conciliatory in his relations with the War Department, as he and his aides had worked to develop a combat air arm worthy of the name. Reason had finally prevailed in the matter of Air Force control of its own bases, but G-3 had refused to let the matter lie, and its organization section had come forth with a new plan that was more unwieldy than the original corps-area arrangement. Andrews had had enough. No doubt his response to Simonds on the G-3 recommendation contained input by Knerr and others, but this was Andrews with the gloves off. He pointed out that the report on the service test of the GHQ Air Force had been submitted on March 1, 1936, and now, nearly five months later, comes this G-3 study that would accomplish nothing.

> It avoids or postpones such fundamental issues as Tables of Organization, increased personnel requirements in the GHQ Air Force, in the Air Corps, Quartermaster Corps, Signal Corps, Ordnance Dept., Medical Dept., etc., and allied problems which must be solved to accomplish the War Department programs of 980 airplanes welded into a complete, integrated, M-day unit of our National Defense.

He then went to the heart of the matter, and in a single page laid out the reality of what was going on as he saw it:

> When the War Department abolished the Office of Assistant Secretary of War for Air, approved the Drum Board and the Baker Committee's reports, created the GHQ Air Force to be built up to 980 airplanes, it thereby established certain definite War Department policies in connection with our air defense and assumed definite and clear-cut responsibilities to Congress and the people of the United States for this air defense.
>
> A great deal of time has elapsed since these things have taken place. The G-3 division of the General Staff during this time has been in effect the agency of the War Department to plan and recommend action to effect these policies. This headquarters had been unable to discover in this study or in any other action by the organization section of the G-3 division, or in conferences in this section, any cooperative effort on the part of the section or any plan or intention of a plan in the G-3 division, other than in the training section on current training questions, intended to solve the many problems in connection with the development of the GHQ Air Force, or the Air Corps as a whole.
>
> The Commanding General GHQ Air Force feels that he would be deficient in the performance of his duty if he delayed any longer in bringing this to the attention of the Chief of Staff. It has been apparent that fear has been entertained by the G-3 section that the Air Corps, including the GHQ Air Force, might grow to over-shadow other elements of the War Department and that this is in part responsible for the attitude of this division, but whatever the reason concrete results have been inadequate. It is the considered opinion of this hqrs that the greatest ultimate advantage

would accrue to the Army if for the time being the War Department
would make its major effort in National Defense in the maximum development of air power. Air power development is a national defense necessity
and is demanded by the people of this country. It can be made the spear
head of appropriations. Ultimately the dominant influence in National Defense will inevitably accrue to the service having the greatest functions and
responsibilities, and the budgetary means to meet those responsibilities.

If the War Department is to seize its opportunity to take a commanding
position in National Defense for the Army, it must do so now. Further
delay will surely result in air defense passing to the Navy, or a separate Air
arm.

That the memo shook up a few bodies was apparent. Embick, Chief of
War Plans, called Andrews and told him his reply on the G-3 plan would
be passed on to General Hughes and his staff *minus* the two critical paragraphs. The offending words were studied by Malin Craig, who had no
comment. Major Robert E. Eichelberger, in Craig's office, suggested in a
letter to Andrews that he might wish to discuss the expunged material with
Craig at some point.[4] Whether he did or not, and it could be assumed that
he did, the hard-nosed approach by Andrews marked a change in attitude
by the Air Force Commander. In their discussions, Craig had always considered him reasonable and willing to cooperate, but less so once he had
returned to Langley and met with his staff.

In this case, Andrews was speaking for himself, and he continued to do
so on every level affecting his command. "I am not at all happy about the
apparent Air Corps trend now-a-days," he wrote Robbie Robins, "but will
do my best to advocate to the War Department the things we think necessary for adequate defense."

He was referring in part to Woodring's decision to cut research funds for
the XB-15 and divert them to buying spare parts; a continuing failure to
get funds to purchase oxygen equipment; a War Department decision to
buy five autogiros out of Air Corps Research and Development funds, the
craft to be used by other branches of the service. On this last, Robins had
told him that the proposed cut in the R & D program was so severe that it
would mean curtailing the Materiel Division's four major undertakings:
Projects A and D, a multiseater fighter and a new, twin-engine attack
plane. Robins had suggested that money be transferred out of procurement
funds for R & D. "If we don't do this," he said, "the equipment coming
out in 1938 and 1939 will not be furnished with the proper propellers, instruments, and other accessories." As Robins saw it, the situation was the
most serious they had faced in a long time.

Andrews agreed the whole thing sounded suicidal, and he promptly sat
down with Westover to see if something couldn't be done. Hap had assured him the money would be forthcoming for the oxygen equipment, but
so far there had been no result. Oscar knew the situation and so did Hap,

firsthand, but Andrews said that neither his pursuit nor his bombardment units were capable of high-altitude operations, and with an important test coming up they'd better be, and no alibis. It was up to them to see that his planes were properly equipped.

In a later letter to Hap, he made a pointed observation on personnel: "Found March Field in bad shape, due to two changes in Wing Commanders, change in Group Commanders, change in practically all Squadron Commanders and almost every other key position within less than a year. Personnel policies that cause such a situation certainly need to be remedied."

In an attempt to eliminate the continuing factionalism between Westover's headquarters and Andrews' own, it had been agreed back in February that Harvey Burwell would come up from Langley and act as a part-time liaison officer, cooperating on personnel and other changes. But the attempt didn't work very well. Harvey was probably not the man for the job, and he was pretty thoroughly ignored, not even able to get secretarial help.

In August there were two significant political deaths, one no surprise and the other totally unexpected. In the latter case, big John J. McSwain, sixty-one years old, overworked and overweight, was stricken on August 5 while riding on a train. An honorary colonel in the South Carolina National Guard, he had been spending some time with a Coast Artillery regiment at Fort Moultrie and was on his way home to Columbia. The Air Corps knew it had lost a very staunch supporter in the Congress.

Assistant Secretary of War Woodring was quoted as saying that the late Chairman of the House Military Affairs Committee had made a "very substantial contribution to national defense" and that his death "was a national loss." Malin Craig declared that McSwain's death came as "a great shock to the officers and enlisted men of the army."

Oddly, the New York *Times* obituary, which described the Congressman as an advocate of peace seeking to join the Army and Navy departments into a single unit, made no mention whatever of his interest in and efforts on behalf of the Air Corps. Airmen would have to be content with their own knowledge of that.

Three weeks later, Secretary of War George H. Dern, who had been seriously ill for some time, died, on August 27. Henry Woodring had been Acting Secretary since early spring, but it was not expected that he would be chosen by FDR to succeed Dern. When, by September, the President had made no announcement, it was generally accepted that he would wait until after the November election before making the appointment. However, near the end of the month he was informed by his Attorney General, Homer S. Cummings, that it was the law that he fill a cabinet post within thirty days. Roosevelt promptly wired Woodring that he was going to name him Secretary of War on a "temporary" basis. Woodring was happy to accept. He had been filling the number one and two slots in the War Department off and on for a long time, and holding on to them was just fine with him.

Significantly, on the day that Dern died, Hap Arnold telephoned Andy to tell him that a meeting was scheduled for the next day at 10 A.M. in Simonds' office. They would be making a major determination on the military characteristics of aircraft. Woodring believed procurement should be based on the missions assigned rather than by engineering possibilities and capabilities. The same kind of thinking prevailed in the War Department, particularly G-4, under General Spalding.

To both Andrews and Arnold the signal was a large red flag, and Andrews wrote to Delos Emmons, "My main mission in Washington Friday is to appear before a War Department Committee to defend long-range bombardment. I think that the fate of the Army Air Corps in national defense is going to hang on this conference, and I feel that we have enough data to convince the War Department that we must continue with long-range bombardment."

Hugh Knerr was on leave, but at this conference, which included the backfield of the General Staff, Westover and Arnold appeared with Andrews. They were pushing for eleven B-15's and fifty B-17's, but they were pushing against a stone wall. The original thirteen B-17's would be procured but, as they were soon to be informed, until the situation warranted, there was no need for more of the types of bombers they were advocating and no more need for experimental types of the Project A and D types. The General Staff preference, particularly sought by General Embick, was the Douglas B-18. It fulfilled all the needs and requirements of coast defense. Its purchase could be justified from the point of view of cost, maintenance, operations and so forth. Westover was directed to concentrate his efforts and funds on "acquiring aircraft of reasonable performance rather than nothing as a result of reaching for an ideal." Behind this reasoning stood Henry Woodring, the promised figure of 2,320 planes dancing like sugarplums in his head. Little more than a week before the meeting, he told Hap Arnold that in his opinion the new-type combat trainer should be classed as a combat plane.

Shortly after the conference, Andrews took Malin Craig aboard his DC-2 command ship and together they made a tour of the Air Force and its major bases. There was no doubt that Craig was genuinely impressed with what he saw. He returned to Washington at the end of September, referring proudly to his Air Force. But because of a decision by Woodring to save money, his Air Force was not going to be permitted to participate in the annual All American Air Races, in Miami.

Andrews had asked him to get the Secretary to reconsider. The Navy would be there. The public reaction would be unfavorable if the Air Force failed to show. He knew Westover and Arnold were in favor. He suggested a composite group of attack, bombardment and pursuit be entered. He felt that a change this year of long-established policy would be a mistake.

Craig's reply was that he didn't think the Secretary would change his mind, having given the "impression that we gain nothing by going down there in a professional or official way." For the first time in a long time, the Air Corps was not represented, Woodring believing the cost was not worth the public relations. The Navy, quite naturally, thought otherwise, and its growing air arm was well represented.

When it came to the Navy, before the year had ended, joint air-sea maneuvers were conducted off both the East and the West coasts. In both instances the targets selected for practice bombing were not ships but target sleds four feet by eight. Neither Andrews nor anyone involved on the Air Force side felt that the maneuvers offered a real test of what long-range strategic bombardment was all about, nor was the public let in on the action.

However, an action concerning the Navy at the Air Corps Tactical School brought Oscar Westover to Maxwell Field in somewhat of a heat. Larry Kuter had given a lecture on the vulnerability of the Fleet to aerial attack. One of the students, a Navy lieutenant commander, sent word of Kuter's heresy to the Chief of Naval Operations, who hit the ceiling and notified the Secretary of the Navy. He in turn contacted Woodring, who was quickly on the phone to Malin Craig. Craig passed the word to Oscar, and prompt and angry inquiry was flashed to Maxwell Field. Who was the guilty party? Charges would be preferred.

Gentleman John Curry, known for the fine stable he kept and the graciousness of his style of living, also had a backbone. He sent his own message to Washington. He was running the School. He was responsible for what was said. Who was going to prefer charges over what?

Westover came in person to deliver a lengthy lecture to staff and students alike. They would cease and desist from any course of inquiry that did not hew to the lines of curriculum as laid down by the War Department. Loyalty to the top was again his steadfast theme. The assembled listened and, after his departure, proceeded as before in their examination of all things aerial.

In a similar type of move, Oscar raised Andrews' hackles. In late November he sent out a memo to every Air Corps field officer on the subject of excessive drinking. The reason for the memo was that several officers had had to be suspended from flight duty because of too much imbibing. A great deal of umbrage was taken by those who did not mix booze with flying. In a letter to Westover Andrews said: "I wish you had talked it over with me first. I would have advised against it. The great majority of our officers are temperate in their habits and there is a substantial number of teetotalers. These resent the implication that their conduct has not been exemplary. They consider that the receipt of such a letter implies this. . . ."

To Delos Emmons, Andrews confided that Westover had "no authority to send that letter out as the discipline in the GHQ Air Force is my responsibility. . . ."

Westover showed his offending memo to Craig, who said it seemed perfectly all right to him. If nothing else, the incident indicated sensitivity not only on the subject of too much alcohol but also on the line of authority.

It was along this line that Andrews forwarded a proposal to Simonds on how a resolution could be found in reorganizing the Air Corps. Following background on past methods of organization, he listed the major components of the Air Corps as: a. GHQ Air Force; b. air units of overseas departments; c. air observation units and detachments assigned to corps-area commanders; and d. establishments under the direct control of the Chief of Air Corps. This last included the four branch schools and the training center, the Materiel Division, including the depots, and the Air Corps Board. In addition to the direct control of these units, the Chief of Air Corps had jurisdiction over all Air Corps activities in technical matters and items of Air Corps supply.

Andrews cited the basic Air Corps directive AR 95-5 in that the Air Corps Chief would "control and supervise expenditures of funds appropriated for the support of the Air Corps" and "will select and recommend to the Secretary of War types and designs of aeronautical equipment."

Then he got to the heart of it:

No where does it appear that he has any responsibility for the combat efficiency of Air Corps tactical units,—the only reason for the existence of the Air Corps. However, he is charged with control of the Air Corps Tactical School and of the Air Corps Board, which agencies, by long custom formulate air tactics and methods of employment.

We find then a unique situation in which one agency has control of the funds, selects the equipment and personnel, and prescribes for the combat efficiency of those units; while other agencies are responsible for the results, but have no authorization in securing for themselves the means whereby they may accomplish their results.

The foregoing paragraphs are designed to show clearly yet briefly the inherent complex nature of the Air Corps set-up. It indicates the necessity for a simple yet effectively organized method of superior coordination and control. For a number of years, an Assistant Secretary of War for Air endeavored to accomplish this coordination. It appears that this office will remain vacant. Some method of superior coordination and control is necessary and an attempt to supply such control has been made by placing the GHQ Air Force directly under the Chief of Staff.

He then detailed three possible solutions, beginning with the statement: "It is believed impossible and undesirable for the Chief of Staff personally to coordinate the various activities of the Air Corps." The solution, he felt, that would answer the complexities best was to create an Air Division of the War Department General Staff, coequal with the existing divisions. It would be similar in structure to the Navy's Bureau of Aeronautics. In describing the makeup of the Air Division, which would handle all matters pertaining to the Air Corps, Andrews stated: "The creation of an Air Division of the War Department General Staff would do much to allay the clamor for a separate Air Corps. It would weld the Army Air Corps firmly together as an integral part of the Army and would prevent the recurrence of such conditions as made it necessary to separate the tactical units from the Chief of the Air Corps. If legislation is necessary, it is believed that the Congress would readily grant it." Cogent as Andrews' proposal was, it received absolutely no attention at the time, was filed away and forgotten.

The attitude in the War Department worked its own effect on Andrews' thinking, as did so much else. Because of the rotation policy, which he had also been trying to change, in an eighteen-month period there had been five officers to command his three wings, thirteen officers commanding his nine groups, eighty-eight officers to command his thirty-nine squadrons, with four additional officers due for relief as group commanders in the near future.

By the end of the year he also knew he was probably going to be losing Conger Pratt, who had become a permanent brigadier general of the line, the first airman to be so appointed. On the possibility, he went back to Malin Craig to see if he could move Clagett into the vacancy. Harry had written him a letter saying he had considered contacting the President to ask for an investigation of his demotion. He had thought better of it and, separately, he and Mary had gone to see Senator Morris Sheppard instead. Clagett was sure Hap Arnold was the man behind the gun, and he hoped Andy would move to make a recommendation before Arnold got the jump and recommended someone else. He also believed that after Mary had shown Senator Sheppard Andy's previous letter of praise, the Senator might wish to see him also. But, in any case, Sheppard would contact Craig, and he didn't think the Chief of Staff would buck the Senator unless Arnold had gotten there first.

Craig responded to Andrews' recommendation that he would consider it when the time came to fill the job.[5]

On yet another vacancy Andrews suffered defeat. Major Clements McMullen, who had assumed Kenney's job in July, was ordered to take the next class at Fort Leavenworth, and Andrews went back to Craig to try to have Kenney returned to his staff. He reminded the Chief that he had told him he could have Kenney back at the end of the year. Andrews had already discussed the move with Westover, whom he knew was in the process of making up the slate for personnel changes. Westover had been noncommittal, saying he had received no instructions.

"I need Kenney very badly," wrote Andrews, "and I know that he would be willing to come back to work with this headquarters immediately . . . without taking any of the academic leave . . . so I also request that his orders direct him to report to this headquarters immediately after the close of the school year." Andrews also was able to suggest an officer who was well qualified and anxious to take Kenney's position at Fort Benning.

That didn't matter. Andrews was not going to get George Kenney back under any circumstances, even though he kept trying for the next six months.

While he was attempting to do so, Hugh Knerr was forced to take sick leave for overwork. "I have hypertension, whatever that is," he told Andrews, "and I have to relax or go bust." He didn't want to be a useless luxury. However, before he took his leave he and Andrews came up with a plan for a special coronation flight of the first B-17. It was anticipated the Boeing would be ready for its flight to Langley shortly after the new year. The idea was to give the maiden flight full exposure in the press, make it a real event. Knerr had also talked the idea over with Arnold and then reported that he didn't think Hap would give the plan much support. Andrews would probably have to sell the idea to Craig and Woodring on his own. Further, Knerr believed that Arnold was planning to grab the first three B-17's and, with himself in command, pull off his own stunt, unless Andrews beat him to it. He admitted that the thought might be a figment of his suspicious nature, and it was. But the bone of contention caught in the throats of the two commands was not about to go away.

Twenty-one

End Run

In June 1936, General Walther Wever, Chief of Staff of the Luftwaffe, was killed in a crash at Dresden, Germany. The accident's cause bore marked similarities to the crash of the Boeing-299 at Dayton. Wever, not a very experienced pilot, was flying a Heinkel He 70, the fastest aircraft in the German service at the time. It was not the plane's speed, however, that killed him, but his failure to activate a lever in the cockpit to unlock the controls. Like Major Pete Hill's, his flight was brief, but in retrospect it would be difficult to measure the effect of his death on events to come.

Wever and his staff subordinates were firm believers in the bombing strategies of Douhet and Mitchell. The former's *Command of the Air* had been published in Germany in 1929, and some of Mitchell's articles in 1933. Fostering these theories was the remembrance of the deadlocked trench fighting on the Western Front, the belief that Germany had lost the war at home and not on the battlefield, and the conviction that in any future war Germany must win quickly because of limited industrial production.

Toward this end, at almost exactly the same time the Boeing B-17 was being prepared for use by GHQ Air Force, the Junkers 89 and the Dornier Do-19, both four-engine long-range bombers pushed by Wever, were under construction in Germany. There were three prototypes produced of each model scheduled for production, but after Wever's death Goering, who was not a strategist in any sense of the word, rejected the whole concept. Wever was later recognized as one of the few really brilliant officers in the Luftwaffe. After he was gone, gone, too, was the possibility that Hitler would have a powerful long-range strategic bomber force—a "Ural bomber," as it was called—at his disposal. Gone, as well, were Wever's adherents, moved out of their positions by Goering's appointees.

Lieutenant Colonel Truman Smith, the U. S. Military Attaché in Berlin at the time, gave a quick glimpse of the prevailing atmosphere in a letter to Benny Foulois: "Colonel [Ernst] Udet is now prancing around in a colonel's uniform designing pursuit tactics, and some new mysterious air corps office or other opens up almost every week in the Tiergarten quarter.

"In Pomerania between Hamburg and Berlin the new air fields in various states of completion dot the landscape, so that in the air you are hardly ever out of sight of one."

Truman Smith considered the Luftwaffe to be one vast training school, but in July, a month after General Wever's death, civil war erupted in Spain, and soon thereafter school was out for some of Hitler's pilots. By October, Soviet tanks and planes were in combat supporting the Loyalist (Republican) side. A few weeks later the German Condor Legion, commanded by General Hugo von Sperrle, was dropping bombs on Madrid on behalf of General Franco's Nationalist forces. The Legion consisted of four bomber and four fighter squadrons, forty-eight planes in each. There was also a mixed squadron of seaplanes and reconnaissance and experimental aircraft. Supporting the airmen were antiaircraft and antitank units.

For three days in November the capital was bombed. Over one thousand of its citizens perished. It was the first major Western city to be so tested. The Germans were out to examine clinically Douhet's theory of breaking the will of the people by methodically attempting to set the city afire district by district, with the emphasis on destroying public buildings such as hospitals and the telephone exchange. At one point, twenty thousand people were forced to live on the streets, but the effect, instead of creating panic, inspired only hatred and a determination to fight on. *No pas-ar-án!* was the drumbeatlike cry of the beleaguered.

Its sound and strategical meaning failed to make much of an impression on bomber theorists at Maxwell Field and elsewhere. This may have been so partly because there were very few qualified air observers on the scene, not only in Spain but also in China when the Japanese attacked there. The reason was steeped in a policy that adhered to the belief that a ground or naval officer was just as competent to judge aerial combat as any form of warfare. But, very probably, competent air observers in either arena of combat would have changed few believers. Most of the bombing equipment being used did not measure up to U.S. standards, or so it was believed. That the bombing planes—German, Russian, Italian, French— required fighter escort, else attackers on both sides suffered heavy losses, made no more impression on Air Corps thinkers than the results of the bombing of defenseless cities.

The political effect of the war in Spain was to bring a tightening of the U. S. Neutrality Act, passed in May of 1935 as a result of Mussolini's invasion of Ethiopia. The Act made it illegal for Americans to sell or transport arms to belligerents once the President declared that a state of war existed.

Both political parties were in favor, and both in the fall of 1936 were occupied with the presidential election. When, in August, the Glenn L. Martin Aircraft Company approached the State Department on selling eight B-10's to the Loyalist Government, the proposal was flatly turned down. But because the Act said nothing about civil wars and its character was more moral than binding, by the end of the year partisans on both sides were out to work their way around it.

Roosevelt, having been reelected by an overwhelming majority, moved quickly; with near-total congressional support, a resolution was passed to ban all arms shipments to Spain. A Gallup poll indicated that 94 percent of the American public was opposed to any involvement in the war.

U.S. military men by and large saw the war as a battleground where the dictatorial powers of Europe were testing their armament and contesting their respective ideologies—Fascism against Communism. Germany, Italy and Russia became heavily involved, while Britain and France impotently skittered around the edges of the murderous conflagration, their International Non-intervention Committee incapable of meaningful action. The French sent aircraft to the Loyalists, British merchants shipped arms through to both sides, and all sides, including the non-Spanish belligerents, outdid each other in their diplomatic duplicity.

The fear in Europe and America was that Spain was the bloody proving ground for more to come. Certainly that was how Hermann Goering saw it. In December he told a group of his air officers, "We are already in a state of war." With the new year, "all factories for aircraft production shall run as if mobilization had been ordered."

A few days later he was to say to a gathering of industrialists, "The battle we are now approaching demands a colossal measure of production capacity. No limit on rearmament can be visualized. The only alternatives are victory or destruction."

Lieutenant Colonel Truman Smith and his assistant, Air Corps Major Arthur Vanaman were not privy, of course, to these predictions by the Reichsmarshal. Instead, Smith was "tremendously impressed with the great thoroughness of the preparations, even if as yet I doubt if they [the Luftwaffe] have many first-line tactical air units in existence. The whole German air corps," he observed, "is still a school." He said he had not fully made up his mind, but in observing the Luftwaffe's maneuvers he was coming to believe "that antiaircraft and artillery and pursuit work must come in the closest cooperation in the defense of cities."

The reports of both Smith and Vanaman, as well as all U.S. military attachés, were not made available to the Air Corps or the Air Force, their study restricted to G-2 and select members of the General Staff. Only through unofficial correspondence and nonmilitary observers did U. S. Army airmen get a glimpse of what was going on in Germany and elsewhere during this critical period.[1]

Conversely, Smith's opposite number in Washington was Lieutenant General Friedrich von Boetticher, the German Military Attaché. For entirely different reasons, von Boetticher's reports to Berlin had an effect that, like Walther Wever's death, would be difficult to measure on events to come.

Andrews had gotten to know the seemingly genial, heavy-set officer during the summer of the Balbo flight. They had met at the Chicago World's Fair, when von Boetticher was there on a visit. Blond, round-faced, extremely gracious, the German officer spoke excellent English, and Andrews invited him to visit Selfridge. Von Boetticher, having established the contact, continued to cultivate it, particularly after Andrews became Commander of GHQ Air Force.

Not knowing they had already met, Major William F. Fruhoff, at the War College, sent Andrews a note of introduction to the General, maintaining, "You will find Gen. von Boetticher, who is a very close friend of mine, a charming gentleman and outstanding soldier and I'm sure your contacts will result in mutual benefit."

Von Boetticher, who had been the German Military Attaché since 1933, had other friends and acquaintances, who included Generals MacArthur and Craig as well as acting Secretary of War Woodring. His job, of course, was not only to make friends but also to gather political and military information, which was passed directly to Hitler. Through his many contacts and the U.S. press, he was able to obtain a clear enough picture of American thinking on the accelerating pace of the arms buildup in Germany and elsewhere. But his bigotry got in the way of any objective appraisal. He was as anti-Semitic as Hitler, and so all his reports were colored by the contention that the Jews ran everything in America, including the military and the presidency.

Hitler believed without question everything his Military Attaché sent on the subject, for von Boetticher's blindness matched his own. That America was too weak and decadent ever to fight was what he wanted to hear. It was to be a fatal misjudgment, generated in part by the isolationist circles into which the General was welcomed. And so, three years before the Wehrmacht invaded Poland, Hitler's man in Washington was busy getting himself invited to Langley, writing a letter of condolence to Andrews when a Captain Carl Gimmler, whom he had previously met, was killed in a crash, and sending back to Berlin reports that the major reason why the U. S. Government was doing so little to arm itself was due to Jewish influence.

While war raged in Spain and pilots from a host of countries engaged in daily aerial combat, flying the aircraft of at least six nations, Andrews' thoughts were concentrated in two areas: the long-anticipated arrival of the first four-engine Boeings and forthcoming action by the new U. S. Congress. With the former, his hope was to have two groups of the aircraft by 1938, one on the East Coast, the other on the West. That G-4 of the General Staff and Secretary of War Woodring were already opposed to the idea was just one more reason why Andrews' interest in the seventy-fifth Congress had sharpened.

In the House, Representative Lister Hill, of Alabama, had been named to take McSwain's place as Chairman of the Military Affairs Committee. Hill, a strong Roosevelt supporter, was also impressed by Henry Woodring and thought that FDR should keep him on as Secretary of War. That the Air Force Tactical School had moved to Maxwell Field was largely Hill's doing. Walter Weaver had sold him on the idea that to convert what was a desolate, run-down air depot into a really handsome, up-to-date post dotted with white stucco buildings made good political sense. PWA funds had gone into the reconversion; it had afforded a good many jobs, and Hill was a popular man with the home folks.

Long-time Chairman of the Senate Military Affairs Committee was Senator Morris Sheppard, of Texas. At sixty-one, he was a power in the Congress. In 1935 he had worked hard for the passage of General MacArthur's much sought bill on Army promotions, and before that, when Roosevelt was considering dropping MacArthur as Chief of Staff, Sheppard had told FDR he thought the General should be kept on.

Both Hill and Sheppard were looked upon by airmen as being supportive of their needs, but neither was a congressman who would boldly champion a separate air force unless the President was so inclined. Instead, it was Representative Mark Wilcox, as a member of Lister Hill's committee, who had taken up the cause where McSwain had left it.

Although doubt remains as to just who in Benny Foulois's office authored the McSwain bills of 1934 on a separate air force, there can be no doubt as to who were behind the bill introduced by Mark Wilcox before the Military Affairs Committee on January 18, 1937: Frank Andrews and Hugh Knerr, with Knerr acting as an intermediary between the Air Force Commander and the Florida Congressman.

Wilcox's bill, H.R. 3151, proposed "To create an Air Corps under the Secretary of War, to be known as the United States Air Corps," which would "consist of the regular Air Corps, the National Guard units and personnel while in the service of the United States and the Air Corps Reserve."

The bill's purpose, as Andrews was to state later, "would recognize Air Power as being on an equal footing with military and naval power, and would charge the United States Air Corps with the responsibility of establishing and maintaining the Air Power of the United States, and with providing the aerial defense of the United States and its possessions. . . . The Chief of Aviation, generally speaking, would be placed on an *equal* status, under the Secretary of War, with the Chief of Staff of the Army. . . ."

It was an oblique variation on a well-tried theme, and it illustrated how far Andrews' thinking had shifted in his two years of commanding the GHQ Air Force. More than that, it showed that, like others before him, he was willing to risk career and future for a belief. He was no wheeling-dealing politician. He put it on the line the way he saw it. Up to a point he was willing to accept rejection and make the best of it, but that point had been reached.

That Knerr had an influence on his thinking there is no doubt, but Knerr knew better than anyone that after everything was said, it was Andrews who thought it out and made his own decision. Likewise, although George Kenney was drafting some of Andrews' addresses before the War College and the Tactical School even after he had departed from Langley Field, the final result was Andrews'.

His thinking on the Wilcox bill had evolved out of the two-year struggle to build an air force and the growing belief that under the strictures of the General Staff—its failure either to come forth with its own plan or to accept one by Andrews or Westover to end the divided air command, plus its failure to push new aeronautical development in a world growing increasingly dangerous—a new course of action must be employed.

Andrews had first gotten to know Wilcox when testifying on the Congressman's bill to increase the number of air bases. Following the passage of the bill, contacts with Wilcox had centered on the inability to obtain the necessary funds to build the bases, and then on Andrews' outlook on frontier defense, leading to ways to break the Air Force loose from the political and economic bonds that retarded its development.

Wilcox was a guest at Langley on numerous semiofficial occasions, but Andrews was careful not to meet with him in Washington, using Knerr as a go-between, principally in Florida. Andrews had planned to join them there on one occasion, but bad weather had prevented the meeting. Wilcox was anxious to come up with a bill that would enlist strong congressional support, knowing that the War Department would oppose anything that followed the McSwain line of an independent air force in or out of a department of defense. When Knerr went on sick leave, he and Hazel journeyed to Florida, and there the two men held a long session. Knerr then sent his notes of the meeting along with a copy of a paper he had written a year past that combined Andrews' and Knerr's thinking on the concept of "an Air Corps within the Army." The paper, Knerr said, could be extended also from the viewpoint of a separate department of air.

Then he offered the opinion: "I personally think an effort for a completely independent Department of Air Defense will certainly result in turning land-based aviation over to the Navy, lock, stock and barrel to be operated by the Marine Corps. In other words, our effort to establish an Air Corps in the Army, would be taken over by the Navy and the same thing done within the Navy. So our only chance is exactly where it was when the effort was made last year. However, if you think a completely separate Air Defense Department can be accomplished, I defer my judgment to yours.

"Before you make up your mind to any course of action I know you have your eyes open to the fact that your chance of official survival is practically nil.

"Wilcox says that the opposition to their [House Military Affairs Committee's] big Navy program needs but some nucleus to insure growth of a more reasonable Navy program and that a convincing argument in favor of Air Forces is all that is needed to start it. However, somebody will get hurt if handled by the Air Corps because it will make the President mad."

In making up his mind, Andrews agreed that trying to follow the McSwain line of creating an independent air force in or out of a department of defense would be unworkable, not only because of fears that the Navy would take control but also because such a bill would have no chance of passage. He was looking for the possible, and the idea of a coequal air force within the Army was an extension of the previous plan he had submitted to the Chief of Staff, to which Knerr referred. He had made the direct approach without any reaction. Now he had gone the forbidden congressional route, he and Knerr all but drafting H.R. 3151.

About the time Wilcox was offering his bill to the House Military Affairs Committee, Andrews was responding to George Kenney's latest query on, When can I come back to you and what's new?

On the first question, Andrews reported that Mike Kilner had told him not to be impatient and that he remained confident that General Craig would stick to his promise. So George should be patient too.

On the second, they were still getting opposition to the "big ships," but he hoped that actual performance would wipe that out. A crew for the second B-17 had been sent to the Boeing factory for transition training.

A month later he would admit to Kenney, "There is still some influence working against your return to GHQ Air Force. What it is, I don't know, but I believe General Craig is going to send you back to us. I had a talk with him today about it again and although he did not definitely commit himself, he left me with the impressions that he would take care of it."

He said nothing in either letter about the Wilcox bill, and it can be assumed that only Hugh Knerr knew his part in the proposed legislation.

Canadian-born Cy Caldwell, a former RFC bomber pilot and well-known aviation and military-affairs writer, offered his own, unsolicited opinion to Andrews on the Wilcox bill. He did so also sending Andrews an article on how in the next war Germany's only way to defeat England was through the air, that the Germans must win quickly or lose surely. "The political and economic history of the world is being re-written," predicted Caldwell. "Sea power is being delegated to the position occupied by the bow and arrow. There will be sea forces, but they will be naval air forces based on the sea—on the carrier."

Confident that he knew all there was to know about the Corps and its personnel, he foretold Andrews what the future held. "Now about yourself and the part you are to play in the American Air Force. I pick you to command it—though you may never do so, for you may die before we have an Air Force. But in my vision of the United States as a world air power, I see you in command—right now I don't see you as commanding anything. You are merely an auxiliary air employee of the ground army. You can't do a thing the General Staff doesn't like—and they don't like anything that shatters the illusion that the infantry is the Queen of Battle."

Caldwell then gave a free evaluation of the way he viewed certain officers. "They [the War Department] promote [Conger] Pratt—and then send him back to the ground to prove to him where he heads in—and Susie [Sadie] will see that he (isn't that his charming and clever wife's name?) will see that he heads that way—toward higher ground in the Army—which is not necessarily higher in the air.[2] Susie comes of an old ground army family—she knows where the butter and jam is spread on the bread. I have always given her credit for half of Pratt's brains—the best half. He sits on the other half. I look upon Pratt as a very efficient equipment officer, and that's all. An ideal man for a supply department, even when it's an Air Force. Danforth is a grape-fruit grower and fisherman—that tells the whole story about Danny. Arnold is a good man—he will be of use to you. So will Colonel [Tony] Frank."

And then came the advice. "Don't let yourself be led into political back alleys, such as this Wilcox bill—to put the Air under the Secretary of War. No good. You will find yourself junior to the whole General Staff when you go into his office. In theory you and the Air will be co-equal with the Army. In practice you cannot be. As you have brains, I needn't labor that point—simply recall your own conference with your superiors on the G.S. [General Staff]—the bill won't change that at all. They'll outweigh you as the ton does the ounce . . . so don't commit political and military suicide before the time comes. Time effects all changes—leave it to Time, build up the present Air Force. . . ."

He was offering Andrews a private and confidential pipeline to say what he wanted under the by-line of Cy Caldwell, not only in *Aero Digest* but also in *Collier's Magazine*. Public opinion must be formed, but he really was the only aviation writer who could be trusted. He would be coming down for a visit in the spring, but because he was conspiratorial he and his wife and two children would stay at a hotel so that it would not appear that he was Andrews' guest.

In his somewhat formal response, Andrews agreed that in time "air power will be predominate, not only in this country, but in all nations of the world, although I do not think five years will do it. It will take another big war in Europe or somewhere, to impress on the minds of the people of the world the value of the airplane. . . ."

On the Wilcox bill, he said he had in no way committed himself and at the present time had no intention of doing so. "As a matter of fact," he observed, "I strongly suspect that it has already run up against administration opposition. As you probably noted, the Military Affairs Committee, when Wilcox tried to get them to start hearings of the full committee on the bill, stalled around and did nothing."

And then Andrews opened the door just a crack into one reason for his private moves, when he informed Caldwell, "I have no intention of committing political or military suicide, but my position in this temporary-rank scheme is so precarious that I might get sniped off at any time."

It was a dangerous poker game he was playing, but, then, gambling and danger were ingredients of his profession. He had always ignored the latter and made light of the former, whether dealing with the General Staff or a deck of cards.

Andrews' appraisal of administration reaction to the Wilcox bill was reflected by Secretary of War Woodring and the General Staff. In a four-and-a-half-page letter to Congressman Lister Hill, Woodring explained why "I am unalterably opposed to any change therein, and recommend strongly against the enactment of H.R. 3151." In detailing his objections, Woodring cited four previous attempts in the past eleven years to offer similar or correlated proposals. The first, in 1926, a bill to create an Air Corps, under the Secretary of War, to be known as the United States Air Corps, he said was very similar to the Wilcox offering.

The bill had been introduced by Congressman Jonathan Wainwright and was commonly known as the Patrick bill, named for General Mason Patrick. This was the legislative attempt that had gotten Hap Arnold thrown out of Washington. Declared Woodring: "The views of the War Department . . . are as sound today as they were when written eleven years ago."

The views were uniform. Major General Stanley D. Embick, who had become Deputy Chief of Staff, supplanting Simonds at the end of the previous May, was automatically against approval. He was seconded by all the assistant chiefs of staff. Colonel T. P. Lincoln, heading up G-2, Military Intelligence, added to his rejection that he did not like the idea of having separate air attachés. Military Intelligence, he pointed out, already had air attachés stationed in the major European countries. He didn't add that none of their reports were available to the Air Corps.

Glasser, of G-1; Hughes, of G-3; and Spalding, of G-4 all signaled their rejections, as, of course, did Malin Craig. Woodring added in his letter to Lister Hill that if hearings were held, the War Department would want to be given the opportunity to present its views in greater detail.

There was a slightly different twist in the blanket rejection, for Oscar

Westover joined it, submitting a fifteen-page report to Major General A. W. Brown, the Judge Advocate General, recommending that the Secretary of War report unfavorably on H.R. 3151. Instead, he suggested that the portion of the bill placing supreme command and control of the Air Corps and the Air Force under the Chief of Air Corps in time of peace be taken out of context and made effective by the issuance of a War Department directive. Westover had not given up on what he saw as the best way to go, but the same General Staff reasons for rejecting his thinking remained in lockstep, steadfast, unmovable against *any* fundamental change.

Hugh Knerr had framed a broad series of questions and answers on the bill for use if the time ever came when the Air Force was free to speak out on it. The final question was: *Why is it that bills of this nature periodically crop up? Is there a small group of agitators in the Air Corps that are responsible?*

And the answer:

> The reason is obvious. It is the same reason that accounts for Lassiter Committees, Morrow Boards, Baker Boards, Howell Commissions, etc. The process will continue until positive action is taken to provide this country with adequate Air Defense. Until such action is taken and so long as there is an Air Corps there will always be a small group of Air Officers who place loyalty to their duty and loyalty to their subordinates above their own personal fortunes.
>
> Recall that a Chief of the Air Corps [Patrick] at one time presented over half of this bill. Ten years have gone by and the idea is as vigorous as ever. If nothing is done meanwhile you will have it before you ten years from now.

While the Wilcox bill was ruffling feathers in the War Department, Andrews was seeking clarification of other General Staff moves. He had heard unofficially that the War Plans Division was at work preparing "another balanced Air Corps program," and he informed Craig he was working very much in the dark until he knew what the plan was.

Since the advent of the long-range bomber, the term *a balanced Air Corps,* like *unity of command,* had risen to the fore. What lay behind concentration on the terms was not just a buildup toward the magic figure of 2,320 aircraft but a political and military determination by Secretary Woodring and the General Staff to prevent the Air Corps from putting too many of its limited eggs in the B-17 basket.

As Assistant Chief of Air Corps, one of Hap Arnold's major responsibilities was procurement. Back in September he had carried the ball on bulling through final approval to go ahead on Project D, the XB-19, which had been hanging in limbo for a year. But a balanced Air Corps meant just that to him, and he sat in the middle, between the War Department resistance to any increase in long-range bombardment aircraft and a growing desire on Andrews' part to greatly increase their numbers. Following the talk Hap had had with Hugh Knerr on Andrews' proposed coronation flight of the first B-17, Arnold had called on Malin Craig, urging that an additional twenty B-17's be added to the projected 1938 budget.

Craig listened and was willing to submit the figure to Woodring. The Secretary of War promptly said no. It wasn't just the cost, although that would have been enough, but the Boeing plane was an unproven entity. The first one had crashed. No one knew what the second might do. There were no data on it, no proof of what its performance might be, and its declared use did not fit the political and thus strategic thinking of Henry H. Woodring.

Internally, Andrews met with additional rejection. Once again he had asked Oscar "for a little return cooperation in giving . . . what assistance" he could "in getting Kenney sent back to the G-3 job."

Westover sent his regrets, saying he could not consistently recommend a change in Kenney's assignment.

Neither could he do anything to see that Henry Clagett was restored to commanding a wing. When Conger Pratt relinquished command of the 2nd Wing, Andrews moved Jerry Brant up from the 3rd to take his place, planning on Clagett's assuming command at Barksdale. Instead, he was informed that Malin Craig had selected Colonel Frederick L. Martin for the post.

Martin was a vintage pro who had been placed in command of the 1924 Round-the-World venture, and like everyone else of his stage and age had served since in many capacities. Once the selection had been made, Andrews accepted it and got on with the job at hand, which at the moment was the long-awaited arrival at Langley of the first of the Boeing B-17's. In this case, there were no qualms over the arrival being in the best of hands.

In July 1936, Bob Olds had assumed command of the 2nd Bomb Group, the premier bomb group of the Air Corps. It was to be the only unit in the Air Force equipped with the B-17, and no officer in the entire military air arm was better qualified to prepare for and command B-17 operations than Olds. The men who served under him considered him the toughest, most demanding, most tireless tactical commander to be found, and they didn't wish to serve under any other.

First Lieutenant Curtis LeMay, who had become operations officer of the Group's 49th Squadron, felt he didn't really have a picture of what the Air Force was all about until Olds took over as CO. Then he learned fast. You'd better be at work by seven o'clock, and you'd better know your stuff. LeMay's tiny office was in a protruding section of a hangar, with Olds's office directly above. The lean, whipcordlike CO would always stop by LeMay's desk and make an inquiry or two. The Lieutenant promptly learned that he'd better have the answers. At their first morning meeting,

Olds wanted to know if LeMay had maps for routes to some far-distant points. The answer was a puzzled "No, sir."

"Your planes can fly that far, can't they? Don't you think it would be a good idea for you to know the best way to get there?"

The same questions were true with regard to weather, and for the first time in his career, LeMay, who had become one of the Air Corps's very few navigators, began to perceive what a bomb unit could and should be—that prior preparation in all areas of flight was essential to the flight's success.

How much was riding on the B-17's success was well known to Hal George, who was now Olds's Executive Officer. Andrews had called them in and told them this was one airplane that could not be cracked up, not even in a minor way. A single accident would put an end to what procurement there was. So much as a broken tail wheel, and the General Staff would be sure to announce not only that there was no tactical or strategical purpose for the four-engine bomber but also that the Air Corps's best pilots couldn't fly them. Andrews stressed that the plane's pilots had to be proficient in all kinds of weather and that crews must be prepared to fly the B-17 anywhere in the world, over land or sea.

Olds needed no urging. George considered that his boss had a brilliant mind and with it the capability of grasping all the complexities of a situation and making accurate decisions. Training pilots in his Group to be proficient in blind landings and takeoffs was such a situation. At the time, blind landings and takeoffs were not exactly standard procedure. There was no equipment for such practice, but Olds managed to borrow some from an airline and put the machine shop to work improvising. He then issued a short memo to his pilots that said that "Nobody will be cleared to fly a B-17 two months from this date unless he holds a certificate over my signature stating that he is qualified in instrument landings and take-offs."

Flying instruction began around the clock in the gerrymandered equipment and in the Link trainers. Olds led the pack. All the pilots held instrument ratings to begin with, but by the time the CO's signature was put on their certificates they were qualified to fly when the birds would refuse the opportunity.

March 1, 1937, was the great day when the first B-17 arrived at Langley Field, piloted by Major Barney Giles and Captain Cornelius Cousland. They had departed Seattle, Washington, on February 9 with Lieutenant William O. Senter as navigator and a flight crew of three sergeants. There have been landings and inspections at March Field, where Andrews had been on hand to fly the bird, and at Barksdale, Louisiana. Not exactly a coronation flight, it was nevertheless a historic one. This was the aircraft Andrews and his people had long awaited, a concept brought home, a reality now in being. In a message to Craig, Andrews made the point that the B-17 had arrived without mishap to be greeted by a large gathering, including newspaper reporters, photographers and newsreel cameramen. "It is an imposing-looking airplane," he observed, "and handles very nicely in the air and on the ground." He stressed to Craig that he was confident that any multiengine pilot would be able to check out in it quickly. He didn't add: under Bob Olds's wary, watchful eye. Curt LeMay knew exactly what that meant. Because he was junior in rank and junior in bomber experience, it was going to be quite some time before he got his hands on the controls of a B-17, but in the meantime, he was a much sought-after navigator. March 1 was an important day in the minds of Army airmen who were looking to the future.

On a strictly personal level, March 1 was also the day Andrews' father came down with a bad cold. Andrews had been planning to invite his father to join the family in a Key West vacation. Since he would be flying Jeannie down later in the month to join her mother and Josie, who had been enjoying the sun and sea since January, he would pick his father up at the new Nashville airport.

On the night of the third, Andrews' sister Josephine telephoned to say that their father was in very serious condition. In Washington on official business, Andrews took off from Bolling Field at midnight and flew directly to Nashville. He arrived early in the morning. Shortly thereafter, seventy-nine-year-old J. D. Andrews, a man of many parts and interests, passed away.

When his son returned to Langley, he found trouble from on high awaiting him. He had written an article for the *Army Ordnance* magazine titled "The GHQ Air Force as an Instrument of Defense." In it, he had detailed the tenets of strategic air power. He had given aerial support of ground troops their due, and credit to the power of the Navy, but he had stressed the aircraft as a new weapon of war needing only to overcome weather, antiaircraft defenses and distances. That he underestimated the effects of the first two and in the third neglected to mention that in overcoming distance lay the additional barrier of enemy interception, were errors in theoretical belief, but they were not the points with which Malin Craig took issue.

The article, said Craig, "is not in harmony with the program now being considered by the War Department to balance requirements of the Air Force and the Army Air Corps." He suggested that Andrews defer publication until he had received the program and looked it over, at which time he could "prepare a suitable revision."

This was the balanced program Andrews had made earlier requests to see, and when Craig sent it to him, and he responded, the Chief of Staff was pleased that Andrews was favorably impressed with the plan. However, on the essential point underlying strategic air power Craig observed:

I have noted your views with reference to additional defense measures for the Panama Canal. The governing idea in the preparation of the Air Corps Balanced Program was to provide a force adequate to carry out the role of the Air Corps and one which would at the same time be practicable, realizable, and non-provocative. There were many cogent reasons for the omission from the program of the facilities recommended in your letter. In fact, to have included those establishments . . . would have been most unwise and would have jeopardized the entire program.

This was in answer to Andrews' suggestion that the air defense of the Canal on the Atlantic side required the use of bases in the Caribbean, and with "those bases, make the Caribbean as much an American lake, as the Italians made the Mediterranean an Italian lake in the Ethiopian war." He felt that for the reinforcement of the Canal, Chapman Field, outside of Miami, should be the principal staging point.

It was a question whether Craig's mention of being "non-provocative" referred to the fact that the Navy considered the Caribbean its own lake, or that establishing air bases in Cuba, Puerto Rico and the Virgin Islands would arouse the French and the British. In either case, where Andrews' thinking made good strategic sense, the excuse for rejecting the proposal, aside from economic considerations, indicated the kind of hidebound timidity that prevailed in the War Department.

On the twentieth of March, Andrews got away from it all for a couple of weeks. He packed Jeannie on board his Douglas command ship and, with Sunny Williams along to fly the plane back, took off for Key West to join Johnny and Josie. His plan was to drive the family home when the leave was over.

The island they vacationed on was Long Key. It boasted a Coast Artillery installation, Fort Taylor, and with it a few rough vacation cottages that could be rented. The usual method of arrival was by a lengthy ferryboat ride from Matacombe Key. The ferry was little more than a barge with a noisy diesel engine and a toilet that never worked. Hugh Knerr knew, because he and Hazel had taken it coming down to join the Andrews.

There wasn't any better place to conspire, either deep-sea fishing or sitting in the shade of a palm tree watching the waves roll in or swimming out and daring to roll in with them. Jim and Catherine Fechet were also on hand. Fechet loved the water, but no matter how hard he coaxed he couldn't get Hazel to join him, because she was afraid of barracuda.

Barracuda! Nonsense, proclaimed the General. Not a barracuda within a hundred miles. He was paddling about in the water off the end of the wharf where she was sitting. When she spotted the big, arrowlike fish easing up toward the General, she screamed. Afterward, the question remained, Who moved the faster, Fechet or the fish? It had to be the former, because he emerged unscathed, although breathing a bit hard, his suntan momentarily off color.

But Andy and Hugh had fish of another kind to catch, and the fishing had not been good. They already knew that there would be little or no support for the Wilcox bill. Wilcox was hopeful that he would be able to get hearings under way in a month or so. He claimed to have considerable support, but even if he did, they had to face facts. Andrews was going to have to give the Chief of Staff an opinion on the bill. He already knew what Westover's reaction had been. In a letter to George Brett he had confided, "I am afraid that I am making very little progress on my idea of a 'Chief of Aviation,' not even getting much support higher up in the Air Corps." So what chance did they have of the bill's passage? Precious little.

On the issue of aircraft, before the year was out they might have ten B-17's; the other three would be going to the Materiel Division for further testing toward modifications. Ten four-engine bombers, and their objective was to convince the General Staff and their own compatriots in Westover's office that another ninety-eight should be procured over a three-year period, enough to equip two groups. Already they knew that the Douglas B-18 was an inferior aircraft, that as a first-line bomber it did not measure up. Yet, as planned, it was to be the backbone of the GHQ Air Force based on the balanced-program figures. In that regard, Secretary of War Woodring's directive foresaw that the approved number of 2,320 aircraft would be reached by June 30, 1940.

At root was really only one fundamental question to be answered. Did they continue to press for what they believed was essential against all odds, or did they back off, fall into line and accept the War Department's position?

There was really only one answer.

As Hugh Knerr later wrote: "We decided to fight it out. The result was disastrous."

But first there was a minor personal disaster at Long Key. Both Josie and Jeannie came down with measles. This meant that although Knerr returned to Langley, Andrews remained, and while he and Johnny were waiting for their daughters to recover, Major General Charles Kilbourne, who had retired in January, arrived on the key with his wife for a bit of rest and relaxation.

Kilbourne at sixty-four could look back on a military career of nearly forty years. He was highly decorated, having won the Congressional Medal of Honor as a young lieutenant in the Philippine fighting. Then, during the World War, he had been awarded both the Distinguished Service Cross and the Distinguished Service Medal, "attesting to physical courage in battle and coolness as a staff officer under the pressures of war." That he had not seen eye to eye with air-power proponents in their efforts to be free of War Department control was natural enough. That he could not perceive the use of strategic air power as it was viewed by Billy Mitchell and his followers was certainly in keeping with the General Staff norm. However, he genuinely believed the Army had tried to help the Air Corps, that he could present a long list of very positive sacrifices made to gain even such limited advances as the Air Corps had made. He knew the results had been totally inadequate, but he blamed that on meager military appropriations chopped down further by the Bureau of the Budget, not on General Staff intransigence. Or so he told Andrews when Andrews brought up the Wilcox bill and they talked it over.

In the autumn of his life, with no military axes to grind—he was soon to become the Superintendent of the Virginia Military Institute—Kilbourne saw the bill as an instrument that could only produce further contention. He looked upon Andrews as a man who would honestly stand by his convictions regardless of personal interest, and he told him, "Your personal interest lies in supporting the status quo, especially if the bill fails of passage." But he added that he didn't think anyone could blame Andrews for his attitude, although if he gave the bill his backing he would be in a tough spot.

Kilbourne felt the fundamental error lay in not being able to complete the Baker Board recommendations, and had that been done there would be no support for the Wilcox measure. He admitted sadly, in view of what had been accomplished, that he had no basis for confidence that the Baker Board goal would be reached. Still, he believed that if the Wilcox bill passed, it would not have the desired effect. It would only bring greater competition and acrimony, greater struggles for the appropriation dollar.

Andrews appreciated hearing the older man's views as a result of their chance meeting on a sunbaked Florida isle, but Kilbourne's admitted distress over unfulfilled resolutions of the past and his pessimistic outlook on things to come did nothing to change Andrews' mind.

Upon his return to Langley, Andrews contacted Malin Craig and made a date for April 24 to talk over the Wilcox bill with him. Craig said he'd be glad to discuss it and then added, "You may laugh when I tell you that I have never even seen it. I talked a little about it to Follett Bradley, who apparently knew its whole history and gave me the impression that he is not in favor of it, as now written at least."

Craig's impression was not altogether correct on two scores. Bradley did not know the whole history of the bill, and he was generally in favor of it. When Hugh Knerr had passed around his own evaluation of the legislation, Bradley had written to Tony Frank that he was surprised and glad that Knerr was apparently reconciled to building up the Air Force within the framework of the War Department. Bradley agreed that it was the only way they could get anything done, short of a war. Maybe if things bogged down in a war, separation would come, but for now he agreed the course proposed was the best one.

In his own, step-by-step critique of Knerr's appraisal, Bradley made note that current press releases, "probably inspired by the War and Navy

Departments," on the war in Spain and operations in China were handing out the line that air power is a myth. A rebuttal should be written, although he was too busy to do so, suggesting instead that Tony or one of his assistants might take it on.

News of the bombing of the Spanish town of Guernica should have made the writing of a rebuttal on the effects of air power unnecessary. On April 26, Guernica was all but obliterated by the systematic attack of German aircraft striking in waves, using machine guns and thousand-pound incendiary bombs. The strike was purposely made on the afternoon of a crowded market day, and over a third of the town's population of seven thousand were killed or wounded in an assault that lasted for nearly four hours. Some years later, Goering would admit that Guernica, which was a military target because it was a communications center not far from the front, had been considered a testing ground by which German airmen could observe the effect of their bombing tactics. The immediate effect was that the area fell to the Nationalist forces without further resistance.

Although the grim success of German bombing tactics in Spain might seem far removed from possible congressional action on a change in Air Corps command structure, there was, of course, a relationship in the comparative uses and command of air power in war and peace.

Andrews spent several days rewriting and revising the statement he proposed to make before the House Military Affairs Committee should the Wilcox bill "ever come up for hearings." Behind the closed door of his den, the family could hear the drone of his voice as he rehearsed. Then, on the twenty-fourth, he went to Washington and showed Craig his official statement, which he would present to the Adjutant General. It was four pages long, and the first part gave his understanding of what the bill was all about: that "The United States Air Corps under this bill would be authorized to have its own laws and regulations," that it "would create a United States Air Corps in the War Department under the Secretary of War, on practically an equal status with the Army."

He then listed the fundamentals in the bill with which he concurred, giving in two pages a precise account of the development of air power and what its capabilities must be in the future. It was on the future that he summed up:

> I do not believe that any balanced plan to provide the nation with an adequate, effective air Force of modern type, can be obtained within the limitations of the War Department budget, and without providing an organization individual to the needs of such an air Force. Proposed legislation to establish such an organization has periodically appeared before the Congress in the past and will continue to appear until this turbulent and vital problem of National Air Defense is satisfactorily solved.
>
> Detailed study and analysis of H.R. 3151, in connection with our country's needs for Air Defense, has led me to the conviction that in general

principle, this bill, subject to special and detailed revisions, which I am prepared to suggest, provides fundamentally, the organic framework upon which can be built the Air Defense of the nation.[3]

What Malin Craig had to say to Andrews after he read his official position is not known. However, Henry Woodring had already sent his personal rejection of the bill in all its parts to Congressman Lister Hill.

Behind the scenes, Wilcox had been working to set up a meeting to bring both Andrews and Knerr in contact with the President. He had been unsuccessful for several reasons, the principal one being that Roosevelt refused to "consider reforms or reorganizations within the War Department unless and until it has received the full consideration of the Department, and he has been advised as to the Department's attitude."

Wilcox had submitted to Hill a list of the names of officers that he was anxious to have testify, with Andrews and Knerr heading the list. On the same morning that Andrews was discussing the Wilcox bill with Craig, a division chiefs' meeting was held in the Office of the Chief of Air Corps, presided over by Hap Arnold. At it he announced that hearings on the bill were scheduled for April 28, and that all those present should be prepared to express an opinion if called on. Shortly thereafter, a telephone call came from Colonel W. E. Cocheau, of G-3, informing Arnold that the Chief of Staff had decided to appoint Westover to be the War Department representative at the hearings. It was not difficult to see the sagacity of Craig's selection. What better way to blunt Andrews' testimony in favor of the bill than by having the Chief of Air Corps speak out against it?

The move wasn't necessary, for Woodring had been in touch with the President, and on the twenty-seventh, Charles West, Roosevelt's personal representative, passed the word to Lister Hill that FDR wanted the hearings postponed.

The word in Hap Arnold's office was that the postponement would be more or less indefinite, at least until June. Should action be taken then, Craig had changed his mind on Westover as the War Department spokesman. The reason for the switch, as Arnold put it, was "the fact that certain questions may arise wherein it is shown that the Chief of the Air Corps made certain recommendations of the War Department proper and no action was taken in which case the C/A.C. would have to straddle the fence so to speak."

Although the War Department was not aware of the degree of input by Andrews and Knerr into the Wilcox bill, Deputy Chief of Staff Major General Stanley Embick came to learn of Knerr's efforts to solicit support for it, and he reported the transgression to Malin Craig. Embick had been called on by a Mr. Dudley, Secretary of the Chamber of Commerce of Sacramento. Dudley informed him that several days past he had received a telephone call from Colonel Knerr, at Langley Field, asking for an appointment. Dudley agreed, and Knerr had arrived at his hotel with a Mr. Stetson, Secretary of the Air Defense League. Knerr informed Dudley that the purpose of his visit was to discuss the Wilcox bill. He said he was strongly in favor of it and read a lengthy statement that Dudley understood would be presented at the hearings. The gist of it was that efficiency could not be attained by the Air Corps under the present organization, that such progress that had been made was due to GHQ Air Force, but that efforts by the Air Force toward greater improvement were being continually thwarted and nullified by the Secretary of War, the Chief of Staff, the Chief of Air Corps and the General Staff.

Dudley was shocked by what he heard, which he felt was, in effect, accusing the "War Department of mal-administration and hostility to the Air Corps." He equated the charges with those made in bygone days by Billy Mitchell and his followers, and he believed that if such a statement was made public it would be harmful to the national defense.

The result was that as soon as Knerr and Stetson had departed, Dudley called Embick for an appointment. Embick in turn sent him over to relate his story to Oscar Westover.

Whoever had recommended that Hugh Knerr try his brand of salesmanship on the Chamber of Commerce Secretary from Sacramento had made a large blunder. More, Craig and Embick must have reasoned that Knerr would not have made his solicitation without the approval and knowledge of Andrews.

Whereas April became a month of stillborn congressional hopes, which included the failure of the House to appropriate funds to build an air base in Alaska, May was a month of successful testing, wherein for the first time the entire bomber strength of the Air Force was assembled in California for maneuvers. The Muroc bombing range was designated the city of Los Angeles. B-10's and B-12's made up the attacking force. One group would act as decoy, drawing off the pursuit, while the main force would move in and bomb the target.

At the conclusion of the exercise, Andrews made a statement to the press and was interviewed on a coast-to-coast radio hookup. He pulled no punches. He said air defenses were weak, that if the United States was attacked on two fronts it would have the forces to defend only one. Three hundred planes and three thousand men had taken part in the three weeks of mock combat without injury to anyone, he said, but his salient point was that, three years after the approval to build up a force of one thousand combat planes, he could still muster only about a third of that number.

During the maneuvers, Andrews had written to Knerr and asked him if there was anything doing on the Wilcox bill. There was not, but, again, there was a great deal doing on procurement, and Andrews returned to the

East to engage in the first of a number of fruitless battles to try to change War Department thinking on the types of planes to buy.

In that same month, Roosevelt named Henry Woodring to be his Secretary of War on a permanent basis. The Senate quickly confirmed the long-dangling appointment, and then the only question was, Whom would FDR name as the Assistant Secretary? The question that occupied Woodring's thinking more at the moment was the matter of aircraft procurement, with particular regard to the type of bombers to be purchased in the next two years.

Arnold, in a meeting with Robins and Echols, had cited three major conditions affecting purchases: Woodring's demand for numbers, balanced-program figures and distribution, and available funds. Since G-4 of the General Staff, headed by sixty-year-old Brigadier General George R. Spalding, a career officer in the Engineers, was the controlling War Department force, its position was in line with Woodring's. *Numbers* was the watchword.

On June 1, Andrews presented to the War Department a detailed analysis on why all heavy bombardment purchases should thenceforth be in the four-engine class. His position was that the B-17 could be utilized equally for long-range reconnaissance and for heavy bombardment. It would serve a dual defensive purpose, like no other bomber. Further, it was a far safer aircraft than any twin-engine model, offering not only greater flexibility of action but also greater economy. He pointed out that the twin-engine bombers now on contract would equip two full-strength groups and three reconnaissance squadrons by 1939 and would be in the front line of national defense until 1946. "Unless four-engine development is encouraged now," he warned, "we shall be hopelessly behind the state of aeronautical art before then and shall be equipped with outmoded, outclassed and obsolete bombers," adding that only a four-engine bomber could reinforce the Hawaiian Islands.

Two days after Andrews' submission, Arnold noted in his daily log that an abstract of proposals for bombardment competition would be awarded to the Douglas Company for 177 B-18's. This was passed on to Woodring for approval. The next day, Andrews' plea for no more twin-engine bombers arrived in Hap's office. He remarked that "in view of the fact that such a policy comes under the policy of the War Department as to the proper role and requirements of GHQ Air Force, I communicated with Col. [James E.] Burns, giving him an outline of the letter, and invited his attention to the fact that this should be considered prior to the decision on the Evaluation Competition for bombers." The final decision, he added, must be made prior to June 15.

On the fifth, G-4 returned Andrews' letter to the office of the Chief of Air Corps, asking for remarks and recommendations. The letter was given a number-one priority, and Robins, at the Materiel Division, was so informed.

Apparently support for Andrews' recommendation was not all that forthcoming from Westover's office. When Andrews called and talked to Hap a few days later, Arnold tried to tell him they had no real selling point, because they were short of data on the B-17, that what was needed was service-test information. In any case, Woodring had already approved the purchase of the B-18's, and the next opportunity for buying B-17's would come as a result of the multi-engine competition just completed. Hap asked Andrews to send in all available test data. Andrews replied the trouble was that he had to come to Washington and justify everything he did instead of his information being accepted for what it was worth.

A week later, the two sat down in Hap's office and Andy spent a half hour telling his former wing commander what a terrible mistake had been made in buying the Douglas instead of the Boeing. Arnold later recorded: "Andrews made such statements as 'These planes are obsolete before they are bought. We will have them for seven years. They haven't the firepower. We should buy planes based upon firepower, not based on numbers.'"

According to Arnold, Andrews refused to take into account considerations such as balanced-program figures, evaluation of two-engine bombers, possible tactical needs for such a plane and, most important, objection to the four-engine bomber by the General Staff, Secretary of War and possibly the President. Wrote Arnold, "All these were thrust aside with the wave of the hand."

The objection to the B-17 was summed up by Deputy Chief of Staff Embick when he stated:

> Defense of the sea areas, other than within the coastal zones, is a function of the Navy.
> The military superiority of a B-17 over two or three smaller airplanes that could be procured by the same funds remains to be established, in view of the vulnerability, air-base limitations and complexity of operation of the former type . . . If the equipment to be provided for the Air Corps be the best adapted to carry out the specific functions appropriately assigned to it under Joint Action there would appear to be no need for a plane larger than the B-17.

No words better illustrate the caliber of thinking opposing Andrews' position. The ominously quickening pace of international developments seemed to have no appreciable effect on the General Staff aeronautical view or that of the Secretary of War.

By year's end Andrews had made no fewer than six major attempts to change the War Department's attitude on the matter. His written proposals were followed by meetings with Craig, with Westover, with Spalding, with

Woodring, all to no avail. In August, the Secretary of War rejected the Air Corps recommendation that thirteen more B-17's be purchased.

Statements by Andrews such as the following made no impression:

All great nations, including our own, are developing pursuit equipment with speeds up to around 400 m.p.h. A bombardment airplane of the performance characteristics of the B-18 type is at the mercy of such pursuit planes in aerial combat. To continue the equipment of our units with airplanes of low performance handicaps national defense and is without justification; particularly in view of the fact that airplanes of greater performance are available. The B-17, for example, by supercharging its engines . . . can be brought up to a performance of over 200 m.p.h. at 25,000 feet. It is noteworthy that the recent British mission to the United States did not purchase the B-18 type of airplane, probably on account of its lack of speed.

I therefore strongly recommend against the purchase by the United States Army of additional tactical airplanes of inferior performance.

In a meeting on the subject in Woodring's office with Malin Craig and key officers of the General Staff present, Andrews predicted that a day of reckoning was going to come and someone was going to be found guilty of the failure to go for quality instead of quantity. Knerr, who was also present, observed that the remark angered Craig. It was also pointed out that Group Captain G. C. Pirie, who headed an RAF purchasing mission, considered the B-18 an inferior aircraft. On the other hand, as the British Air Attaché, Air Commodore T. E. B. Howe, informed Andrews, "I am most grateful to you for showing me the four-engine Boeing bomber, which looks most impressive and gives one an interesting idea of what may be expected in the future.

"We were both most interested and impressed with the demonstrations of flying in various formations by the B-17 bombers under Colonel Olds' leadership."

The impression was such that Claire Egtvedt mentioned to Hap Arnold that the British appeared to be in the market to buy thirty-five of the Boeings. His company was hurting over the Air Corps's failure to procure the B-17. He said that both the Russians and the Chinese had been making inquiries on purchases.

In this same period there were other pressures besetting Andrews and his staff. An attempt to get some inkling of procurement plans for 1938 and the estimates for 1939 was blocked by Westover. As always, he refused to release anything officially or unofficially until the War Department gave approval.

Earlier, Craig had finally informed Andrews that he couldn't have George Kenney back. The reason was that a new regulation now prevented the return of an officer to a previous station. There was no doubt in either Kenney's or Andrews' mind that the long delay in reaching the decision had been due to waiting for approval of the regulation. Andrews believed that at root he was to blame for what had happened, telling George, "I am in some way responsible for your having been detailed to Benning." He then moved Mike Kilner into the G-3 position and made Jan Howard the Chief Inspector.

In another area, Andrews had learned that the Norden bombsight was a more advanced and accurate instrument than the one being perfected by Sperry. He was anxious to have his B-17's equipped with the Norden and so communicated to Robins, who informed him that they were proceeding with both companies but that it was necessary to get permission from the Navy to deal with Norden directly.

To obtain the necessary service-test data on the B-17's, Bob Olds had taken his squadron complement of eight planes to Orlando, Florida, and was putting his crews through their paces when a great opportunity arose to show the world, if not the General Staff, what the plane could do. The Peruvian Government sent an invitation to the U. S. Government to participate in an aviation conference at Lima. Andrews immediately recommended that a unit of B-17's be dispatched to attend.

Woodring blocked the proposal. According to Westover, his reasons were based on "some questions concerning the servicing and fueling of the planes down there as well as the use of funds at the time in view of the stringent demands for economy."

In his response to Westover, Andrews noted among other things that the Navy was sending the aircraft carrier *Ranger* with seventy-eight planes. He added with a tone of cordiality, "Why don't you and Arnold hop in a plane someday and come down and have luncheon and play golf with Brant and myself?" Their differences were official, their personal relationship congenial. One way to soften the bumps was on the fairway.

Andrews' proposal for the Lima flight was refused, but by the end of August there were sufficient data of a dramatic quality on the B-17's capability—data for the General Staff to study and the Navy to try to hide in Davy Jones's locker.

Since 1931 there had been no joint maneuvers between the Air Corps and the Navy in which the former attempted to bomb the latter's ships. The 1931 effort, referred to as the *Mount Shasta* incident, had turned into something of a debacle for the Air Corps, and the War Department had not been anxious to pursue such exercises further. Neither, for that matter, had the Navy, particularly since the state of the art had come a long way since Billy Mitchell's day. Air Corps pilots believed, correctly or not, that a goodly portion of the War Department's continued reluctance to permit its planes to go after the Navy's shipping was in deference to the Navy's wishes.

In the fall of 1936, the President changed all that. He wanted to know for his own private information just what Army land-based bombers could do against a fleet. In the 1936 exercise he didn't find out, because although Air Force units and Navy elements carried out a joint exercise, it did not include the acid test. That was reserved for the exercise to come.

Before the parameters for the next test were ironed out by the Joint Army-Navy Board, Andrews submitted a list of considerations to Malin Craig that he felt should be weighed before a decision was reached. The principal one was that the exercise should be held on the East Coast because of facilities, good communication, and the possession of a small, radio-controlled boat that the bombers could use for practice. Additionally, the West Coast at the time of year selected for the exercise was usually fogged in, making it inferior for training purposes.

On July 10, 1937, after a meeting with Roosevelt, Craig and Admiral William D. Leahy announced in a secret memo that the exercise would be held on the West Coast. The target would be the battleship *Utah*. The area involved would cover a tract of over one hundred thousand square miles, extending three hundred miles outward from the coast, westerly from San Francisco and Hamilton Field in the north, and to the south on a parallel line out from San Pedro Bay. The time set for the contest was from noon August 12 to noon August 13. No bombs could be dropped after dark on the twelfth or after noon on the thirteenth.

As Delos Emmons, who had been named as Commander-in-Chief of Air Forces, put it: "A more unpropitious time for this mission could hardly have been selected. We can expect unfavorable fog conditions off the coast of southern California during the month of August. The probabilities are that this fog will extend out to the limit of the problem."

To Andrews and his staff the exercise was a stacked deck. Their forces, the Blue, would be made up of thirty B-10's, four B-18's of the 7th and 19th Bombardment groups and seven of Bob Olds's B-17's. There would also be thirty Navy patrol planes to be used for scouting, but the Blue defenders would have no ships except for rescue. The attacking force, the Black, would supposedly consist of two battleships, one aircraft carrier and nine destroyers, all represented by the *Utah*. The bombs to be dropped were fifty-pound Navy water bombs, never used before by the Air Force.

Further, Navy reconnaissance planes could not start looking for the *Utah*, whose purpose was to attack the West Coast somewhere within the target area, until noon of the twelfth. As soon as they located the *Utah*, they were to flash the word to the Army planes, which would then fly out and attack the ship.

Emmons established two chains of operating bases, one along the coast and the other farther inland, because he knew the chances were that the coast fields would be fogged in. His B-10's had a radius of action of only three hundred miles. On one score, the Air Force was able to hoodwink the Navy. When the latter asked how fast the Army planes could fly, instead of giving the correct air speed, the airman's response was 190 mph. Otherwise, it had been reasoned, the Navy would keep the *Utah* so far out to sea that there wouldn't be time for search and attack.

From the point of view of Lieutenant Curtis LeMay, the *Utah* mission began when Colonel Olds's B-17 unit began practice-bombing targets in the shape of battleships. The crews were enthusiastic about clobbering the targets but they realized they were pretty useless when compared to the real thing because they were stationary. Even going after the radio-controlled boat did not really afford what was necessary, but they worked to perfect their precision. On that score, it was LeMay who was going to be called on to show how well he had perfected his. He was named the principal navigator, in the lead plane, flown by Captain Caleb V. Haynes. Andrews and Olds would also be on board.

In their practice bombing they had been using sand-filled, powder-charged bombs. Upon reaching the Coast, they were anxious to try out the Navy's water bombs, but somehow the Navy suddenly found it had an unexpected shortage, and it wasn't until several days before the exercise was to begin that Emmons was able to lay hands on a supply for his bombers.

High noon of August 12 found the entire West Coast socked in solid with fog, extending outward an estimated two hundred miles. The crews waiting at Oakland, at Monterrey, at Hamilton, could only sit and wait . . . and wait. There was absolutely nothing to go on, but Lucky Emmons guessed that the *Utah* would enter the target area at the northwest corner, make a feint at San Francisco and head for San Pedro Bay to attack the aircraft factories around the Los Angeles area.

At 3:37, Navy patrol planes signaled that they had located the *Utah*. She was three hundred miles southwest of San Francisco, but by the time the message was forwarded through the 1st Naval District Headquarters to Emmons, at Hamilton Field, it was after four o'clock. The B-17's and the B-18's scrambled. There was no point in the B-10's taking off; they had neither the speed nor the range.

A few minutes after Olds's squadron was airborne, a Navy patrol plane signaled the *Utah* was forty miles farther east than first claimed. Three additional reports confirmed the sighting. LeMay made his calibrations, and when the flight was over the designated site, Haynes took his Fortress down through the fog while the other six ships stayed on top, circling. They broke out at about seven hundred feet and there was no ship in sight —just a large amount of empty sea and the light fading fast. LeMay swallowed hard on that one and began giving headings for a square search pattern. Nothing. And then it was dark and they had to go home.

Bob Olds came back from the cockpit and looked down at the Lieutenant. "Are you sure you knew where that boat was supposed to be?" he asked.

LeMay stared back at him and said that he was. He took a celestial sighting that indicated to him that his error could have been no more than a mile or two from the exact location given. To prove that he knew what he was talking about, he told the unsmiling Colonel the course to San Francisco and the exact time they would be over it.

Olds said, well, they still had tomorrow morning, and then added, "I want the *Utah*. You'd better find it for me."

Ten minutes before his estimated time of arrival, LeMay went up to the cockpit and stood between Haynes and Olds and waited. With the fog, they'd never see the city, but they'd see the glow of its lights, that is, if San Francisco was where it was supposed to be. It was, right on the money! Olds crumpled up the slip of paper LeMay had given him with the ETA and said, "By God. You were right. Then, why didn't we find the *Utah?*"[4]

Andrews had the answer. "It turned out," he said, "that all these reports by Naval patrol planes were incorrect. They were not corrected by the Navy until nine-thirty that night." Just a small error, said the Navy, only one degree off. Just a matter of sixty miles! No one in the Air Force, from Andrews on down, was about to believe for a second that the error hadn't been intentional.

The B-17's couldn't get into Oakland that night. Nor could they land at Mather Field, for its runways were too short and it had no runway lights, so they put down at Sacramento Municipal Airport.

It was eleven o'clock and the crews worked on their planes to have them fueled and ready for the morning. They got their food at a hot-dog stand and then lay down on the hangar floor for some much-needed sleep.

It was Olds who woke LeMay, saying, "Get up, LeMay. Let's go have a cup of coffee." Over the coffee he informed the Lieutenant that he had been right all along. The Navy had admitted its error. From then on, Bob Olds never questioned his chief navigator's accuracy.

Andrews soon arrived with the announcement that again the Coast was fogged in, that the Navy planes that were supposed to keep contact with the *Utah* during the night had been unable to do so and would not be able to take off and go hunting for some time. By nine o'clock, with no word on the *Utah*'s location, Emmons decided to get his planes airborne. Andrews climbed aboard with Olds, and Haynes took off. Once he had the Fortress trimmed up and cruising, the big North Carolinian made an announcement in his deep, heavily accented drawl. "If I'm going to command this flight, I want to be in command. If too many people give commands, everything will be screwed up."

Andrews laughed and said, "That's all right. Come on, Bob."

After his passengers had departed the cockpit, Haynes latched the door, but he'd no sooner done so than Andrews came through to him on the intercom. "There's just one order I want to give you," he said. "You are not to fly below one thousand feet."

It was not until eleven o'clock that a Navy plane signaled the *Utah*'s location. Had they remained on the ground until the position report arrived, there would have been no chance to find the ship, and had they followed the exact course calibrated by LeMay from the information given, they still wouldn't have found the target.

At ten minutes of twelve, Haynes contacted his radio operator. "Sergeant, did you hear that order that I was not to go below a thousand feet?" he asked.

"Yes, sir," came the reply.

"Can you tell whether I'm above or below a thousand feet?"

"Yes, sir. There's an altimeter right here."

"If there's an altimeter right there, Sergeant, I can't very well go below one thousand feet, can I?"

The Sergeant got the message. "Oh, I see what you mean, sir." He then swung around, his body blocking the instrument, and began a conversation with the two ranking officers. Haynes passed the word to his squadron and down they went through the fog, spreading out in a line abreast in a last try at what appeared to be the impossible. They broke into the clear, and there, in what LeMay described as "the greatest happenstance in the world," lay the *Utah*. She was 285 miles from shore, her crew sprawled all over the decks, sailors taking their ease. The air crackled with the excited chatter of the B-17 crews. Even the crews of the B-18's, which wouldn't arrive in time to attack, were cheering. Below, on the battleship, the scene was one of disorder, a wild scattering of seamen making for hatches.

LeMay, who was now up in the nose section with bombardier Lieutenant Doug Kilpatrick, saw the first water bomb hit the *Utah*'s deck, a joyous sight. In the last five minutes of the exercise, three direct hits were made on the battleship. There were many near misses, and had they been real bombs they would have gutted her.[5]

Caleb Haynes, with a cheering Bob Olds, climbed his Fortress back on top of the fog deck and, with his jubilant squadron, headed for home. LeMay had never seen anyone grin like General Andrews. He had completely forgotten his orders to Haynes about staying above a thousand feet. They were all literally floating on air, but Curt had made another observation, which he called to Olds's attention. By his calculations, from the point of interception to the point at which they crossed the California coast, he proved that once again the Navy had given them a position report on the *Utah* that was one degree, or sixty miles, off the mark. Had they remained on top of the fog instead of going down to have a look, not only would they have arrived too late at the location given, but also there would have been no *Utah* there.

Aside from the Navy's seeming inability to give accurate position reports, its brass attempted to wipe the bomb water off its collective face by maintaining the successful attack hadn't proved a thing. It had been a sneak attack and the *Utah* had had no chance to maneuver.

Since the exercise had been laid on by order of the President, Andrews said, if the point hasn't been proved, get your ship out in the open sea, and let it take all the evasive action it can. We'll bomb from all altitudes. Olds angrily put the challenge more bluntly. "All right, God damn you! Get out from under those clouds. We'll bomb you from altitude and see what happens!"

The Navy was boxed in on this one. If the exercise had not proved what the Air Force could do, what had it proved?

Andrews' request that the *Utah* make itself available for another day of tests had to be granted.

This time the weather was clear.

The *Utah* took evasive action. The B-10's, the B-17's and the B-18's dropped their water bombs from eight thousand to eighteen thousand feet. There were thirty-seven direct hits on the battlewagon, about 23 percent of the total dropped, enough to have sunk her many times over.

Although the Navy could do nothing to discount the action except to insist that its results be kept secret, a report on the previous day's exercise was notable for its omissions. A confidential memo from the Senior Naval Commandant at San Francisco to the Chief of Naval Operations dated August 31, 1937, noted: "The *Utah* was intercepted by 7 B-17's and 3 B-18's of the 2nd Bombardment Group on August 13, 1937. The airplanes arrived over the *Utah* at 1155 hours. The exercise ended at 1200, and the planes returned to bases. The ceiling over the *Utah* was 400 feet. No bombing was possible."[6]

In an account he gave at the Army War College in October, Andrews described Joint Air Exercise No. 4, as it was called, and drew attention to the superiority of the B-17 over all other GHQ Air Force bombers. It was his constant theme now, wherever and to whomever he spoke. "The air power of a nation," he concluded his lecture, "is essentially what it actually has in the air today. That which is on the drafting board are the statistical tables of resources and manpower and can only become its air power five years from now, too late for tomorrow's employment."

Malin Craig sent a letter of commendation to Andrews and his staff, praising the efficiency and dependability of GHQ Air Force, and in a personal note to its commander reiterated, "Again, this old swivel-chair general is proud of you and your staff and the work you did in connection with Coastal Frontier Defense, Joint Exercise No. 4."

The secrecy of the action, determinedly insisted upon by the Navy and agreed to by the War Department, came unstuck when newscaster Boake Carter, somehow in possession of evidence, stated on his nightly broadcast: "I've got some pretty important news. The Air Corps can not only find battleships with Flying Fortresses, it can put bombs on them, enough bombs to sink them. What's more, it can put bombs on a towed sled that's only about one third the size of a ship. I know what I'm talking about. I've got photographs right here in front of me to prove it."

The Navy brass were furious over the leak and informed the Army brass. Craig had specifically stated to Andrews that "there will be no publicity about the matter that can be avoided, as the President stated the exercise was for the information of himself, the Secretary of War and the Secretary of the Navy." Now the American public had been let in on the secret, and there was outrage in high places. GHQ Air Force was queried in no uncertain terms: What did it know about the leak? Who was the culprit? At Langley, straight-faced officers shook their heads. No one had a clue. But Curt LeMay would never forget the grin on the Air Force Commander's face when they had clobbered the unsuspecting *Utah*.

Twenty-two

Gyrations

In September 1937, Andrews had occasion to write a personal and confidential letter to Oscar Westover on an article the latter had written titled "The Army Is Behind Its Air Corps." Although he praised Westover on his writing skill, he took gentle but firm exception to the tone of the article, which gave all credit to the War Department for everything the Air Corps had accomplished, "failing to mention faults and deficiencies in past and present policies."

It was his experience, he said, "that every reform and progressive step forward has been forced through, in the face of lack of interest, inertia, and in some cases, active opposition of the War Department."

More, the article would have been strengthened if Westover had pointed out that numbers of planes meant little without men to man them, and that numbers of planes without regard to type were equally valueless.

He reiterated that the backbone of air defense must be the long-range bomber, and he knew that Oscar agreed with him on the point, as did most of the officers in the Air Corps. Then he made his key observation:

> The Air Corps until recently prided itself in securing the best equipment obtainable regardless of expense, but the recent procurement of bombers illy suited to carry out the Army's mission in coast defense is the first time that an inferior airplane has been produced when a superior one was available.
>
> If and when the War Department truly recognizes the need for adequate personnel and for suitable bombers, . . . and sincerely fights for their provision, then and only then, I think we can say, without reservation, that "the Army is behind the Air Corps."

Within his evaluation of the *status quo* outlook of the War Department, he also believed that it was drawing "just exactly the wrong conclusions from air operations in Spain and China." From his own information, he believed these operations clearly indicated that only if a nation had a strong bomber force could it protect itself against air raids on its cities and industrial heartland, because a bomber force would be able to knock out an enemy's air bases and be in the position to retaliate. To Delos Emmons he stressed, "We have got to sell this idea to the War Department. It is a big job and I confess we don't seem to be making much progress at the present time."

Hap Arnold also considered the problem extremely serious and discussed it with Hugh Knerr. They had to do something to convince the War and Navy Departments that air power in Spain and China was not being properly employed. Knerr felt it was a matter of poor bombing equipment, particularly bombsights.

On air operations in Spain and China, Delos Emmons told Andrews that he had positively begged Oscar Westover to have someone make a thorough analysis of such operations, not just covering the type of equipment being used but also showing the way it was being misused. Such a study, he said, should be submitted to the War Department so that the General Staff would not continue to draw the wrong conclusions.

There appeared to be a consistency in the drawing of wrong conclusions that defied efforts of clarification on many levels. The B-18 was a prime example. George Brett laid his finger on an internal weakness that had permitted selection of the aircraft in the first place. He said that it had come about because the Materiel Division did not fly experimental aircraft, yet was in the position to recommend production. Emmons fully agreed, and complained to Andrews, "I can't understand how an airplane like the B-18 could have been approved for production when it is so full of gross bugs." On one engine, it could fly only five hundred miles. It was poorly armed and was a sitting duck for interceptors. It lacked any means for towing targets, and since the units equipped with it had no other type of bomber to substitute, combat crews could not work on their aerial gunnery. Andrews confessed he was very disappointed with the aircraft; it would never be able to reinforce the Hawaiian Islands.

A special board headed by Westover and joined by Arnold met with Emmons at March Field to see if they could not come up with "some means of making the plane more satisfactory." Andrews asked Emmons to emphasize to both Westover and Arnold the need for a "common commander for the service and combat elements of the Air Corps." He had thought Westover had been sold on the idea after the May maneuvers, but at the present there was no indication of interest on anyone's part. He realized the Wilcox bill was never going to get to the hearing stage. Therefore, with the chance of wide exposure on the need for change scuttled, Andrews' only recourse was to try to bring pressure to bear by internal support and by the force of reason. As he said to George Brett, "Every great nation has found it necessary to develop a special type of organization in order to build up an adequate air defense. I thought that the Wilcox Bill offered a very practical solution. . . . However, as this bill seems to have no chance of ever coming to a hearing, we must seek some other solution, which I believe should be a Deputy Chief of Staff for Air. . . ."

It was simply another variation that got nowhere, as did his suggestion that GHQ Air Force headquarters be moved to Washington in order to afford closer communication. There were many in the War Department who felt that Langley was much too close as it was.

Andrews' two-pronged thrust of reorganization and procurement of the best was unceasing. The B-17, he insisted, was the most efficient and most advanced airplane for reconnaissance and bombing ever built. But it was not the ultimate. They must look forward. The Boeing XB-15 had been completed in the fall of 1937. Its weight, it was found, was too great for its four 1,000-hp engines. It could manage a top speed of only 190 mph, and its gas consumption impeded its range. But an attempt to construct an improved and modified version was rejected.

As long as Andrews' efforts were kept in the proper military channels, no one on the General Staff became overly exercised. After all, he had every right to push for what he believed he needed to improve his command. But when the article Malin Craig had asked him to withhold and rewrite earlier in the year appeared in the November-December issue of *Army Ordnance,* it raised a hue and cry all the way up into Woodring's office. It wasn't as though he had said anything different from what he had been saying months on end, but now he had said it semipublicly, even though *Army Ordnance* certainly had a limited readership.

In attempting to reprimand him by citing the displeasure of the Secretary of War, issue was taken not over his development of strategic air power, not over the need he stressed for thinking ahead five years, not over the influence of aviation on any wars of the future, not over the failure to build the air bases authorized two years past by the Wilcox bill,[1] not even over the shortages in men and planes, but over the fact that because of these shortages it had been necessary at times to use reserve officers and even air cadets "as airplane commanders with far less flying experience than is required of even a copilot on a transport line."

This, declared the Adjutant General's Office, was in violation of specified requirements. The complaint was more a reflection of embarrassment over the hard reality of Andrews' measured, concise and unadorned account of what was what. It was patently clear where the responsibility lay, and Henry Woodring didn't like the inference one bit.

The demand for an explanation by Andrews could also be viewed in the nature of a warning.

There were other warnings of a different nature but far more obvious. In July, Japan had provoked open warfare with China. On December 12, the U.S. gunboat *Panay* was some twenty-seven miles up the Yangtze River from Nanking. Her decks were painted with large Stars and Stripes to show that she was a neutral ship in a war zone. Twelve Japanese bombers flew over her and dropped their bombs. In the unprovoked attack, the *Panay* and three Standard Oil supply ships were sunk. Three U.S. sailors died; others were wounded. Japan made a formal apology and paid requested indemnities. The incident was quickly put aside.

During the summer, President Roosevelt had announced the appointment of Louis A. Johnson as the new Assistant Secretary of War, under Henry H. Woodring. At the time, Woodring indicated no objection to the appointment, but by the end of the year a running feud between the two was an open secret, supplying choice gossip on the Washington cocktail circuit. Johnson, like Woodring, was a staunch member of the American Legion and had been its National Commander in 1932. An equally staunch and ambitious Democrat who had also served as his party's floor leader in the West Virginia House of Delegates, Johnson had, during the campaign of 1936, organized the Veterans Division of the Democratic National Committee. The offer of the post of Assistant Secretary of War was in payment for political services rendered, but, again, like Woodring, when first offered the position, Johnson had said no.

Jim Farley had talked Woodring into accepting the job; now he did the same thing with Louis Johnson. Only, in this case, according to Johnson, he did so on the promise that Johnson would soon be taking over the top position in the War Department, because FDR was going to drop Woodring. Farley later denied the promise, but Johnson assumed office with two thoughts in mind, which he expressed freely and frequently. He would be Secretary of War within a few months, and as Assistant Secretary his responsibilities made him answerable specifically to the Congress and the President and not to Woodring.

Since Johnson's main tasks centered on military procurement and economic mobilization, to ignore the Secretary of War in pursuing goals along both paths could only produce an unnecessary conflict within the War Department. It was a conflict that grew increasingly bitter, catching Malin Craig in its pressure cooker. Roosevelt, of course, knew all about it and could have put a stop to it but did nothing. He had a penchant for playing his appointees off against each other so that he could make the final decision, keeping them and everyone else in doubt. At one point, Woodring, lacking whatever it took to face the President, actually asked Craig to do so for him in order to explain the unworkable situation. Craig refused. This was a civilian battle, not a military one. Woodring then went to *Army and Navy Journal* publisher John C. O'Laughlin and asked him to intercede. Cal was a good friend, but he didn't think such a move would be very smart. He did talk the impasse over with Press Secretary Steve Early, who called Johnson "a horse's ass" and said bluntly, "If Woodring had any guts he would ask the President to relieve him or Johnson." But the Secretary of War didn't want to chance anything like that. Johnson, for his part, tried to prevail on Malin Craig to keep his mouth shut on various plans he was advancing. This the Chief of Staff refused to do. His boss was Woodring, and his loyalty was to him. How would Johnson like it, he asked, if the roles between himself and the Secretary of War were reversed? Johnson had no answer to that.

Quite naturally, the Woodring-Johnson feud had its effect on the Air Corps and its leaders. At first Johnson was viewed as being on a par with Woodring in his understanding of air power. The concerns of Andrews, Arnold and Delos Emmons over the War Department's misreading of the meaning of air combat in Spain and China centered on Johnson's openly expressed opinion that he thought the use of the airplane in war had been greatly overrated. Andrews had a chance to size up the Assistant Secretary at a late-fall 1937 meeting on the everlasting issue of the balanced program, looking forward to 1945. Present were both secretaries, Malin Craig, Embick, all the assistant chiefs, Westover, Andrews, Arnold, Robins, Knerr and a host of lesser lights from the General Staff and the Air Corps. Only Andrews and Knerr were present to represent the Air Force.

General Spalding, Chief of G-4, presided but had little to say. Westover presented a projection that was a compromise based on keeping to the 2,320 aircraft figure and staying within the limitations of the budget. His program envisioned two heavy bomb groups in the United States and one in the Hawaiian Islands. Andrews agreed with it, knowing it was the best that could be hoped for under conditions imposed. He then said again that he believed that there should be only two types of bombers: the heavy and the attack. Johnson then spoke up and asked Andrews to please submit to him a program along those lines. Andrews said he'd be glad to, and added that he believed a board should be appointed to consider the whole issue of air defense before they committed themselves to another long-range program. This drew Woodring's ire. He said he thought the General Staff was better able to come to a decision on the question than any board. The conference ended, as Andrews expected it would, with no decisions made on anything. Woodring announced there would be another session on the subject, but he gave no date.

Shortly after the New Year, Andrews received some encouraging news from Emmons. Johnson had stopped off briefly at March Field and Delos had asked him if he had taken the time to read the General's statement on long-range bombers. Johnson replied that he had and was very much impressed. Emmons then asked him if he planned to push the matter. The Assistant Secretary said he would do all he could, but it was his opinion they would never get any long-range bombers as long as Woodring re-mained as Secretary of War.

Emmons also informed Andrews that he had written a long letter to Hugh Drum, in the Hawaiian Islands, on the need for long-range bom-bardment aircraft, and enclosed a paper with charts prepared by Follett Bradley. "It may do some good in educating him," wrote Delos. Andy replied that anything that could be done to educate Drum would be worth the effort. He didn't add that at the moment he was more interested in educating Louie Johnson.

Andrews' own knowledge of European developments had been en-hanced through a private meeting with Charles A. Lindbergh at the Metro-politan Club, in Washington. In late 1935, the Lindberghs had moved to Europe because of unceasing harassment. Their lives had been made miser-able by a combination of irresponsible news reporters, criminals and luna-tics, their home having to be constantly under guard. Thus, when Lind-bergh and his noted wife returned to the United States late in 1937 for a few weeks' visit, they attempted to slip into the country quietly and make themselves inconspicuous.

Andrews received a telegram from the noted flyer suggesting a luncheon meeting. The decision to talk to Andrews most probably came from the suggestion of U. S. Military Attachés Colonel Truman Smith and Major Arthur W. Vanaman, stationed in Berlin. Both officers knew Andrews well.

Through an invitation by Hermann Goering, Lindbergh had been per-mitted to inspect the Luftwaffe, as had no other outsider. He was greatly disturbed by what he saw developing, and it can be reasoned that his pur-pose in wanting to see Andrews was to give him the benefit of his observa-tions.[2] Andrews was, of course, intensely interested in all aspects of for-eign air power and, through Hap Arnold, received yet another helpful briefing, this one written on bomber action in China.

The worry over growing misconceptions concerning the value of air power in the two ongoing wars in Spain and China had moved Arnold to appoint Lieutenant Colonel Gerald E. Brower, Air Corps G-4, to start compiling all the data that could be obtained in both theaters. Andrews contacted Brower, who sent him what he had on air combat losses. From this information, made up from eyewitness reports by various sources and Japanese news agencies, it appeared that Japanese bomber losses without fighter escort were catastrophically high. These were twin-engine types looking, said one U. S. Air Corps Reserve officer in China, like Lockheeds. They carried guns in the nose, tail and waist of the plane. There were twenty-one in one raid, their target being the Chinese main air base near Shanghai. Half of the bombers were shot down over the target by inter-ceptors, and the reporter believed that nearly all the rest were eliminated in their return to Formosa, the rumor circulating that only one plane arrived safely.

This kind of information did nothing to change the thinking of Andrews and his staff and those who studied similar information at the Air Corps Tactical School and in the Air Corps. They saw the Japanese losses as a combination of inferior equipment and poor tactics, and since the Japanese were claiming much heavier losses by the Chinese while indiscriminately bombing Chinese cities, it was difficult to obtain an accurate picture.

On one point Andrews was sure. Circumspection must be used in making public statements on all aerial military matters. "The present times are full of dynamite, nationally and internationally," he wrote to Jerry Brant. "We must exercise extreme caution to be sure that we do not step on the toes of any foreign nation, or of high officials in our government. To do so might easily result unfavorably to our program for development of air defense.

"The international situation is such that unusual care must be exercised that foreign nations do not learn of the performance characteristics of any of our new equipment beyond that which cannot be kept secret."

This latter War Department sensitivity was exhibited when, in January 1938, the XB-15 made its maiden transcontinental flight to the West Coast and return to Langley. There was considerable news publicity concerning the country's biggest bomber and its performance. Deputy Chief of Staff Embick was promptly on the phone to Andrews. The War Department wanted no further publicity that might reveal performance data to anyone. A perfectly understandable wish, but in the minds of the suspicious, like Hugh Knerr, the *anyone* would also include those in Congress and elsewhere who might be inclined to see value in further development of aircraft that Embick and his General Staff colleagues saw no use for.

The point was soon to be reproven. Woodring had finally called for another conference on the procurement program for 1940–45. Andrews, by this time, as well as other Air Corps planners, recognized that they appeared to have a sympathetic ear in Louie Johnson. It didn't matter how much of the Assistant Secretary's support was based on a desire to oust Woodring or whether he was genuinely sold on Andrews' position. That his support was there was all that mattered. On January 25, Andrews met with him at his office, and they went over the presentation that the Air Force would make at the conference. Johnson was willing to back it on all points.

The same cast was present as had been at the November gathering. Previous cooperation by Westover, which Andrews had greatly appreciated, had included the sending of two officers to Langley to discuss the Chief of Air Corps's proposals. At the conference, Andrews praised Westover's

program, termed it "excellent in most respects." But—there were certain
points of difference.

They were the same old differences: two types of bombers, more aircraft, more personnel, the relocating of GHQ Air Force headquarters in
Washington and not way the hell out in Scott Field, Illinois, the crying
need for greater aeronautical development in order to keep pace. Andrews'
amplifications were pointed. To Woodring, to Craig, to all of them, he
made it clear that the 31st Bombardment Squadron, of the 7th Bomb
Group, now ordered to the Hawaiian Islands, was going by boat because
the B-18 couldn't safely fly that far. If the squadron had been equipped
with B-17's, there would have been no problem, not even the costly necessity of a permanent change of station.

"Airplanes that can be flown to theaters of operation, continental, insu-
lar possessions, or foreign, in which this country might become engaged,
and which can be transferred by air from one theater to another, have an
obvious national-defense advantage with our limited shipping and in our
particular geographical situation." So, too, he explained the importance of
air-base location and particularly Air Force Headquarters. He had recommended repeatedly that it be moved to the capital, for the obvious reason
of efficiency and close communication. He made it plain that he devoutly
hoped that there might be a broader understanding of the application of
air power.

Certainly there was elsewhere. The world's major nations recognized
that the aircraft was a weapon of war that had brought into being a new
and entirely different mode of combat. "It is another means, operating in
another element for the same basic purpose as ground and sea power—the
destruction of the enemy's will to fight. It has already become one of the
vital agencies of all major countries to ensure in peace the continuation of
national policy, and to gain the victory in war." The words had been worn
so smooth with use, repeated so often, written, spoken, recited, submitted
in memos, proposals, programs *ad infinitum,* that it was a wonder Billy
Mitchell's ghost didn't rise up in their august midst and cry, "Woe! Woe!"

Once again Andrews tried, on a more earthly plane, to enlist their attention and interest in things to come. "It should be remembered that the
largest bombardment plane we now have available for procurement has
only a one-thousand-mile tactical radius of action and a bomb load small
in comparison with what could be built." Future capabilities, he said, were
just beginning to be realized, and the struggle for strategic air bases and
effective air fleets was well under way and becoming intensified through fast
technical development. It was the Army's responsibility to get with it. He
thought the program he was recommending as a compromise was actually
too conservative and not progressive enough. Outside of Hugh Knerr, and
possibly Hap Arnold, there wasn't another Army officer present willing to

agree with his conclusion. As for the Assistant Secretary of War, he was willing to concur so long as Woodring didn't.

And suddenly this negative factor won a widely acclaimed victory for Andy Andrews and his Air Force. It began with rejection by Woodring, Craig and the War Department of a request by the State Department that the Army send six B-17's to Buenos Aires, Argentina, on a goodwill flight to aid in that country's inaugural celebration of a new president. As he had with the invitation for B-17's to attend the Lima, Peru, aviation conference, Woodring was quick to say no. It would cost too much. There was nothing to be gained, a risk to be taken, etc.

It was said that Johnson got wind of the turndown from a newsman who covered the War Department, but he might also have gotten it from Langley. He went to see Malin Craig for verification and then asked Craig not to tell Woodring what he had in mind. Craig refused to go along, saying his loyalty was to the Secretary. Johnson, unfazed, made an appointment to see the President and thereupon sold FDR on the value of the requested flight. Roosevelt liked the idea, and much to Woodring's chagrin, telephoned him and said he wanted an acceptance.

Bob Olds, who had started off the year by getting in dutch with the War Department, particularly General Embick, because of a story he had given the Washington *Post* after completing a record-breaking transcontinental B-17 flight, was chosen by Andrews to command the 5,600-mile mission. Again Lieutenant Curt LeMay was to be the chief navigator. Word had come down to Andrews from the White House, probably via Pa Watson, that *if* the planes could arrive in Buenos Aires the day after they left the United States, an impression would be made not only in Argentina but in other countries as well.

Olds timed his takeoff from Miami for one minute after midnight, giving himself and his crews a full forty-eight hours to meet the request. They landed at Lima, Peru, to refuel, took off under zero-zero conditions and made world headlines, arriving over Buenos Aires the next day.

Craig was duly impressed. He thought that Olds and his men were doing a splendid bit of flying, but the Chief of Staff would be relieved when they got back to Langley safely. Westover had brought in a plan to have the planes return to Bolling for a big, hoopla reception. Craig said no, and Andrews fully agreed that more favorable reaction would be attained if the squadron returned to its home base as though the mission had been in the nature of a routine undertaking. Craig assured Andrews that the War Department would be sure to take official cognizance of the mission "with appropriate commendation and decoration and wide publicity."

Like the pungent smell of garlic, a certain delicious irony capped the widely heralded and publicly acclaimed venture. Woodring, who had been so completely opposed to the flight, took full credit for its success, thereby further infuriating Louie Johnson. Months later, in his annual report,

Malin Craig would draw attention to the event, calling it "a demonstration of speed, range and navigation accuracy unexcelled by any military planes in the world."[3] At the time, he personally escorted Bob Olds to the White House for a congratulatory meeting with the President. Before they arrived, the Secretary of War allowed that he was going to tell Roosevelt that there wouldn't be so much antagonism to the President's proposed $800-million naval shipbuilding program if the Air Corps received the one thousand planes the program envisioned giving to the Navy. At the meeting, however, Woodring said nothing, possibly because Admiral Adolphus Andrews was seated next to the President throughout.

While the War Department enjoyed the headlines, after they had faded there was no change in general attitude. Nor was there any change in Andrews' effort to have some say in who his principal commanders would be. A final attempt to get George Kenney back as CO at Hamilton Field failed. Through the process of normal rotation, Andrews was losing Jerry Brant, and once again, without success, Andrews made an attempt to have Clagett returned to a wing commander's post. Lieutenant Colonel Arnold N. Krogstad was selected instead, certainly a well-regarded pro. The loss of Hugh Knerr by the same method of rotation was something else again. The change of duty, of course, had been anticipated, and as in Jerry Brant's case Andrews asked Craig if he could talk over with him what Knerr's next duty would be, but when Knerr's next assignment was announced, the message from G-1 of the General Staff was clear enough. Payment had come due for the lean, blue-eyed rebel. He was being sent to Fort Sam Houston, Texas, as Air Officer for the VIII Corps Area, exactly the same job that had been given to Billy Mitchell when he had been exiled to San Antonio.

For an officer of Hugh Knerr's capabilities, it was an assignment in limbo, and to make the message absolutely clear, he was placed in the same dingy office Mitchell had been given. There was a picture of Billy on the desk, and in the corner an open-top latrine used by the clerks working in the outer area. Knerr was also demoted back to lieutenant colonel, but he refused to complain and, instead, enjoyed the flying he was able to get in covering the Southwest. With time on his hands, he wrote a book, *The Student Pilot's Primer,* which was illustrated by his daughter-in-law Sally and published by Van Nostrand. He kept in close touch with Andy on developments, but the frustration of being literally immobilized slowly began to eat at him. As Chief of Staff of GHQ Air Force the extent to which he had overworked, combined with the intensity of his nature, was such that he developed sciatica. It was traceable to his 1923 crash in the West Virginia mountains, and the pain became so severe that he had to fly standing up in the cockpit of the old wreck he had been assigned. Word got back to the flight surgeon, Andy Smith, who not only grounded Knerr but hospitalized him.[4]

The removal of Knerr from his staff completed a change of all the original members. Tony Frank became Chief of Staff. Lieutenant Colonel Ralph P. Cousins was G-1; Major James P. Hodge, G-2, Follett Bradley having become CO of Moffet Field, in California. Mike Kilner was now G-3, and Jan Howard, who was Chief Inspector, would soon become G-4, with Joe McNarney assigned to VI Corps Area.

There was to be yet a new Air Force staff position: aide-de-camp to the Commanding General. Second Lieutenant Hiette S. Williams, of the Air Corps Reserve, was appointed to that duty by his new father-in-law, for on February 9, 1938, the wedding bells had rung at Langley. Josie Andrews and Sunny Williams plighted their troth, Delos Emmons sent the recipe for the wedding punch, and although Andrews lost a daughter, he gained an aide who continued as his copilot.

In the shifting of command assignments, General Hugh Drum had returned from the Hawaiian Islands to become the Commanding General of VI Corps Area. He and Delos Emmons had been corresponding, Emmons forwarding Andrews' position papers to him. Lo and behold, the General, who five years before had denigrated the Balbo flight and took a dim view of strategic air power visionaries, began to sound a bit like one himself. "The increased range and efficiency of the bomber has materially affected the depth of the battle area and opened the way for a restudy of future strategical plans," he declared to Emmons, and then added, "Of course, with the increased bomb load, a restudy should be made of the potential objectives for the large bombers. . . . There can be no question as to the need for the large, long-range bomber, which can be used for both reconnaissance and destruction. . . ." He concluded with the observation, "I fear some columnist may twit the Army because we are shipping bombers to Hawaii, while the Navy flies theirs."

Such a change in Hugh Drum's evaluation of air power must have brought a chuckle from Andrews and Emmons, because the General wasn't aware that he'd changed his thinking at all. But the fact was that Drum now seems to have gained a greater capacity to grasp the reality of the changes airmen had long been predicting than did most of his confreres on the General Staff. Andrews was anxious to nurse the shift. Drum's name was already being mentioned as a successor to Malin Craig.

In March, harsh reality of a less theoretical sort shook Europe and grabbed the attention of the American public. Hitler had bluffed his way into occupying the Rhineland just two years past. Now, in a bloodless yet ruthless coup, he annexed Austria, dissolved the Austrian Government, and made the territory a part of the German Reich.

The President kept silent. In January, through Under Secretary of State Sumner Welles, he had secretly sounded out British Prime Minister Neville Chamberlain on the possibility of a summit meeting to be held in the United States over which he would preside in an attempt to discuss the underlying causes of major European differences. Chamberlain had been less than receptive and Roosevelt had dropped the idea. Now as noted he was trying to prevail on Congress to authorize a naval shipbuilding program costing an estimated $800 million.

Andrews, for his part, was hoping that the results of the South American flight would have a favorable effect on his airplane proposals and the acquiring of enough personnel to bring into being the oft-cited balanced program. He told Emmons that although the War Department had disapproved the Hawaiian flight, the mission to Buenos Aires demonstrated everything such a flight would have proved.

There were follow-up letters of congratulations from air-minded congressmen, and in all his responses Andrews stressed that "had the 31st Bombardment Squadron been equipped with the large bomber, it could have flown to Hawaii and then when required elsewhere flown there also. Large bombardment airplanes, we also believe, are more economical per ton of bombs delivered, than small ones."

Although Andrews' strategic concepts were manifestly focused on the long-range bomber as forming the centerpiece of his Air Force, he was fully cognizant of and interested in pursuit development, as any air commander would be. At the time, his interceptor squadrons were equipped with the Boeing P-26, the Seversky P-35 and the Curtiss P-36. The latter was the newest and best, a low-wing, all-metal fighter powered by a 900-hp air-cooled Pratt & Whitney engine, with a top speed of about 270 mph. The French Government was interested in the plane and dispatched Michel Détroyat, one of its most noted test pilots, to have a look at it and, if possible, test-fly it. Détroyat approached the War Department, and Craig referred the matter to Westover. Air Corps policy was that no foreigner should be permitted to fly any Army planes unless there was reciprocity.

A meeting was held at the French Embassy, at which Détroyat was informed by Colonel Harrison H. C. Richards, G-2 of the Air Corps, that he was welcome to visit either Dayton or Buffalo and look the plane over but that without the permission of the Chief of Staff he would not be permitted to fly it. The French pilot said he was not interested in visiting either location unless he was given a chance to put the P-36 through its paces. The matter went back to Craig, who was not about to make a yea-or-nay decision and instead arranged an appointment with the President. He went to it armed with a memo showing all the reasons why the Air Corps was against foreign pilots flying U. S. Army equipment. FDR decided that Détroyat should be allowed to test-fly the P-36 under certain ground rules. The flight must last no more than twenty minutes. Utmost secrecy must prevail. An Air Corps pilot would bring the plane to an outlying field, where the Frenchman would be waiting. Any instruments that were of a confidential nature—even the altimeter if it was a clear day—should be removed from the aircraft's panel. Hap Arnold was given the task of going to Dayton and making the arrangements with Materiel Division Chief Robins. He was furiously opposed to the whole idea for obvious reasons plus the fact there would be no reciprocity involved. Roosevelt was permitting the flight as a special favor to the French.

Air strength, or a lack of it, was becoming an increasingly important subject, on which politicians and others were beginning to speak out.

One was Assistant Secretary Johnson. He had given an address to the American Legion in Chicago comparing the strength of the major nations. *Time* magazine had criticized his figures. Johnson called Arnold and asked him if he would supply him with additional background. Hap gave the job to Lieutenant Colonel Ira Eaker.

Eaker was now the Assistant Chief of Air Corps Information, having completed the course at the Command and General Staff School the previous June. Glad to be back in Washington, glad to be working again under the command of Arnold—the two were privately collaborating on a new book—Eaker's writing and thinking capabilities were being put to good use.

The paper he prepared for Hap to give to Johnson offered the latest G-2 figures on air comparisons and, as he noted, were almost a year old. The point said much about the effectiveness of Army intelligence. The numbers Eaker quoted included planes of all types and not just combat aircraft. The British headed the list, with more than fifty-six hundred. Italy and Russia were next, with approximately forty-one hundred plus, Germany and the United States were about equal, in the thirty-two-hundred bracket, with France and Japan at the bottom, in the twenty-three-hundred range. If nothing else, the figures were widely misleading, because, as Eaker pointed out, they lumped together everything that could fly.

Captain Eddie Rickenbacker had followed Assistant Secretary Johnson on the speaking platform at Chicago and made the claim that the United States needed one hundred thousand pilots and thirty thousand aircraft. The figures were so far beyond anything that could be attained either economically or through existing production facilities that Eaker said they just couldn't be justified on either a commercial or a military basis. Eaker also drew attention to the differences between quantity and quality. To take one alone as a measurement of strength would be misleading, he said. The swiftly growing interest in comparative global air power was but another measure of increasing international tension. Unsubstantiated word came to Arnold that the Japanese had somehow managed to lay their hands on the design plans for Larry Bell's new pursuit, the XFM-1.[5]

During this particular period, Andrews was in the Caribbean and Florida, observing joint maneuvers in which the 2nd Wing participated. Through them, it was determined that the P-35 at altitude was much too slow to do anything against the B-17. Then, to remind Andrews that all work and no play was debilitating, he received a challenge from Fred Martin. It was issued by the 3rd Wing CO and his Executive to Andrews and Kilner. No quarter would be asked or given, it said.

"The principal thing prompting this challenge," wrote Martin, "is my belief that nothing will do you more good than to leave behind the trials and tribulations of your office and spend an entire weekend as my guest, making sufficient allowance in time that we might gorge ourselves on golf to the extent of three 18-hole matches, in order that a proper decision may be obtained." He suggested that Andy and Mike take off of a Friday so as to have all day Saturday and as much of Sunday as necessary for recreation, "during which time we will refrain as much as possible from talking shop."

Andrews took up the gage, which he said "sounded dangerous." Mike would not be able to come along, because he was too tied up getting ready for the May maneuvers.

Danger of another sort reared its dirty political head when Colonel Davenport "Johnny" Johnson was summarily relieved of his command of the 7th Bomb Group, at Hamilton Field, by order of the Adjutant General. Andrews was informed of the action by a telephone call from the War Department and flew up to Washington the next day to find out what it was all about. Both he and Emmons considered Johnson to be one of the better Air Force group commanders.

What it was all about was an effort to be cooperative that, for reasons of political vindictiveness, was used to demote a valuable officer and transfer him out of an important command. The action did not personally concern Johnny Johnson at all but the official he had attempted to aid, High Commissioner to the Philippines Paul V. McNutt. McNutt, a vigorous, handsome politician with a mane of prematurely white hair, was a former Governor of Indiana. Previously, he had been mentioned as a possible running mate for FDR. Now he had presidential aspirations of his own, and he had returned from the Philippines to test the political waters. In Los Angeles and San Francisco he had made speeches before large, enthusiastic audiences, and he wished to move on to Denver for still another. He asked Johnny Johnson if the Air Force could fly him there. The Colonel could see no reason why not; McNutt was a high-ranking U.S. functionary. To double-check, he sent a wire to Air Corps Headquarters. When there was no immediate reply, Johnson okayed the flight and the plane with McNutt took off. A few minutes later, Ira Eaker was on the phone from Washington. When the details were explained to him he said he thought there would be no problem. Ira was wrong.

Henry Woodring disapproved permission for the flight, and since it had already taken place, Colonel Davenport Johnson was in violation of orders from the Secretary of War and must be punished.

Johnson's Wing Commander, Delos Emmons, was nonplussed. So were the high-ranking ground brass in the area. General Simonds got on the phone and talked to Oscar Westover, saying the whole thing was totally unjustified. Westover said he would take the matter up with the Chief of Staff immediately. But Craig was caught in the middle, and it didn't matter what either Simonds, Westover or Andrews had to say. Woodring was adamant. Andrews tried without success to get Colonel Johnson's orders suspended until an investigation could be undertaken.

Emmons wrote him: "We will take no action that will put the War Department in an embarrassing situation. Johnson accepts his orders as a soldier should. If there is any come-back on me, I will accept it without complaint. . . ." Delos was not one to buck the tide, and he advised Andrews not to either. Instead, Andrews went back to the War Department and expressed himself plainly. "The whole business is very un-American. It happened once before, in Clagett's case, and I wish we could find some way to put a stop to it," was the sum and substance of his position.[6]

Emmons' position to Andrews was, "Only agree, won't you, that we had better soft-pedal the affair until the circumstances that caused the situation to be developed are overcome?"

The circumstances could not be overcome as long as Franklin Roosevelt was President, for it was by White House order that Woodring had issued the turndown. Soon after FDR had taken office, he notified the Air Corps that military planes would no longer be made available for anything but official business—supposedly as an economy move. Very quickly, this edict became highly selective: those congressmen or functionaries whom the White House favored were permitted to travel aboard military aircraft, to make a speech here or pat a baby's bottom there. Naturally, it was assumed that McNutt, as High Commissioner to the Philippines, would be so favored. Apparently not, even though the next presidential election was a far piece down the road.

In a letter of regret to Johnny Johnson, Andrews gave a hint of his own future: "Please be assured of my full understanding and sympathy. Although I will not be in the GHQ Air Force much longer myself, I hope to see you back in it as soon as possible."

With regard to the future, Andrews had some frank and personal words for Oscar Westover. They arose out of the appointment of a four-engine bomber board and who would be on it. Andrews had appointed as his representative Bob Olds, who in his estimation was better qualified to serve than any other officer. He sincerely hoped that Westover understood his position in the selection of boards. "You know, I realize, the problems and responsibilities of this job of mine," he wrote in a private note, "and unless I can insist upon, and obtain, the necessary power and authority and support to handle the job properly, then it should be given to someone to whom these things can be entrusted. I am telling you this as my friend, and not as the Chief of the Air Corps. We have had official differences of view

in the past and undoubtedly will in the future. I hope this question of selection of board members from the GHQ Air Force is not one of them, but
if it is I want you to understand that I shall do all in my power to prevent
this, or any other difference, from affecting our personal relations. I rate a
friendship as one of the most valuable things in life and as we grow older
new friendships don't happen often, so I want to treasure those I now possess." So did Oscar Westover.

A matter now arose of such a secret nature that a memo sent from the
Secretary of War's office to Hap Arnold, who was Acting Air Corps Chief
while Westover was away, was withdrawn because of its sensitivity by
Colonel James E. Burns, Johnson's military aide. It concerned the anticipated visit of a British purchasing mission coming over to visit U.S. aircraft factories with an eye toward "determining what types [of planes]
would be suitable and dates of delivery possible."

Having been directly involved in Détroyat's visit and opposed to his
flight-testing, Arnold was concerned at what appeared to be a far more extensive approach by the British. He thought it was a matter for the President, Secretary of State or Secretary of War to decide.

And so he noted, in the *Confidential and Secret Diary* kept by the office
of the Chief of Air Corps. The British mission, he felt, raised questions the
Air Corps could not answer:

"For instance, are we going to show favoritism to Gr. Britain over other
countries?

"Are we going to assume that Gr. Britain is an ally of ours now and will
always be an ally?

"Should we upset existing policies relative to export of our latest aircraft?

"In other words, can we, should we, must we show Gr. Britain our latest airplanes such as the B-17, B-15 or the Bell Fighter?

". . . there are so many questions of policy relative to this, and it is
liable to put us in the position as an aid to certain nations in war by virtue
of furnishing munitions and thereby endanger our neutrality. . . ."

This was April 1938, and Arnold's concern was reflective of a general
Air Corps attitude toward the sale of hard-to-come-by aircraft to foreign
buyers who might or might not be allies for the long run. This even included Canada, anxious also to look over U.S. military aircraft, with a
desire to purchase in mind.

At a preliminary meeting with Embick and Burns, it was determined
that a conference would be held prior to April 19, when the Canadians
would be arriving. At it they must have the dates of release for foreign sale
as well as recommendations from Westover upon the effect such release
would have on deliveries to the Air Corps. Some planes should not be
released at all, "due to their having certain implements, gadgets, or accessories which were far in advance of any developed in foreign countries."

The Navy was equally concerned over the idea of foreign sales, and Admiral Arthur Cook, of the Navy Department, contacted Arnold to say that he thought it was absolutely essential that the Army, Navy and State Department get together and come to some definite policy before the arrival of the British mission. The Admiral said he had gotten wind of the Air Corps permitting the Frenchman Détroyat to test-fly the P-36, and he wanted to know if the same procedure was going to be adopted for the British.

Following some preliminary huddles with Embick and Johnson, all concerned paid a visit to the State Department and sat down to reach a decision with Sumner Welles. He informed them of the official attitude as approved by the White House. Foreign sales should be encouraged of those aircraft previously released, and there should be no hesitation in showing later types that might be released in the near future. All countries must be placed on the same basis, no one any more favorite than the other. It would be proper to encourage the visitors to buy from those plants now about to run out of work, particularly Lockheed and Vultee. The only major objection was to the release of manufacturing rights to any foreign country. It was clear that the Administration's willingness to permit purchase of U.S. military aircraft was motivated as much by economic considerations as political.

In the course of what followed, it was no wonder that airmen gritted their teeth and shook their heads. Woodring approved Boeing's request to sell B-17's to the British, willing to have the Air Corps's best piece of aeronautical equipment sold to a foreign power but unwilling to authorize additional procurement of it for the Air Force. Oddly, when Boeing requested that members of the British mission get an inspection flight in the Fortress, Deputy Chief of Staff Embick said no.

Essentially, the British said no, too. They were interested in buying two hundred twin-engine Lockheed 14's (Hudsons) with gun turrets installed for reconnaissance purposes, and two hundred North American basic trainers. As Group Captain Pirie had previously informed Andrews, the RAF had no interest in purchasing the Douglas B-18. Its performance didn't match what they already had, and as for the B-17, they decided they didn't want it, either. In a few months they would have a bomber of their own, having, he said, far superior characteristics to the Boeing. This was the twin-engine Manchester, out of which the four-engine Lancaster would be developed, but, like its U.S. counterpart, it was produced before the war in very short supply. However, in this case, the British lack of bomber production was for reasons that were both strategic and economic.

In 1936 Sir Thomas Inskip had been appointed British Minister for the Coordination of Defence. He saw that Britain was losing the bomber race to Germany, and he reasoned that if there was war, Hitler would have to win quickly, but if the RAF could hold off the Luftwaffe, then in time Germany could be defeated. He believed "the role of our Air Force is not to deliver an early knock-out blow . . . but to prevent the Germans from knocking us out." He was not inclined to accept the Douhetian-Trenchard dictum that there was no defense against the bomber. He foresaw that the new RAF fighters, the Hurricane and the Spitfire, coupled with the secret development of radar, could defend England from bomber attack. He sold his argument to the British Cabinet in the same way that Henry Woodring reasoned over Air Corps purchases: two small planes for the price of one large one.

Inskip was supported by Air Marshall Sir Hugh Dowding, who in 1937 had been made Chief of RAF Fighter Command, a post considered by Trenchard and other bomber proponents as second-rate. Dowding's single-minded belief was that his job was to defend England from air attack, and he devoted all his energies and talents toward that end. Events would show he was the right man at the right time in the right job. But, at the time, the revolutionary change in air policy brought cries of outrage from the RAF's bomber theorists, Trenchard declaring that the radical shift in strategy could defeat England in any new conflict. No doubt Andrews would have agreed with him, as would all the bomber advocates and most Air Corps planners. The conflicting viewpoints were understandable, and geography had a strong influence on both.

Bomber dogma in the Air Corps was riveted to the graceful lines and performance capabilities of the B-17 and what further advances could be projected from it. The geographic components remained the same, with the additional factor that the Air Corps's best interceptors did not measure up as a formidable bomber deterrent.

The French, on the other hand, following Détroyat's evaluation, began negotiating with the Curtiss Company to buy the P-36.[7]

In the thorny area of future aircraft, Westover asked Andrews to come up with a proposal. The decision on procurement was still hanging fire. Andrews waited until he had all his wing commanders assembled, so that a joint position could be presented. His submission noted that all experimental bombardment funds for fiscal 1940-41, as proposed by G-4, would be used only for the development of a supercharged cabin on a B-15-type aircraft. Beyond that, "It [the funds] practically prohibits the continuation of progressive development in bombardment airplanes until 1941 at the earliest. . . ." Once the Project D, XB-19 had been delivered there would be no follow on. Andrews and his commanders did not think the delay was justified. Already there was indication that a 250,000-pound plane was practicable, and until world stabilization of size had been reached, the United States could not afford to find itself lagging behind. "Our only safe course is to develop equipment to the limit of possibilities in funds, engineering skill and materiel."

The joint recommendation to Westover was to utilize fiscal-year 1938 funds to solve the problems involved in the building of a military pressure cabin to be adapted later to a production bombardment airplane; start at once a secret project for procurement of design data for a 250,000-pound bombardment airplane, for the purpose of determining the practicability of bombers of greater range and large weight-carrying capabilities.

The fate of Andrews' recommendations was noted by Hap Arnold. G-4, he learned, in late May had received a memo from Embick with Woodring's approval saying in effect there was no need to purchase anything larger than the B-17, since there was no military requirement for such an aircraft. G-4 informed Arnold that it was willing to put up a fight for the Air Corps position.

Two weeks later, Arnold reported in the office diary that although it appeared that characteristics for the four-engine bomber for 1940 would be approved, the entire question of a larger bomber would be put up to the Joint Army-Navy Board. Remarked Arnold, "If this subject is put up before the Joint Board, God help any further development in large airplanes by the Army Air Corps because in my opinion the big plane for the Army Air Corps hasn't a friend on the Joint Board."

While Westover and Arnold fought the procurement battle, all of Andrews' energies were centered on the May 1938 maneuvers of his Air Force. These were the most extensive maneuvers the Air Force had held to date. Headquarters was established at Mitchel Field. The three hundred aircraft employed were stationed on nineteen airports, only four of which were military posts. They extended from Schenectady, New York, to Aberdeen, Maryland, and westward to Harrisburg, Pennsylvania. All three wings were involved, a total complement of nearly three thousand officers and enlisted men participating. The scenario this time foresaw an aggressor attacking with planes, ships and troops, aiming to capture industrial territory in the Northeast. The Fleet was busy in the Pacific, tangling with an "Asiatic power," so it was up to the Air Force to prevent a landing along the eastern seaboard.

Andrews explained to Congressman Paul Kvale what the action was all about officially: ". . . a form of exercise designed primarily for the training of commanders and their staffs." And later he cited as the "most spectacular problem" the blackout at Farmingdale, Long Island. Farmingdale's citizens had cooperated and, at a fixed time, had turned off all lights in the community. Searchlights probed the sky. Bombers sought out the Republic Aviation factory, supposedly dropping their bombs from fifteen thousand feet. Interceptors rose to the attack, and the 62nd Coast Artillery Regiment blasted away at the attackers caught in the beams of the searchlights. It was a grand display of semimake-believe, with participants and audience caught up in the excitement and drama of *it can't happen here, but should it—we're ready*. But the real drama lay elsewhere, in a remarkable display of navigation and pilotage that, instead of bringing enlightenment, was to result in even greater restriction.

Twenty-three

Andrews' Folly

As Chief of Information for the maneuvers, Ira Eaker had been assigned to Andrews' staff at Mitchel Field. In this case, there would be no blackout on the news of the exercise. The Navy was not participating, not about to put the *Utah* or any other ship on public display against Andrews' bombers. Therefore, Andrews had insisted on, and the War Department had approved, full coverage by the media. It was Eaker's job to oversee the manner in which the coverage was handled, and he brought with him as his aide Reserve Lieutenant Harris Hull.

Hull, who had missed out on Hap Arnold's flight to Alaska, was not inclined to miss out on much concerning the Air Corps. He knew it was Eaker's practice to sit down with the members of his staff in order to surface their thinking on whatever the issue might be. Hull, as a newsman and Air Corps officer, was anxious that the Air Force capture some headlines that were meaningful. He thought he might have an idea of how it could be done, and he took his thinking to Eaker. Now Chief of Air Corps Information, Ira listened to his young assistant, smiled his cagey smile, and went looking for Andrews.

The way the maneuver had been set up, the Black forces were somewhere out in the Atlantic, and the Blue, defending, forces hoped to locate the attackers before they could launch an air attack from their carriers. But there were in actuality no ships approaching the U.S. coast, so the reconnaissance flights to find the enemy would simply be operational exercises in flying out to sea and back home again. Or would they? Lieutenant Hull knew that the Italian liner *Rex* was making its crossing to New York. What about designating her as the attacking fleet? How far off the coast could Bob Olds's B-17's pick her up?

Mike Kilner, Jim Hodges, Tony Frank, George Kenney—who had finally managed to get permission to attend the maneuvers as an observer and ex-officio adviser—thought it was a helluvan idea! So did Andrews, who immediately agreed with Eaker that the press should be brought in on the flight. Eaker also suggested that the Air Corps's top photographic officer, Major George W. Goddard, be ordered down from Dayton posthaste. Photographic evidence of contact between the bombers and the famous ocean liner, far at sea, would make dandy page-one material. Beyond that, it might send a message, if not to Henry Woodring and the War Department, then to the White House, where its occupant was having difficulty pushing his naval buildup program.

To undertake the mission, approval was needed from the War Department and both approval and cooperation from the Italian steamship line. There was no problem from either source, the former through Craig missing the overall purpose of the flight, and the latter only too happy to cooperate in gaining worthwhile tourist publicity.

It was decided that three of Olds's B-17's would undertake the interception. They were temporarily based at Harrisburg, Pennsylvania, where Wing Commander Arnold Krogstad had set up his headquarters. On Wednesday, May 11, the crews of the three aircraft were ordered to bring their planes to Mitchel Field. Bob Olds, with Andrews' approval, had named Major Vincent J. Meloy as commander of the theoretical attack. He would fly in Captain Caleb Haynes's Fortress, Number 80, and once again, that made Lieutenant Curtis LeMay the lead navigator. Captain Cornelius "Connie" Cousland was the pilot of the second B-17, Number 81, and George Goddard would be riding in the copilot's seat, armed with his trusty Graflex. Captain Archibald Y. Smith would be at the controls of aircraft 82. On board Haynes's ship would be a three-man NBC team—two technicians and an announcer. With Cousland would be riding C. B. Allen, of the New York *Herald Tribune,* and Harris Hull, who this time was not going to be left behind. The word was that A. Y. Smith would be hosting noted New York *Times* correspondent Hanson W. Baldwin. One couldn't ask for more complete coverage. But one could ask for a better break on the weather. It was downright rotten: wind, rain, no visibility and no indication that it was about to improve.

As soon as the three B-17's landed at Mitchel Field, Lieutenant LeMay was handed a radiogram from the *Rex* giving her noontime position. By his calculations the ship should be about six hundred miles off Sandy Hook at noon the next day, which was to be the approximate time of intercept. Given the weather, which the forecasters said would be worse at sea, given the time element and the fuel supply, finding the Italian ship was going to take a combination of expert navigation, flying skill and a smile from somebody up there above the overcast. At least that was how LeMay and his two fellow navigators felt. The Air Force was going public on the mission, and if they missed their target, they might just as well try to swim home.

That night, no follow-on position report was received from the ship. LeMay could not sleep. The dawn's gray light found him comparing the scene and situation with the airmail days. He was present when Andrews greeted the press and radio people. The passengers were outfitted with parachutes and told about emergency procedures for ditching. LeMay and his fellow navigators fled to the Weather Office, but the way the rain was pelting down it was a wasted trip. There was no further news from the *Rex*. No one answered the phone at the steamship company. LeMay was faced with the crucial question by Haynes, What would their ETA be over the *Rex?* Takeoff was scheduled for eight-thirty. While the grim-faced Lieutenant got down to his calculations, staff members huddled with Andrews, wondering if the whole thing should be called off. There could be no margin for error, but everything was based not on the geometric niceties of longitude and latitude but on a flock of wind-blown *ifs*. *If* the *Rex* held to her course, *if* she maintained a constant speed, *if* the weather permitted—*if* the god-damned ocean weren't so far and wide!

LeMay gave his estimated time of arrival over the *Rex* as twelve twenty-five, but he felt like running off somewhere and hiding. General Andrews said a few words to the crews in his soft, unhurried voice. Briefly, he emphasized the importance of this mission. It was doubly important that Americans become aware of the capability of the B-17 and the expertise of its crews, he said, and when he wished them good luck he had his eyes on Lieutenant LeMay.

The three planes had begun to taxi out for takeoff when a piece of good luck did catch up. Haynes slowed down. There was a pounding on the hatch. The crew chief opened it and a sergeant shoved in a wet piece of paper; then they were rolling again. The paper was delivered to LeMay where he hunched over his charts, soggy due to a leaky window. But suddenly that didn't matter, for a position report had come in literally at the last moment from the errant *Rex*. He went to work on new calculations and realized the ship was farther out than first anticipated. The revised ETA would be later than the arrival time he had given Haynes, at which time, unknown to LeMay, the NBC announcer would start broadcasting to millions of American listeners on a coast-to-coast hookup.

For four hours the three planes bore through a conglomerate mass of bad weather. It was vitally essential that they stay together, and because there was no telling how high or thick the overcast was, they stayed low, trying to keep visual contact. Low meant practically skimming the wave tops at times. Haynes led, Cousland off his left wing, Smith the right. As the big planes heaved and bucked their way through a swirling shroud, visibility was often nonexistent, and radio communication was their only linkage.

Another reason for staying low was so that LeMay and the other two navigators could measure drift from the wave tops. In this way, they could calibrate their actual speed. Now and again they would break into patches of clear air, and then it was possible to get a more accurate reading. As a result, LeMay realized that they were facing a stronger head wind than originally forecast by the weather experts. Since having enough fuel to get back home was the most crucial factor of all, throttle settings could not be advanced to compensate for the additional wind. This meant there would be no extra time to bisect the *Rex*'s course ahead of its approach and then turn and fly a search pattern along the ship's course. LeMay had already figured this safety margin into the estimates he had given Haynes. From his new estimates he saw there would be no margin for a search, and in fact he did not see how they could make contact by the given time.

Vince Meloy came struggling back to where he was assimilating the unyielding mathematics of the problem. Until this moment LeMay had not known the real significance of finding the *Rex* before but not later than twelve twenty-five. When Meloy informed him of the broadcast, he was literally thunderstruck. Millions of listeners, and if they didn't find the *Rex!*

Everything had been based on LeMay's calculations. He grappled his way forward to the cockpit and, clinging to the back of Haynes's seat, he shouted in the pilot's ear that his calibrations had been changed. All he could get in response was the growled observation, "You're the navigator, Curt!"

The towering wall of another cold front lay ahead, and while LeMay made his careful way back to his soggy charts, Haynes instructed the other two planes to move off his wings to give them all plenty of room as they slammed into the weather barrier. Ten minutes and a furious pounding later, they emerged into bright sunlight, a new world of blue above and shimmering sea below. They went up to five hundred feet, then a thousand. They spread out in a line thirty miles apart. The minutes flew by in the slipstream. No one had to say anything. They all knew how much was riding on making an exact interception with an object less than nine hundred feet long in the immensity of an empty ocean.

At twelve twenty-one LeMay noted that the sun had been blotted out by a squall. Two minutes later they were coming into the clear again, and just two minutes after that it was Cousland who spotted the *Rex* dead ahead. "There she is! There she is!" he boomed out, his clarion cry picked up by radio on the other planes. Then it was wild shouts and cheers as Haynes firmed up his flight and they headed down to circle the ship.

To Curt LeMay, *"It was all a movie. It was happening to someone else, it wasn't real, wasn't happening to us.*[1]

But it was happening, and to photographer George Goddard, the *Rex*, with her red-white-and-green-striped funnels, was the most beautiful ship he'd ever seen. Unlimbering his camera, he watched passengers come flocking out onto her decks to wave and cheer.

The news of the contact was flashed to Mitchel Field. Meloy then established communication with the *Rex*'s skipper, Captain Cavallini. The Captain, with fine Italian courtesy, invited them all to come down for lunch. They were sorry to have to decline, particularly because they were famished. Also, it was picture-taking time. Goddard directed the other two planes to circle the *Rex* at smokestack level, and then, while Cousland held a steady course, he went to work with his Graflex. He was to write of the moment: "When word came through from GHQ at Mitchel to proceed back to base, we circled the liner a final time. On the afterdeck we could see a group of passengers set apart and waving furiously. Later we learned they were Americans, and though we couldn't hear them, or the *Rex*'s whistle over the sound of our engines, they were singing *The Star Spangled Banner. . . .*"[2]

Throughout the brief but dramatic encounter over seven hundred fifty miles from shore, Navigator LeMay stared down at the ship, still not quite sure it was there, but aware that the radio listeners back home were not being spun a fairy tale.

However, the excitement was not over. The weather they had flown through to reach the *Rex* was waiting for their return, magnified in its intensity. They lacked the fuel to fly around the virulent thunderstorms blocking their way, and so they had to fly straight through them. LeMay never knew a rougher four hours of flight, and at times he wondered if they were going to make it—altitude varying from six hundred to six thousand feet, the instruments unable to keep up with the violent, bone-jarring gyrations.

To Goddard, "The turbulence at times was so violent that the B-17 would drop sickeningly, hit what seemed to be solid cement and then shudder and shake from nose to tail, the props of its four Cyclones chewing impotently at the air. It didn't seem that any altitude afforded us relief."

Hailstones the size of golf balls hammered at them, and they were afraid the plexiglass cockpit windows would go. Lightning crackled and spat along the wings. It did not seem that any aircraft could withstand such a structural beating. "White-faced and taut," Goddard wrote, "we had the weird experience of trying to secure a large gaggle of golf clubs that were scattered from the bomb bay to the cockpit. They were Cousland's, who never traveled anywhere without them. They'd been in a large bag behind his seat and when we hit the turbulence they and the bag had taken flight. No one could have painted such a scene, and I only wished I could have photographed it, but I was too busy trying to pick a niblick out of the air before it clouted me for a hole in one."[3]

The three planes had separated to make the return, and the crew of each could not help but wonder if the other two were going to get through. At one point, Cousland's ship lost an engine to ice, but with heat and mixture control they got it going again. And finally, on the tag end of a late-spring afternoon, they broke free of the weather. "The far distant rim of land, smoky in its curtain of haze, welcomed us home," wrote Goddard. "It was not the sailor but the flier home from the sea."

They had been airborne for nearly eight hours, covered over fifteen hundred miles, located their pinpoint target and returned. Finding the *Rex* was the obvious dramatic news. Only the crews knew what the B-17's had withstood to carry them through their mission. Cousland could show LeMay and others the damaged nose and leading edges of the wings of his aircraft, dimpled and hail-scarred. But a few years down the road, B-17's would be returning to home base far more gravely damaged, by man-made elements. Yet the flight had been somewhat in the nature of a test, proving that the Flying Fortress was a tremendously rugged aircraft, able to withstand enormous punishment. This point was totally unrecognized in the War Department, its meaning being lost outside the Air Corps.

Nothing was lost, however, in reporting the success of the flight. It was page-one across the country and around most of the globe. The headline in the New York *Herald Tribune* read: *FLYING FORTS, 630 MILES OUT, SPOT ENEMY TROOP SHIP—8 Hour Reconnaissance Locates Liner Rex at Sea; Generals elated by success of defense.*[4] Next to the by-line account, by C. B. Allen, was George Goddard's epic photograph of Haynes's and Smith's B-17's winging past the ship at mast level. It was a photograph that was reprinted on the front pages of more than eighteen hundred newspapers, magazines and periodicals.

The New York *Times* was to declare that the performance of the Boeings on the first day of the mimic war was "one from which valuable lessons about the aerial defense of the United States will be drawn, and one which already has furnished . . . a striking example of the mobility and range of modern aviation."

Andrews, Westover, Arnold and all the Air Corps were elated over the exhibition. Andrews was careful to stress, however, to the press, to the members of the House Military Affairs Committee attending the maneuvers, and to whomever he wrote or spoke to on the subject, that the flight had been a routine operational performance.

What followed is clouded in controversy and contradiction. From where Ira Eaker sat at a staff meeting the next morning, a telephone call interrupted the session. Eaker assumed the caller was General Craig, for he heard Andrews say, "I'd like to have that in writing, sir." And then, after he had hung up, he told the assembled he had just received orders that henceforth all over-water flights by the Air Force were to be restricted to a distance of one hundred miles off the shore.

In the moment of stunned silence that greeted the news, none of those present had to be told what had happened, and nothing could ever be said or written to change their minds on it. To their way of thinking, the Navy had acted, determined to prevent what it saw as an invasion of its own area of operations. The Air Corps reasoned that Secretary of the Navy Claude Swanson had called Secretary of War Woodring to protest the *Rex* mission and that Woodring in turn had contacted Malin Craig, who had given the order to Andrews. There is no written evidence to support this widely accepted joint assumption. Further, it was Tooey Spaatz, then the Executive Officer of GHQ Air Force for the maneuvers, who received the call from Malin Craig and passed it on to Andrews. Spaatz did not believe the order had been put in writing. That it had been issued, there was no doubt in anyone's mind.

Of its validity, Hap Arnold wrote: "Somebody in the Navy apparently got in quick touch with somebody on the General Staff and in less time than it takes to tell about it, the War Department had sent down an order limiting all activities of the Army Air Corps to within 100 miles from the shoreline of the United States."[5] Lieutenant Curt LeMay believed the order was simply a repeat of an earlier edict given to Bob Olds's 2nd Bomb Group after a unit of B-17's had flown over elements of the U. S. Navy conducting maneuvers several hundred miles off Bermuda.

Of that restriction, Colonel Hal George related that Colonel Olds had called a meeting of his officers and read the War Department instructions. He then said to them, "Hereafter, as the War Department General Staff says, don't schedule any flights more than one hundred miles from the shore. However, there is nothing in these instructions to prevent you from following the shoreline from Long Island to Miami provided you don't get further away from the shoreline than one hundred miles. As a matter of fact, that restriction will provide better training, because to keep not more than one hundred miles from shore will require damned good navigation." He added that his instructions would not be published, because he didn't want the War Department to respond by issuing "some more instructions, saying we can't fly out of sight of land."

Olds's caution probably dated back five years, to the fall of 1933, when he had served under Benny Foulois in Washington and the fight over coastal defense was burning bright. Foulois and most of his staff believed that they were being sold out on the issue by the General Staff in order to keep Army-Navy harmony. Some poet laureate had written of the situation in the Washington *Star:*

> Mother, may I fly out to sea?
> Yes my darling daughter,
> But keep your eye on the land and me,
> And hurry away from the water.

Actually, before the advent of the B-17 the Air Corps was willing and, in fact, often insistent that its off-shore activities be limited first to one hundred miles and then later to three hundred miles out to sea—the change due both to the increasing range of its aircraft and the type of exercise intended. Hap Arnold had shown what his B-10's could do when he had brought them from Juneau to Portland over nearly a thousand miles of water in the summer of 1934. But that had not been a joint Army-Navy maneuver, nor had Andrews' record-breaking flights from the Caribbean. In Joint Coastal Frontier Defense Air Exercise No. 1, in 1936, Andrews had specified that, since the exercise was the first in a series, to minimize "danger to personnel and materiel" it be conducted not more than one hundred miles offshore. Caleb Haynes and his B-17's claimed to have found the *Utah* 345 miles at sea—although the limitations had been set at three hundred miles—but except for the leak through broadcaster Boake Carter, that encounter had been kept secret. When Andrews later spoke to the Newport News Rotary Club, following the May maneuvers, he stressed that the *Rex* mission was a repetition, on a smaller scale, of the Buenos Aires flight, a repeat, in fact, of the kind the 2nd Bombardment Group and other GHQ Air Force units had been doing over land for the past year.

Until the *Rex* incident, the record showed that the Air Corps had tailored its over-water flights to suit the need, always bearing aircraft capability and crew proficiency in mind. Following it, Andrews in a letter to Knerr wrote that the situation with regard to the strategic mission was getting progressively worse, "and we have no court of appeal that I can think of." He had some confidential information on the matter that he was not at liberty to divulge as yet, he told Knerr. Possibly the information referred to the new restriction, and in spite of the past record it was new because it had come as a blanket order and not specific to a particular exercise.

Some months later, in November, when the next joint exercise was being set up, 2nd Wing Commander Brigadier General Arnold Krogstad noted he had received verbal instructions from the Chief of Staff of the Air Force "that War Department orders had been received which restricted Army participation to within a line one hundred miles offshore."

A month later, the Navy proposed in a tentative draft of the exercise that it be "so conducted as not to require Army aircraft to fly more than one hundred miles to seaward."

At this, Andrews strongly objected, pointing out icily that a one-thousand-mile weapon was being reduced to the operating range of one hundred miles. The restriction was removed for the particular exercise. It was removed by Malin Craig, for, apparently, it was he who had ordered it in the first place. He had issued the restriction in a fit of pique, not having been informed in detail of the *Rex* mission until after it had been set in motion. His displeasure had to stem from a sensitivity toward stirring the Navy's ire and also because of the position he was in between the feuding

War Department secretaries: Woodring and Johnson. He had acted, obviously, with no forethought of what reaction his order would bring.

The General had been protecting his rear, and in doing so he set off a misunderstood chain reaction that, aside from inhibiting Air Force flying proficiency, hardened the attitude Army airmen had toward the Navy—a condition that could do neither service any good, and in the end could only harm the nation.

Whether Craig's original order was the result of a complaint by Henry Woodring because the Secretary knew that Louie Johnson was in favor of such flights, is not known. What is known is that as long as the Chief of Staff had "timely request" for long-distance over-water flights, he had no objection to them in maneuvers with or without the Navy. But, in the meantime, he had created an uproar, and even though he was willing to rescind the order, it became entangled in the General Staff bureaucracy and would not come unstuck.[6] As late as the end of February 1939, Hap Arnold was writing to the Assistant Chief of Staff, G-3, recommending, along with Andrews, that "the present limitation of one hundred miles on flights to sea by aircraft under his [Andrews'] control is too restrictive." He recommended that "the restriction be removed in the interest of adequate training in aerial navigation."

The grinding folly of the affair did offer a touch or two of ironic humor. Admiral Cook, who served on a joint aeronautical board with Hap Arnold, made a discreet and delicate inquiry. What would the Air Corps's attitude be, he wondered, if two naval officers were to apply to attend the Corps's navigational school? Hap managed to keep a straight face and replied in all probability the request would be approved. The Admiral then sadly admitted that he fully realized Army air navigators were far in advance of Navy navigators. Arnold noted: "What a fall for the Navy—their one hundred-year inviolability of Navigation principles fallen by the wayside."

There was no better example of the running competition between the two services than their joint participation in national air races and meets. Johnson was hot for the Air Corps showing its stuff. The War Department agreed that the airmen could participate in three events each year, but the Navy was not anxious to be involved at all. Only reluctantly did the Chief of Naval Operations agree that the Navy would take part in a single yearly event.

When the War Department informed the Navy that movements by GHQ Air Force from one place to another did not constitute an *exhibition,* the Navy responded that when they moved their fleet around and launched their planes from carriers, that didn't constitute an *exhibition* either. As Arnold saw it, with his flair for painting the situation in bright colors, ". . . from now on it looks like we are in for just one big competition between the Army and Navy, and it will be a competition to see who can give the biggest exhibit throughout the United States. Further, in such a race the Army Air Corps is bound to be the loser. . . ." He didn't explain why, but the previous flight by Olds and his B-17's to Buenos Aires, the *Rex* incident and a forthcoming B-17 flight to Bogotá, Colombia, hardly indicated that his prognostication was correct.

All the competition indicated, as Admiral Cook was frankly to point out, that "the Army distrusted the Navy and the Navy distrusted the Army."

Distrust was not special to the relations between the Army and the Navy that summer of 1938. Its wormwood within the War Department manifested itself in various ways, and none more debilitating to Andrews than the fate of the B-17. One clear illustration of the attitude within the confines of the War Department and mentioned in the press was that his unceasing drive to have the plane purchased in substantial numbers was snidely referred to as "Andrews' Folly."

On a summer's day, Andy Andrews and Hap Arnold had lunch together at the University Club in Washington. The two friends had much to talk over, and away from the confines of the office they could let their hair down. In the spring, Congress had approved $37 million for the procurement of additional bombardment aircraft in 1939 and 1940. The approval had been followed by General Embick's famous statement that the military superiority of the B-17 over "two or three smaller" planes was yet to be proved. Still, through the combined salesmanship of Andrews, Westover and Arnold, Woodring had tentatively agreed to the purchase of twelve B-17's in 1939 and sixty-seven in 1940. At the conference where Woodring had presented his and the War Department's thinking, Johnson had raised an objection to the plan and Woodring had told him, "We all know you are opposed to it."

"Yes, I'm opposed to it," agreed Johnson and asked that Andrews be permitted to speak. Andrews had made his plea for more B-17's, which did nothing to change Woodring's mind. "This is still the program," he reiterated.

"With all due respect to your office," responded Johnson, "there is a statutory responsibility involved. This is *not* the program until the Commander-in-Chief approves it." Thereupon, he had beaten Woodring to the punch, taking the proposed figures to Roosevelt and asking that more heavy bombers be included. The President was occupied with his plan to build up the Navy and its obsolescent air arm. As a result, he was in favor of Woodring's plan, which meant no further buildup of B-17's or more-advanced heavy bombers.

Strengthened by FDR's approval and anxious to speed up the numbers game, Woodring decided on the basis of cost to further reduce the number of B-17's to be procured. He started by canceling the purchase of three, declaring that the funds would go toward additional light and medium bombers. The War Department was in favor of the cutback, for light and medium bombers, of course, meant close support of the troops.

When Andrews and Arnold sat down to break bread, both were aware that the latest rumor was that Woodring was about to cancel *all* heavy-bomber procurement for 1939 and 1940. Regardless of what he might say, Arnold's belief in the paramount use of the heavy bomber did not match Andrews'. There were fundamental reasons for his position. He had not been in on the B-17's inception. He had not attended the Air Corps Tactical School, where Mitchell's bombardment theories held sway. When he was CO at March Field, in 1933, a special effort had been made from Foulois's office to enlist his support in the development of the plane. There was no definite assurance that it had been gained.

Following the wing commanders' conference at Langley Field in April 1935, Arnold had written a long series of questions to Andrews about the new plane—what it would take to maintain and fly it. The questions indicated he was not all that sold on the aircraft. His interest at the time was focused on the Norden bombsight and improving its use. Once he became Assistant Chief of the Air Corps, he was exposed almost daily to the thinking of Oscar Westover and Malin Craig. He admired and liked both men; the latter's influence, he knew, had brought him to his position. Often, it was his own thinking that had a positive effect on the Chief of Staff's attitude regarding air matters. But his outlook on what the Air Corps needed in the way of aircraft was more nearly matched with West-over's, because from where he sat he had to contend with all aspects of aerial procurement: what was economically available and what could be bought with it. Like Andrews, he was anxious and worried about Wood-ring's and the General Staff's inability to understand the vital importance of research and development into more advanced long-range bombardment. But, like Westover, he believed in traditional balance, and therefore, although he had come to grasp fully the importance and capability of the B-17, he was not for it beyond the numbers advanced by the Office of the Chief of Air Corps.

Out of a proposed complement of one thousand combat planes, Andrews wanted about 25 percent, or 244, to be four-engine bombers. In the summer of 1938 his Air Force had its original thirteen, and Arnold was working as best he could to have that number increased. Both men had come to realize that while Louie Johnson was a force to be utilized, the Assistant Secretary was mercurial by temperament and therefore not very dependable, nor was he very well informed. His desire to replace Wood-ring influenced his every move, and although both airmen would have been happy to bid the Secretary of War a fond farewell, neither was all that sold on Johnson.

In July, when Woodring finally went on a much-needed vacation, there were a couple of prime examples of Johnson's bull-in-the-china-shop activities. One concerned the proposed flight to Bogotá. Arnold first learned of the possibility when Malin Craig called him to ask if some of the planes could be sent to Colombia for that country's presidential inauguration, on August 7. Hap made pointed note, "wanting to use but not wanting to produce," and called Andy at Langley. Andrews' response was that Bob Olds could lead a squadron of three B-17's out of Miami, flying direct to Bogotá and returning via the Canal Zone.

Westover recommended that B-10's might be used instead of the Boeings. While this suggestion might have seemed odd on the surface, most probably it was a result of a donneybrook that had involved Bob Olds in May. Some months after the Buenos Aires flight, word had come from the State Department that Olds, while in Argentina, had made some very critical statements to a newsman there on the qualities of the B-10, saying that it was in fact obsolete. Since the Argentine Government had arranged to buy a number of the bombers, Olds's supposed quotes made banner headlines in Buenos Aires and brought both embarrassment and complaint from the State Department. Olds was summoned by Woodring and Craig to explain, and he denied categorically ever having said anything attributed to him in disparaging the B-10.

That should have been sufficient, but apparently not to Johnson, who was extremely interested in selling the Army Air Corps image, as well as some of the aircraft it flew, to foreign buyers. He blamed Olds for the blowup. Possibly he was influenced by Woodring's having accepted Olds's disclaimer. Because the 2nd Bomb Group Commander was not hesitant in speaking out forthrightly on aerial matters and moving quickly, with Andrews' approval, to take advantage of showing off his B-17's, Johnson decided the Colonel was "the most difficult officer in the Air Corps" with whom he had to deal. He told Craig and Arnold together that Colonel Olds caused him more trouble than anyone, failing to explain in what specific regard.

Now, as acting Secretary of War, he directed that Colonel Olds *not* be allowed to command the flight to Bogotá. Previously, when Olds had been called on the War Department carpet, Deputy Chief of Staff Embick had asked Arnold if the Colonel was the only officer qualified to command the B-17's. Arnold had told him Olds was the only man to date who had the necessary experience.

It was Westover who telephoned Andrews to give him Johnson's directive. Andrews protested. The choice of flight commander was his responsibility, not that of some civilian appointee who really knew nothing about what was involved. Later, he called back to advise that if his protest was not accepted, he was recommending Major Vincent Meloy as the commander, but this would require a complete rearrangement of the crews.

Meloy was a very competent officer, and the flight went off as planned, but Johnson's action was an example of his political judgment by personal dislike without any regard for the military requirements involved.

Hap Arnold had frequent occasion to joust with and outmaneuver the Assistant Secretary. Shortly after the Bogotá mission, he took a letter to Johnson seeking procurement authority for a new attack bomber. Johnson immediately informed him that the plane as proposed wouldn't fly fast enough. There were two other planes in the world of the same type, he claimed, that had a higher speed. Arnold explained this was a twin-engine plane, and so far as was known, there weren't any planes anywhere in the world that could equal it. Johnson then implied that he had already discussed the matter at the White House. Arnold patiently pointed out that for the past eighteen months, the Chief of Air Corps had been discussing the need for such a plane, and that this was the first time they had seen their way clear to procure it. Since this implied that the blockage had come from Woodring, Johnson did an immediate about-face and said he would take the plan to the White House at once—and that Arnold should stand by in case his presence was needed there.

The Assistant Secretary then confessed he wanted to let his hair down over the whole matter of procurement, because he felt too much emphasis was being placed on large bombers, a complete shift from his previous stand. "Have you seen any communication from the Chief of Air Corps since instructions came from the War Department to the Secretary forcing the issue relative to large bombers?" he asked Arnold.

Arnold replied that General Westover was in favor of large bombers in limited numbers because they had a special place in the scheme of things, but he was also in favor of having a large number of smaller bombers for close-in support of the troops. He then asked Johnson if he would do all in his power to expedite approval of the new twin-engine plane.

Johnson, again looking over the letter relative to the plane's procurement, asked, "Who else knows about this? Have you been to see Secretary Woodring? Have you been to see General Craig?"

Arnold assured Johnson his office was the "first port of call"; however, he admitted, the plane's military characteristics were now in circulation in the War Department.

The next day, a Saturday, Johnson telephoned Arnold's office from the first tee at the Columbia Country Club. He said that Westover was to be informed that the President had approved the Air Corps's plan for procuring attack bombers with maximum performance.

Johnson's grandstand play from the first tee had to be for the benefit of his golfing partners and anyone else within range of his voice. If nothing else, it illustrated the caliber of his self-importance. However, it was Arnold's handwritten note on the call, citing *"max. performance"* of the attack bombers that was significant.

Andrews had been protesting that the Douglas B-18 was an inferior aircraft since its acquisition by GHQ Air Force. Both the Air Corps and the Materiel Division were slowly coming around to recognizing that his criticism was valid. Even though Westover and Arnold had concurred in Woodring's recommendation that seventy-eight of the B-18's be purchased in 1939, Arnold had sat down with the chiefs of the division and said he wanted answers "to discover the flaw in that airplane." He instructed Supply Division to write a letter to Dayton to find out why the bomber didn't have sufficient power for takeoff—which was only one of its problems— caused most probably because it did not have controllable-pitch propellers. Whatever the cause, the need for a twin-engine attack bomber with superior performance had become paramount in Arnold's mind as well as Westover's.

In the meantime, the Air Corps was stuck with the B-18, and in an effort to improve on the performance of its next production, Arnold called a conference. To it he summoned from his San Antonio exile Hugh Knerr as well as Andrews' G-4, Lieutenant Colonel Jan Howard. With other officers from Dayton and Washington, it was decided that a five-man board, with three members from the Air Force, would go to the Douglas plant to work out new provisions for the plane, the B-18A. The provisions would, indeed, bring improvements—the Air Force had recommended sixteen changes—but the plane in any form simply did not measure up as a frontline combat aircraft, and never would.

In his written comments on the situation, Arnold privately indicated he had a chip-on-the-shoulder attitude toward GHQ Air Force criticisms. Langley had had the plane for a long time. Why had Andrews' staff been so late in registering complaints? At the conference he had called, Arnold later declared they "had torn to pieces" the GHQ report. But the fact was, Andrews and his staff had been sending in reports on the B-18's lack of performance for a long time, and after their recommendations were supposedly torn to pieces, evidently they were put back together again, for the board went to the Douglas plant to see to the changes.

Arnold's sensitivity to criticism over such a serious matter was a common enough trait, particularly in a general officer. A lousy plane had been, and was still being, purchased in large numbers, and the office of the Chief of Air Corps—in spite of Woodring and the General Staff—was in some measure to blame. Robbie Robins, at Dayton, had complained often enough that due to the restrictions of the Bryden Board, laid down in 1936, the Materiel Division's engineering department had not been able to keep pace with foreign aircraft development. He felt whipsawed between the need to keep costs low by adhering to outdated procedures, and aircraft manufacturers who were not interested in seeking contracts with the Air Corps because of the attempt to stabilize design in a field whose technology was in a steady climb.

Arnold had always felt that the P-26 was a poor pursuit. He now felt the same way about the Seversky P-35, and not much better about the Curtiss P-36. The B-18 was simply the latest example of a system that had gone for second best because it was less expensive and it was possible to rationalize use by fitting it into the accepted War Department strategic mold, to which Oscar Westover faithfully adhered.

Within the confines of his position, Arnold had become an astute politician, although he would have been loath to admit it. In trying to attain the possible against entrenched odds, he had had to learn how to maneuver both ends against the middle in order to attain limited goals. Caught in the glue pot of military/congressional politics, which he detested, he had learned to play the game of compromise, even though the results continued, as they had for twenty years, to be disappointing. He maintained a penchant for lacing his accounts of boneheaded actions by others as well as any criticism directed toward himself with wry exaggerations. Of major importance in the circumstances, his sense of humor remained intact.

So did Andrews', but after nearly four years of command at Langley Field, there was a great deal going on in Washington that he was not privy to until it was too late for him to bring what little influence he had, to bear. Had there been closer cooperation between his office and Westover's, had there been agreement on what the Air Corps's strategic plan really was, had there not been divided bases of power and authority between administration and operation, then at least on the matter of aircraft procurement there could have been more cohesion and a united front presented. As it was, although Andrews was fully aware of the contentions over the B-18, because it was supposedly the backbone of his Air Force, he knew nothing officially at the time about Arnold's efforts to sell a more advanced twin-engine bomber. Privately, Arnold, who came down to Langley for a summer visit, filled him in. But there were other, equally important matters during those months about which he remained in the dark.

He did not know that on July 2, Arnold had been sworn to secrecy by Craig and told that it looked very much as though the GHQ Air Force was going to be placed under the Chief of Air Corps within a month. Craig wanted no one to know of the intended change except Westover, but he wanted Arnold to have a memo drawn up that would outline the items the Air Corps office believed should be in the directive. Who had instituted the intended change and why was not explained, nor the fact that the major shift was not carried out as proposed. It was a change Westover had long desired and Andrews had come to believe was the only solution to the joint administration. Yet Craig did not wish to bring him in on it, despite the fact that such a shift would directly involve him.

Nor did Andrews know that the rumors of June, concerning the cutting off of all B-17 procurement for 1940–41 by Woodring, had by the end of July come much closer to reality. Word from the Chief of Staff to Westover at the end of the month indicated this was the Secretary of War's plan.

Andrews was not informed at the time, either, that the Joint Board, as Arnold predicted, had recommended against any bomber more advanced than the B-17. Whatever inside decisions he was kept in the dark about, he was only too aware of the results. In July he wrote Knerr: "We are steadily losing ground on proper organization and equipment. I have no hope of improvement during the present administration. . . ." The purchase of seventy-eight B-18's as M-day units he saw as "a crime against National Defense."

In still another area of procurement, Arnold had supported Andrews, doing much to carry the fight in the War Department for a new substratosphere long-range bomber, the YB-20, a modification of the XB-15. Andrews, using as arguments foreign aircraft development and the time lag of five years between acceptance and use, attempted to change General Staff minds. Earlier support had faded, and Woodring's decision to cancel funds previously allocated for the 1938 purchase of two XB-15's, so more B-18's could be bought, was the bottom line. Worse, Westover refused to back either Arnold or Andrews in promoting the new substratosphere bomber, because the XB-15 had not lived up to its performance expectations and he felt he could not justify the additional expenditure of funds. Assistant Secretary Johnson managed to add his weight by informing Arnold there was no military requirement for a substratosphere bomber and that all development should be restricted to the type of plane designed for close support of the troops. He was, of course, parroting the Joint Board conclusion.

On August 9, with Westover on leave, Arnold was acting Chief. He received a decision from the War Department that began: "The development of the Pressure Cabin Bomber with military characteristics as stated in letter, Office of the Chief of Air Corps, dated May 10, 1938 . . . is not favorably considered." The decision went on to reiterate: "Experimentation and Development for F.Y. 1939 and F.Y. 1940 will be restricted to that class of aviation designed for the close support of ground troops and the protection of that type of aircraft such as medium and light bombers, pursuit or other light aircraft. . . ." Arnold noted: "It is thought that the Chief of Air Corps should fight this decision."

He might have added that it was a position that he and many thinking airmen had been fighting since the days of Billy Mitchell, and that in eighteen years of trying, it now appeared to have come full circle, a perfect 360-degree turn into nowhere.

The following day, in accordance with instructions from the Adjutant General, Arnold submitted a new purchase program. The old program for 1939 had projected the procurement of over one hundred light bombers and no heavies, but in 1940, sixty-seven B-17's had been included. This last figure had now been dropped altogether by Woodring's order.

Again, Andrews was not privy to the draconian decision. But at a small dinner given by Eugene Meyer at his home, Andrews had explained to the Washington *Post* publisher where the major error lay which could produce such a decision. "Our National Defense budget is not scientifically constructed with proper balance between air, ground and sea requirements," he told Meyer, "and there is no adequate machinery to do so."

The President, he went on to say, "has no trained organization to coordinate national defense and direct war operations. Congress has a Military Affairs Committee and a Naval Affairs Committee but no Air Committee. But, there again, where is the coordination?" he asked.

Before they sat down to bridge, he admitted to the *Post*'s publisher that he believed every major power in the world was better organized for war than the United States.

If there was one bright spot for Andrews in that frustrating and discouraging summer, it involved another fortuitous meeting, a meeting whose importance was immeasurable and whose results were to have a lasting and positive effect. He made note of it in a letter to Delos Emmons on August 2. "I am planning to visit the West Coast, during the 1st Wing maneuvers, with General George C. Marshall, the new War Plans Division man, and [Lieutenant Colonel] Earl Naiden, my G-3. . . ."

George Marshall, at fifty-eight, had been brought to his General Staff post through the combined efforts of Craig and Embick. Embick had taken credit for recommending that Marshall be made Chief of the War Plans Division preparatory to taking over as Deputy Chief of Staff when Embick would move to a corps-area command. Woodring liked the idea, because having Marshall as Deputy Chief would offer an alternative to Hugh Drum's open desire to become the next Chief of Staff. For once, Johnson approved of Woodring's move.

Andrews did too, for although he did not know Marshall, he knew about him, through his late father-in-law, Major General Henry T. Allen, whom Marshall had served at one time as Chief of Staff.

Directly after Marshall assumed his new duty, Andrews invited him to come to Langley and flew to Bolling Field to pick him up, and when the day was over, flew him back home again.

Marshall wrote his thanks, saying, "I thoroughly enjoyed my day, I think I learned quite a bit about the problem, and will look forward to some further meetings when I have better coordinated my thoughts with the information available."

It was at their Langley meeting that Andrews invited Marshall to accompany him on what would be a comprehensive inspection trip including both Air Force and Air Corps installations. In response to the plan, Conger Pratt wrote from Maxwell Field: "Certainly am glad that you have established this contact with Marshall and that you are going to be able to take him around and show him some things. He is a fine fellow and I am sure can be of great help to the Air Corps as well as the War Department as a whole. Give him my best and here's hoping you have a fine trip."

They did, and Marshall would later relate to General Pershing in capsulized form the itinerary covered:

> I flew from here [Washington, D.C.] with General Andrews of the GHQ Air Force to Selfridge Field, Chanute Field, Minneapolis, then on to Billings, Montana, and across Yellowstone Park to Spokane where the Air Force had concentrated. From there I flew over to [Fort] Lewis where another element of the First Wing had concentrated, did the Boeing airplant at Seattle [where B-17's were being made] and stopped at Vancouver Barracks for the night where another wing of the 1st Group was concentrated.* From the Northwest I flew to San Francisco, Sacramento, Los Angeles—the airplants there—Denver, San Antonio, Barksdale Field, Shreveport and home. Altogether I had a very interesting trip professionally and a most magnificent one personally.

The two men took to each other, and during their nine-day association a meaningful bond was formed. Noted for his retiring nature, Marshall was open and voluble with Andrews, later referring to their journey as the point where his education on air power began. The tour, he told Andrews, had been of tremendous assistance, and it gave him a perspective, as it were, against which he could sort the facts in the days and months ahead. He thought it was just as well that he had known so little at the time, rather than having started out with too many preconceived notions. And upon his return to Washington, he wrote Andrews, thanking him for "the splendid trip" and saying he had enjoyed every moment of their association.

Of all that Marshall had seen, the visit to the Boeing plant, where replacement models for Andrews' B-17's were under construction, made the deepest impression. Andrews had purposely taken him to Seattle not only to show him the big Boeing but also all that was involved in its production.

Yet at the very time they were being conducted through the plant, Hap Arnold was engaged, as ordered, in preparing the new procurement program, devoid of all B-17's.

In the course of their tour, the two officers discussed the deepening crisis in the world and how it might affect the United States and their own needs. In Europe, since his uncontested take-over of Austria, Hitler had been bringing increasing pressure to bear on Czechoslovakia over the issue of that country's Sudetenland territory, where more than 3 million Germans lived. The Sudeten Nazi Party, headed by Konrad Henlein, on instructions from Berlin, was demanding autonomy from Czechoslovakia.

---

* He got it backwards. He meant another Group of the 1st Wing.

Britain, France, Italy and Russia were all involved in the maneuverings, which, it appeared by early August, might be leading Europe into war.

The conflict in Spain, now more than two years old, was being won by Franco's Nationalist forces with the aid of German and Italian ground and air units. In the Far East, the Japanese were making deep inroads into China, and public opinion polls in the United States indicated that more than 40 percent of the population believed the country could not stay out of a major war if one came.

Against this sharpening international background, Andrews guided his perceptive and open-minded guest through the ambit of all that was embodied in making a clenched fist out of air power—everything from training to precision bombardment.

Earlier, Andrews had attempted to have Westover arrange it so that he could take Marshall to visit the Douglas plant where the XB-19 was under construction. Since they were going to be at March Field on a Sunday, Oscar didn't think the idea was practicable, because the plant worked a forty-hour week. This did not deter Andrews, and he later wrote a letter of thanks to Donald Douglas, president of the firm, for giving them the opportunity to make an inspection tour. They also visited the Northrop factories, and there was no doubt in Andrews' mind of the positive impression gained by Marshall.

So, too, the Air Depot at Duncan Field, San Antonio, Texas, where Lieutenant Colonel Henry J. F. Miller commanded. Andrews' point in having Marshall see the depot was to emphasize the high degree of mechanization needed to maintain air power. He wanted Marshall to realize that in running the Air Corps there were many problems that were different from the other branches. Marshall, he felt, got the message.

At March Field they were the guests of Billie and Delos Emmons, and this was another opening Andrews wished to make. He was anxious for the War Plans Chief to meet as many senior air officers as possible—Martin, Brant, Yount, Bradley, etc.—Air Force and Air Corps commanders.

On the major deterrents to the building of Army air strength, Andrews was utterly frank with his guest, recognizing "his intelligent interest." Sure he was the type of man who wanted to have all the evidence before making a decision, Andrews held nothing back. Both men were selfless in their work, perceptive of the others' qualities and unhesitatingly open. There can be no doubt from all that followed that the nine-day briefing Frank Andrews gave George Marshall was of enormous benefit to the future Chief of Staff and, eventually, to the cause of U.S. air power.

However, as August passed into September and Europe moved swiftly toward the brink of war over Hitler's designs on Czechoslovakia, the advance of U. S. Army air power had been stopped. Andrews learned of Woodring's decision to halt the purchase of all B-17's through 1941. For once, he felt a sense of despair, and later expressed it in a letter to

Marshall. He knew that his tour of duty as GHQ Air Force Commander would terminate on the first of March, 1939, and he confessed, "I have only a few months in this job of mine, and I will be glad to get out of it, for as it works out, I carry the responsibility and very little authority. I don't even know who my principal assistants are until their selection is announced. There is no future in it, and it is like sitting all the time on a powder keg. . . ."

Woodring's scrapping of the plan to buy more B-17's was actually but one of two extremes. The other was that in the effort to build up the Air Force, planes were starting to come off the production line but personnel to man and service them were not.

Wrote Andrews to the Adjutant General in early September: "With the continuing delivery of airplanes, the personnel situation in the GHQ Air Force will become so critical in the near future, that definite corrective action must be taken." Under the existing personnel policies, Andrews informed the General Staff, by July 1, 1939, the Air Force would be short nearly nine hundred officers and more than two thousand enlisted men.

"The airplanes to be assigned the GHQ Air Force cannot be operated in the face of these shortages," he stated. "Unless additional personnel is assigned, a number of airplanes will deteriorate in storage and reserve."

Once, there had been a shortage of both planes and men. Now only the latter would prevail. But, overproduction of the former, fundamental differences as to type were at the base of all else. On this last, the entire concept of U.S. strategic air power lay in the balance. Even Westover came to see it from a practical point of view. He recommended to Woodring and to Craig that procurement of the B-17 not be halted and that procurement of the YB-20 be reinstituted. He told them that unless the Air Corps was able to maintain its continuity of effort and progress, it would be operating on a hit-or-miss basis of doing business. If he was not thinking in strategic terms, he was at least thinking in technical and economic realities that were sound and could be applied strategically should the need arise. His recommendations failed to make any more headway than Andrews' and Arnold's.

The public, of course, was totally unaware of the internal problems that gripped the Air Corps. The September Cleveland Air Races, in which Army and Navy air units excited the crowds with their flying proficiency and the look of their smartly aligned aircraft, offered no hint of weakness. Quite the contrary. It was a week of splendid aerial activity, followed shortly thereafter by an equally brave aerial display on the West Coast, this one tied in to the American Legion convention being held in Los Angeles.

With both Woodring and Johnson highly touted American Legionnaires, Air Force participation had, as Andrews put it, turned into "pretty much of a political football." Back when arrangements for the convention were

being made, Johnson announced that the entire Air Force would participate. Woodring, upon returning from his vacation, cut that idea down to
size. Participation, he said, would be limited to the 1st Wing, Olds's B-17's
and the B-15.

It became even more limited when Olds, evidently on his own, invited
"the Aviation Authority to go with him to Los Angeles in the B-15."

This angered Johnson, who was anxious to issue his own invitations to
those who would ride in the big bomber. By chance, Arnold bumped into
both Johnson and Craig in a Munitions Building hallway, and they asked
him just what units were going to be represented in California. Hap told
them, and Craig announced the B-15 was not to go.

As Arnold noted, "Rumor has it, as a result, orders issued direct to
Andrews that Olds not to go to L.A."

Andrews went, in any case, and on the nineteenth of September, he and
Westover attended the American Legion luncheon, at which Woodring
spoke. It was a Monday, and whatever impression the legionnaires had
gained watching Delos Emmons' 1st Wing perform on the weekend, it was
overshadowed not by Woodring's reassuring words but by a single
thought: Was there to be war in Europe?

The Czechoslovakian crisis was fast coming to a climax, and as far as
the public knew, the possibility of war had never been closer. Hitler's
Wehrmacht was poised to march, his bombers prepared to strike. Prime
Minister Chamberlain had met with the German dictator for three hours at
Berchtesgarten and had returned to London on Saturday. On Sunday Daladier the French Prime Minister, crossed the English Channel to discuss
with Chamberlain Hitler's demands over the Sudetenland. The public did
not know that the abandonment of Czechoslovakia was under way, only
that negotiations were continuing and that France had a treaty with the
Czechs to come to their aid in case of attack. News headlines and radio
bulletins heightened the tension.

In that hour, what Andrews or Westover was thinking as they sat listening to the Secretary of War describe the might of U.S. air power is not
known. Since youth—for more than thirty-five years—they had known each
other, air power the central point in their lives. Regardless of differences in
temperament or differences in the manner in which they viewed the use of
air power and its administration, there was an unspoken bond between
them, established during their West Point days and strengthened during all
the years of flight that lay between then and now.

Both had remained very active pilots, Westover, in his AT-17, spending
much time in the air flying to inspect bases and installations around the
country, Andrews seemingly ever on the go in his DC-2. Now, sitting together at the crowded luncheon—Westover short of stature, round of face,
mustached and intent, Andrews relaxed, full-featured and strikingly handsome, his pelt of fair hair going to gray—they represented the best of what

the military air arm had to offer. Even in the disparity of their views, there lay strength, for their goal was a common one, and in the bedrock of their actions none could ever doubt their dedication or their integrity. They were at that moment the ranking elite of a small, close-knit aristocracy of airmen. Knowing all they knew, not knowing what was coming in Europe but worried for obvious reasons, they had no premonition that this was to be their final meeting.

The next day, Andrews flew East. Westover continued his inspection tour. On Wednesday morning he took off from the Lockheed Air Terminus at Burbank, California, with his long-time crew chief, Technical Sergeant Sam Hymes, who had been promoted five days previously. It was only a short hop to the Vultee plant where Westover was to look over a new basic trainer. He was in good spirits, and it was noted before he departed to return to Lockheed that he made his usual methodical inspection of his aircraft. The Lockheed field, though long, was narrow, set behind the aircraft factory. A crosswind was blowing and he flew over the field to check its direction. Then, with flaps and gear down, he began his final approach.

He had done it thousands of times before; the procedure was routine, and so was his pattern until he began his turn onto final. There was turbulence, tricky winds from off the nearby mountains, thermal currents rising from the sunbaked earth. Oscar Westover, at fifty-five, was not all that sharp a pilot. He stalled the plane in the turn and it whipped into a spin. When he saw he couldn't pull out, his last act was to shut off the power to prevent fire on impact.

Major George Goddard happened to be flying in the area at the time, testing a new photographic color process with a fellow pilot. They were heading for the Lockheed terminal when they saw, close to it, a black plume of smoke rising into the sky, with flames flaring beneath. They knew it meant a crash, and they landed quickly, arriving at the scene of the accident just as the ambulance sped away, siren wailing. They were too late, and so was the ambulance. Major General Oscar Westover and Technical Sergeant Sam Hymes were dead.

There were others who had seen the Air Corps Chief spin in. One was Major K. B. Wolfe, the Air Corps Lockheed representative. He had been standing at the window watching the General make his landing approach, planning to have lunch with him. Even before would-be rescuers reached the scene, Wolfe was on the phone to Washington. With the four-hour time difference it was five o'clock, and Hap Arnold was at home. In fact, he was in the back yard washing his car when Bee called to say he was wanted long distance from Los Angeles.

After K.B. stated the purpose of the call, he gave it to Hap straight. "I can see the plane burning. There is no hope."

Bee saw the color drain from her husband's face. For a moment, she thought he was going to faint. When he hung up, he sat down and put his head on the breakfast table, unable to speak.

The shock of the news was all-engulfing, the loss enormously sad. Westover's son Lieutenant Charles Westover was a recent graduate of Kelly Field; his daughter Patsy had been her father's gracious and dependable social aide during his wife's illness. It would be Hap and Bee's task to break the news to them before they suffered the blow of hearing it on the evening news.

In a letter to Andrews the next day, Malin Craig expressed the feeling that ran throughout the Air Corps and the War Department: "The tragic death of Westover has completely floored me. He was as conscientious and honest a man as I ever knew in my life and his ability and loyalty were simply superb."

Aside from the personal tragedy of it, the question was swift to arise in Washington and Langley and throughout the Air Corps: Who was going to replace him?

Twenty-four

The Turning Point

Since 1926, the line of succession to becoming Chief of Air Corps had followed a consistent course. Fechet, as Assistant Chief in the Office of the Chief of Air Corps, had replaced Patrick; Foulois had followed Fechet; and Westover, Foulois. There had been doubts over the appointments of Foulois and Westover for reasons stated, but in the latter case the War Department's decision to bring in Brigadier General Henry H. Arnold as Assistant to Westover had assuaged much of the criticism on the choice. Now, with Westover's unexpected death, it seemed apparent that Hap Arnold would succeed him.

During Westover's many official trips away from Washington, Arnold had been acting Chief, and with Westover's final departure he assumed the temporary title and waited for word from the White House of permanency in a position that but twelve years past he would have equated with his ability to fly to the moon. But, in the days directly following, there was no word. Instead there were rumors, some very ugly.

For three years Arnold had been dealing with the War Department, the Navy, the Congress and a host of influential aviation-minded individuals from aircraft manufacturers to magazine editors. Woodring approved of him, and his earlier mentor, Malin Craig, was a golfing partner. Further, he had been adroit enough not to ruffle too many feathers on the General Staff. Nevertheless, within the Congress and within the Air Corps ranks there was support for other officers whom the War Department might recommend to Roosevelt for the top air job.

Frank Andrews headed the list. He was senior to Arnold in both grade and rank. So was Jim Chaney, who headed the Training Command, and Conger Pratt, at the Air Corps Tactical School. Both had friends in high places. Augustine Robins was a classmate of Hap's and had received his appointment of temporary brigadier general at the same time. However, only Andrews was really in contention.

Andrews, with his record as head of the Air Force and the knowledge that his tour as its commanding general would be ended the first of March, had to be considered. Certainly, neither Woodring nor Craig preferred him over Arnold, with whom they knew they could work and feel comfortable. But Louie Johnson's attitude had to be figured into the equation due to his close connection with the White House. His position on the succession was difficult to predict, because there was no telling how much his ambivalence was focused on his desire to get rid of Woodring or on whom he had spoken to last.

As Arnold had found, on one day the Assistant Secretary of War might announce that he didn't want to buy any more pursuit aircraft because the Air Corps should develop an automatically controlled torpedo plane, and the next he would be giving approval for the purchase of a twin-engine bomber, reversing the policy of the Secretary of War for reasons that were more personal than judgmental. In this regard, whenever Woodring had to go out of town he had his aide standing by to call him should Johnson attempt to pull a fast one. If the call came, the Secretary of War's personal pilot was on hand to fly him back to the capital. Woodring tried to make every cabinet meeting, so that Johnson could not sit in for him. Roosevelt had once told Jim Farley he would never appoint Louis Johnson Secretary of War.[1]

Even so, as the situation stood, Johnson had much closer access to FDR than did Woodring. Directly following the American Legion convention in California and Westover's death, Johnson had written a glowing letter of congratulations to Andrews on the fine showing of GHQ Air Force. Andrews answered thanking him and saying that he wanted to meet with him again in Washington, which indicated that they had done some talking at the convention.

Stripped of his political motivations and desires, Johnson was at heart an internationalist, whereas Woodring was an isolationist. The President's thinking was far more in keeping with his Assistant Secretary of War, particularly during the fateful week that led Europe to the abyss of war, ending with Hitler's victory at Munich. Whether Johnson would have aligned Hap Arnold with Woodring is not known, but he was inclined at the moment to support Andrews' positions, impressed by the cogency of his reasoning and the force of his personality. He may have expressed the opinion to Roosevelt that Andrews, not Hap Arnold, should be the next Chief of Air Corps.

There were others of a like mind. One was Walter Weaver. As base commander at Langley Field, he remained one of Andrews' closest friends and confidants. When a flying proficiency board had grounded Weaver for failure to keep up his air time, Andrews had literally stormed up to Washington and notified the decision makers in the office of the Chief of Air Corps that, unless Weaver was given an opportunity to regain his standing, he would demand an investigation that would ground most of the air officers in the capital. In the end it was physical infirmity that took away Weaver's flight pay, but there was nothing he wouldn't have done to aid Andy, and to his way of thinking that meant aiding the Air Corps. Unbeknownst to Andrews, he was in quick touch with General Pa Watson, at the White House, putting in a strong pitch for his friend. Watson was ever and always a favorite of the President's, and as military aide could bring a certain amount of gentle persuasion to bear. Bluff, good-natured, with a sense of humor Roosevelt appreciated, Watson's opinion might count for something in the choice of Westover's successor. Press Secretary Steve Early was also an Andrews booster.

In the end, FDR's thinking on the appointment had to be influenced by a number of factors. One of them came very close to finishing Hap Arnold's career. It is not known who the perpetrator was: whether the story was manufactured in the White House or whether it was fed to Cal O'Laughlin, of the *Army and Navy Journal,* by someone who disliked Arnold. The rumor was that Brigadier General H. H. Arnold was a drunk. It was being passed about that when he had served in the Hawaiian Islands he had frequently been seen getting soused in Honolulu. Since he had never served in the Hawaiians, the claim lacked credibility to begin with, but the poisonous tale was being circulated, and Arnold knew it had to be dispelled swiftly or his chance to be Chief would be gone.

When he learned what was apparently blocking his appointment, he sat down with Tooey Spaatz and Ira Eaker in the privacy of his home to talk over what could be done to correct the lie.

Although Spaatz had not served with Arnold for five years, Hap had borrowed him from Andrews and his job as Executive of the 2nd Wing, to head a flying evaluation board. However, the association of more than twenty years did not depend on proximity. The two got together by serving on boards, during maneuvers and whenever a suitable occasion arose, social or otherwise. Theirs was the kind of friendship that picked up where it left off at any time, anywhere. And when Hap wanted some cool, steady advice, it was to Tooey that he turned.

The relationship established between Arnold and Eaker during the March Field period had matured more firmly with Eaker's posting to Washington. The Eakers were frequent guests of the Arnolds, and to eleven-year-old David, the Arnold offspring still remaining at home, they were practically Uncle Ira and Aunt Ruth.

Officially Eaker, as Chief of Information, did a voluminous amount of writing—articles, speeches, background papers for Westover, news releases for the press, and ghost pieces for retired General Fechet and others—on matters close to Air Corps needs. His thinking was always out in front, his ear to the ground. In this last, his contacts with the press were extremely broad, and he had picked up the ill wind of the rumor concerning Arnold's drinking habits.

The question was, should an attempt be made to kill the story by having Ira either write a denial for publication under someone else's name or should the straight story be given to some of his news contacts? Both approaches were dangerous, in that they could attract much greater attention to what at the time was not widely known. No one had to tell Arnold he was apt to act on impulse, particularly when aroused. He might not be the audacious rebel of a decade past, but his temper could bring the wrong response. Bereft as he was over Oscar Westover's death, the clear, unflappable thinking of Tooey, combined with Ira's caution, were what he needed most.

To Spaatz, the course of action was plain and direct. Hap's efficiency report spanning the years would give the lie to the rumor. Place it in the right hands at the White House. And whose hands were those?

Harry Hopkins, relief specialist and WPA Administrator, was reported to be FDR's most trusted adviser. Arnold didn't know him, but back in May, Hopkins had indicated a desire to help the Air Corps in the construction of new air bases. Colonel Chafee, Arnold noted in his diary, "volunteered the information that Harry Hopkins couldn't possibly do anything he said he was going to do." The Colonel was incorrect, and it was apparent simply from reading the newspapers that if anyone had Roosevelt's trust, it was Hopkins.

Spaatz and Eaker jointly urged Arnold to make an appointment to see him. This he decided to do, not sure the appointment would be granted, due to the Czech crisis, or if it was, what manner of individual he would find the lean, rawboned Iowan to be. He found him to be extremely attentive and particularly interested in Air Corps matters. Arnold, for his part, said if there were any doubts about his sobriety, his record spoke for itself. And, indeed, it did.

Harry Hopkins was to become one of Hap Arnold's closest friends, but at the moment he quickly scotched the dangerous rumors and lent his influence to Arnold's confirmation.

Roosevelt's announcement, on September 28, 1938, that Hap Arnold was the new Chief of Air Corps brought an immediate wire from Andrews. "The GHQ Air Force assures you of its sincere good wishes and support and that goes for me personally."

That Mike Kilner was to be Assistant Chief was greeted with equal approval, although Andrews was naturally sorry to lose him as his new Chief of Staff. His replacement came in the nature of a gift from Arnold that indicated Hap's desire to work with the Air Force. "After talking over the matter with Mike," he wrote Andrews, "I reluctantly agreed to George Brett coming to you for duty on your staff. Reluctantly because I really need him very much as an Executive Officer. Cheerfully, because I appreciate the fact that you need him probably more than I do."

The appointment of Arnold and Kilner came a day before the climax of the most momentous month in twenty years of European history. It had begun on September 5 with the French canceling all Army leaves and calling up reservists to man the Maginot Line. By the twenty-second, the British were issuing thousands of gas masks, testing air raid sirens and digging trenches in the London parks. On the twenty-seventh, following Roosevelt's appeal to the heads of state of the four major European powers to keep negotiating, Hitler replied that war or peace was up to the Czechs. Actually, it was up to the British and the French, and on the thirtieth, Chamberlain returned from Munich, having agreed to give Hitler the Sudetenland, believing the paper he grasped spelled "peace for our time."

Two and a half weeks previously, on September 12, Roosevelt, in his private railroad car, at Rochester, Minnesota, accompanied by Harry Hopkins and aide Howard Hunter, heard Hitler's famous Nuremberg speech. The President could understand German, and it was at that point he concluded that the German dictator was bent on war. The result was that he secretly dispatched Harry Hopkins to go West to investigate the capacity of the aircraft industry to build military planes. Wrote Hopkins: "The President was sure we were going to get into war and he believed that air power would win it. . . ."

At the time, neither Arnold nor Andrews was aware of Hopkins' mission. It was, however, directly after his return that Arnold called on him in hopes of gaining his support. He found the rumpled, unprepossessing social worker very keen on talking aviation buildup.

Unknown to those involved, the Arnold-Hopkins talks that followed ran parallel to the earlier Andrews-Marshall talks. Their substance would be joined and in joining create a force like an unnoticed change in temperature beginning to work on the solid ice of winter. Andrews had given Marshall a nine-day indoctrination on the perquisites and strategies of air power. Arnold followed a similar course with Harry Hopkins. Both recipients were perceptive, open-minded, anxious to know.

In mid-October, Marshall was named Deputy Chief of Staff. Andrews wrote him a long letter of congratulations, enclosing the talk he was scheduled to give at the War College. He also brought up the inanity of Air Force training flights being limited to within one hundred miles of the coast. He stressed the fate of the four-engine bomber. "In every test or exercise we have ever had, including the Fort Bragg exercise just completed, this plane stands out head and shoulders above every other type; yet for 1940 and 1941 our estimates do not include a single one. For the support of the Monroe Doctrine on the American continent such a plane would be of inestimable value. In the control of the three important defiles in the world, Singapore, the Mediterranean, and Panama, one of which is our own bailiwick, the large-capacity plane is easily the outstanding weapon. . . ."

This was the letter in which he signaled his discouragement, but he said he hoped that in the few remaining months he had left to serve he would be included in the discussions and conferences on future plans and policies for the development of the Air Force.

Marshall, in his reply, assured him that as he became more informed on War Department conditions and policies, he would want to "talk over matters with you. . . ." He then went on to reveal, "The more I go along here the more fortunate I think it was that I had that fine trip around the Air Service with you. . . ." He had been down to Maxwell Field, where Conger Pratt had arranged for the Tactical School's senior members to give a four-hour demonstration that Marshall said he found most informative.

He then admitted that he had made "a little use of the Air Corps for family reasons. . . ." On the day of Oscar Westover's crash, a devastating hurricane had swept up the eastern seaboard, wreaking havoc and causing many fatalities. Mrs. Marshall and her sister Mollie were vacationing on Fire Island, a thin strip of sand dunes adjacent to the ocean side of Long Island. Marshall, hearing that Fire Island had been submerged, and gravely worried about the safety of Mrs. Marshall, was flown to the site. From the air he could see that their cottage was still standing. His pilot landed him at Mitchel Field, where they transferred to a training plane better able to put down on the sand amid the debris. He found his wife and Mollie safe, having survived a harrowing experience, escaping to high ground in water up to their waists in the teeth of the hurricane. The new Deputy Chief of Staff was getting a practical as well as a theoretical indoctrination into the uses of air power.

Harry Hopkins had also been looking into the sad state of America's defenses, and he was very critical of the manner in which the War Department had been parceling out WPA funds to build them. Secretly, he saw to the transferring of several million dollars of relief money toward the purchase of machine tools to start manufacturing small-arms ammunition. Although this move had nothing to do with the Air Corps *per se,* it was the first step, however small, in the direction of enlarging defenses beyond what had been prescribed.

In the meantime, nothing had changed the thinking of Secretary of War Woodring. In early October, by error, a memo signed by Woodring's aide, Colonel Shulz, arrived on Arnold's desk, outlining the method of procurement for the coming year. Purchases were to be decided solely on the basis of economy. While Arnold's Plans Section was studying the memo, the Adjutant General called up to say that the memo should have been routed to G-4. In passing it back, Plans drew up its own memo, "outlining the seriousness of any such step, which would not permit us to buy modern pursuit planes in 1939. . . ." It pointed out that European nations all had pursuit planes that would do well over 300 mph while the best the P-35 and P-36 could do was less than 300 mph. The same reasoning applied to bombardment.

Several days later, a Colonel Phillips from G-4 stopped by the Air Corps offices to see how Arnold and Kilner felt about having a Technical Committee meeting on 1939 procurement. Arnold told him that, as he knew, the Technical Committee was now meeting on bombardment, but its decisions wouldn't change anything anyway, because all it could do was make recommendations. Phillips was also informed that Secretary Johnson had been convinced that the Air Corps must have the best in pursuit and that present models were four to five years old.

On the issue of bombardment, Arnold and Kilner had already had a go-round with four officers from Woodring's office intent upon exercising the option to buy more B-18's. Arnold told them that if they insisted, they could be sure that the whole Air Corps, including the Assistant Secretary of War, would be severely criticized, particularly because Andrews was "violently opposed to buying any more B-18's."

A few days later, on October 15, the new Chief of Air Corps called on the Secretary of War to do battle on bombardment procurement for 1939. Arnold explained that it was impossible to think of buying anything other than modern bombers. More than a lack of money was at stake—it was a lack of time. At first Woodring failed to see the point, but Hap kept at it, and finally the Secretary, noted for his own speechmaking capability, was swayed by Arnold's powers of persuasion. He said all right, they would go ahead as soon as possible for a new bomber in 1940. Arnold explained that before a move could be started, someone had to make a decision on whether it was to be a two- or a four-engine bomber, long-range, large or light. Woodring's response was that there were not sufficient funds to buy large bombers.

Three days later, Mike Kilner was notified by General Spalding that Woodring had made a positive and final decision—following earlier final decisions and appeals of them by Andrews and Arnold. There would be no four-engine bombers bought in 1939—period! The ninety-one bombers to be purchased would be twin-engine B-18's. The Air Corps was to proceed with procurement at once. The Secretary wanted the full complement of 2,320 planes by 1940.

Spalding indicated his own faulty understanding of military aviation when he suggested that he was in favor of buying only a few B-18's and acquiring around two hundred attack planes in order to save money. Kilner explained to him that such a move would be very unfortunate, for it would mean bombardment units without aircraft, and attack outfits with aircraft they couldn't use because they didn't have the personnel to man them. Furthermore, bombers were considered the backbone of the Air Force. Well, couldn't attack planes be used in bomber squadrons? the

General wanted to know. Kilner, with infinite patience, said no, that wasn't quite the way it worked, but what would be more preferable would be for the Air Corps to hold onto its money and ask the Congress to extend the time of required expenditure, so that a new bomber could be bought in 1940. The General doubted very much if the Secretary of War would take to that idea; it would delay reaching the fixed goal of 2,320.

At this juncture Arnold's daily log reported:

> Bombardment program for 1939 one mess with Secretary of War approving all procedure, Asst. Sec. of War demanding action at once, General Staff hands off because it is a "hot potato" and nobody wants to do anything but everybody wants to get something done yesterday!
>
> Conference with Sec. of War this date with a view of having procedure for 1939 procurement and 1940 development for bombardment plane resulted in no action. Will see Secretary of War again tomorrow. Letter from Asst. Sec. of War which must be answered by 11 o'clock tomorrow requests information as to whether or not all circular proposals for 1939 procurement will be in his hands by Dec. 1. The reply will be "yes."

The Assistant Secretary of War's knowledge of aircraft was on a par with General Spalding's suggested use of it, which only added to the mess referred to. Johnson had approved the twin-engine attack-bomber project, believing it was the same type of plane as the Spitfire and the Messerschmitt and comparable to them! A memo was prepared for him outlining the status of the Air Corps's two developing fighters, the Bell P-39 and the Lockheed P-38, and how it was expected their performance would measure up to the British and German fighters. The memo then carefully pointed out for him the Air Corps's plans for procuring attack bombers and the differences between such aircraft and pursuit. Arnold personally gave the memo to Johnson, explaining it in detail and cautioning that the information was secret.

Quite obviously the situation in the War Department was confused. Since mid-October, Roosevelt had been speaking out on the need to increase U.S. air defenses. At the same time, he was preoccupied with the upcoming November congressional and gubernatorial elections. It was just at the moment of Woodring's final decision on what planes were to be bought that the President asked him to go out on the campaign trail and bring in the votes for the Democrats. Woodring departed Washington, and Hap Arnold did also, flying down to Langley to talk over developments with Andrews.

Arnold was a major general now, and at fifty-two he had attained his position as Air Corps Chief through a mix of talents and drives. From the dogmatism of Billy Mitchell's cause, he had, with passing years, become pragmatic in his approach. Aerial independence was still out there in the blue somewhere, but not before a real air arm had been built. After three years in Washington, holding down the number-two spot, he knew the realities, knew how to accept them, because either you accepted them and learned to operate within the system or you got out. Operations was what he loved most, but he knew that those days were gone. A chief had to administrate, and it must be reasoned that before his appointment those who had served with him and with Andrews must have considered how it would have been had Andy been made Chief, and Hap head of the GHQ Air Force. As Andrews had been pointing out for nearly four years, the Air Force Commander had little clout in the decision making, but oh my, he did have tactical units to train and utilize, and that Hap would have loved dearly.

In approach, Andrews' thinking encompassed the best way to utilize and employ the total force. Arnold was more inclined to seek the use of the specific. Both were intellectually curious, Arnold scientifically so.

A move he made not long after becoming Chief was a prime illustration. He invited several noted scientists to a private meeting at the National Academy of Sciences, in Washington. One of them was Dr. Theodor von Karman, a Hungarian-born aerodynamicist who, as an assistant to Robert Millikan at Cal Tech, had first gotten to know Arnold when Hap commanded at March Field. Also present at the meeting was Professor Jerome Hunsaker, director of the MIT Guggenheim school. Arnold told them he had two problems that were very urgent. One was the designing of windows in a bomber that would afford good visibility at high altitudes and speeds. The other was the need to develop rockets to assist heavy-bomber takeoff. Although runways were being lengthened, Hap was worried that they couldn't be lengthened that much for future equipment, and to his way of thinking rockets might be the solution.

Hunsaker said, "All right. We'll take the bomber-window visibility problem and Karman can take the other job."

He did not say "rocket" because at the time it was considered a bad word within the scientific community. No less a personage than Dr. Vannevar Bush had commented to Millikan and von Karman jointly, "I don't understand how a serious engineer or scientist can play around with rockets."

Hap Arnold had ten thousand dollars for von Karman to play around with to develop rocket-assisted takeoff and rocket use. The doctor went back to Cal Tech and retained a handful of graduate students who were experimenting with small rockets. They were referred to as the Cal Tech Suicide Club. Under von Karman's guidance they became the GALCIT project—Guggenheim Aeronautical Laboratory, California Institute of Technology. This was the beginning of what was to become the Jet Propulsion Laboratory, and to von Karman's way of thinking, if it had not been for Hap Arnold's interest and his ability to stimulate interest in others, jet propulsion in the United States "would have come much later."

Nothing better summed up the newness of the idea and the reach of Arnold's thinking than a question Major Benjamin W. Chidlaw asked von Karman at the time the contract was let. Chidlaw, a specialist in aeronautical engineering, was handling research and development for Arnold and would later be chief of the experimental engineering branch overseeing jet propulsion. He asked von Karman, "Doctor, do you really believe the Air Corps should spend as much as ten thousand dollars for rockets?"

This was one instance of Arnold at his best. He knew how to take an idea and run with it, which in turn sometimes illustrated Arnold at his worst. Often, in the running he ran alone, ignoring his staff, issuing contradictory or conflicting orders and generally upsetting the machinery of command. Andrews, on the other hand, knew what he wanted, delegated the authority through his chief of staff and kept a well-balanced eye on the results.

Withal, neither man could tell what the future would bring, but they perceived its shape and within the confinements of the system they struggled to meet its demands before events overtook them. They were not always right. Not infrequently, they exaggerated the effects of actions taken by the War Department, the Congress and the White House, but nevertheless, having endured so long the restrictions brought by economics and public attitude, they kept their foresight intact.

For Andrews, Hap's drop-in at Langley was "very pleasant, but too brief. . . ." There was so much to talk over. The big question was, What next in Europe? Larry Bell, of Bell Aircraft, had recently spent six weeks there looking over aircraft development, principally in Germany. Both Andy and Hap had read his report, which closely paralleled a report they had also read sent by Major Arthur Vanaman, Assistant Military Attaché in Berlin. In their minds, Britain and France had knuckled under to Hitler at Munich for one reason only: German air power. Presently Arnold was to write in a secret memo to Johnson: "The total strength of the German air forces is probably somewhat in excess of 10,000 fighting planes. Production capacity is believed to be in excess of 12,000 fighting planes per year."

Neither he nor Andrews could know that these figures, due to clever German manipulation of invited observers such as Charles Lindbergh and Larry Bell and the military attachés on the scene, were greatly inflated.[2] In view of his meetings with Hopkins and Marshall, in view of the President's starting to speak out on the need for air power, Arnold was growing confident that a change was in the wind. Events in Europe were demanding it. He told Andrews that before Woodring's campaign departure, the Secretary had suddenly asked to see figures on a revised estimate for Air Corps appropriations embodying an increase—over the 178 originally recommended—of one thousand aircraft! Of course, in the estimate, no four-engine bombers were to be included, and to Andrews this indicated the utter blindness of such an increase. But, to Arnold, it wasn't that bad. Of the increase, nearly half the recommended total was for training planes, for, as Andrews had stressed to Marshall, tactical planes were no good if you didn't have trained pilots to fly them.

Arnold returned to Washington knowing he and Andrews would operate well together. They had done so in the past under somewhat different conditions of command. Whoever had instigated the rumors denigrating Hap's character, he knew with certainty that Andy Andrews would have played no part in it. He didn't really give a damn who had. No man could rise to his position and not make enemies along the way.

On his return he took time out to write his daughter a funny letter. Lois was now Mrs. Ernest Snowden. Nearly a year ago, in December, she had written to her mother from California, where she was working for the Consolidated Aircraft Company's in-house magazine, that she had met the man of her dreams. He was a pilot, too, but as a graduate of Annapolis, he was a naval aviator! It was all very sudden. Pop and Bee were momentarily bowled over, but they recovered quickly enough. Hap promptly borrowed some money from a bank so he could give his daughter away properly. In the process, he had gained a son-in-law whom he liked immensely . . . even if he was a Navy pilot. There was still one bottle of brandy left to present the newlyweds from the case bought so long ago, and in one of the first letters he wrote to them, he had told them that they could weather anything as long as they kept their sense of humor. His own letters proved his was still in fine form.

His description of his daily routine ran:

> We get up in the morning. I get breakfast, I go to the office. I meet an endless number of visitors. I may or may not have lunch. I leave the office. I drive home through the usual afternoon traffic congestion. I finally reach home. I find the family at home or away. We have dinner. We go to bed. We sleep. We wake again. We get up—Repeat as often as you wish or until you get tired, then you know just how I feel.

And on sports:

> We went to the football game. It was Navy playing Princeton. I had a buck bet on Navy. I wanted Navy to win. I was pulling for Navy. I hoped for the best. I didn't surmise that they would lose their sense of proportion. I figured that with such excellent backfield men as Wood, Lenz, Cook, the Navy could not pull many boners. I did not count upon the inexperience of youth. I know now that Navy should have made one more touchdown. I know that Princeton has a heavy slow dumb team. I know that Princeton has only one real backfield man, Mountain. I know that he caught the whole Navy team flat-footed upon two occasions. I didn't win that dollar.

He and Ernie had previously agreed on the bride price for Lo. Like the

Indians, they figured it in horses, one hundred of them, which were actually
the plastic horses tied around the neck of a bottle of White Horse Scotch.

> I know something else. I haven't received any horses for quite some
> time. I want to know whether the girl is worth any more horses. I want to
> know whether a bargain is a bargain or is it a Versailles Treaty. I want
> some horses.

And on the general family situation, he told them:

> I am glad that you are coming East. I am glad that I haven't any more
> troubles than I have. I am glad that Bruce is at Annapolis.[3] I am glad that
> Ernie and Lois are in the Navy. I am glad that Hank is not holding down
> the exact bottom of his class. I am glad that some of the women that I
> know don't live in this house. I am glad that I don't have to associate with
> politicians after working hours. I am glad that I live in Maryland. I am
> glad that I received a letter of congratulations from my daughter.

The story of the bride price became known throughout the Army and the
Navy, and the father of the bride began receiving contributions from interested friends everywhere. Consequently, he dispatched a message to Lieutenant Ernest Snowden, USN, stating:

> Sir:

> This is to inform you that as a result of the generosity of your friends, and
> in a slight measure to hard work on your part, your debt of 100 horses for
> my daughter has been paid in full. The last shipment of horses, coming
> unbeknown to you and from an unknown source, contained 36, which,
> with the exception of one which unfortunately had a broken leg and had
> to be shot, were the finest-looking, best-formed, healthy pintos I have re-
> ceived to date. It is a shame that you did not have a chance to see them
> and know the kind of horses that other people send me to help you pay
> your debt. In any event, it is paid in full. . . .

That was the lighter side of it that veiled the side that occupied most of
his waking hours. Directly after he wrote to his favorite daughter and her
husband, Arnold received a letter of interest from Charles Lindbergh, living at the time in Normandy, France. Lindbergh had just returned from a
month of test-flying both French and German aircraft. He had learned that
if war had come in September, the French were without a single modern
pursuit plane to defend Paris, and nothing they could put in the air was as
fast as the newest German bombers. In Germany he had been given the
royal treatment, permitted to see the Luftwaffe's latest equipment and to
fly the ME 109. At a stag dinner at the American Embassy, Goering had
taken him by surprise, presenting him, by order of Hitler, with the German
Eagle, one of Germany's highest decorations. General Erhardt Milch,
Deputy Minister for Air under Goering, suggested the Lindberghs might
want to spend the winter in Germany, a move they seriously considered.

Upon his return home to the Côtes-du-Nord, Lindbergh had written to Arnold emphasizing the need for the Chief of Air Corps to have firsthand information on German aircraft production. Arnold couldn't have agreed more; three months previously, he had been slated to undertake such a mission, but the War Department had decided it would cost too much. Now, because of the political situation, Woodring had ruled such a visit would not be advisable. In his reply to Lindbergh, Arnold wrote: "No one realizes more than I do the tremendous advances made by the Germans in aviation. . . ." He wanted to go into the whole production question, and he hoped that until such time as a mission could be sent over, Lindbergh would send back whatever information he could obtain—which he did.

While Arnold was receiving firsthand information on German air strength and French lack of it from Lindbergh, Roosevelt had received a similar report from U. S. Ambassador William C. Bullitt, who had returned from his Paris post in October. Bullitt told the President that the French believed the only way they and the British could build up their air strength rapidly was by buying as many aircraft as possible from U.S. manufacturers.

It was directly after his meeting with Bullitt that FDR told a press conference the time had come to have a new look at the country's defenses—or lack of them. Bullitt had been accompanied by Jean Monnet, Prime Minister Daladier's representative, and on October 25 they met with Roosevelt and Secretary of the Treasury Henry Morgenthau, Jr. The upshot of their discussion was that the President assured Monnet the U.S. aircraft industry would be prepared to sell the French two thousand planes, divided between bombers and pursuit. At the time it was a promise that no one in the War Department, including Woodring, knew anything about. Although the assurance did not exactly fly in the face of the Neutrality Act of 1937, it was given without due regard to important manufacturing considerations and protective Air Corps regulations that sought to prevent any foreign country from obtaining military aircraft—Army or Navy—until they were anywhere from six months to two years old.

This was not the first unilateral move Roosevelt had made to aid the French. Michel Détroyat's secret flight of the P-36 and the French decision to purchase one hundred of the planes under the name Hawk 75-A was an earlier case in point. But it was the first time he had secretly circumvented U.S. neutrality, committing U.S. foreign policy to aiding an ally, offering nearly as many combat aircraft as the Air Corps had been struggling to obtain in total for five years. At the time, his Secretary of State, Cordell Hull, was not informed of the meeting or the decision.

On that same day, in a follow-up to his promise to Monnet, Roosevelt called in Louie Johnson, who was now acting Secretary of War, to discuss a plan to greatly increase aircraft production. Johnson was to head a three-man board whose job would be to figure out how a major expansion could take place. Johnson's fellow thinkers were Assistant Secretary of the Navy Charles Edison and Aubrey Williams, Harry Hopkins' Deputy Administrator of the WPA. Roosevelt gave no figures as to how large an expansion he had in mind, but several days later Johnson brought him a preliminary plan that forecast the production of thirty-one thousand planes within two years, and twenty thousand yearly after that. The cost would be nearly a billion dollars.

To most airmen, but not Hap Arnold, the figures were outlandish, beyond belief. When Johnson came to him and asked that he submit his own program on what was really needed, he called a meeting of his air staff to sound out thinking. Thinking had been forced to deal with paucity in every respect for so long that the best the assembled could come up with was the figure of fifteen hundred combat planes to fill U.S. requirements around the world. Arnold was shocked. He tried again, stressing the locales and missions of the Air Corps, and finally got the total up to seventy-five hundred. He left the number pinned to the conference-room easel as a reminder, in later months, to think big.

In his own presentation to Johnson, Arnold was comprehensive, and he injected the long-suffering objective of air equality. In recommending a council to advise the President on the needs of national defense, he included, with representatives of land and sea forces, the State Department and industry,—air forces.

The importance of strategic air power was also laid down:

> The phase of the air problem which parallels the Navy mission is the attack of distant strategic objectives. It is obviously not easy to estimate requirements for this purpose. One formula that might be applied is that the strength of this part of our air force should be at least equal to the air force that any European or Asiatic country, or combination thereof, could bring to bear on any part of the American continents. This would mean we should be able to match their strength by forces in being and be able also to reinforce at an equal rate.

When it came to the numbers, Arnold flew the four-engine bomber back into perspective. This was after he had given the comparative figures on the German, British, Italian and Soviet air forces:

> . . . We feel that a reasonable round-number objective for the strength of the Army Air Force at this time would be 7,000 planes, including fighting and non-fighting types and a supporting production capacity of 10,000 per year. *Sufficient of our planes should be of the long-range bomber type so that our strength in this class as to performance and numbers will be second to none.* . . . (italics added)

Also within the proposal was contained the possibility of German bombers having the range to fly from bases in West Africa to South

America. As soon as Johnson read the information, he called Arnold and asked for more specifics. A memo was quickly drawn up, which was hand carried to Johnson's aide, Colonel James Burns. It said that Germany was trying to regain control of its former colonies of Togoland and the Cameroons. The distance between Togoland and Natal, Brazil, was twenty-six hundred miles, and from the Cameroons, three thousand. The Luftwaffe was believed to have seventeen hundred Heinkel He-111 bombers, with a maximum range of thirty-three hundred miles.[4] "It is impossible for these planes to make the trip carrying any bombs. In order for them to be effective, a base would have to be prepared in Brazil by German sympathizers or German agents," concluded the memo.

Five years previously, the Drum Board had pooh-poohed the idea of enemy aircraft being able to threaten the continental United States. Now the possibility was suddenly to loom large and to become of genuine concern. The Panama Canal was the most vital ocean link the United States possessed. A half dozen bombers secretly based in a Latin American country could pose an enormous threat to its operation.

On Monday, November 14, 1938, at 2 P.M., a meeting took place at the White House that, to Hap Arnold's way of thinking, was the most important in his career and in the history of the Air Corps. Like other recent meetings held by Roosevelt, it was a secret one, so secret, in fact, that FDR didn't bother to notify his Secretary of War that it was going to take place.

Woodring had recently returned, worn out from the campaign hustings. He had met with the President, discussed the political situation and then gone home to rest for a couple of weeks. The White House did not wish to disturb him; Louie Johnson could manage in his place.

At the White House conference with Johnson were Secretary of the Treasury Morgenthau; Harry Hopkins; Robert Jackson, the Solicitor General; Herman Oliphant, General Counsel of the Treasury; Malin Craig; George Marshall; Colonel Burns; Pa Watson; Captain Daniel Callahan, naval aide to the President; and last, but certainly not least, Hap Arnold. He knew something of what was coming because he had met that morning along with Craig and Marshall at Oliphant's office to discuss revisions of the procurement laws to make the purchase of aircraft easier.

That afternoon, Roosevelt did most of the talking, and to Arnold his words were something out of a dream. The Commander-in-Chief did not want to hear about ground forces, he wanted to talk about air forces. Hitler would not be impressed by anything else. He told the assembled he was going to ask Congress for twenty thousand planes, with a production capacity of twenty-four thousand. That, he felt, would be scaled down to about half, but he believed the numbers could be reached in two years by utilizing private industry plus seven new plants the government would construct.

While the President's announced program to build air power sounded like the promised land to Arnold, to Craig and Marshall, particularly, it was lacking in practicality, for FDR did not wish to be bothered with the basic problems of men to fly the planes and ordnance to make them worth flying. He talked instead about getting planes to aid France and England and planes to protect the hemisphere. Such a deterrent would discourage aggression, while an Army would not, he said. Everyone else had very little to say. This was true of General Marshall. The President turned to him, and addressing him by his first name—which Marshall did not like—he said, "Don't you think so, George?" And the General replied, "I'm sorry, Mr. President, but I don't agree with that at all." FDR gave his Deputy Chief of Staff a startled look, and that was the end of the conference. The others in the room were sure that was also the end of Marshall's career in Washington.

What Marshall and Craig and the uninvited Secretary of War believed was that there must be a balance to the buildup of military forces, and that while the country should have an air force second to none, the deplorable state of the Army must be faced and corrected. A handful of figures told the story of abject neglect. There were 12,500 semiautomatic rifles on hand out of an intended minimum of 227,000. Of fifteen hundred 75-mm guns authorized, the Army had 141. Of 3,750 required 60-mm mortars, the Army had one. There were no 155-mm howitzers, although fifty-five were authorized in the Protective Mobilization Plan; light tanks numbered thirty-six of an intended 244, and of eleven hundred medium tanks there were less than a third that many. To overcome the shortages would take a year. To build an Army of no more than five divisions would take much longer. At Munich the Wehrmacht could muster ninety fully equipped divisions. This was why Marshall had told Roosevelt he did not agree with the idea of putting all the eggs in the air-power basket.

From where he sat, Arnold was to call the November 14 meeting the Air Corps Magna Carta. If the analogy is apt, like the Magna Carta, the results were in for some very heavy weather. The major significance of the meeting, however, and Roosevelt's new-found resolve to have an air force worthy of the name, was that the desire—the impetus for air power—came from the White House, the seat of power. This had never been so before, and it made all the difference, for although the President would change his mind on numbers and costs, would weather-vane in one direction or the other, listening now to Johnson, now to Woodring, would lock horns with an isolationist-minded Congress and make decisions that would grind the teeth of air planners, he had started a momentum, and the momentum once started could not be stopped, orchestrated as it was by a determined leader and the onrush of climactic events.

Hap Arnold made some changes too. Directly following the White House conference, he sent a wire to Langley Field, appointing a secret three-man board to be headed by Tooey Spaatz with the support of Colonels Joe McNarney and Claude Duncan. He had borrowed Spaatz from Andrews' command before. Now he was borrowing him permanently. Shortly, Tooey would be made Hap's chief of plans, his number one henceforward.

At the moment the three-man board reported, they were in the dark as to their call; they only knew they had been summoned and that they were in Washington without provision for being paid per diem. They soon forgot that detail when Arnold brought them into his office and said, boys, this is it. The grin was wide and the sparkle in his eye must have been an inner reflection of the sun rising after twenty years of overcast.

Their assignment, he told them, was to draw up an expansion plan that would total ten thousand aircraft in a two-year period and the personnel to man and maintain them. Since there were 2,320 planes already in being or under order, the need was for an additional 7,680, two thousand to be constructed in two government-owned plants and the balance by the aircraft industry. It was a whopping big job, the biggest peacetime reorganization ever, and they had a month to accomplish it. There would be no publicity. There would be no time for comedy. The President had appointed a board to rewrite the procurement laws so that split contracts could be awarded. On costs, they were to present Arnold with a defense of the necessary budget so he could defend the plan before Congress. Were there any questions?

Whether there were or not, it would take some time for them and for the Air Corps to adjust to having reached a turning point. In the six weeks following Munich, a 180-degree change in course had been made. It would bring to the Air Corps all the upheaval and explosiveness of a revolution.

Twenty-five

Chain of Command

Between the September of Munich and the September of Poland, a year later, the Air Corps knew sudden and wrenching change. It was as though an aircraft with a 65-horsepower engine were to be converted almost overnight to two thousand horsepower, and in the metamorphosis its pilot, its equipment, its maintenance and its base of operations were to be updated accordingly. It simply couldn't be done with any degree of smoothness, for aside from the mechanics of what was required, a multitude of forces were at work—many of them at cross-purposes.

The elections of November had not gone anywhere near as well as predicted within the Administration. In the House of Representatives the Democrats' margin had been cut from 244 to 93 and in the Senate from 58 to 46. Munich had only tended to strengthen isolationist appeal, and the new Congress was laced with hostility toward the idea of greatly enlarged appropriations for the Army and Navy.

The original costs submitted to Roosevelt for the air buildup he had signaled at the November 14 meeting had been considerably reduced by December. The figure of $2 billion was more than even he had bargained for. He wanted planes, and he failed to understand all the additional military support that was necessary to have them. It was Johnson who was quarterbacking the sudden leap skyward, but it was Woodring, Craig and Marshall who continued to insist there must be balance. In war, they argued, an air force without an army might not be an air force at all.

Although the Assistant Secretary of War's bailiwick was procurement, the sidetracking and often exclusion of the Secretary of War from the decisions made in Johnson's office was a continuing and unnecessary element of divisiveness. Even before the mid-November meeting, Johnson had been in command of the developing plans for expansion. It was to his office that the General Staff, the Navy, the Air Corps, the WPA and the PWA sent their representatives. What was discussed there was heady stuff, and when the $2-billion-plus figure was presented to Roosevelt by Malin Craig, it also included substantial increases for the other Army branches.

The President gave his State of the Union message to the Congress on January 4, 1939. In it he stressed the paramount need of air power in a world that had grown hostile and whose boundaries had shrunk, but for the Army, the amount he asked of Congress was about a quarter of the original total: $552 million, with $300 million of that allotted to the Air Corps. The number of planes had been reduced from ten thousand to fifty-five hundred, with a 50-percent reserve force.

In spite of the paring down of costs and numbers, the figures Roosevelt cited for the expansion were grist to the opposition mill. It was going to be no easy matter to get such a vast program approved. There would be some hot and heavy questioning by congressmen who saw the proposed buildup as another way to tax the hard-pressed citizenry. The fact that such a program would stimulate business and bring jobs to an economy that was still in the grip of a depression, with over 11 million of the work force still unemployed, was a strong point in the Administration's favor, but it was not a uniformly accepted point, because it indicated that it took a military expansion program to get the country out of its economic doldrums—relief agencies such as WPA and PWA notwithstanding.

Aside from the politics tied to the cost of his program, there were two major considerations influencing Roosevelt's thinking. The first concerned German-Italian inroads into Latin America. There was genuine State Department and military apprehension over the possibility—as Arnold had outlined to Johnson—of German air power gaining a foothold in the hemisphere. Secretary of State Cordell Hull had previously expressed these fears to FDR, and the major diplomatic reason for the White House authorizing Andrews' B-17's to make goodwill flights to various South American capitals, over parochial War Department objections, was a psychological borrowing of the Navy's ability to show the flag in trouble spots and, in so doing, convey the message of strength to friend and potential foe alike. In this case, it was showing the plane—the swiftness with which it could be on the scene and the might it was capable of delivering. The ability of the Air Force to make these flights, for the B-17 to be able to land at the Panama Canal, had greatly upgraded the Air Corps's mission of coastal defense. Roosevelt spoke about it in his address to Congress. He made clear that the five-year-old Drum-Baker board recommendations were obsolete—no longer applicable—a point on which Andrews had been trying to convince the War Department since the advent of the B-17. The mission of the Air Corps was to be no longer tied just to the muddy waters of coastal defense but enlarged to encompass the broad parameters of hemispheric defense. This did not mean that the Navy would relinquish what it considered its prerogatives in both areas, particularly because its air arm was slated for a similar increase, but Army airmen had now been given a presidential mandate—a purpose they had heretofore been somewhat paranoid about.

The second consideration influencing presidential thinking was the plight of French and British air power, particularly the former. As Lindbergh had noted and Ambassador Bullitt had reported to Roosevelt, the French Air Force would be absolutely no match against the Luftwaffe. This, said Bullitt and others in the know, was the major reason why Prime Minister Daladier had knuckled under to Hitler on the sellout of Czechoslovakia. Ambassador Joseph Kennedy, home from his post in London, had told FDR that when Chamberlain had gone back to see Hitler the second time, he was prepared to tell the Fuehrer that if he insisted on annexation of the Sudetenland, there would be no recourse; France and England would go to war against him. As Arnold reported Kennedy's words, spoken at a later conference in Woodring's office, "Mr. Hitler told Mr. Chamberlain that he would welcome France and England combining because, with the aid of Italy, he could lick the hell out of them." That the Ambassador's account of what was said at Bad Godesburg was not exactly accurate was far less important than the impression it made on Roosevelt and others who heard it. Kennedy also reported that England had fewer than five hundred modern aircraft at the time of Munich. He said he had attended a dinner at which there were a large number of RAF officers present. "It was just like the Last Supper," Kennedy recounted, "because none of the flying officers there ever expected to be back again, because the majority were flying planes with maximum speeds of one hundred fifty miles an hour."

Bullitt's figures of ten thousand German combat planes came, he said, from French intelligence, whose agents appeared to be peppered all over the Reich. French aircraft production had sunk out of sight, and the French Air Ministry knew that, should war come, within a month they would lose all their experienced airmen, and there would be no one to replace them.

That German air strength was greatly overestimated by the allies and by individual observers was understandable, but there was no overestimation of the sad state of French and British defenses. If the Munich Pact was a sellout to the threat of aggression, it was one that bought badly needed time. No one knew how much time there was. Ambassador Kennedy thought there would be war by spring, and Roosevelt was out to do all he could to help the French prepare before that time came.

The earlier French purchasing mission had returned in December desperately anxious to purchase in large numbers the latest and best military aircraft the United States had to offer, particularly pursuit and attack models. The French had a direct line to the White House through Baron

Amaury de La Grange, a member of the Chamber of Deputies, who had known FDR since 1910 and was married to an American. They had become friends during the First World War, when the Baron was a captain in the French Air Force and Roosevelt was Assistant Secretary of the Navy. It had been a year before, in January 1938, that La Grange had described his country's aerial plight, and in the twelve-month period events had simply sharpened the need not only for purchases but also for speed—this last because the U. S. Neutrality Act would prevent the sale of armament to countries engaged in war.

The rigors of these two presidential concerns, hemispheric defense and aid to France, were veiled in secrecy, concealed within the proposed plans for expansion, which in turn had to be translated into action on a multitude of levels both military and civilian. It was a highly charged atmosphere in which conflicting aims and ideas were bound to clash. Beneath the surface of grandiose projections lay difficult realities.

When Roosevelt first spoke of ten thousand planes in two years, the Boeing production line was equipped to turn out thirty-eight B-17's annually. The production lines of other aircraft firms—Martin, Douglas, North American and Curtiss—were comparably geared.

When Louie Johnson called his own secret meeting, a week after Roosevelt's, he summoned to it heads of the U.S. aviation industry. They numbered thirteen—seven aircraft manufacturers, two makers of aircraft instruments, two of powerplants and one of propellers. The thirteenth was the representative from the Aluminum Company of America. Mass production was the order of the day, but it was going to be a long while before the assembled could—and in some cases would—tool up.

On the announced plans of how much of everything was to be produced, there were cogent differences of opinion. Lieutenant Colonel Russell L. "Maxie" Maxwell, an Ordnance officer and good friend of Andrews' on Johnson's growing staff, sent Andy an informal address he had given at the Aberdeen Proving Grounds explaining what he termed the "Extraordinary Defense Program."

In his reply to Maxwell, Andrews took issue with the percentage of planes that would be held in reserve. "If it takes three months to train an artilleryman and ten months to build him a cannon, then you have got to have a reserve of cannon," Andrews agreed. "But, when it takes a year to build an airplane and from two to three years to train the crews to operate and maintain that airplane, then there is not quite such a big argument for a reserve of airplanes, particularly where aeronautical advancement in types is as rapid as it is today. We cannot afford to equip the air force of tomorrow with the airplanes of yesterday."

The point was well taken, but Andrews was not thinking in terms of attrition once the fateful M Day came. No one knew what that percentage would be, and no one would know until the conflict was joined. In the meantime, the question of how much to hold in reserve was debated by the planners. So was the question of personnel to fly and maintain a force that had been programed by Spaatz, McNarney, Duncan and others—a force aimed at more than doubling in two years the yet unreached goal of 2,320 planes.

How was it going to be done? How could the Air Corps graduate ten thousand pilots a year? The Air Staff pondered the problem. The Training Command chewed on it. If Randolph Field worked three shifts, it might turn out fifteen hundred pilots annually. The only answer was to build more Randolph Fields.

"Yes, but how long would that take?" asked Arnold. And then he came up with an idea that bowled over his staff and flew in the face of his training experts. In so doing, he exhibited once again the freedom of his thinking and his willingness to gamble even against the advice of his experts. Spaatz was opposed. So were McNarney, Eaker and almost everyone else.

Like Johnson calling in the heads of the aviation industry to enlist their cooperation, Arnold called in the operators of the leading civilian flying schools to enlist theirs. He made them an offer. He wanted each to set up a flight-training school for air cadets—all the facilities, housing, food, training. He visualized that what would be needed was not ten thousand pilots but eventually one hundred thousand. The flight instructors at the civilian schools would take indoctrination training at Randolph. The Air Corps would supply the schools with training planes. West Point graduates would serve as supervisory personnel at the particular schools in order to establish discipline and a code of honor.

The civilian operators were as bowled over by the proposal as were Arnold's subordinates, who did not believe it was possible to train a military pilot outside of a military reservation. A considerable amount of risk was involved on both sides—economic for the operators, in the results for Arnold. The former figured each would have to put up an initial investment of two hundred thousand dollars to establish a primary flight-training school. Where was the money to come from? Arnold didn't have a nickel, but the banks did. If they'd borrow what they needed, he was sure that he could get it back from Congress through a supplemental appropriation. Once they were in business, payment to the operators would be on the basis of students graduated, with a lesser amount for those who washed out. There it was. What did they think? There were ten of them present, and like Hap, they liked challenge. Also, they trusted him. They said yes, and neither they nor the country nor the Air Corps was ever to regret the gamble.

Few developments worked that well in the galvanic reorganization that was afoot, and although 1939 would always be remembered as the year in which World War II began, for Hap Arnold and Frank Andrews it would be remembered for other upsetting reasons as well.

It was a year in which history came close to repeating itself for Arnold, and he was nearly thrown out of Washington again—permanently!

It was a year for Andrews in which the treatment he received paralleled that of Billy Mitchell. With Arnold, the problem was twisted around the President's desire to help the French. With Andrews, it was the determination to shut him up by exile or to force his resignation.

Following Roosevelt's assurance to Jean Monnet, head of the French purchasing mission, that the U.S. aviation industry would be prepared to sell them two thousand aircraft, the French industrialist returned home to report to Daladier. The Prime Minister was concerned about the U. S. Neutrality Act, but Monnet assured him that Roosevelt had promised that, should war come, he would work for the Act's repeal; and if he couldn't obtain it, arrangements could be made to send U.S. planes to Canada, where they could be shipped to France. Only Ambassador Bullitt and Secretary of the Treasury Morgenthau knew of the President's assurances. Neither Cordell Hull nor Henry Woodring nor any other Administration officials were informed of them.

In mid-December Monnet returned heading a new mission, carrying approval by the French Defense Council to buy one thousand planes. The Secretary of War learned of the mission's intent not from the President or the Secretary of State but from General Marshall, who had, in turn, been notified by Hap Arnold. The Treasury Secretary had requested that the Chief of Air Corps grant permission to the French to inspect the latest models of pursuit and attack bombers now on the production line.

Arnold's reaction to the request was just as negative as had been his attitude toward Détroyat's test-flying the P-36. Then he had listed his reasons, which centered around the danger of disclosure and no assurances that the disclosures could be kept from hostile hands. He recognized the economic advantages to Curtiss and the jobs it would bring, should the French buy the aircraft, but he, like most military men, took an extremely dim view of passing on the latest and the best equipment to others, even if they were considered allies. Not only had this attitude been fostered by years of having too little, and the fear that even an expanded production line turning out aircraft for the French could well delay and even curtail Air Corps production, it was also influenced by political and military convictions. At the time, neither Arnold nor Andrews, nor any Air Corps planner, was on record as favoring direct involvement of the United States in a European war, should the worst happen. There was no good reason why they should be, and many good reasons why they should not. Coastal defense, hemispheric defense, defense of possessions, yes, but Europe, no. When Richard Aldworth, Chairman of the New Jersey State Aviation Commission, had written to Arnold and Andrews on the need to set up training schools for aircraft mechanics, he had remarked, "We both know there is another war coming up, just as sure as death and taxes."

Andrews in his reply summed up for Arnold as well when he wrote, "I hope we can avoid this war that seems to be brewing in Europe. . . ." Secretary of War Woodring did too. Directly after he learned from Marshall of the Monnet mission, he met with Morgenthau and Under Secretary Sumner Welles to discuss its intent. Since Hopkins was about to be confirmed as Secretary of Commerce, the President had turned to his Secretary of the Treasury to handle the delicate and secret negotiations with the French. Morgenthau was an internationalist, and would have few qualms in skirting War and Navy department policies in selling aircraft to the French. Never mind that the assignment was not in the normal realm of the Secretary of the Treasury's affairs. The Department had a procurement division capable of handling whatever the French bought. Before they bought anything, the mission wished to inspect and test-fly the latest Curtiss pursuit and the Martin F-166 and Douglas B-7 attack bombers.

Woodring said absolutely not. Testing of the Curtiss YP-37, XP-40 and the Seversky AP-4 in competition had not even taken place. The bombers were still under construction. War Department policy, agreed to by Roosevelt, blocked the sale of military planes to foreign buyers until a year after the second production model came off the line.

On top of that, Woodring expressed Arnold's fear that any large order by the French would interfere with Air Corps orders. Like Roosevelt, Morgenthau believed that France had become this nation's first line of defense, and he was as single-minded in his desire to aid Monnet and his group as Woodring and the War Department were to prevent what was intended. The meeting ended in stalemate. The two cabinet members did not care for each other. In a meeting that followed the next day, both Arnold and Marshall were present. They agreed that the French should be permitted to buy more P-36's, but they were against any inspection or sale of the Martin and Douglas bombers. Arnold pointed out that if the French were allowed to go ahead and buy a thousand planes, it would take a year and a half to produce them and, in the process, Air Corps production was bound to be hurt.

Morgenthau, who could be waspish, impatient and arrogant all at the same time, said he had his orders from the President and, whether the War Department liked it or not, he was going to carry them out. Woodring felt he had his orders too, but he was far more inclined to compromise, stimulated by his uncertain position with the White House. He warned Morgenthau that if Congress found out what was going on, there would be hell to pay. He said all he was trying to do was protect Harry.

"I don't want to be protected!" Morgenthau shot back. But that he was worried there was no doubt, for when he informed Monnet that Woodring had finally agreed to the inspection of the Martin, he told him that the whole U. S. Army was against his move. "I am doing it secretly, and I just cannot continue forcing the United States Army to show planes which they say they want for themselves."

But the French were determined too. By year's end, they had decided to buy another one hundred P-36's and sixty Martins. However, they had not been able to inspect the Douglas, because the War Department had dug in its heels. Morgenthau had a way to get around it—make him responsible for the purchasing of all aircraft. Then he could sell whatever was needed.

Roosevelt wasn't willing to go that far, so on January 16, he called in Woodring, Johnson, Morgenthau and Bullitt to get things straight. Shortly after the session, Arnold received a telephone call from Louie Johnson. The result was that three days later he sent a coded message to Delos Emmons.

> Deliver following message to K. B. Wolfe—Captain Kraus U. S. Navy and three members of French Mission will arrive Santa Monica or Mines Field Saturday night. Authority granted for them to inspect Douglas Attack Bomber less secret accessories, fly in same and open negotiations with Douglas Company. Keep this office informed of action taken. Above information to be held as confidential. Make arrangements for plane demonstration Sunday. Expedite.

Arnold was furious. A few days later, as he put it, "All hell broke loose."

On January 23, French test pilot Paul Chemidlin had taken off with a Douglas pilot, John Cable, to feel out the DB-7. Something went wrong and the plane crashed in a Los Angeles parking lot. Cable was killed, Chemidlin plus ten persons on the ground were injured—most of them by flying debris—and a number of cars were demolished. The Frenchman was rushed to a nearby hospital. In an effort to maintain the secrecy veiling the flight, Douglas officials said he was a mechanic named Schmidt. The ruse didn't work. The press quickly penetrated Chemidlin's alias and Douglas was faced with admitting that he was a pilot and technical adviser with the French mission.

The story was reported nationwide, and it brought an immediate outcry from isolationist-minded congressmen: *What the hell was a French pilot doing on board an experimental model of our latest bomber!*

It was a question presently put to Hap Arnold. He had been testifying in executive session before the Senate Military Affairs Committee on the needs of the Air Corps in the coming year. In so doing, he was defending the President's plan for expansion. Senator Bennett Champ Clark, of Missouri, a leading proponent of America first, had known about the French mission because, in strictest confidence, Woodring had leaked the information to him. Clark could not speak out, for fear of revealing his source, but with the story of the crash emblazoned by the press, he was free to expose the secret presidential dealings. He did so, joined by some of his fellow committeemen, questioning Arnold, putting him in a ticklish spot. Arnold was forced to admit to the senators that the French mission had been given a free hand in the face of War Department objections.

What was a French aviation expert doing on a secret bomber? Clark wanted to know.

"He was out there under the direction of the Treasury Department with a view of looking into the possible purchase of airplanes by the French mission," Arnold replied. War Department authorization had been given, but—

The senators were not interested in that line. They wanted to know: Does the Secretary of the Treasury run the Air Corps? Does he give orders about Air Corps procurement? Arnold answered as best he could, but Clark and the Committee decided to call both Morgenthau and Woodring to get their testimony on what was now seen by White House opponents as direct interference in the affairs of Europe, violation of America's neutrality and secret diplomacy that contravened U.S. policies.

For Arnold, the battle took on a more dangerous complexion. Morgenthau saw a transcript of Arnold's testimony and was most unhappy. He said it was the Chief of Air Corps who had given authorization for the fatal flight, and stomped over to the White House to complain. Arnold believed what Morgenthau was really angry about was the Air Corps's original attempt to block him from giving wide-open access of Air Corps equipment to outsiders. It may have been that, but more likely Morgenthau was upset by having to testify on a matter that had become a hot political issue, concerning the essential international question of the day.

The next presidential conference Hap Arnold was called to attend he felt might be his last. Present were the top Army and Navy brass, Woodring and Johnson and others. Roosevelt was not in a gracious mood. The attack on his policies revealed by the Douglas bomber crash had not subsided. This had done nothing to sway his determination to help the French, but it was bound to present problems in his plans to modify the Neutrality Act. Morgenthau had apparently placed much of the blame for lack of military support on Arnold, and Roosevelt went after him, by indirection. As the Commander-in-Chief, he lectured to the gathering on the need for cooperation in furthering foreign sales of aircraft. Previously, in private, he had reiterated to the entire Senate Military Affairs Committee his belief: "Our first line of defense is in France." Now he made it clear he wasn't going to see that position endangered by resistance from the War Department or the Air Corps. On the matter of giving testimony to congressional committees, he was equally displeased. This was particularly true of War Department witnesses, and then, as Arnold described it, ". . . looking directly at me he said there were places to which officers who did not play ball might be sent—such as Guam."[1]

Arnold returned to his office knowing he had lost the confidence of the

President at what he considered the most critical period of his career. He felt his hours in Washington were numbered—that it wouldn't be a train ride to Fort Riley this time but a slow boat over the horizon.

Roosevelt's displeasure stemmed not only from his secret dealings being made public—which he claimed were not secret at all—but also from the effect the publicity had in delaying French purchases. The outcry in the Congress and the furor in the press made it possible for Arnold's office, with the War Department's blessing, to keep the new P-40 out of French hands.

Just prior to the Douglas crash, a competition had been held at Wright Field between the P-40, which was essentially a souped-up P-36 with a 1,600-hp Alison liquid-cooled engine; the P-37, another version of the P-36, with the cockpit set far back to accommodate an Alison engine with a turbine-driven supercharger; and the P-4, Seversky's update on the P-35. In the competition, the P-37's supercharger failed to work properly, the P-4 caught fire and its pilot had to bail out, and in a sense, the P-40 won by default, as had the B-18 when the test B-17 had crashed. Whatever its merits or demerits, the P-40 was the best fighter the Air Corps had, and Arnold, with his neck already in a noose, was anxious to prevent the French from laying hands on it.[2]

The same attitude did not prevail with regard to purchases of the Martin bomber, for as Mike Kilner put it, "The actual placing of this new type of airplane in production, at an early date, would provide a basis for further development." As Assistant Chief of Air Corps, he recommended that the sale be approved. At the same time, Johnson had okayed the sale of the Douglas bomber as well. In all, the French ordered 655 aircraft, including one hundred more Hawk 75's (P-36's), forty Chance-Vought dive bombers and two hundred trainers. But the crux of the purchases, as viewed by Arnold, was that the French order would take every single Martin and Douglas bomber slated for production in 1939. Where did that leave the buildup of U.S. air power, which he saw as his primary responsibility and the greatest challenge of his career?

He had no way of knowing that most of the French orders would never be filled, due to the politics involved in both the United States and France, the slowness of production, and then the German defeat of France. Instead, the winter months of 1939 were as cold for him politically as meteorologically. Shut off from the White House—neither Woodring nor Johnson offering to mediate with a congressionally engaged President—he concentrated on controlling the whirlwind of change that was beginning to gust through Munitions Building corridors.

There were many outside agencies involved, not the least of which was the State Department. An indication of how it, as well as the Treasury Secretary, had an effect on the change was evidenced by a piece of legislation authored by the Department concerning the sale of aircraft and antiaircraft guns to "other American Republics." The proposed "legislation was so far-reaching," wrote Kilner, "that it would have permitted the use of our appropriation to furnish South American countries with our aircraft. Also, it involved furnishing any of these countries that so desired, any secret plans or secret developments that we might have. . . ."

The War Department went to work on "such phraseology as would render innocuous some of the more drastic features of the legislation. . . ." It was believed by Arnold the idea for the legislation had come from the White House.

He had one friend there, and that was Harry Hopkins, but Hopkins had not only become Secretary of Commerce, he had also become very ill and consequently was only infrequently with the President. Still, Arnold felt he had Hopkins' support, and that if anyone was preventing his dismissal, it was Harry. As to what was going on in the numerous high-level meetings with Roosevelt concerning military development, George Marshall kept Arnold informed. It would be a long nine months before the Chief of Air Corps was invited back into the inner sanctum.

During this wintry period, while Hap Arnold was excluded from presidential view, Frank Andrews, as Air Force Commander, was unable to supply any real input into expansion plans until after they were announced. Then it was a matter of reaction. He was disappointed in Roosevelt's program, not because of numbers or the amount of money to be allocated but because it lacked balance with regard to types of planes, the necessary personnel to man them, and the bases to maintain them. He could foresee a battle ahead to get that balance, and he believed Arnold was in a position where he would have to put up a fight or accept something the whole Air Corps would know was wrong. He was sure Hap would fight, and he stood ready to join in with him. This he did.

On Monday, January 16, sporting a bow tie, he was the featured speaker at the National Aeronautic Association Convention dinner in St. Louis. In the crowded Jefferson Hotel ballroom, he supported the things the President had said before Congress, but he went even further, declaring the United States was a fifth- or sixth-rate air power. He revoiced his theme that the Air Force must be built up around the long-range bomber, and everything he said, he had said many, many times before, but not before such a high-powered audience and a press that was now anxious to quote his every word.

"Our air forces are inadequate . . . the entire Army combat force in continental United States . . . numbers only slightly over four hundred fighting planes." The statement was reprinted in major newspapers across the country. So, too, was his statement on air strategy, which the press reported as a good offense being the best defense—a turnaround from the War Department axiom of defense. "There are only two ways of stopping an air attack," he said. "One is to prevent the hostile air force from getting close enough to launch an attack; the other is to stop the aggressor nation from even planning the attack, through fear of retaliation."

He spoke about the vulnerability of U.S. targets: the Panama Canal, the northeastern Atlantic seaboard. "Eighteen power plants," he pointed out, "supply 75 percent of all electric power for the New York metropolitan area. It would not take a large force of bombardment planes to accomplish the destruction of these plants."

In the Q and A period, Andrews answered the question of what constitutes an effective air force by saying, "This means equipping our Air Force with enough airplanes of sufficient range and bomb-carrying capacity to enable them, from available bases, to reach any locality where an enemy might attempt to establish air bases, either land or water." Yes, geography played a part in determining the makeup of an air force, but the side that had the longer-range weapons was the side that had a distinct advantage.

That what he had to say made a strong impression was evidenced by the press reaction. A year before, when he had attended the same convention with Oscar Westover and Westover had spoken, little notice was taken of what was said. Now, with Austria annexed, the Munich Pact seen as surrender, and Roosevelt's call to build air power, Andrews' words were in line with a developing administration policy. Alarmist to some, they must have been fully acceptable at the White House. The same could not be said of the Secretary of War's office. Andrews' address flew in the face of Henry Woodring's accomplishments, not to mention those of the War Department. It was one thing for the President to say that an expansion program must be launched. It was quite another for the highest-ranking officer in the Air Corps to announce publicly that the United States was a sixth-rate power in a world slipping toward war. It was Woodring, with General Staff agreement, who had put a halt to long-range bomber production; and now it was Andrews who was stating in clear, concise terms that an air force, to be meaningful, must be built around the long-range bomber. In so doing, he had ennunciated the present lack of balance in personnel, training and bases. Again, an old complaint, but theretofore delivered essentially behind the closed doors of the War Department, at the War College or in official documents.

To the Secretary of War, all that Andrews said for public consumption must have seemed like a finger pointing at his door. Oh, yes, Congress was to blame, and the White House, and the Bureau of the Budget and a good many others as well. But if U. S. Army air strength was at such a low ebb, a good share of that blame was a reflection on Henry Woodring's stewardship—or so he and some members of the General Staff were bound to conclude Andrews was claiming before the most important aviation body in the country. Worse, what the Air Force Chief had to say was in direct contradiction to what Woodring had already said on air strength in his report to the President at year's end:

> There has been provided a powerful defensive arm in the form of the General Headquarters Air Force. In creating this extremely important arm, it was necessary to do more than merely procure increased numbers of airplanes. A balanced air force had to be established in personnel, ground installations, training and supplies. In the gradual development of this air force, we constantly strove to keep abreast of rapid development of aviation equipment and technique and simultaneously provide military aircraft of unexcelled quality. Considering our initial deficiencies, it is my opinion that we have builded wisely and well in developing our General Headquarters Air Force. The efficiency attained by that force in a few short years of its history is a most noteworthy achievement. We have a substantial framework for the extension which now appears essential.

Whether anyone in the White House or outside it bothered to draw a comparison between what Andrews had to say and what Woodring maintained, is not known, and in view of the President's call for a huge expansion, is not likely. But Woodring and some of those around him had to be sensitive to what could have been seen as an attack on his credibility, particularly with Johnson leading the expansion attack.

Following his St. Louis address, Andrews received a considerable amount of congratulatory mail, and in responding to Sherman Ford, a former bridge-playing partner at Washington's Metropolitan Club, he said, "Don't worry about my making the front page too often. My detail as GHQ Air Force Commander expires on the first of March, and unless I am reappointed, my publicity value will become nil, which suits me okay. . . ."

For some months, he had been discussing with Malin Craig the question of his next post. Craig had appeared to be genuinely interested in helping Andrews to assume a command that would be commensurate with his abilities. There were two major possibilities: command of Maxwell Field and the Air Corps Tactical School, or the Air Corps Training Center at San Antonio. Of the two, Andrews favored the latter because, as he remarked in a letter to Tony Frank, who was commanding the 18th Wing, in the Hawaiian Islands, "my past experience and my knowledge of the requirements for the product there should qualify me to be useful. . . ."

He said he had never been able to follow Westover's reasoning in the personnel setup at Maxwell Field, and he thought the job should go to a general officer of the line. For himself, he wished to remain with the Air Corps for the rest of his service. This was his position until Roosevelt issued his call to arms to the Congress. After that, he began to think in terms of being reappointed Air Force Commander.

When Senator Lister Hill wrote to him saying he was delighted to hear that Andrews was to take over at Maxwell Field, Andrews replied that he had received no information on his next assignment from the War Department and added, ". . . although I appreciate the importance of the Tactical School to the Air Corps, I had rather hoped to be allowed to continue to handle the GHQ Air Force during the coming proposed expansion. . . ."

He had written to the Senator on January 21. Ten days later, Kern Dodge, President of the Air Defense League, a pro-Air Corps civilian organization, was testifying before the House Committee on Military Affairs. He said: "The GHQ Air Force was the result of the wish of your committee toward a more comprehensive aviation policy. Its work is outstanding in spite of the fetters that have hampered it from its beginning, with General Andrews, whom I have never talked to, surrendering his rank in a very short time. Thus we lose his knowledge in aviation and he returns to a subordinate position. Did such a condition ever face the development of battleships, of infantry, of artillery and other branches of defense? No! They were put in charge of men who spent their lives in perfecting these various sciences."

That others on Capitol Hill at this time, both in the Congress and testifying before military and appropriations committees, were interested in Andrews' value and his future, was apparent. Malin Craig was able to tell him that whatever his next duty and rank, a bill was to be passed assuring his retirement as a major general. On February 6, Senator Robert R. Reynolds, Democrat of North Carolina, entered Andrews' St. Louis speech in the Congressional Record and spoke at length on the floor of the Senate as to his caliber as an air officer.

In all his correspondence to friends military and civilian, Andrews was saying: "Unless I can hang on to my present job, I would like very much to have as my next station the Air Corps Training Center."

That there were air officers anxiously waiting in the wings to assume Andrews' job, was natural. Jerry Brant was one who wrote Hap a personal letter stating the reasons he hoped he would be chosen. He was fifty-nine, and he believed he had as much command experience as anyone in the Air Corps. He then listed his credits, one of which was that in four Army-Navy joint exercises he had headed the defending forces and the Navy had never gotten to first base.

Hap responded saying the "makes" were announced while he was on the West Coast. Jerry had been considered but "only sorry that you were not one of the lucky ones finally chosen but that's the way it goes in this man's Army—one wins, one loses."

Arnold's letter to Brant was written on February 3. As of the next day, Andrews said in a note to Emmons he hadn't a clue as to who his replacement would be, but he figured it would be Delos or Jim Chaney—now commanding at Mitchel Field. He was still hoping he would be retained as he prepared to conduct a joint Air Force-Coast Artillery maneuver to be observed by interested members of Congress. Shortly thereafter, he learned that Delos Emmons was to be the new GHQ Air Force Commander, and so, as he said in a letter on February 16 to retired Colonel James A. Mars, in San Antonio, he was still hoping to be assigned to the Training Center, but he hadn't heard whether he was going to succeed. It was on that same day that he flew up to Washington with his son-in-law copilot to find out.

What he found out was that he was indeed going to San Antonio but not to head anything, really. He was going to be Air Officer for the VIII Corps Area, the job they had given to Billy Mitchell and then, a year ago, to Hugh Knerr.

No longer a major general, not even a brigadier, he was being reduced in rank to colonel. In a letter to his son Allen, he described his duties as handling the Reserves and the National Guard. Whoever in the War Department had decided on the duty, Secretary of War Henry Woodring approved it—possibly suggested it.[3]

Hugh Knerr's immediate reaction was that the move had been made to force Andrews' resignation, and that he should resign. Andrews had no such thought in mind. He accepted the unexpected assignment calmly and philosophically. Both he and Johnny liked San Antonio. They had many friends there, and one had written to tell Johnny that polo was very much in vogue. Andrews wrote to his brother Billum in Miami, hoping to get down to see him and Ashlyn before leaving. He had no written orders as yet. Sunny Williams did. As of March 1, he would no longer be his father-in-law's aide. He was being transferred to the 2nd Wing. As a pilot, that was fine with him, but he would miss his boss.

There were many others who would miss him too, and said so, but none said it better than Devon Francis, of the Associated Press:

> I want you to know with what regret I learned about the finish of your tour as the head of the GHQ Air Force. I need not tell you that the newspaper men who have had the opportunity of "covering" you bear you an affection seldom bestowed on a news source. But that is understandable. Few men have the patience, kindliness and confidence which you have shown us. For myself as an individual and as Aviation Writer for the Associated Press, and as President of the Aviation Writers Association, I wish you Godspeed and success in any new undertaking.

A week after he had learned of his next duty, Andrews commanded the aforementioned final maneuvers, and Malin Craig sent him a note of congratulations saying the bombing had been conducted with admirable precision and effectiveness. But what Craig had to say about the orders he was to sign on March 1 transferring Andrews into what was meant in the minds of some as aerial oblivion, he did not put on paper.

On March 9, Johnny Andrews wrote in her diary: "Women of the post gave me a diamond watch and officers [gave] Andy a lovely silver box—all had their names on it. Cocktails at the club for our farewell." And the next day: "Grand send off with troops and music—left post at 9:15." All was very matter-of-fact, with no feelings of bitterness or remorse on Johnny's part. The real feelings were expressed by those who bid the Andrews farewell. During the ceremony many officers and their wives wept unashamedly. It was difficult to accept the wretchedness of small minds in high places who, for whatever reasons, would cast from command at a time of great need an officer considered by his air colleagues as the most brilliant in their midst.

A reflection of the need made headlines a few days later when Hitler, escorted by Colonel Erwin Rommel and an advance guard, drove into Prague and Hradcany Castle to take the surrender of Czechoslovakia's President, Emil Hacha. The elderly Czech official had given up his country under the threat of aerial bombardment.

As for Andrews, he and Johnny took some leave in Key West. His friend Richard Aldworth had cried out, "Why, oh why, did you not let some of us who love you know that your detail was up so that we could have at least endeavored to do something about it, or is there anything we can still try to do?"

And he had replied, "I expect to enjoy my tour in San Antonio, but trust that I won't be there too long. I will be looking forward to the time when I can get back more in the center of things because I believe there is a lot to be done and that I can be of some help."

He was optimistic when they put him in the same office Knerr and Mitchell had occupied, the one with a latrine in it for the clerks. He didn't say anything. But when Hugh Knerr saw to it that the story got back to the Chairman of the Senate Military Affairs Committee, Morris Sheppard, and Senator Arthur Vandenberg, they wouldn't believe it. Sheppard let it be known he was going to drop in on Andrews to verify the insult.

Of all the messages Andrews received following departure from commanding the Air Force, none was more important than the one from the Deputy Chief of Staff:

April 1, 1939

Dear Andrews,

I did not find the opportunity to have a talk with you at the time of your final relinquishment of command of the GHQ Air Force, and since then things have been happening so rapidly that I must admit I lost track of the matter.

You know full well that you created, more or less out of whole cloth, a splendid tactical command, but I should like to add my appreciation of this fact, along with the statement that I think you gave a remarkably fine exhibition of leadership in the development of discipline, tactics, and general progress, in organizing and rounding into practical shape the GHQ

Air Force. I should like to have been able to look forward to some more agreeable and highly instructive trips with you like that of last summer—which, incidentally, was of tremendous assistance to me this winter. Not that one can acquire an intimate knowledge of the Air Corps in nine days, but it gave me a perspective, as it were, against which I could sort the facts I collected during the following months. It was probably just as well that I knew so little at the time, rather than having started out with too many preconceived notions. However that may be, it was a thoroughly delightful and highly informative experience.

With warm regards

> Faithfully yours,
> (Sgd) G. C. MARSHALL

What Hap Arnold had to say about Andy Andrews' new assignment is not known. Those who were not all that fond of Arnold, such as Hugh Knerr, believed that if Hap felt his own position was threatened, he would take action to remove the threat. Because of Morgenthau's complaint to Roosevelt, Arnold's position was in jeopardy at the very time the decision was made in the War Department to ship Andrews to Texas. But the axing of Andrews came from much higher up, and those who knew Arnold best, such as Andrews himself, knew he was no more responsible for the move than Andrews had been to defame Arnold's character by maintaining he was a drunk.

When the bill was passed authorizing that Andrews, upon his retirement, would be rated a major general, it was Hap who sent a telegram giving him the news. In a reply thanking him, Andy wrote that he hoped "plans for carrying out the new program are working okay. If anything new comes up, will you please have someone drop me a line? I like to keep up to date on what's happening. . . ."

And a great deal, of course, was happening. Before his departure as Air Force Commander, Andrews had written to both Tony Frank and Follett Bradley on the ever-pressing question of command organization. He felt that with the expansion program, the existing division between Air Corps and Air Force would have to cease or the program would flounder, with the War Department not all that anxious to cooperate in a buildup that failed to give its other branches an equal share of the appropriation pie. With the Air Force under the Air Corps and not the Chief of Staff, he believed the program would have a better chance of success. This was the setup Oscar Westover had long wanted, Arnold had supported, and Andrews, in failing to obtain greater Air Force autonomy, had come to accept as the only feasible course to follow. Now Arnold was able to follow it with Marshall's support—and Craig's acquiescence possibly influenced by his desire to retire.

Delos Emmons, as the new GHQ Air Force Chief, was not inclined to make a very strong stand against the change. It was more judicious to accept and try to adjust accordingly. The move made in April did not work all that well, because, at best, the staffs were bound to conflict and duplicate efforts; and in the present atmosphere there was no time for a proper system to be engineered out of a highly complicated amalgamation. The move, however, was the end result of efforts begun by Oscar Westover. It was not a motion toward autonomy but one toward a stronger Air Corps position. The experiment would not be a long-lasting one. Events would work still another variation. But all such changes emanated from the same essential need, whether voiced, written, thought or concealed: the need to be an independent service. Arnold was not seeking it at the time. His mind was on how to make things work as they were under complicated circumstances. But, like a will-o'-the-wisp, it was always there.

In that same month, aboard the liner *Aquitania,* Charles Lindbergh was coming home. His wife, Anne, and the children would be arriving later, all depending on what he found he might be called on to do in the coming summer. A few days before his departure, the British had declared they would support Poland against aggression, and he believed the decision might lead quickly to war. When, on April 7, Italian troops entered Albania, he asked himself not only what France and England would do, but most important, America. "Are we on the verge of the world's greatest and most catastrophic war?" he wrote in his diary.[4] He didn't know, but his return home was to see if he could be of help in what must follow.

As noted, he had been corresponding with Hap Arnold since Munich, sending back information on European air strength. Ambassador Kennedy had said that Lindbergh's report on the Luftwaffe's strength had done much to influence Chamberlain's decision at Munich. Even though the report had been sent to G-2 of the General Staff, Arnold had not been given the opportunity to see it, and so he started his own line of inquiry.

Two days before the *Aquitania* arrived, Lindbergh received cables from Arnold and the National Advisory Committee for Aeronautics, on which both he and Hap served. Arnold asked that Lindbergh call him as soon as the ship docked, and NACA wanted him to come to a meeting on April 20.[5] It was obvious that his services were going to be well utilized.

Once the *Aquitania* had tied up to the pier, Lindbergh had a difficult time getting away from the waiting men of the press. He considered his arrival a barbaric way to be greeted in a civilized country. From Englewood, New Jersey, the home of his mother-in-law—Mrs. Dwight D. Morrow—he telephoned Hap Arnold, who was at the Fountain Inn, in Doylestown, Pennsylvania. Bee and Hap were on their way to West Point to visit their son, Hank. The Inn had a special meaning, for Hap had taken his West Point entrance examination there thirty-five years before. When Lindbergh told him the problems he was having avoiding the press, Arnold suggested he meet them at the Thayer Hotel, at West Point.

The following morning, a bright spring Saturday, Lindbergh drove up the Hudson in his De Soto unobserved by anyone. He arrived at noon to find that Arnold had arranged with the hotel manager to have their lunch served behind the closed doors of the dining room. It was a long lunch, in which little was eaten, but there was a great deal of food for thought—the returned authority on air power giving the Chief of Air Corps the kind of personal briefing letters can't provide. They spent the remainder of the afternoon watching a track meet and a baseball game—West Point playing Syracuse. The three of them sat in the stands unobserved by the press, some of its representatives close by. Hidden amid the cheering cadets, they talked aviation.

Early the next week, with Lindbergh in Washington to attend the NACA meeting, Hap invited him to his home. There he asked a question: Would Lindbergh come on active duty? His expertise was needed, particularly in devising ways to improve the efficiency of aeronautical research. The following day, he accepted and was now officially a colonel. Arnold took him to meet Secretary of War Woodring, and although the state of European aviation was the main topic of conversation, it was apparent to the Colonel that Woodring was very anxious that he not testify before any congressional committees on air matters. It was not Lindbergh's desire to do so, but he couldn't help wondering what Woodring's political motives were. Undoubtedly Arnold could have told him that his testimony would only reveal how unprepared the United States was in the air.

Lindbergh came away from the meeting with a somewhat unfavorable impression of the Secretary of War, and when, a week later, Woodring telephoned and asked him to be his guest at a Chamber of Commerce dinner in Kansas City, his feelings hardened. Here he was just beginning important work for the Air Corps, and because of his fame the Secretary was trying to use him for political gain. He declined the invitation.

Lindbergh's important work commenced with an inspection tour of various bases and factories. George Brett had been shifted from duty as Chief of Staff of the GHQ Air Force to head the Materiel Division. Lindbergh spent time with him and his staff at Wright Field, examining a variety of aircraft, including the Bell P-39. At Buffalo, he looked over Curtiss aircraft developments with Burdette Wright, whom he had long known. Shortly thereafter, he flew across the country in his assigned P-36 on the same sort of inspection, going all the way to March Field, where Jake Fickel now commanded. In California, he talked to the major aircraft manufacturers—Douglas, Ryan, J. H. Kindelberger of North American, Gross of Lockheed—and at Cal Tech with the Millikan brothers of the Guggenheim College of Aeronautics. Of all the production lines he observed, that of "Dutch" Kindelberger impressed him the most.

Upon his return, he sat down with Mike Kilner, Tooey Spaatz, Earl Naiden and engineering specialist Major Al Lyon to talk over the proposed revisions for the Air Corps's new five-year plan. Their talk focused on the amount of money available for research and development and the need to get more. Other meetings followed, and as a result, Arnold designated the group the Kilner Board. The task of the five was, as Arnold put it, "To revise the military characteristics of all types of military aircraft under consideration in our five-year program." Quite simply, but not all that simply done, the Board was to come up with a plan that would assure uniform and far-reaching progress in all phases of military aircraft, equipment and weapons development. A key requirement was to get a well-funded research and development program under way. To Arnold, the Kilner Board's contributions in that hectic time were immeasurable, and Lindbergh's input was greatly appreciated and valued.

In his mind's eye, Arnold had a picture of the brilliant and unassuming aviator being tailed through the halls of the Munitions Building by clerks and unabashed newsmen, going about his work with thoroughness and skill.

To Lindbergh, Hap Arnold was the best Chief the Air Corps had ever had. He liked his style, his smartness, his genuineness, his solid enthusiasm. Their working relationship would last less than six months, but they understood and appreciated each other.

While the Kilner Board was seeking to come up with a program that would assure a consistent line of air development in the years ahead, the usual counterforces were busily engaged in the War Department. To Arnold, many higher-ups on the General Staff, particularly Henry Woodring, remained fixed on the idea of numbers of planes and not types. Supply the French, supply the British, supply the South Americans with all they wanted—sell them thousands of planes. But, as for the Air Corps, B-17's cost too much.

One way Arnold got around the thinking was to include a flock of inexpensive light aircraft along with a handful of B-17's. Because the total number fitted in with the cost figure, he got approval. Since the Army could use the Piper Cubs for liaison duty, everyone was happy. But it was not the right way to build an air force.

There were others, such as Hugh Knerr, who didn't think so either. Being a civilian had not dampened his ardor over the issue of air power and the U.S. lack of it. In fact, now that he had greater freedom of action, he was busy writing articles for major magazines attacking the Navy and the War Department while calling for an independent air force. He had half hoped Andrews would join him in retirement, and they had kept in close touch. At the time Arnold was enlisting Lindbergh's services, Knerr was foxily attempting to catch FDR's attention. In so doing, he had supplied both Lindbergh and Senator Morris Sheppard with a copy of the material he had sent to the President. The material had been more or less smuggled into the White House through a friend of Eleanor Roosevelt's, Mrs. Walter Lloyd Bender, of Newport News, Virginia.

Lucy Bender, also a mutual friend of Knerr and Andrews, had written to Andy on April 30 telling him what she had done, enclosing a copy of the letter she had written to Eleanor along with Colonel Knerr's submission. She confessed that she didn't know how Lindbergh would take the matter, but at least he would have the facts. She was sure the President would read them too.

In closing, she exhibited her feelings for Andrews: "God bless you, and direct you in the fight you are now having," she wrote. "You know that I and your friends would lay down our lives for you anytime. As ever, Lucy." Lucy Bender was right about FDR seeing Hugh Knerr's three-and-a-half-page indictment.

In a follow-up to Andrews, Knerr described the uproar his effort produced. Through one of the White House secretaries, he had learned that FDR and Eleanor had had quite an argument over the paper, that the President ordered three copies of it be made. He then turned on Pa Watson and said he had been lied to. He was in quite a temper, said Knerr. The next day, Roosevelt had called in Senators Sheppard and Vandenberg, Cordell Hull, and Congressman Andrew J. May, Chairman of the House Military Affairs Committee. According to Knerr, the material "was dissected by them and a statement made that something was to be done at once." Knerr allowed how "Arnold is furious and trying to find out for certain that I am in back of it. . . ." He was informing Andrews to be on his guard, because he believed he would be approached without being aware of what the approach was all about. For himself, he wasn't going to do any more for the present. That Knerr had acted without Andrews' knowledge there is no doubt, but Andrews would have concurred with most of what his former Chief of Staff had written in his paper titled: "Air Power for the United States."

The first page, in eight brief paragraphs, outlined the standard airman's belief in the importance of air power to national defense. Except for the final paragraph, there was nothing to which the President in his present frame of mind would take exception. But, in coming to the setting up of GHQ Air Force, Knerr wrote:

> From the beginning, the constant opposition of the War Department to every innovation of organization, equipment and method had to be overcome. Willing generous assistance was lacking.

The accusation set the tone for what followed. Knerr cited the General Staff, and particularly G-4, for its failure to build real air power. He stressed numbers being sought instead of quality.

Arnold would probably have been in agreement with most of the points Knerr made until he declared that the staff of the Commanding General of GHQ Air Force had been "scattered to the four corners of the country" for contesting the War Department view. It had been replaced, he maintained, by General Staff officers. Certainly Knerr and most probably George Kenney had been given this treatment, but Knerr had served in his job for three years, and he, like everyone else, had to expect rotation in a personnel-short Air Corps. His next two paragraphs were what had to infuriate Arnold:

> Recently, the Air Force has been placed under the Chief of the Air Corps where formerly it was directly under the Chief of Staff of the Army. Under the Chief of the Air Corps convenience of administration always took precedence over efficiency in operation. Air Power cannot be developed in an organization where the supply and administrative responsibility controls the operating responsibility.
>
> Since the Air Force has come under the control of the Chief of Air Corps tactical units have become so depleted in experienced personnel ordered out to supply and administrative functions that their functioning is considered unduly hazardous. As a consequence the morale of the Air Corps is at a low point.

Aside from Knerr not being in a position to know what the morale of the Air Corps was, he was exaggerating to make his point. Arnold, and anyone else who was in the know, could argue with conviction that the demands for experienced personnel were coming from everywhere in the Corps and everywhere they remained shorthanded.

Knerr's criticism of the Air Force, now under the Air Corps, was perhaps the key point of his paper, for from it he made his recommendation. It was the same recommendation Andrews had made first to the General Staff and then incorporated in the Wilcox bill:

> Create a United States Air Corps within the War Department, with a Chief of Staff for Air, reporting directly to the Secretary of War.

With the proposal went a separate promotion list for officers, a separate budget and a seven-grade pay system for enlisted men.

Artful and determined, Hugh Knerr with his talent for overstating the case—he blamed the crash of the Douglas B-7 on the War Department—had finally penetrated to the seat of power. His attempt created a momentary furor in the White House, brought men in high places running, and had a poor effect on Hap Arnold's digestion. But, in the press of more momentous considerations, his thrust was put aside, and although not entirely forgotten, it was shelved.

One momentous consideration was the choice of Malin Craig's successor. Who, official Washington wanted to know, was going to be the next Army Chief of Staff? In view of the deteriorating hopes for peace in Europe, the answer to the question was of singular importance.

In military circles it had been known for two years that Hugh Drum, now commanding the II Corps Area, had been actively running for the job. He had no compunction in seeking General Pershing's important support in recommending him for the post. It had been thought by some that when George Marshall, only a brigadier general, had been made Deputy Chief of Staff, through Malin Craig's recommendation (and with the unusual agreement of both the Secretary and the Assistant Secretary), Craig saw him as his successor. But, in the fall of 1938, Woodring let it be privately known that his man was Drum.

As for Craig, he was looking forward to getting out from between Woodring and Johnson, on the one hand, and a President, on the other, who didn't seem to appreciate the need to build an Army as well as an Air Corps (although he was pleased Roosevelt had scaled down his original armament plan). Shortly after the first of the year, Craig had let it be known that he would be retiring sometime soon.

When Cal O'Laughlin, of the *Army and Navy Journal,* asked Woodring if he had determined who would be the next Chief of Staff, the Secretary said he had not. And when O'Laughlin mentioned the possibility of Marshall, the Secretary said there was no precedent for it, and he doubted Roosevelt would make such a choice. But both men knew the selection would be made by the White House and not by the War Department.

In mid-February, O'Laughlin informed General Pershing that Roosevelt was "pretty sore at Woodring and Craig" due to their opposition to the French plane proposal. Since Marshall was backing both of them, he doubted Roosevelt would want him as his Chief of Staff, and that meant Drum's chances were much better. At the same time, rumors began to pop that Roosevelt was finally going to dispose of Woodring and replace him with Johnson. Secretary of the Interior Harold L. Ickes had referred to their feuding as "the holy show." While the President was vacationing, Johnson received a telegram from Pa Watson saying, "I talked with you-know-who, and he says matters will be arranged in a short time." They would not, and neither would matters on who would replace Craig.

In March, Roosevelt was urged, because of the gravity of the international situation, to announce his selection soon. At a cabinet meeting soon after, he told Woodring he was going to give him as good a Chief of Staff as the Chief of Naval Operations he had given Secretary Swanson (Admiral Harold R. Stark).

In mid-March, O'Laughlin wrote to General Pershing that although Marshall would be made a major general, on Craig's retirement he would not become the Chief of Staff. It appeared that Drum would—although nothing was certain.

At the beginning of April, Roosevelt asked Craig to send him the records of the outstanding general officers. There were eight in all. By the fifteenth, no announcement had been made. Hugh Drum had come to Washington for the Gridiron Dinner and in the hope that lightning might strike while he was in town. He had a long talk with Malin Craig that encouraged his hopes. Craig believed that Marshall had been eliminated because of objections by Pa Watson and Steve Early, or so O'Laughlin reported to Pershing.

And then came the end of the month, and it was no longer springtime for Hugh Drum. Roosevelt announced that General George C. Marshall would be the next Chief of Staff. Major support for Marshall had come from Harry Hopkins, Johnson and, finally, Woodring—for the Secretary had gotten tired of pressure being exerted by those pushing Drum. How much influence this support had in helping Roosevelt to make his decision is not known, but it would later be considered as the best military appointment he ever made.

There was some wry humor lurking in the War Department and White House corridors before the choice was made. Marshall knew that Johnson was for him as well as Woodring, but he didn't want either to know, for fear that if one found that the other was in favor, both would withdraw their support, and that if the President learned Craig was a Marshall man, FDR wouldn't be.

In a letter of condolence to Hugh Drum, O'Laughlin wrote: "Everybody here cannot understand why you were not appointed Chief of Staff. It was expected and there is considerable soreness over the way you were jumped. The Press of the country has paid practically no attention to Marshall's selection."

Evidently, Drum didn't either, for when he failed to send routine felicitations to Marshall, O'Laughlin remarked that there had been some comment on his oversight. "I am sure you will take an early opportunity to show that you can take it," he suggested.

In San Antonio, Andrews heard the news and sent an immediate letter of congratulations. At the time it arrived, Marshall was on a flying trip to the West with Arnold, and he had hoped to see Andrews, but they had missed connections. At Christmas, Marshall had written to him: "The Deputy has no Christmas stationery, only a collection of headaches; but your greetings and good wishes brought a very pleasant touch." These headaches had not grown any less, and in his reply to Andrews, whom he addressed not as Colonel but Major General, he told him he was taking off for Brazil and would not return until about the first of July.

It was a trip that had been some months in the making and was evidence of how seriously the Administration and the War Department had come to take what was conceived as a threat of foreign infiltration into Latin America. For many years there had been German and Italian economic and political interests in the major Latin American countries. In January, at Johnson's request, Marshall, working with the War Plans Division, had come up with a two-year program to bring five divisions up to combat strength. Since the President had not asked for an increase in Army ground forces, the Congress didn't go along with the plan, but one of the reasons Marshall gave for making the proposal was the threat posed by dictatorships to the south that might align themselves with Hitler.

Roosevelt was not only aware of the threat as a military possibility but also as a political ploy to use against his isolationist detractors. If it could be shown that the hemisphere was vulnerable—the inviolability of the Monroe Doctrine endangered by attack from the air or the ground—then those who were opposed to his call for preparedness could not preach the cause of "no foreign entanglements."

At the time Roosevelt was calling for air power, Marshall quietly asked the War College to make a study on what would be needed to prevent the take-over of Brazil and Venezuela by Nazi subversion and sabotage. The Joint Army-Navy Planning Board was also formulating plans for stronger defenses of the Panama Canal as well as other U.S. possessions. On the diplomatic front, Secretary of State Cordell Hull had attended the Inter-American Conference in Lima, Peru, at the end of 1938 and had returned with the Declaration of Lima, in which the republics of America agreed to come to the aid of each other in the event of either direct or indirect attack. As Hap Arnold had informed Louie Johnson, Brazil, because of its geography, was the most likely spot an invader would seek to gain a foothold on by air.

When it was learned that the Brazilian Army Chief of Staff had been invited by the German General Staff to Berlin, the Brazilian Foreign Minister, Oswaldo Aranha, suggested that the U. S. Chief of Staff might come to Rio de Janeiro, and after his visit, invite his counterpart to come to Washington. Malin Craig had been slated to make the trip, but with Marshall named to succeed him, it made sense that Marshall go instead. He learned of his mission while on his westward swing with Hap Arnold. His hope to see Andrews during that swing was based on more than a desire to exchange pleasantries, but now what he had in mind would have to wait until his return.

In the meantime, while Marshall went to Brazil and established a firmer military relationship with that country, Andrews packed his wife and daughter into the family car and headed for a look at the Grand Canyon. They were living in a small apartment just off the post, there being no available quarters at Fort Sam Houston to house them. Life in San Antonio was low-key and relaxed, and even though it was a backwater, out of the stream of duty for Andrews, he adapted easily, his equanimity unruffled. It was as though inside he knew perfectly well that, with all that was going on, an officer of his experience and capabilities could not be sidelined for long. He had the patience and fortitude to get the most out of his exile, making his calls around the VIII Corps Area in a single-engine photographic ship, enjoying the southwestern scenery and the many friends he and Johnny had in the area. It was a relaxing but unnecessary interlude, payment for his refusal to knuckle under.

Marshall had told him he anticipated returning from Brazil by the first of July, but he was back a bit sooner. When he returned, Malin Craig was eager to begin his final leave before his official retirement, and Marshall became the acting Chief of Staff.

Whether the wire of June 30 Andrews received was the result of Marshall's first act upon taking over is not known. But there is no doubt that it was the first appointment the new Chief of Staff made in the chain of command. The telegram from the War Department read: *The President has submitted to the Senate your nomination for appointment as Brigadier General. Wire acceptance.* Andrews actually received it standing on the edge of the Grand Canyon. And then the congratulatory wires and letters of joy poured in from all over the country and around the world. In volume and content there was nothing to compare with the genuine outpouring of feeling. It was as though even officers in the Navy and Marines had been ashamed of the way the Army had treated one of its best. Hap had called him on the twenty-ninth—the first to announce the glad tidings that Andrews was to be made a general officer of the line, and on that same day, Lieutenant Colonel Alexander Day Surles, old friend and War Department spokesman, wired: *What was due to happen has happened.*

Jake Fickel and his wife, Hester, perhaps summed up the announcement best: *We were delighted to hear of your appointment to a line star. I am sure it is the most popular appointment that has been made and all ranks of the Air Corps are cheering.*

To Hugh Knerr, Andy's "make was a ray of sunshine in an evil world." Along with his joy, Knerr offered the only sour note. He figured the appointment to a line officer, which meant that Andrews was now available for duty in any service branch, was to get him out of the Air Corps and force his retirement. Andrews told his friend, "George Marshall doesn't work that way." He had sent his own telegram of thanks to Marshall for his confidence, and to George Brett he confessed, "The appointment came as a complete surprise and I have no idea where I will be sent." He didn't find out until the middle of July. The assignment couldn't have been more fitting, not only from the point of view of his qualities but also as to its underlying meaning to the Air Corps. He was to be Assistant Chief of Staff for Operations and Training—G-3 for the entire United States Army. He was the first airman to have attained such military heights in the War Department power structure. And to his friends who had served him in the GHQ Air Force, there had to be a certain amount of delicious irony attached to the news. It had been G-3 that had dragged its feet and had attempted to obstruct him in the formation of the Air Force, and once he had formed it, the Division had failed to give the necessary support on every level. Now he was to head G-3.

The two-week delay in announcing the appointment was the result of the battle that Marshall had fought to gain War Department acceptance, particularly behind the closed door of the Secretary of War's office. But when the time came to make the designation public, there were no voices of protest raised. If there had been, they would have been drowned by the cheers.

Andrews made his return to Washington during the last weeks of peace in Europe to take on what was to be a mammoth job. If he looked back on the previous twenty years of struggle to bring military air power into proper national perspective, he could see that quickening events in Europe and the Far East were doing more to influence a buildup of U.S. aerial strength than all the reasoning that had gone into the arguments of the past two decades. Now all that mattered was the end result, and Andrews would be directing his energies toward that result in the training of all U. S. Army branches while working against the clock and years of lost time.

Aerial strength would come now because political demand would coalesce with aeronautical development and industrial know-how. But what of aerial independence? Twenty years of contesting, of actions above and below board, years of carrying on a crusade for an air arm coequal with the Army and the Navy, the issue remained as contentious and unresolved as in Billy Mitchell's day. Andrews was no longer in a position to carry the fight. Knerr was a voice on the outside. Kenney, Weaver, Bradley, George, Olds, all the others who still had their eye on the dream—none were in a role in which they could rally strength to the cause. Only Arnold could do that, and his attention at the moment, as Chief of Air Corps, was fixed almost exclusively on the nuts and bolts of expansion and all the ramifications that were involved in such growth.

Quite naturally, he was enormously pleased at Andrews' becoming G-3. Andy in the War Department could only be of great benefit to the Air Corps and its plans. Toward that end, he had been building his own close-knit command structure, working to put some of the old pros in the right slots: Tony Frank in Hawaii, Bert Dargue in the Canal Zone, Jakie Fickel at March Field, Jim Chaney at Mitchel, Robbie Robins commanding the Training Center. For closer coordination, George Brett and his entire staff of the Materiel Division would be moved from Dayton to Washington. As for his own staff, Arnold had made Ira Eaker his right hand—his Executive. Tooey was to be his Chief of Plans, Bart Yount his Chief of Training.

It was Eaker who saw in the arrangement a remindful scene: Rockwell Field following war's end, he the Acting Adjutant working with Arnold, Spaatz and Yount to disassemble an air force. And now here they were, twenty years later, working furiously to build one. At that moment in time they could only guess what the future held. General Marshall had recently given a briefing to ten new officers on his staff. His subject was "Yours Is a Wartime Assignment," and that, too, was the way the thinking went in Arnold's office. Understrength as they were in men and equipment, for reasons over which they had had no control, they were prepared in every other way as professionals.

Neither Tooey Spaatz nor Ira Eaker would know that down the road a short way they would be called to high command and, in commanding, put to the test the long-held concept of strategic air power in the crucible of total war.

On September 1, 1939, Hap Arnold and George Marshall were winging their way to the West Coast in separate aircraft. The radio operator on Arnold's plane received word of Marshall's official appointment as Chief of Staff. Hap sent a message of congratulations into the blue.

Already, in the blue over Poland, Hermann Goering's Luftwaffe had struck, and on the ground Hitler's Army had marched. Peace was no more.

Author Acknowledgments

This book could not have been written without the cooperation and assistance of many interested contributors of information. Most of those whom I interviewed played active roles in the events related within these pages, and I extend my thanks and gratitude to them for helping and guiding me along the way.

From the outset and for more than two years, Lieutenant General Ira C. Eaker gave unstintingly of his time and energies in lengthy sessions that covered his relationships with the other principal characters in this book, including the part he played during the period involved. More than detail, however, he added important evaluation.

I confess I was a bit wary of interviewing General Curtis E. LeMay, having heard he was on the abrupt side. I found him just the opposite, found he was ready to help in every way—ready to explain, to offer answers to my questions recalling the old days that had been the early days for him. My only regret was that we ran out of time.

So, too, with Lieutenant General Elwood R. Quesada. We had two sessions, and I would like to have made it two dozen. His input flew along with his enthusiasm. A day spent with Major General Haywood S. Hansell filled a hefty notebook with his comments and left me with some aching fingers. Inadvertently, I learned how he acquired the nickname "Possum," for when I asked him its derivation, he dryly replied, "I suppose someone thought I looked like one." Hardly an apt description for an officer and gentleman who supplied me with a great deal of valuable material.

The late General Laurence S. Kuter, forced off the golf course by an unexpected Florida gale that blew in my favor, gave me the benefit of his concise and extensive recollections and evaluations of men and events between the wars.

When I met Major General Frank O'D. Hunter, in Savannah, Georgia, his bushy mustache a vivid white, he greeted me saying, "I don't know why you want to question me. I'm senile." If General Hunter is senile, then we should all be so afflicted; it was a grand encounter, in which his memories took us back to other days. And this, of course, was true of my interviews with Generals Bernard A. Schriever, Earle E. Partridge, Theodore R. Milton and Jacob E. Smart, as well as Major General Thomas C. Darcy and Brigadier Generals George W. Goddard and Harris Hull. Each added his imprint and contributed greatly to my knowledge.

I regret that I did not personally meet with Lieutenant General Harold Lee George, but my queries were brought to him through the good offices of Lieutenant General John B. McPherson, and his answers added new life to old conflicts. Neither did I have the opportunity to meet with Major General Richard A. Grussendorf, former Chief of the Office of Air Force History. But through correspondence he acquainted me with some of the finer points concerning his former CO Major B. Q. Jones.

For insight into the personality of General "Hap" Arnold, I had the invaluable and generous aid of his son Colonel William Bruce Arnold. No one could have been more willing to offer remembrances of his famous "Pop." This same degree of interest and assistance goes for Mrs. Jean Andrews Peterson and Allen Andrews, whose father, Lieutenant General Frank M. Andrews, was one of the most brilliant and far-seeing air officers of his time. Both aided me with personal interviews and a continuing correspondence.

Two very gracious ladies named Ruth, Ruth Spaatz and Ruth Eaker, answered my inquiries about their noted husbands—a somewhat dangerous course of questioning to pursue, but enlightening and a pleasure all the same.

Colonel John E. Whiteley, who has written a most informative account of his career as an Air Corps pilot and officer, supplied me with a wealth of inside details of GHQ Air Force and a better understanding of the personalities who staffed it.

Mrs. Patricia Power Esty, an Army Air Corps "brat" of the late twenties whose father was an engineering officer at Wright Field, helped to fill in the picture on the distaff side, whereas aviation leader C. R. Smith knew the old pilots and the bold pilots and the politicians, too. In this last area, Duke University history professor I. B. Holley, Jr., spent many hours on my behalf extracting needed congressional correspondence. Colonel and Mrs. Joseph Viner combined hospitality with humorous anecdotes of the glory days when the saddle vied with the cockpit. Finally, there was Major General Benny Foulois—not recently but twenty years ago. Back then, I had an idea to write a book of this nature, and I went to visit him at his home in Ventnor City, New Jersey. We spent a day together while he regaled me with some of the highlights of his life. I kept the notes made that day, and like all the days and hours of interviews that followed with the above named, they were of great value to the research and writing of this book.

In that regard, anyone who has done so knows that gathering documentation can be extremely difficult without the guidance of experts. I was fortunate to have such guidance throughout my research, and my special thanks go to Major General John W. Huston and those members of his staff of the Office of Air Force History who assisted me. Air Force History archivist Dr. George Watson helped to ease my way through mazes of microfilm, while Dr. John T. Greenwood and his colleagues were ever ready to obtain answers and make cogent suggestions. David Schoem, Chief, Support Division, gave needed support to my endeavors.

At the U. S. Air Force Academy, at Colorado Springs, Colorado, Colonel Alfred F. Hurley, Head of the History Department, and members of his staff were equally helpful in their efforts to offer their own perspective on my research. My particular thanks to Majors John F. Shiner and John R. Tate for their thoughts and to Academy Archivist Duane Reed for his capable direction.

My thanks as well to the staff of the Manuscript Division of the Library of Congress, who recognized the scope of my research and cooperated in aiding me. Dr. Forrest Pogue, noted author and military historian, also gave valuable guidance to my line of inquiry.

Of all those who contributed their time and knowledge to this effort, to Air Force Historian Thomas A. Sturm I offer my special thanks. From the very outset, he volunteered his aid and encouragement and was never too busy to lend a hand. He corrected errors in my first draft, smoothed rough spots in the second, and joined my sigh of relief at the final landing—a friend indeed.

My faithful manuscript typists, Jeri Olsen and Suzanne Gibbons, deserve a special citation not only for their ability but also for their good humor in having to translate my early-Sumerian script into English.

And finally but foremost, my wife, Susan's, infinite patience and unflagging support, not to mention hours of researching, typing, proofreading and general care, were necessary to the entire project.

Notes

OVERVIEW

1. Douglas MacArthur, *Reminiscences* (New York: McGraw-Hill Book Co., 1964), pp. 100–6; also cited in D. Clayton James, *The Years of MacArthur,* Vol. 1 (Boston: Houghton Mifflin, 1970).

ONE

1. There were five Arnold children: Elizabeth, Tom, Henry, Clifford and Price.

2. Dr. Arnold recovered.

3. H. H. Arnold, *Global Mission* (New York: Harper & Row, 1949), p. 15.

4. Between 1909, when the Army bought its first plane, and the end of 1913, twenty-four officers were qualified as pilots, eleven were killed in training, and seven others were subsequently killed in crashes.

5. The Army had no set of standards to qualify its pilots, so it adopted those established by the Fédération Aéronautique Internationale, whose representative in the United States was the Aero Club of America. Capt. Chandler was appointed by the club to pass on the qualifications of Army officers. F.A.I. certificates were the first Army wings. In 1912, Army pilots were given the rating of military aviator and awarded winged badges to wear. A year later, Army pilots were authorized by Congress a 35 percent increase in pay, with a maximum complement of thirty officers; flight pay was later increased to 50 percent over base pay. In 1914 the number of officers was doubled with the creation of the Aviation Section of the Signal Corps. There were now two aeronautical ratings: junior military aviator and military aviator.

6. Mauborgne later became a Major General and Chief of the Signal Corps.

7. Dargue was temporarily transferred to the Field Artillery.

8. Glassford was removed from his command after an investigation by the

Army Inspector General's Office. He retired at the mandatory age of 65 in April 1917.

TWO

1. Because of production failures at home, Pershing had contracted with the French and Italian governments for six thousand planes by June 1918. Again because of failure by the U. S. Aircraft Production Board to supply promised materials and tools to build the planes, the contracts were canceled. As a result, until late in the war, U.S. pursuit units were stuck with flying older French planes in combat. The famous Nieuport was one of them.

2. *Global Mission,* p. 69; before war's end, over twelve hundred DH-4s were shipped to the Western Front.

3. It was strategic only in the sense that control of the air was attained, not that bombing attacks were aimed at targets far behind enemy lines. Mitchell's air power was tied directly to "facilitating the advance of the ground troops." Alfred F. Hurley, USAF, *Billy Mitchell, Crusader for Air Power* (Bloomington: Indiana University Press, 1975, p. 36).

THREE

1. The original spelling of the name was Spatz (pronounced "Spahtz"). In 1938, Tooey had the name legally changed to Spaatz. He did this at the request of his wife and three daughters, because all too often the pronunciation of the name was incorrect and came out as *Spats.*

2. A year after the event, Post Office Department pilots were following the routes blazed by Mitchell's fliers. It was Mitchell who established the first airways system in the United States, overseeing the plan in 1919.

FOUR

1. While Mitchell was the principal U.S. champion of air power, he was not alone in his theories. In Europe, other rebels against military tradition had been speaking out. In England, there were Air Marshals Trenchard and Sykes and their supporters, who, during the war, had pioneered strategic concepts. Similar thinking was stirring in France and in defeated Germany. The foremost theorist on the subject, however, was the Italian General Giulio Douhet. He saw mass bombardment as the method to achieve victory. Douhet's book *Il Dominio dell'Aria (The Command of the Air)* detailed these theories and included the need for a separate military branch for air. In 1918, British airmen had gained independence with the creation of the Royal Air Force. In 1922, a year after Douhet's book was published, Benito Mussolini subscribed to the idea and established a separate air arm under his newly formed government.

2. Gen. Patrick had said there were twelve hundred planes fit for duty. Mitchell later changed his figure to nineteen pursuit planes, which was hardly an exaggeration.

3. Elizabeth Trumbell was Mitchell's second wife, whom he married in October 1923. He and his first wife, Caroline Stoddard, were divorced in 1922.

4. Rep. Charles F. Curry, Republican (California), served as a commissioned officer in the Air Service during the World War.

5. The committee was named for Rep. Florian Lampert (Republican, Wis.).
It recommended air independence.

6. Later Sen. Hiram Bingham (Republican, New York).

FIVE

1. King was later to write one of the finest efficiency reports Arnold ever received.

SIX

1. Military officers on duty in Washington at the time were required to wear a uniform only one day a month, for the purposes of inspection.

2. The course from San Antonio was down the coast of Mexico to Veracruz, then across the Isthmus of Tehuantepec, and with the exception of a major stop at France Field, Canal Zone, following the Pacific coast of Central and South America to Valdivia, Chile, then eastward to Bahía Blanca, Argentina, then north to Asuncion, Paraguay and doubling back to Montevideo, Uruguay. The return followed the Atlantic coastline of South America to Trinidad and Puerto Cabello, Venezuela. The final leg, encompassing the West Indies, finished at Langley Field, Virginia.

3. The *San Antonio* had had to remain in Santiago, Chile, for repairs.

4. The flight of the *Bremen* was, of course, headline news, considered on a par with Lindbergh's crossing. When word crackled over the wires that the crew was safe, other airmen rushed to bring aid. Duke Schiller, a Canadian bush pilot, managed to land his ski-equipped Fairchild on Greenly Island before Fred Melcheor had made his parachute jump from Eaker's Loening amphibian. Schiller flew Fitzmaurice to Quebec while Koehl and Von Huenefeld and Melcheor waited for equipment to repair the plane to arrive by plane and boat. The attempted takeoff was a failure, and again the *Bremen* was in an unflyable condition. The three decided to leave it. In the meantime, Herbert Bayard Swope, editor of the New York *World,* got into the act. Naval Commander Richard E. Byrd was preparing for a new polar flight in a Ford trimotor using skis. Bernt Balchen and Floyd Bennett were his pilots. Both were in Canada, both down with the flu. Swope talked Byrd into letting the pair effect a dramatic "rescue" of the *Bremen*'s crew using the trimotor. By the time they reached Quebec, Bennett was so ill he was rushed to a hospital. Lindbergh, hearing the news, made a dramatic flight from Detroit to Quebec, carrying serum through a vicious snowstorm. In spite of his mercy flight, Bennett died. Balchen and Fitzmaurice went on to Greenly Island, landed on the ice and brought off the others. A gala reception followed in Washington, where the three crewmen were decorated by President Coolidge, after which they went to Arlington Cemetery to attend Floyd Bennett's funeral, an unnecessary loss in an effort to capture headlines.

5. For the comfort of the crew, the *Question Mark* carried a small stove and two wicker chairs.

SEVEN

1. Gen. Pershing had asked Foulois to write a report on the record of the Army Air Service in France during the war.

2. Spaatz and Adamson, in concert with Davison, were quick to come up with ideas to demonstrate Air Corps capabilities. At the time, they had a scheme going to fly a squadron of P-12s across the country, giving demonstrations of formation and combat flying over major cities. Fechet liked the idea but had to reject it because of lack of funds.

EIGHT

1. The first officer in charge of the school was Maj. T. DeWitt Milling. His Assistant was Maj. William C. Sherman, an early proponent of long-range bombardment. McNarney was an instructor.

2. A partial breakdown of Air Corps assignments during 1932: four officers assigned to the office of the Assistant Secretary of War for Air; five to the War Department General Staff; forty-five to the Office of the Chief of Air Corps (reduced in 1933 to thirty-six); four Army War College; four Army Industrial College; fifteen instructors and thirty students Air Corps Tactical School; sixty-two plus fifteen students Materiel Division Wright Field; sixty-two Selfridge Field; forty-seven Fort Crockett; twenty Crissy Field; eight Rockwell Field Depot; seven Assistant Military Attachés—United Kingdom, France, Germany, Italy, Chile, two Cuba.

3. The 11th Bombardment Squadron, of the 7th Bomb Group, operated out of an emergency field at Winslow, Arizona, in making the drops. It was awarded the Mackay Trophy. The Squadron's CO, 1st Lt. Charles H. Howard, was killed in a crash in 1936.

4. The B-12 was a B-10 with Pratt & Whitney engines.

5. A good case in point was the P-26, which Arnold came to consider an unsatisfactory plane for tactical purposes.

6. In 1929 Elmendorf maneuvered a squadron up to 28,500 feet, breaking the world's record for combat-formation flying. The previous record had been 17,000 feet.

7. Lt. Woodring was also Capt. Ross G. Hoyt's crewman in the refueling of the *Question Mark,* in January 1929.

8. Bombardment, pursuit and attack aviation were designated as air forces. Observation, photographic and reconnaissance were designated as air services. Both forces and services were under corps area direction and control.

9. *Global Mission,* p. 141.

10. Udet was a pilot in Baron von Richtofen's famous "Flying Circus" squadron. When Richtofen was killed, Hermann Goering was selected to take his place. Goering and Udet became friends. When Goering became Air Reich Marshal under Hitler, it was Udet who sold him on developing a dive bomber, using the Curtiss Hawk as a test model. Out of this came the JU-78, the "Stuka." In 1936 Udet was made Director of the Technical Department of the Air Ministry. He committed suicide on November 17, 1941.

NINE

1. When Gen. Allen's tour was ended, the people of Coblenz gave him a stone plaque thanking him for his service to them. When the General died, in 1930, the plaque became the footstone at his grave.

2. The Rapallo Treaty, of 1922, signed by Germany and Russia, contained a secret clause that permitted the Germans to occupy an airdrome in Lipetsk, two hundred miles southeast of Moscow. Using forged passports, wearing civilian clothes, more than two hundred German officers traveled to Lipetsk, while in the same period aircraft engines and parts as well as ammunition and tools were smuggled aboard Russian ships to Leningrad. At Lipetsk, while training the Soviet Air Staff, German airmen began developing the strategy and tactics for what would later become the Luftwaffe.

3. Italo Balbo was not only Mussolini's favorite airman, he was also one of the Italian dictator's four top leaders. A World War fighter pilot, he had joined Mussolini's Fascist party in 1922 and had taken part in the famous march on Rome. Made Minister of Aeronautics in 1929, he was imbued with the importance of air power. In 1931 he had gained world attention by leading a flight of eight Savoia Marchettis of a less advanced design on a flight from Rome to Rio de Janeiro, Brazil.

4. MacArthur also was against paying for the entertainment of the Balbo airmen.

5. The annual appropriation to G-2 for obtaining military information had been cut from fifty thousand to twenty-five thousand dollars.

6. Thirty-eight officers of these first two classes completed their training on December 16, 1933. These were not, however, instrument-flight training classes per se.

TEN

1. During the 1932–33 Geneva World Disarmament Conference, MacArthur privately urged President Hoover to push for abolition of bombardment aircraft.

2. This idea was suggested to the Lassiter Board by Chief of the Air Service Maj. Gen. Mason M. Patrick.

3. Brig. Gen. Douglas MacArthur, Commander of the 84th Brigade, of the Rainbow Division, later wrote that Pershing's order "precipitated what narrowly missed being one of the great tragedies of American history."

4. The quotes are taken in large part from Gen. Foulois's uncorrected testimony before the Committee, the rest from the corrected testimony, the corrections offering greater clarity but no change in substance.

5. Capt. William V. Andrews, Air Corps; Capt. James D. Andrews, Jr., U. S. Army Corps of Engineers; and the Colonel.

6. Editor and publisher of the *Army and Navy Journal* was John C. O'Laughlin, a long-time newspaperman who had bought the *Army and Navy Journal* in 1925 and become a military "authority" with the right connections. He was a close confidant of Gen. Pershing, now retired, and former President Herbert Hoover. O'Laughlin wrote daily detailed reports to the latter on the Washington political scene, sending frequent copies to Pershing. As a self-

styled expert, his letters to the General carried a deferential, slightly obsequious tone. He was a supporter of MacArthur, and a friend of Hugh Drum and was completely oriented to official Army and Navy thinking. On Army matters his voice was the voice of the War Department. Consequently, he did not hold Benny Foulois in very high regard, nor for that matter any airman. Behind the scenes there was something waspish and snipping about Cal O'Laughlin, a man secure in his highly protected position. His publication carried weight in political and official military circles, and many of O'Laughlin's inside gossipy tips, which he circulated to those with whom he corresponded, proved incorrect.

7. See note 4.

ELEVEN

1. The Aviation Corporation of Delaware held thirteen contracts; United Aircraft and Transportation held six; North American Aviation, Inc., held five.

2. Burdette Wright became president of Curtiss-Wright in 1938. Foulois felt that he and Ruben Fleet, former Air Service pilot and president of Consolidated Aircraft, were adversaries because of their politicking to get contracts.

3. Sec. Baker refused to buy the planes outright, saying that eight would be bought first for experimental purposes in order to get the best type.

4. There were 238 reserve officers serving on temporary duty. Most of them were assigned to flying the mail.

5. Westover was also president of the Spruce Production Corporation, formed by the Army during the World War and used to buy spruce for airplane construction. The corporation remained in being in name if not in fact long after spruce was no longer a major need for aircraft construction.

6. In 1939, Maj. B. Q. Jones transferred back to the Cavalry. It was said he was furious over not being promoted and given a command commensurate with his seniority and experience.

7. If anyone asked the question as to why Lt. Col. Frank Andrews wasn't selected to command the Central Zone, since Selfridge Field was so much closer to Chicago than Galveston, there is no record of it. The answer could have been that Hickam, in addition to being equally qualified, was a friend of Foulois's, whereas Andrews was not.

8. Maj. Clarence Tinker.

9. Capt. B. T. Castor. The route was later shortened to run from Cheyenne to Denver.

10. Capt. Charles T. Phillips.

11. In January 1934, Ocker, the oldest pilot in the Army in point of service, was court-martialed for using improper language to a superior officer. Ocker had been grounded at Kelly Field for weak eyesight. He had gone to another field and passed the eye test. Returning to Kelly, he accused the flight surgeon and CO Lt. Col. Henry "Sue" Clagett of collusion in an attempt to ground him, adding that if Clagett was given more than a cursory examination he, too, would be grounded. Ocker was the inventor of a blind-flying system whose patent Congress bought for one thousand dollars.

12. All told, 1,068 enlisted men and 454 officers, regular and reserve, flying and nonflying, served in AACMO, as did 554 civilian employees.

13. Rickenbacker had been in the aviation business since the early 1920s. In 1932 he had been vice-president of American Airways, one of the properties owned by AVCO, a giant holding company that had been formed by W. Averell Harriman and Robert Lehman. He left AVCO to go with NAA when the former failed to win a proxy fight for control.

14. Capt. Robert Olds, CO of the 2nd Bomb Group, had seen to it that all his pilots were checked out in pursuit planes.

TWELVE

1. Edward V. Rickenbacker, *Rickenbacker, An Autobiography* (Englewood Cliffs, N.J.: Prentice-Hall, 1967), pp. 250–58.

2. Foulois had given the figure of 144 planes needed to do the job, but it took at least 250 aircraft to have the necessary number in service to maintain the routes. It was found that every plane in actual operation would require three additional planes in reserve along the routes at Air Corps stations and depots.

3. In the Eastern Zone there were 139 applications, of which twenty-four were accepted. In the Western Zone nineteen were taken on and in the Central Zone eighteen.

4. The amount of $1,431,650 was not made available until April 3.

THIRTEEN

1. There were twenty pilots on Route 5, and in the first month of operations they flew over twelve hundred hours.

2. Even after the reduction, there were still one hundred two pilots with less than eighteen months' duty who were flying the mail as late as mid-April.

3. On April 8, the Central Zone added a Chicago-to-St.-Paul route.

4. Frazier Hunt, *The Untold Story of Douglas MacArthur* (New York: Devin-Adair, 1956), pp. 148–49.

5. The Baker Board hearings began on April 17, 1934.

6. Lt. Thurmond Wood flew his A-12 into a severe thunderstorm on the night of March 30. He could see the storm and should have known better.

7. The Air Mail Act of 1934 permitted the awarding of contracts for an initial one-year period. If the contractor performed satisfactorily, the contract could be extended indefinitely. Contractors were prohibited from holding an interest in any other aviation enterprise, except airports.

8. In January 1942, the U. S. Court of Claims cleared the airlines carriers and former Postmaster General Walter F. Brown of the charges of fraud and collusion but ruled that Farley's cancellation of the contracts was justified.

FOURTEEN

1. Rep. James had helped draft the Air Corps Act of 1926, which left the loophole permitting the negotiated bid.

2. O'Laughlin was referring to a grand-jury investigation into profiteering on defense contracts, similar to the investigation McSwain was conducting, and apparently O'Laughlin's assumptions were incorrect. The split was over the obvious, not some hidden scandal.

3. Gen. Kilbourne indirectly verified this line of reasoning, advanced by O'Laughlin to Gen. Pershing, when testifying before the Baker Board.

4. In a compromise move, Secretary of War Dern agreed that the Inspector General's office would look into the charges against Foulois by the Rogers Subcommittee. It was a long investigation, the findings not announced until June 1935. Foulois was found guilty of having "departed from the ethics and standards of the service by making exaggerated, unfair and misleading statements to a congressional committee." For this he was officially reprimanded. Rogers and some of his subcommittee were infuriated. Rogers called the punishment "a slap on the wrist" and threatened to block Air Corps appropriations. The low opinion of Foulois's capabilities within the General Staff had not moderated. Drum and Kilbourne were dead set against him. MacArthur was due to step down as Chief of Staff. Foulois knew he was boxed in, knew at the end of 1935 he would have served his four years as Chief, and there wasn't much of any place else to go, unless he commanded the GHQ Air Force. Sick at heart, realizing his remaining could only be detrimental, he announced his departure. Benny Foulois was not a born leader and probably only a fair administrator. He lacked charisma, and the Drums of the War Department looked down on him for his lack of education; he didn't belong in the General Staff coterie. He belonged in the air, and everything he did throughout his career was toward the end of assuring that the nation had a strong aerial defense. His efforts were aimed at assuring that. His ambivalence toward air independence was probably dictated as much by the opportunities and conditions of the time as by the personalities involved.

FIFTEEN

1. The National Recovery Act fixed a forty-hour week. July 14 was a Saturday and thus no one was working.

2. Rep. H. E. Hull served in the House 1915–24. In those days, Benny Foulois said, Hull wasn't just the best friend the Air Service had, he was its only friend.

3. Upon Arnold's return to Washington, he passed the information on to G-2, Army Intelligence.

4. There were a good many points in the Baker Board recommendations that no airman would contest, such as the point that practically all deficiencies in equipment were due to a lack of funds and that the Air Corps should adopt the latest equipment and methods and that Army cargo planes should be converted from commercial types. It was in the area of strategy and the creation of a GHQ Air Force under General Staff control that the major differences lay.

5. The other members of the Commission were Vice-Chairman Edward P. Warner, aeronautical engineer and former Assistant Secretary of the Navy for Air; Albert J. Berres, labor relations expert; Jerome C. Hunsaker, former naval officer in civil aviation; and Franklin K. Lane, attorney with Army and Navy experience.

6. Arnold's efforts to get the awards for his men were not known by them. In March 1935 he again was awarded the Mackay Trophy for having led the flight. A board, consisting of Lt. Col. Arnold Krogstad, Majors Carl Spaatz, Frank D. Lackland and Vincent B. Dixon and Lt. Thomas M. Lowe, made the

recommendation, approved by General Foulois and Secretary of War Dern. At the time, Oscar Westover wrote Bee Arnold that he "was astounded to learn today that he [Arnold] had not received the Distinguished Flying Cross as a result of his Alaskan Flight." He knew that the War Department had not awarded the medal to the other members of the flight and he hoped "some favorable action could be taken on the matter in the future." In 1936, through Westover's efforts and those of the new Chief of Staff, Gen. Malin Craig, Arnold did receive the DFC. Privately, a good many of his Alaska airmen held the award against him, believing wrongly that he had sought it without any effort to see that they, too, were commended.

SIXTEEN

1. Lt. Frank S. Patterson lost the wings of his DH-4 over Fairfield, Ohio. Patterson Field was named for him, and in 1931 the installation became known as Wright-Patterson Field.

2. Andrews and Knerr had first met in 1917, when both had transferred to the Aviation Section of the Signal Corps.

3. Brereton was also an Annapolis graduate who transferred, in 1911, to the Army. He learned to fly in 1916 and had a distinguished war record, becoming Chief of Aviation for the I Army Corps and later Operations Chief on Billy Mitchell's staff. He was a strong independence advocate.

4. One such error was Knerr's long-held belief, until corrected by Air Historian Dr. Murray Green, that Hap Arnold had done nothing to try to get the Alaska flyers the recognition they deserved.

5. Maj. Edwin E. Aldrin was the father of moon astronaut Brig. Gen. Edwin E. Aldrin, Jr.

6. By 1935, after a long internal battle, the RAF recognized the need to mount machine guns in the wings, and when war came in 1939, both the Hurricane and the Spitfire carried eight 7.7 mm wing guns apiece.

7. Later Col. John F. Whiteley, author of *Early Army Aviation; The Emerging Air Force* (Manhattan, Kan.: Aerospace Historian, Kansas University Press, 1974).

8. There was a mix-up on the assignment, and Brett did not get to the War College as soon as expected. He graduated in 1936.

9. The quote is from a memo by Andrews to the Chief of Staff, and inherent in the quote, the strategic role of air power is emphasized.

10. Three months later, Clark Howell informed the House Military Affairs Committee that he was in favor of a separate air force.

SEVENTEEN

1. A West Pointer and career Infantry officer, Simonds had served and commanded in the usual places: the Philippines, China, the western states, West Point, the Canal Zone. During the war, he was a member of the General Staff of the A.E.F. and later Chief of Staff of the II Army Corps on the British front. Following the war, his duties indicated the breadth of his capabilities: commands at the War College, Tank School, War Plans Division and, at the State Department, Military Adviser during the Geneva Conference 1931–32.

2. The sum for the Air Corps for fiscal 1936 was $45 million, a 66 percent increase over the previous year but not nearly enough for GHQ Air Force requirements and only a small part of the War Department's budget request for more than $361 million. Despite the pervasive power of U.S. pacifist forces—historian Charles A. Beard demanded that Roosevelt sack Assistant Secretary of War Woodring for approving the idea of military training at CCC camps—Congress, beginning to take note of the military buildup and aggressive sounds and actions of the major dictatorships, gave Gen. MacArthur almost all of what he'd asked for: $355,500,000. Congress also authorized the buildup from an Army of 119,000 enlisted men to 165,000, plus an increase of cadets at West Point from 1,374 to 1,960. MacArthur steadfastly believed that manpower must come before weaponry. However, he recognized the needs of the new Air Force, and he had gone before McSwain's committee in February of 1935 and presented a proposal nearly doubling the air appropriation, recommending that the additional amount be obtained from PWA funds. At the time of Andrews' speech the increase was pending.

3. Andrews related the details of the episode when he was Caribbean Defense Commander, in 1941, to his aide, Col. Thomas Darcy (now retired Maj. Gen. Darcy).

4. Roosevelt, after indicating to MacArthur he would receive the appointment, later in a fit of pique named Philippines Governor General Frank Murphy to the post and made MacArthur Military Adviser to the Commonwealth.

5. Roosevelt signed the bill into law in August 1935, but as usual, there were no funds to implement it.

6. Andrews' average speed over the entire course was 160.1 mph, and the previous record was 157.3, but the rules of the NAA and the FAI required that a new record must be at least 5 mph faster than the existing one.

7. Roosevelt knew MacArthur wanted Gen. Simonds to succeed him. He purposely waited until Simonds had less than the four years to serve before retirement and then appointed Craig.

EIGHTEEN

1. Douhet's belief in mass bombardment by night had long been rejected by U.S. airmen from both a humanitarian and a strategic point of view. His theories on the supremacy of the bomber in gaining control of the air, of not requiring fighter escort, the totality of the next war, a nation's dependence on its industrial might and a nation's will to fight being broken by aerial bombardment, were all in line with ACTS thinking.

2. After completing their enlisted tour, both men were denied regular commissions, and they left the Air Corps to accept commissions in the Chinese Air Force. They left for China in July 1936, and Chennault joined them a year later, where he formed the famous American Volunteer Group—"Flying Tigers."

3. In 1935 there were six Chennault sons.

4. The two-plane element was found to be a better, more flexible combination for attack.

5. In June 1934, MacArthur maintained the bomber was the most important weapon the Air Corps had.

6. Harold Mansfield, *Vision* (New York: Duell, Sloan & Pearce, 1956).

7. Carl Norden brought out his first bombsight in 1921. In 1928, his Mark II series impressed the Navy and forty of the sights were ordered. In 1931 Norden patents were secretly granted in the name of Norden and Capt. Frederick I. Entwhistle, who worked with Norden on the sight's improvement. The Sperry Gyroscope Corporation was also engaged in producing a bombsight, but the Norden was considered to be a more advanced and accurate instrument.

NINETEEN

1. The General Staff recognized that the overall concept of MacArthur's 1933 mobilization plan of training 4.5 million men within twelve months of mobilization day was unworkable.

2. Ford Island, in the Hawaiian Islands, was established as an Army air base in March 1917. Naval aviation interest did not manifest itself until the fall of 1919, but, by 1923, for all intents and purposes Ford Island and Pearl Harbor had become major naval air installations.

3. In 1936 the Navy and Marine Corps occupied twenty-five air bases in the continental United States from which they operated land-based planes. Formerly, they had been limited to six land bases.

4. Essentially, this was part of the contingency plan on which Andrews had given his testimony on the Wilcox bill.

5. In 1934, the Japanese announced they were breaking a twelve-year-old treaty between themselves, the United States and Great Britain on the size of their fleet, which had been set at the ratio of 3:5:5. After a month in London (January 1936), the Japanese walked out of the conference.

6. In October 1936, in a confidential address at the Army War College, Andrews said: "With respect to GHQ Air Force, the most striking deficiency is the shortage of airplanes. There were originally assigned to the GHQ Air Force a total of 342 combat and cargo planes, of which only 164 were modern. At this time the number of planes has been increased by little and practically no further increase is in sight for more than a year."

7. At the time GHQ Air Force was organized, an Air Corps board that Andrews sat on drew up requirements for officers to be rated as active pilots. Many of the older pilots who were no longer active fliers, such as Walter Weaver, did not meet the standards set.

TWENTY

1. The Douglas was designated XB-19. It made its maiden flight in 1941. Though its gross weight of 160,000 pounds was too much for its engines to perform as hoped, knowledge gained from its engineering was invaluable in the later production of the B-29 and the B-36.

2. In August 1936 Clagett took command of the 1st Pursuit Group, at Selfridge Field.

3. Capt. John F. Whiteley had felt so strongly about Horace Hickam's death that he had commissioned Molly Eglin, wife of Lt. Col. Frederick I.

Eglin, to do a portrait of Hickam from a photograph. To pay for the painting, Whiteley wrote to a dozen of Hickam's close friends, asking for a five-dollar donation. Hap Arnold was one, and for some unknown reason he was furious about the solicitation, maintaining that the officers at Hickam Field should do the subscribing.

4. On July 25, 1936, Deputy Chief of Staff Simonds asked that Andrews' comments on the G-3 study be withdrawn and returned to Andrews personally.

5. In spite of Andrews' efforts, Clagett did not get the wing-commander assignment. There is evidence that Clagett was not popular with some officers at Selfridge Field and that they sought to have him dumped.

TWENTY-ONE

1. Not until 1940, through Hap Arnold's insistence, did the War Department permit the Air Corps to organize its own system of air attachés.

2. Conger Pratt became Commandant at the Air Corps Tactical School and then, as Caldwell predicted, he did go back to ground commands.

3. When Gen. George C. Marshall put through his reorganization of the Air Corps after World War II began, much of what Andrews had proposed went into the change. In early 1939, Andrews had given Marshall a copy of the paper he had presented to Malin Craig.

4. Curtis E. LeMay with MacKinlay Kantor, *Mission with LeMay* (New York: Doubleday & Company, Inc., 1964), p. 146; also author's interview with General LeMay.

5. It was later revealed, although never officially admitted, that one sailor was killed and several others injured in the attack.

6. On December 7, 1941, the U.S.S. *Utah* was sunk with heavy loss of life by Japanese aircraft at Pearl Harbor.

TWENTY-TWO

1. Construction had begun on a base near Tacoma, Washington.

2. Apparently, on this brief visit Lindbergh did not speak to any other high-ranking Air Corps officer concerning Germany's fast-growing air strength.

3. When the department commander in Hawaii requested that B-17's be flown out for maneuvers in March, Craig turned down the request.

4. In January 1939, Hugh Knerr accepted retirement and soon afterward began a one-man publicity crusade to enlist support for the buildup of heavy bombardment aircraft. The villains, in his thinking, were the Navy and the War Department.

5. This was the prototype of the P-39 Bell Airacobra. Shortly thereafter, a Japanese export firm tried to buy the plane, saying Bell could name his own price and date of delivery, offering him 50 percent of the price. Bell refused the offer. In any case, the Neutrality Act would probably have prevented the sale.

6. On April 14, 1938, Malin Craig informed Arnold that he was planning to move Davenport Johnson from Chanute Field back to commanding a group. Wrote Arnold in the daily office log: "Not knowing what politics might be back of such a statement, the undersigned made no comment."

7. This was because their own models did not measure up to either German

or British interceptors, and they believed that with French modifications the P-36 could rank as a formidable fighter.

TWENTY-THREE

1. *Mission with LeMay*, p. 191.
2. George W. Goddard, with DeWitt S. Copp, *Overview* (New York: Doubleday & Company, Inc., 1969), p. 258.
3. Ibid., pp. 258–59.
4. The figures vary on the actual distance from shore that the bombers picked up the *Rex*. GHQ Air Force gave it as 766 miles from Mitchell Field.
5. *Global Mission*, p. 176.
6. Because War Department files over the issuing of the restrictions have turned up missing, evidence of what actually occurred is based on remaining documents and correspondence—including a letter from Gen. George C. Marshall when he was Deputy Chief of Staff, in January 1939, to Gen. George H. Brett when he was GHQ Air Force Chief of Staff—as well as the memories of some of those involved.

TWENTY-FOUR

1. Roosevelt's own ambivalence toward Johnson is apparent, for when he was discussing with Harry Hopkins, Hopkins' possible candidacy for the presidency, he recommended that Johnson be Hopkins' Secretary of War.
2. When Germany attacked Poland, on September 1, 1939, it had a total of about four thousand combat aircraft.
3. Bruce graduated not from Annapolis but from West Point in 1943, whereas Hank graduated in the West Point Class of 1940.
4. On September 1, 1939, the Luftwaffe had a total of 1,180 medium bombers of *all* types. Figures are from the Quartermaster General records of the Luftwaffe.

TWENTY-FIVE

1. *Global Mission*, p. 186.
2. The French did obtain permission to order the P-40, but none were delivered, due to the fall of France, in June 1940.
3. There is no proof that he did suggest it, but Andrews' rank, position, authority and popularity made him an officer whose new assignment was a matter that Woodring would have to have dealt with at least in his acceptance. That he signed the order transferring Andrews to Fort Sam Houston is proof that he approved it. Craig's position in the transfer is clouded. His correspondence with Andrews on the matter, covering a six-month period and indicating that they had discussed it, indicates that he was not out to cast Andrews into exile. Members of the Andrews family were under the impression that the President was behind the move. This does not seem likely either. Roosevelt was calling for exactly the same kind of aerial buildup Andrews had long been advocating at the time of his transfer to Texas. Further, Pa Watson, FDR's aide, was a confidant of Andrews. At Roosevelt's first inaugural, Johnny Andrews had ridden in a special honor guard, and after her husband's appointment as the Com-

mander of GHQ Air Force, they were invited to the White House for tea with Eleanor. None of this had any bearing on Andrews' future status but indicated that there was contact although there is no record of Andrews and FDR having ever sat down to talk. In the end, it was Woodring who was more anxious to dispose of Frank Andrews, and some part of his desire could have been influenced by the fact that he knew Louis Johnson was an Andrews supporter. It is too bad that research has not revealed the provable answer to this question and has left it lying in the realm of assumption and conjecture.

4. *The Wartime Journals of Charles A. Lindbergh* (New York: Harcourt Brace Jovanovich, 1970), p. 175.

5. The National Advisory Committee for Aeronautics (NACA) was established by Congress in 1915. The Committee maintained its laboratory at Langley Field, and four each of its fifteen members were drawn from the War Department and the Navy.

Bibliography

BOOKS

American Heritage. *History of Flight.* New York: American Heritage Publishing Co., Inc., 1962.

Arnold, Henry H. *Global Mission.* New York: Harper & Brothers, 1948.

Blake, George W. *Paul V. McNutt, Portrait of a Hoosier.* Indianapolis, Ind.: Indianapolis Press, 1958.

Borden, Norman E., Jr. *The Air Mail Emergency.* Portland, Me.: The Bond Wheelwright Co., 1958.

Chennault, Claire E. *Way of a Fighter.* New York: G. P. Putnam's Sons, 1949.

Churchill, Sir Winston S. *The Gathering Storm.* New York: Houghton Mifflin Co., 1948.

Craven, W. F.; and Cate, J. L. *The Army Air Forces in World War II. Vol. I, Plans and Early Operations; Vol. VI, Men and Planes.* Chicago: University of Chicago Press, 1948.

Davis, Burke. *The Billy Mitchell Affair.* New York: Random House, 1967.

Deighton, Len. *Fighter, The True Story of the Battle of Britain.* New York: Alfred A. Knopf, Inc., 1978.

Farago, Ladislas. *The Game of the Foxes.* New York: David McKay Company, 1971.

Foulois, Benjamin D. *The Memoirs of Major General Benjamin D. Foulois,* with C. V. Glines USAF.

——. *From the Wright Brothers to the Astronauts.* New York: McGraw-Hill, Inc., 1968.

Goddard, George W. *Overview,* with DeWitt S. Copp. Garden City, N.Y.: Doubleday & Co., Inc., 1969.

Goldberg, Alfred. *A History of the United States Air Force, 1903–1967.* New York: Van Nostrand Reinhold Co., 1967.

Hansell, Haywood S., Jr. *The Air Plan That Defeated Hitler.* Higgins-MacArthur/Longino & Porter, 1972.

Hinton, Harold B. *Air Victory; The Men and the Machines.* New York: Harper & Brothers, 1948.

Huie, W. Bradford. *The Fight for Air Power.* New York: L. B. Fischer Co., 1942.

Hunt, Frazier. *The Untold Story of Douglas MacArthur.* Old Greenwich, Conn.: Devin-Adair Co., Inc., 1954.

Hurley, Alfred F., USAF. *Billy Mitchell, Crusader for Air Power.* Bloomington: Indiana University Press, 1975.

Jablonski, Edward. *Atlantic Fever.* Garden City, N.Y.: Doubleday & Co., Inc., 1972.

James, D. Clayton. *The Years of MacArthur. Vol. 1, 1880–1941.* New York: Houghton Mifflin Co., 1970.

Kahn, David. *Hitler's Spies, Germany's Military Intelligence in World War II.* Macmillan, Inc., 1978.

Killen, John A. *A History of the Luftwaffe.* Garden City, N.Y.: Doubleday & Co., Inc., 1968.

Knerr, Hugh J. "The Vital Era"—unpublished autobiography.

LeMay, Curtis E.; and Kantor, MacKinlay. *Mission with LeMay.* Garden City, N.Y.: Doubleday & Co., Inc., 1964.

Levine, Isaac Don. *Mitchell, Pioneer of Air Power.* New York: Duell, Sloan & Pearce, Inc., 1943.

Krock, Arthur. *Memoirs, Sixty Years on the Firing Line.* New York: Funk & Wagnalls, Inc., 1968.

Lindbergh, Charles A. *The Wartime Journals of Charles A. Lindbergh.* New York: Harcourt Brace Jovanovich, 1970.

MacArthur, Douglas. *Reminiscences.* New York: McGraw-Hill, Inc., 1964.

McFarland, Keith D. *Harry H. Woodring, A Political Biography of FDR's Controversial Secretary of War.* Lawrence, Kans.: University Press of Kansas, 1974.

Mansfield, Harold. *Vision.* New York: Duell, Sloan & Pearce, Inc., 1956.

Mason, Herbert Malloy, Jr. *The Great Pursuit.* New York: Random House, 1971.

Mosley, Leonard. *The Reich Marshal, A Biography of Hermann Goering.* Garden City, N.Y.: Doubleday & Co., Inc., 1974.

Pogue, Forrest C. *George C. Marshall, Education of a General.* New York: The Viking Press, 1963.

———. *George C. Marshall, Ordeal and Hope, 1939–1942.* New York: The Viking Press, 1965.

Price, Alfred. *Luftwaffe Handbook, 1939–1945.* New York: Charles Scribner's Sons, 1977.

Rickenbacker, Edward V. *Rickenbacker, An Autobiography.* Englewood Cliffs, N.J.: Prentice-Hall, 1967.

Roosevelt, Elliot. *Rendezvous with Destiny.* New York: G. P. Putnam's Sons, 1975.

Schlesinger, Arthur S., Jr. *The Age of Roosevelt: The Coming of the New Deal.* New York: Houghton Mifflin Co., 1959.

Sherwood, Robert E. *Roosevelt and Hopkins.* New York: Harper & Brothers, 1948.

Speer, Albert. *Inside the Third Reich.* New York: Macmillan, Inc., 1969.

Thomas, Hugh. *The Spanish Civil War.* Harper & Row, Publishers, Inc., 1961.

Thomas, Lowell; and Jablonski, Edward. *Doolittle, A Biography.* Garden City, N.Y.: Doubleday & Co., Inc., 1976.

Tillett, Paul. *The Army Flies the Mails.* University, Ala.: University of Alabama Press, 1955.

Toland, John. *Adolph Hitler.* Garden City, N.Y.: Doubleday & Co., Inc., 1976.

Twitchell, Heath, Jr. *Allen, The Biography of an Army Officer.* New Brunswick, N.J.: Rutgers University Press, 1974.

U.S. Air Service in World War I. Edited by Maurer Maurer. Maxwell AFB, Ala: Office of Air Force History, 1978.

Whiteley, John F. *Early Army Aviation; The Emerging Air Force.* Manhattan, Kan.: Dept. of History, Kansas State University, 1975.

ARTICLES FROM PERIODICALS

Andrews, Maj. Gen. Frank M. "Our Use of Air Power" (unpublished ms., 1937).

——. "The GHQ Air Force as an Instrument of Defense," *Army Ordnance Magazine,* Vol. 18, No. 15 (Nov.–Dec. 1937).

Bradley, Gen. Mark E. "The XP-40," *Aerospace Historian,* Vol. 25, No. 3 (Sept. 1978).

Crabbe, Capt. William H., Jr., USAF. "The Army Air Mail Pilots Report," *The Air Power Historian,* Vol. IX, No. 2 (April 1962).

Dargue, Maj. Herbert A., USAS. "How Latin America Looks from the Air," *National Geographic Magazine,* Vol. LII, No. 4 (October 1927).

Downs, Lt. Col. Eldon W., USAF. "Army and the Air Mail—1934," *The Air Power Historian,* Vol. IX, No. 1 (Jan. 1962).

Eaker, Lt. Gen. Ira C. "Major General James E. Fechet, Chief of the Air Corps, 1927–1931," *Air Force Magazine,* Vol. 61, No. 9 (September 1978).

Emme, Dr. Eugene M. "The Renaissance of German Air Power, 1919–1932," *The Air Power Historian,* Vol. V, No. 3 (July 1958).

——. "The Genesis of Nazi Luftpolitik, 1933–1935," *The Air Power Historian,* Vol. VI, No. 1 (Jan. 1959).

——. "The Emergence of Nazi Luftpolitik as a Weapon in International Affairs, 1933–1935," *The Air Power Historian,* Vol. VII, No. 2 (April 1960).

Foulois, Maj. Gen. Benjamin D. "Why Write a Book," *The Air Power Historian,* Vol. II, No. 2 & No. 3 (Apr. 1955 & July 1955).

George, Lt. Gen. Harold Lee. "The Most Outstanding Leader," *Aerospace Historian,* Vol. XV, No. 2 (June 1968).

Gorrell, Col. Edgar S. "An American Proposal for Strategic Bombing in World War I," *The Air Power Historian,* Vol. V, No. 2 (Apr. 1958).

Greene, Dr. Murray. "The Alaskan Flight of 1934; A Spectacular Official Failure," *Aerospace Historian,* Vol. 24, No. 1 (March 1977).

Haight, John McVicker, Jr. "France's First War Mission to the United States," *The Air Power Historian,* Vol. XI, No. 1 (Jan. 1964).

————. "France's Search for American Military Aircraft—Before the Munich Crisis," *Aerospace Historian*, Vol. 25, No. 3 (Sept. 1978).

Hart, T/Sgt. John W., USAF. "A Reputation for Courage," *Aerospace Historian*, Vol. 10, No. 2 (June 1969).

Homze, Edward L. "The Luftwaffe's Failure to Develop a Heavy Bomber Before World War II," *Aerospace Historian*, Vol. 24, No. 1 (March 1977).

Hoyt, Brig. Gen. Ross G. (USAF, Ret.). "The P-26," *Aerospace Historian*, Vol. 23, No. 2 (June 1976).

Knerr, Maj. Gen. Hugh A. (USAF, Ret.). "Washington to Alaska and Back —Memories of the 1934 U. S. Air Corps Test Flight," *Aerospace Historian*, Vol. 19, No. 1 (Apr. 1972).

Kenney, Maj. George, USAC. "The Airplane in Modern Warfare," *U. S. Air Services* (July 1938).

Kuter, Gen. Laurence S. "Edgar Gorrell's Concept of Air War," *Air Force Magazine*, Vol. 61, No. 4 (Apr. 1978).

————. "Maj. Gen. H. A. "Bert" Dargue: A Lesson in Leadership," *Air Force Magazine*, Vol. 62, No. 2 (Feb. 1979).

Whitt, Samuel S. "Frank Lahm: Pioneer Military Aviator," *Aerospace Historian*, Vol. 19, No. 4 (Jan. 1972).

Wilson, L. I. "Thanks Tank," *Aerospace Historian*, Vol. 23, No. 2 (June 1976).

Historical studies at the Office of Air Force History, Washington, D.C., from the Albert F. Simpson Historical Research Center, Maxwell AFB, Alabama:

"Organization of Military Aeronautics 1907–1935 (Congressional and War Department Action)," Mooney and Layman, 1944;

"Organization of the Army Air Arm, 1935–1945," Mooney and Williamson, 1956;

"Development of Air Doctrine in the Army Air Arm, 1917–1941," Greer, 1953;

"The Development of the Heavy Bomber, 1918–1944," DuBuque and Gleckner, 1951;

"History of the Air Corps Tactical School," Finney, 1955;

"Development of the German Air Force, 1919–1939," Suchenwirth, 1968;

"Development of the Long Range Fighter Escort," Boylan, 1955;

"Air Defense of the Panama Canal, 1 January 1939–7 December 1941," K. Williams, 1946;

"The Question of Autonomy for the United States Air Arm, 1907–1945," Air University Documentary Study (Parts 1 & 2), 1951;

"Strategic Air Attack in the United States Air Force," Lt. Col. Thomas A. Fabyanic, USAF, Air War College;

"The Army Air Arm in Transition: General Benjamin D. Foulois and the Air Corps, 1931–1935," Maj. John Frederick Shiner, USAF, Ohio State University, 1975;

"Congress and the Concept of Strategic Aerial Warfare, 1919–1939," Victor B. Anthony, Dept. of History, Duke University, 1964.

SOURCES

In addition to the personal interviews cited in the Author Acknowledgments, there were five major sources of documentation for *A Few Great Captains*. They were the Manuscript Division of the Library of Congress, the Office of Air Force History, and the National Archives, all in Washington, D.C.; the Air Force Academy, Colorado Springs, Colorado; and letters, personal papers and transcripts of interviews supplied outside of official source collections.

A supportive source of information was obtained from newspapers and magazines.

At the Manuscript Division of the Library of Congress, research centered on the papers—official, unofficial and personal—of Generals Frank M. Andrews, Henry H. Arnold, Ira C. Eaker, Benjamin D. Foulois and Carl Spaatz. Selected research was also made into the papers of Generals John J. Pershing and William Mitchell, as well as those of John C. O'Laughlin.

Also at the Library of Congress, the applicable congressional hearings and news and magazine stories were digested. Principal news sources were the New York *Times*, the Washington *Times-Herald*, the New York *Herald Tribune*, the Washington *Post* and the Washington *Star*. The principal magazine of reference was *Time*.

The collections examined at the Office of Air Force History were on microfilm and contained to a large extent material from the Albert F. Simpson Research Center, Maxwell Field, Alabama. The focus was on air leaders, air commands and air events between 1910 and 1940. Particular attention was addressed to official documentation encompassing Air Service and Air Corps history for this period, including oral interviews from the American Heritage and Air Force collections. Personal letters from General Arnold to Mrs. Arnold, recently made available to the Office of Air Force History, were of special interest, as were the papers of Major General Hugh J. Knerr and those from the extensive Lieutenant Colonel Ernest L. Jones Collection.

At the National Archives, research centered on the general correspondence files in the Old Military Records Division of the Records of the Army Air Force.

At the Air Force Academy's Special Collections Branch, the papers of Generals Frank M. Andrews, Follett Bradley, Laurence S. Kuter, Haywood S. Hansell and William Mitchell were examined, as was the Arnold Columbia University Oral Collection.

Material supplied by outside sources included newspaper articles, transcripts of interviews with Alexander de Seversky, Elliot Roosevelt and Generals George C. Kenney, Howard Davidson, Hugh Knerr and Carl Spaatz. In addition, personal letters from Frank Andrews to members of his family were made available, as was Mrs. Andrews' diary between 1935 and 1945.

Altogether, approximately ten thousand documents were gathered through research for the writing on *A Few Great Captains*.

Index